Air and Gas Drilling Manual

Air and Gas Drilling Manual

Applications for Oil and Gas Recovery Wells and Geothermal Fluids Recovery Wells

Third Edition

William C. Lyons, Ph.D., P.E.
New Mexico Institute of Mining and Technology
Socorro, New Mexico (Retired)

Boyun Guo, Ph.D.
University of Louisiana
Lafayette, Louisiana

Reuben L. Graham, B.S.
Weatherford International USA Ltd.
Fort Worth, Texas

Greg D. Hawley, B.A.
Redstone Industries Ltd.
San Angelo, Texas

ELSEVIER

AMSTERDAM • BOSTON • HEIDELBERG • LONDON • NEW YORK • OXFORD
PARIS • SAN DIEGO • SAN FRANCISCO • SINGAPORE • SYDNEY • TOKYO

Gulf Professional Publishing is an imprint of Elsevier

GP | P

Gulf Professional Publishing is an imprint of Elsevier
30 Corporate Drive, Suite 400, Burlington, MA 01803, USA
Linacre House, Jordan Hill, Oxford OX2 8DP, UK

Library of Congress Cataloging-in-Publication Data
Lyons, William C.
Air and gas drilling manual / William C. Lyons. -- 3rd ed.
p. cm.
Includes bibliographical references and index.
ISBN 978-0-12-370895-3
1. Air drilling (Petroleum engineering) 2. Gas drilling (Petroleum engineering) I. Title.
TN871.2.L94 2009
622'.3381--dc22 2008046910

British Library Cataloguing-in-Publication Data
A catalogue record for this book is available from the British Library.

ISBN: 978-0-12-370895-3

For information on all Gulf Professional Publishing publications visit our Web site at www.elsevierdirect.com

Printed and bound by CPI Group (UK) Ltd, Croydon, CR0 4YY

Transferred to Digital Print 2011

Contents

Preface

This third edition of the *Air and Gas Drilling Manual* is written as a practical reference for engineers and earth scientists who are engaged in planning and carrying out deep air and gas drilling operations. The book covers air (or gas) drilling fluids, aerated (gasified) drilling fluids, and foam drilling. Further, from the mechanical rock destruction standpoint, the book covers conventional rotary drilling, downhole positive displacement motor (PDM) drilling, and down-the-hole hammer (DTH) drilling.

The first edition of this book was published in 1984 by Gulf Publishing Company of Houston, Texas. It was written primarily for the oil and gas recovery drilling industry and had moderate success, selling out of the initial printing by the late 1980s. The second edition was published by McGraw-Hill in 2001; it added a significant section on the shallow geotechnical drilling industry (e.g., water wells, environmental monitoring wells, mining ventilation shafts, etc.), and also discussed shallow reverse circulation, dual wall drill pipe, and air hammer technologies.

This new third edition is unique in that we have presented the entire engineering material in both the USCS and SI unit systems, including the important equations used for air (or gas) drilling and aerated (gasified) drilling, along with the foam drilling calculations. Solutions based on these equations are given in MathCad™, using both USCS and SI units, providing readers with complete transparency into the solution process for these complicated fluid flow problems. These MathCad™ solutions are presented in the appendices.

The authors would like to thank the editors and staff at Elsevier Science and Technology Books and especially Senior Acquisitions Editor Ken McCombs and Production Project Manager Anne McGee. Their support and assistance during the preparation of this manuscript were invaluable.

The authors would also like to thank the technical staff of the underbalanced drilling group of Weatherford International for their continued support of our efforts to prepare this manuscript. We particularly appreciate the continued encouragement of Jim Stanley.

We encourage engineers and scientists in industry and practice to comment on this book. We also apologize for any errors that might be still lurking in the manuscript. We know our book is not perfect, but we also know that this edition is now complete.

William C. Lyons
Boyun Guo
Reuben L. Graham
Greg D. Hawley

CHAPTER

Introduction and Units

1

Air and gas drilling technology accounts for approximately 30% of the world's land oil and gas drilling operations. The technology is limited to use in mature sedimentary basins. Mature sedimentary basins are older basins that have competent subsurface rock formations that are cemented and are usually uplifted with little formation water remaining in them. Although modern air and gas drilling technology began in the United States in the 1930s, it is presently being used in drilling operations throughout the world's many oil and gas producing geologic provinces. It is important at this time in the development of this technology that the basic principles of the technology be communicated in a manner that all drilling personnel will understand.

Engineers and drilling supervisory personnel need to make predictive calculations in order to make their drilling operations efficient and cost-effective. The prediction calculations for air and gas drilling technology are complicated and will require the creation of calculation computer programs. There are sophisticated air and gas programs available commercially. However, in the tradition of most engineering fields, once the basic outline of the program has been determined, we tend to hand the "care and feeding" of the program over to the computer science department. The authors have chosen to use MathCad™ as our tool to communicate to the readers the details of how air and gas drilling predictive calculations are made. These MathCad™ solutions are very transparent and are written in a sequence that we would do by hand.

The detailed MathCad™ solutions are given in the appendices and the results are summarized in each applicable text chapter. The solutions are presented in both USCS units and SI units. It is assumed that newcomers to this technology will eventually make the choice of whether to use an existing commercial program or to develop their own company internal program.

1.1 OBJECTIVES AND TERMINOLOGY

The objective of this professional text is to familiarize the readers with the basic terminology and operational applications of this new field of air and gas drilling technology. The use of this technology is limited to land drilling operations. There are three subcategories of this technology: (1) air and gas fluids drilling, (2) aerated fluids drilling, and (3) stable foam fluids drilling. This technology is utilized by the industry to fulfill two specific drilling objectives.

Performance Drilling: This type of drilling takes advantage of the low annulus bottom hole pressures that accompany the use of this technology. Low annulus bottom hole pressures usually result in higher rates of penetrations. This type of drilling operation is applied in the upper portions of a well bore above the potential producing reservoir formation. The specific objective of this type of drilling is to drill more rapidly through the upper formations above the reservoir and to ultimately reduce the cost of a drilling operation.

Underbalanced Drilling: Here again, this type of drilling takes advantage of the low annulus bottom hole pressure characteristic of the technology. Underbalanced drilling uses the various bottom hole pressure capabilities of air and gas, and aerated and stable foam drilling fluids to drill into potential reservoir rock producing formations with annulus bottom hole pressures lower than the static reservoir pressure. In this manner, the reservoir fluids flow to the well bore as the drill bit is advanced through the reservoir. Underbalanced drilling operations attempt to avoid damage to the reservoir rock formation so that the reservoir will produce effectively through its life.

1.2 ENGINEERING CALCULATIONS AND UNITS

Modern engineering practices can be traced back to early eighth century AD with the tradition of the "master builder." This was the time of the creation of the measure known as the Charlemagne foot. From the eighth century into the seventeenth century a variety of weights and measures were used throughout the world. It was not until the Weights and Measures Act of 1824 that a complete British Imperial System (BIS) was codified within the British Isles, British Commonwealth countries, and in some of the former colonies of the British. The United States actually did not accept the full BIS. The United States made use of some of the major units within the system and some of the older units that had evolved through the years of colonialism before 1824. This evolution of usage within the United States ultimately became the United States Customary System (USCS) (units that are still in common use today).

The development of some of the basic units that ultimately became part of the present-day System International d'Unites probably began around the time of Louis XIV of France. This system is simply known today as SI units (or metric units). This system became codified by international treaty in France in 1875. Most unit systems today, including the British Imperial System and the USCS, are referenced to the actual existing weights (mass) and measures of the SI units.

These reference weights and measures are kept in Paris, France, for all nations and other entities to utilize. Since 1875 the SI units system has gained rapid and widespread use throughout the world. This system is characterized by its consistent set of units and simplicity of use. SI units are based on multiples of decades or units of tens. All basic weights and measurement units within SI are in increasing magnitudes of multiples of 10, 100, 1000, etc. Nearly all other unit systems in use in international trade and commerce around the world today must be referenced against the SI standard units before they are considered legitimate for legal matters or for international commercial trade.

1.2.1 Physical Mechanics

There are important fundamental definitions of units that must be used to define any units system. These are as follow:

Force is the action of one body on another that causes acceleration of the second body unless acted on by an equal and opposite action countering the effect of the first body.

Time is a measure of the sequence of events. In Newtonian mechanics, time is an absolute quantity. In relativistic mechanics, it is relative to the *frame of reference* in which the sequence of events is observed. The common unit of time is seconds.

Inertia is that property of matter that causes a resistance to any change in the motion of a body.

Mass is a quantitative measure of *inertia*.

This monograph deals exclusively with Newtonian mechanics. Newton's general laws are as follow:

Law I. If a balanced force system acts on a particle at rest, it will remain at rest. If a balanced force system acts on a particle in motion, it will remain in motion in a straight line without acceleration.

Law II. If an unbalanced force system acts on a particle, it will accelerate in proportion to the magnitude and in the direction of the resultant force.

Law III. When two particles exert forces on each other, these forces are equal in magnitude, opposite in direction, and collinear.

Note that the aforementioned original definitions of the just-defined laws by Newton were conceived around the concept of force.

1.2.2 Basic Units and Usage

The USCS is a *gravitational* system, as its units of *length*, *force*, and *time* (i.e., L, F, and T, respectively) are considered fundamental dimensions of the system and all other units, including *mass*, are derived. The SI is an *absolute* system, as its units of *length*, *mass*, and *time* (i.e., L, M, and T, respectively) are considered fundamental dimensions of the system and all others units, including *force*, can be derived.

The reason for this distinction between *gravitational* and *absolute* is the laudable desire that the concept and magnitude of the mass of an object should remain the same regardless of where it is with respect to other objects that would influence it through gravitational attraction. In this manner, the SI would be in accordance with Newton's universal gravitation, which describes the universality of gravity (Newton's universal gravitation is an extension of Newton's general Law II given earlier). The average force of gravitational attraction is shown mathematically in Newton's universal gravitation by

$$F_{gravity} = G\frac{m_1 m_2}{d^2}, \tag{1-1}$$

where $F_{gravity}$ is the gravitational attraction force (lb, N), G is a constant of proportionality (3.437×10^{-8} lb-ft^2/slug^2, or 6.673×10^{-11} N-m^2/kg^2), m_1 is mass 1 (slug, kg), m_2 is mass 2 (slug, kg), and d is the distance between the masses (ft, m).

Newton's Law II can be written as

$$F = ma, \tag{1-2}$$

where F is the force being applied to a mass near the Earth's surface (lb, N), m is any object mass (slug, kg), and a is the resultant acceleration of that mass as a result of the applied force F (ft/sec^2, m/sec^2).

If a mass is on or near the Earth's surface, the force of attraction of the mass to the Earth's mass becomes the special force denoted as weight (assuming that no other forces act on the mass). In this situation, the acceleration term a becomes g, which is the gravitation acceleration of the mass falling freely toward the Earth's center. Substituting g into Equation (1-2) and letting the F terms in Equations (1-1) and (1-2) equal each other, g becomes

$$g = \frac{G m_{earth}}{d^2}. \tag{1-3}$$

Substituting the respective unit system values, Earth's average mass at midlatitudes, and the distance between the center of the Earth and the object near the Earth's surface gives the acceleration term that is used in most practical engineering mechanics problems. Table 1-1 gives the values of g for both USCS and SI. The high accuracy values are given with the commonly used engineering values.

Table 1-1. Acceleration of Gravity

	g (precision)	g (engineer)
USCS (ft/sec^2)	32.1740	32.2
SI (m/sec^2)	9.8067	9.81

Note that the Earth is not a perfect sphere and, therefore, the acceleration of gravity will be slightly different depending on whether the free falling body is at a pole or at the equator. The elliptical form of the Earth dictates that the acceleration of gravity will be slightly greater at the equator than at the poles. For most engineering applications at or near the Earth's surface, the average acceleration of gravity (engineer) terms is used for calculation purposes. Both of these calculations were made using the exact same calculation method. No special term or constant was employed to obtain either of the aforementioned results.

The objective of the third edition of this monograph is to allow engineers and other technologists to carry out the required air and gas drilling calculations in both USCS and SI units. In particular, the objective is for engineers and technologists who presently use the USCS units to learn to be comfortable using SI units. Most examples discussed within this monograph are steady-state flow problems. To facilitate this objective, two minor alterations in SI common usage have been made by the authors to allow for a more unified method of calculation manipulation that are common to the usage of USCS units. This will allow for transparency in the manipulation of both systems that practitioners working in both systems will recognize.

Nearly all equations in this monograph are derived for use with any consist set of units. Further, because most equations in the edition are for steady-state flow, the equations will have few mass (m) or gravity acceleration (g) terms. Therefore, the first alteration from traditional SI usage will be that any SI data or terms that contain *kg* units are to be changed to force units (N) by multiplying appropriately by 9.8 m/sec^2. An example of this first alteration would be the writing of power in SI units as N-m/sec (watt) instead of the SI purist form of the power unit of kg m^2/sec^3 (which is also a watt). The second alteration from traditional SI use will be that fluid flow pressure will be given in *N/cm*^{2} instead of the more SI purist recognized pascal (N/m^2). Regarding the latter alteration, it is very difficult for either the USCS or the SI practitioner to visualize the force per unit area magnitude being applied with pressure to the inside flow area of a 2-in (*50.8 mm*) nominal diameter pipe or applied as stress to a small machine part when the area of the pressure or stress unit is many times greater than the area of the actual flow area or stress area. Therefore, as is done in the USCS for most applications where a pressure or stress in lb/ft^2 is converted to lb/in^2 (psi), the pascal (*N/m*^{2}) will be converted to *N/cm*^{2}.

In essence, the authors recommend analyzing the SI system just like the USCS in that both are treated as an F, L, and T system, instead of treating the SI as an M, L, and T system. Even though this logic may annoy the SI purist, it will be shown that employing the SI mass concept (and the USCS force concept) in this manner is far superior to employing the cobbled-up unit conversions that define force and mass in the USCS as lb_f and lb_m, and the use the associated artificial constants g_c and g_o [2]. The authors have even seen this conversion concept applied to SI units in the form of N_f and N_m with their associated artificial constants g_c and g_o.

It is suspected that this misguided conversion creation was developed by chemical and mechanical engineering professors who were motivated by the desire to have a heat transfer equation that is rationally in consistent force

Table 1-2. Units that Define Mass [1]		
Unit System	**Mass Unit**	**Dimensions**
USCS	Slug	$F\ T^2/L = lb\ sec^2/ft$
SI	Kilogram	$F\ T^2/L = N\ sec^2/m$

(weight) units to correct to mass based equation [2]. Since specific heat c_p in the SI is in units of *kcal/kg-K* and in the USCS is in units of *BTU/lb-°R*, a heat transfer equation for SI calculations would have to be in mass rate of flow $\dot{m}$ and a heat transfer equation for the USCS would have to be in weight rate of flow $\dot{w}$. It is our contention that for those of us who were not brought up using the SI as our primary system of units, it is easier for engineers to carry out calculations in both systems if both unit systems are manipulated in exactly the same manner. This requires use of the kilogram definition given in Table 1-2. It is likely that this kilogram definition will "ruffle the feathers" of some SI purists.

There is one other important change to be made in the application of the SI to the drilling calculations made in this monograph. N/m^2 or the *pascal* (*Pa*) as a pressure unit will not be used. Most engineers have trouble correlating these values to small cross-sectional areas such as a 2-in nominal pipe (~i.d. of 2.0 in, or 50.8 mm). In its place we will use a pressure term unit of N/cm^2. It is only necessary to multiply this pressure unit by 10^4 to obtain pressure in *pascals* or by 0.6897 to obtain psi. The convenience of using this alternate SI pressure term is illustrated by Figure 1-1, where both scales can be placed easily on the same gauge face for easy reading (and reference to one another).

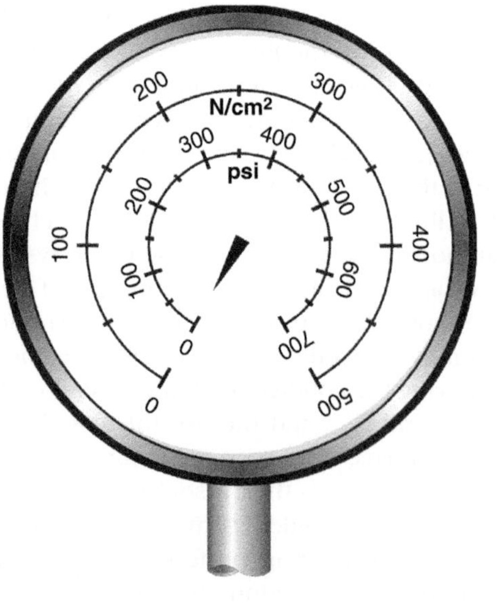

FIGURE 1-1. Pressure gauge with both psi and N/cm^2 units.

REFERENCES

1. Daugherty, R. L., Franzini, J. B., and Finnemore, E. J., *Fluid Mechanics with Engineering Applications*, Eighth Edition, McGraw-Hill, 1985.

2. *SPE Alphabetical List of Units*, SPE Publication, 2006.

CHAPTER

Air and Gas Versus Mud

2

This engineering practice monograph has been prepared for petroleum and related drilling and completion engineers and technicians who work in modern rotary drilling operations. This book derives and illustrates engineering calculation techniques associated with air and gas drilling technology. This book has been written in consistent units to ease application in either USCS or SI. Also, field unit equation use has been minimized in the text. Chapter 1 and Appendix A give definitions of important units and constants and useful conversions for both USCS and SI.

Air and gas drilling technology is the utilization of compressed air or other gases as a rotary drilling circulating fluid to carry the rock cuttings to the surface that are generated at the bottom of the well by the advance of the drill bit. The compressed air or other gas (e.g., nitrogen or natural gas) can be used also or can be injected into the well with incompressible fluids such as fresh water, formation water, formation oil, or drilling mud. There are three distinct operational applications for this technology: air or gas drilling operations (using only compressed air or other gas as the circulating fluid), aerated drilling operations (using compressed air or other gas mixed with an incompressible fluid), and stable foam drilling operations (using compressed air or other gas with an incompressible fluid to create a continuous foam circulating fluid).

In the past, air and gas drilling methods have been a small segment of the petroleum deposit recovery drilling industry. Currently, air and gas drilling methods that utilize compressed air (or other gases), aerated fluids, or foam fluids comprise about 20 to 30% of all operations. There are two separate and unique reasons for utilizing air and gas drilling methods in modern oil and gas deposit recovery operations. These are:

Performance Drilling: Drilling formations above a potential producing formation to generally take advantage of increased rates of penetration of these drilling methods.

Underbalanced Drilling: Drilling of potential producing formations using annulus bottom hole pressures that are below the formation pore pressure. This reduces or eliminates formation damage that could affect follow-on production.

In general, the use of air and gas drilling methods is confined to mature sedimentary basins within mature geologic provinces.

Pneumatic conveying represents the first use of moving air to transport entrained solids in the flowing stream of air. This airstream was created by steam-powered fans that were the direct outgrowth of the industrial revolution of the early sixteenth century. Pneumatic conveying was accomplished on an industrial scale by the late 1860s [1]. The need for higher pressure flows of air and other gases led to the first reliable industrial air compressors (stationary) in the late 1870s [2]. Here again, these early compressors were steam powered. After the development of the internal combustion engines, portable reciprocating and rotary compressors were possible. These portable compressors were first utilized in the late 1880s by an innovative mining industry to drill in mines using pneumatic-actuated hammers for in mine wall boreholes and shaft pilot boreholes [2].

2.1 ROTARY DRILLING

Rotary drilling is a method used to drill deep boreholes in rock formations of the Earth's crust. This method is comparatively new, having been first developed by a French civil engineer, Rudolf Leschot, in 1863 [3]. The method was initially used to drill water wells using fresh water as the circulation fluid. Today, this method is the only rock drilling technique used to drill deep boreholes (greater than 3000 ft, or *900 m*). It is not known when air compressors were first used for the drilling of water wells, but it is known that deep petroleum and natural gas wells were drilled utilizing portable air compressors in the 1920s [4]. Pipeline gas was used to drill a natural gas well in Texas in 1935 using reverse circulation techniques [5].

Today, rotary drilling is used to drill a variety of boreholes. Most water wells and environmental monitoring wells drilled into bedrock are constructed using rotary drilling. In the mining industry, rotary drilling is used to drill ore body test boreholes and pilot boreholes for guiding larger shaft borings. Rotary drilling techniques are used to drill boreholes for water, oil, gas, and other fluid pipelines that need to pass under rivers, highways, and other natural and man-made obstructions. Most recently, rotary drilling is being used to drill boreholes for fiber optics and other telecommunication lines in obstacle-ridden areas such as cites and industrial sites. The most sophisticated application for rotary drilling is the drilling of deep boreholes for the recovery of natural resources such as crude oil, natural gas, and geothermal steam and water. Drilling boreholes for fluid resource recovery usually requires boreholes drilled to depths of 3,000 ft *(900 m)* to as great as 20,000 ft *(6000 m)*.

Rotary drilling is highly versatile. The rotary drilling applications given previously require the drilling of igneous, metamorphic, and sedimentary rock. However, the deep drilling of boreholes for the recovery of crude oil and natural gas is carried out almost exclusively in sedimentary rock. Boreholes for the recovery of geothermal steam and water are constructed in all three rock types. The rotary drilling method requires the use of a rock cutting or crushing drill bit. Figure 2-1 shows a tungsten carbide insert tricone roller cone bit. This type of drill bit uses more of a crushing action to advance the bit in the rock (see Chapter 4 for more details). These bits are used primarily to drill medium hard sedimentary rock.

To advance the drill bit in rock requires the application of an axial force on the bit (to push the bit into the rock face), torque on the bit (to rotate the bit against the resistance of the rock face), and circulating fluid to clear the rock cuttings away from the bit as the bit generates more cuttings with its advance (see Figure 2-2). If the axial force is missing and the other two processes are operating, then the bit will not advance. Likewise, if torque is not present and the other two processes are operating, then again the bit will not advance. However, if circulation is not present and the other two processes are present, the drill string will likely be damaged. This short discussion emphasizes the critical nature of the circulating system.

FIGURE 2-1. Tungsten carbide insert 7 7/8 inch (*200.1 mm*) tricone roller cutter bit IADC Code 627 (courtesy of Hughes Christensen Incorporated).

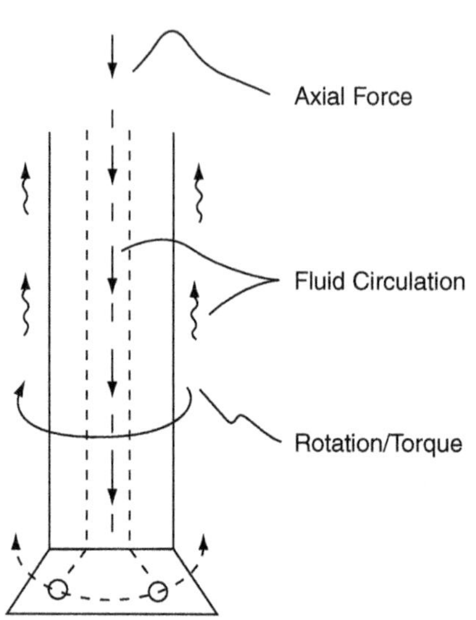

FIGURE 2-2. The three necessary components for rotary drilling.

Rotary drilling is carried out with a variety of drilling rigs. These can be small "single" rigs or larger "double" and "triple" rigs. Today, most of the drilling rigs based on land are mobile units with folding masts. A single drilling rig has a vertical space in its mast for only one joint of drill pipe. A double drilling rig has a vertical space in its mast for two joints of drill pipe and a triple drilling rig space for three joints. Table 2-1 gives API length ranges for drill collars and drill pipe [6].

Figure 2-3 shows a typical single drilling rig. Such small drilling rigs are highly mobile and are used principally to drill shallow (less than 3000 ft, or *900 m* in depth), such as coal-bed methane wells and geothermal hot water-producing boreholes. These single rigs are usually self-propelled. The self-propelled drilling rig shown in Figure 2-3 is a Gardner Denver SD 55. This particular rig uses a range 2 drill pipe.

Single rigs can be fitted with either an onboard air compressor or an onboard mud pump. Some of these rigs can accommodate both subsystems. These rigs have

Table 2-1. API Length Ranges for Drill Collars and Drill Pipe

Range	Minimum Length	Maximum Length
1	18 ft (*5.5 m*)	22 ft (*6.7 m*)
2	27 ft (*8.7 m*)	30 ft (*9.1 m*)
3	38 ft (*11.6 m*)	45 ft (*13.7 m*)

FIGURE 2-3. Typical self-propelled single drilling rig (courtesy of George E. Failing Company).

either a dedicated prime mover on the rig deck or a power take-off system that allows utilization of the truck motor as the prime mover for the drilling rig equipment (when the truck is stationary). Small drilling rigs can provide axial force to push the drill bit into the rock face through the drill string (via a chain or cable actuated pull-down system, or a hydraulic pull-down system). A pull-down system transfers a portion of the weight of the rig to the top of the drill string and then to the drill bit. The torque and rotation at the top of the drill string are also often provided by a hydraulic top-head drive (similar to power swivel systems used on larger drilling rigs), which is moved up and down the mast (on a track) by the chain, cable, or hydraulic drive pull-down system. However, some of these rigs retain the traditional rotary table. Many of these small single drilling rigs are capable of drilling with their masts at angles as high as 45° to the vertical. The prime mover for these rigs is usually fueled by either propane or diesel.

The schematic layout in Figure 2-4 shows a typical self-propelled double drilling rig. This example rig is fitted with a mud pump for circulating drilling mud. A vehicle motor is used to propel the rig over the road.

The same motor is used in a power take-off mode to provide power to the rotary table, draw works, and mud pump. For this rig, this power take-off motor operates a hydraulic pump that provides fluid to hydraulic motors to operate the

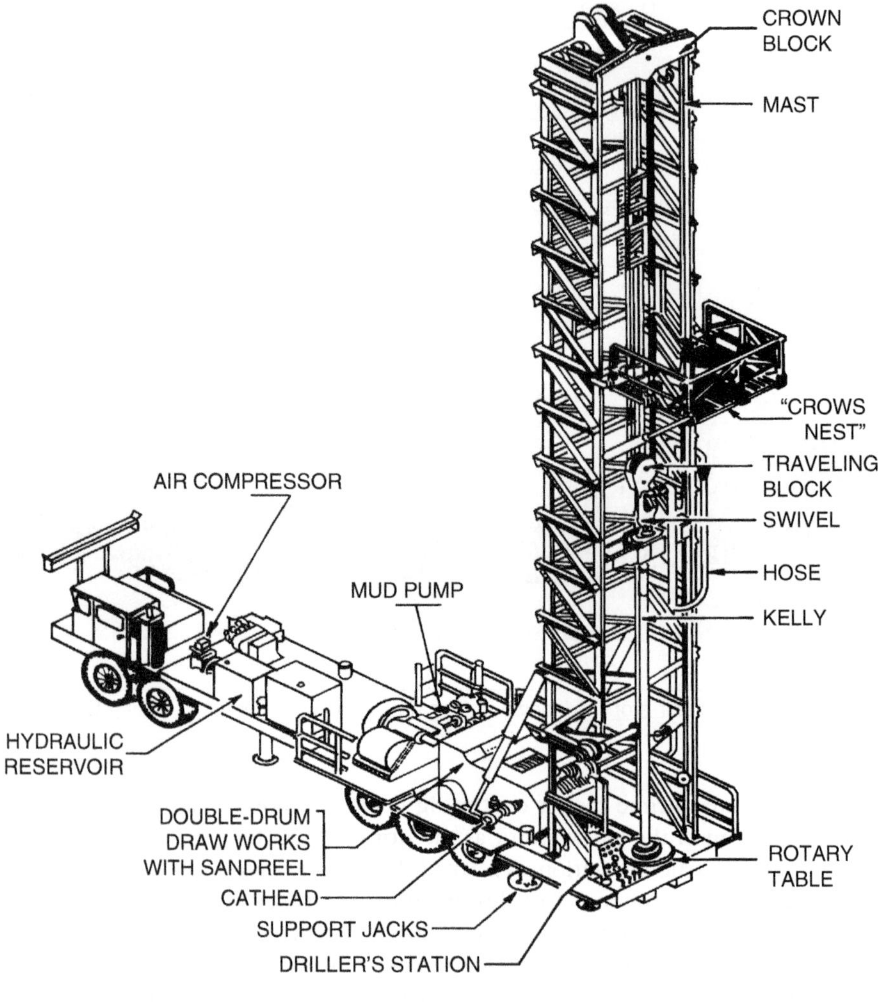

FIGURE 2-4. Typical self-propelled double drilling rig schematic layout.

rotary table, draw works, and mud pump. The "crows nest" on the mast indicates that the rig is capable of drilling with a stand of two joints of drill pipe. This drilling rig utilizes a rotary table and a kelly to provide torque to the top of the drill string. The axial force on the bit is provided by the weight of the drill collars at the bottom of the drill string (there is no chain pull-down capability for this drilling rig). This example schematic shows a rig with onboard equipment that can provide only drilling mud or treated water as a circulation fluid. The small air compressor at the front of the rig deck operates the pneumatic controls of the rig. However, this rig can easily be fitted for air and gas drilling operations. This type of drilling rig (already fitted with a mud pump) would require an auxiliary hookup to an external air compressor(s) to carry out an air drilling operation. The compressor system and its associated equipment for air drilling operations are usually provided by a subcontractor specializing in these operations.

Figure 2-5 shows a new type of triple drilling rig for land operations. This rig is fitted with a power swivel instead of a rotary table. These new Flex Rigs must be assembled at the drilling location.

In addition to having the most modern drilling equipment (automatic pipe loader, single console operation, etc.), the design uniqueness of the Flex Rig

FIGURE 2-5. Helmerick and Payne Flex Rig 3 (courtesy of Helmerick and Payne Incorporated).

concept allows rapid rig-up and rig-down, which minimizes nondrilling time at the location. These rigs have been developed by Helmerick and Payne Incorporated, an international drilling contractor serving both on-land and off-shore drilling operations. These rigs are available in depth capabilities up to 18 000 ft *(5500 m)*.

2.2 CIRCULATION SYSTEMS

Two types of circulation techniques can be used for either a mud drilling system or an air or gas drilling system. These are direct (conventional) circulation and reverse circulation.

2.2.1 Direct Circulation

Figure 2-6 shows a schematic of a rotary drilling, direct circulation mud system that would be used on a typical double (and triple) drilling rig. Direct circulation requires that the drilling mud (or treated water) flows from the slush pump

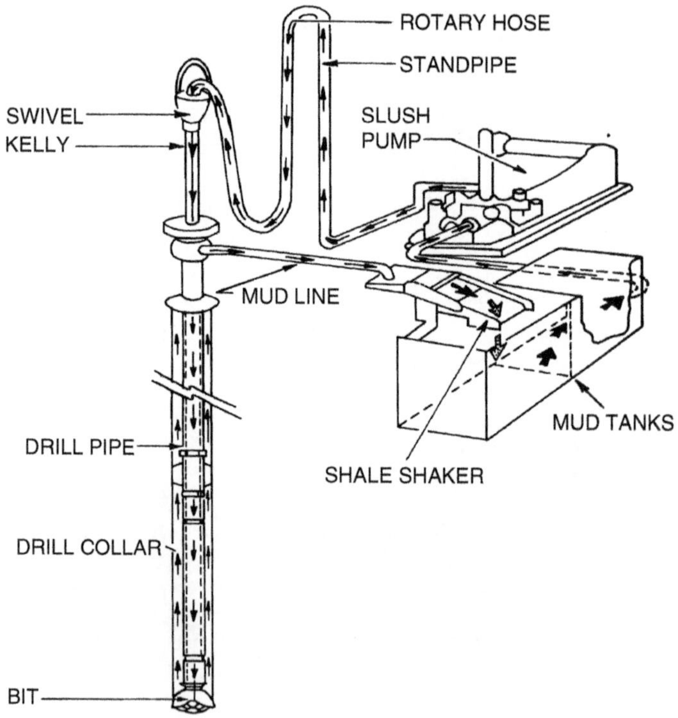

FIGURE 2-6. Direct circulation mud system.

(or mud pump), through the standpipe on the mast, through the rotary hose, through the swivel and down the inside of the kelly, down the inside of the drill pipe and drill collars, and through the drill bit (at the bottom of the borehole) into the annulus space between the outside of the drill string and the inside of the borehole.

The drilling mud entrains the rock bit cuttings at the bottom of the annulus and then flows with the cuttings up the annulus to the surface where the cuttings are removed from the drilling mud by the shale shaker and the drilling mud is returned to the mud tanks (where the slush pump suction side picks up the drilling mud and recirculates the mud back into the well). The slush pumps used on double (and triple) drilling rigs are positive displacement piston-type pumps.

For single drilling rigs, the drilling fluid is often treated fresh water in the mud tank. A heavy-duty hose is run from the suction side of the onboard mud pump (see Figure 2-4) to the mud tank. The drilling water is pumped from the tank, through the pump, through an onboard pipe system, through the rotary hose, through the hydraulic top-head drive, down the inside of the drill pipe, and through the drill bit to the bottom of the well. The drilling water then entrains the rock cuttings from the advance of the bit and carries the cuttings to the surface via the annulus between the outside of the drill pipe and the inside of the borehole. At the surface, the drilling fluid (water) from the annulus with entrained cuttings is returned to a mud pit where the rock cuttings are allowed to settle out to the bottom. The pumps on single drilling rigs are small positive displacement reciprocating piston types.

Figure 2-7 shows a detailed schematic of a direct circulation compressed air drilling system that would be used on a typical double or triple drilling rig.

Direct circulation requires that atmospheric air be compressed by the compressor and then forced through the standpipe on the mast, through the rotary hose, through the swivel and down the inside of the kelly, down the inside of the drill pipe and drill collars, and through the drill bit (at the bottom of the borehole) into the annulus space between the outside of the drill string and the inside of the borehole. The compressed air entrains the rock bit cuttings and then flows with the cuttings up the annulus to the surface where the compressed air and the entrained cuttings exit the circulation system via the blooey line. The compressed air and cuttings exit the blooey line into a large pit dug into the ground surface (burn pit). These pits are lined with an impermeable plastic liner.

If compressed natural gas is to be used as a drilling fluid, a gas pipeline is run from a main natural gas pipeline to the drilling rig. Often this line is fitted with a booster compressor. This allows the pipeline natural gas pressure to be increased (if higher pressure is needed) before the gas reaches the drilling rig standpipe.

2.2.2 Reverse Circulation

Rotary drilling reverse circulation (using either drilling mud and/or compressed air or gas) can be a useful alternative to direct circulation methods. The reverse

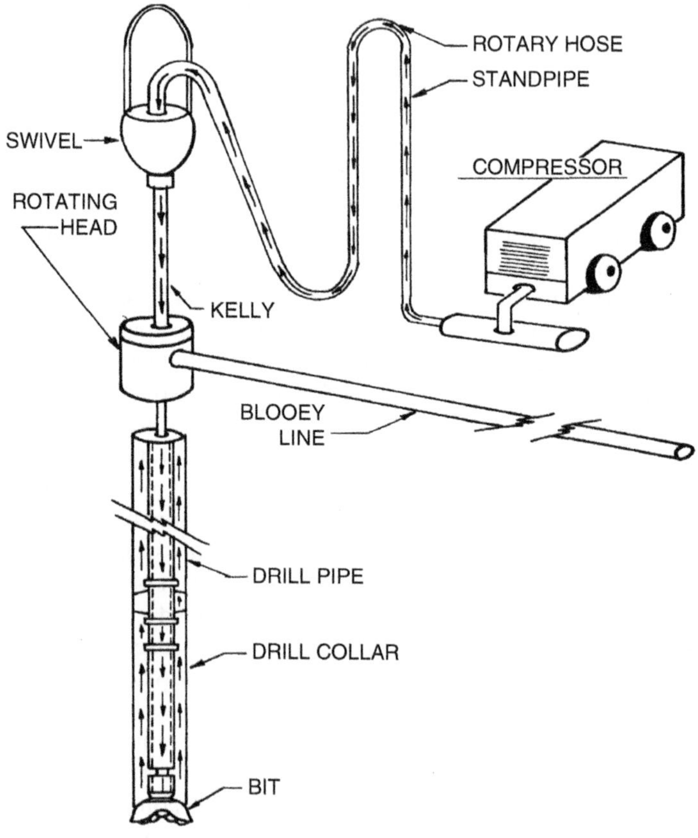

FIGURE 2-7. Direct circulation air system.

circulation technique is particularly useful for drilling relatively shallow large diameter boreholes (e.g., conductor and surface casing boreholes).

In a typical reverse circulation operation utilizing drilling mud, the drilling mud flows from the mud pump to the top of the annulus between the outside of the drill string and the inside the borehole, down the annulus space to the bottom of the borehole. At the bottom of the borehole the drilling mud entrains the rock bit cuttings and flows through the large center opening in the drill bit and then upward to the surface through the inside of the drill string. At the surface, the cuttings are removed from the drilling mud by the shale shaker and the drilling mud is returned to the mud tanks (where the pump suction side picks up the drilling mud and recirculates the mud back to the well).

Reverse circulation can also be carried out using air and gas drilling techniques. Figure 2-8 shows a typical application of reverse circulation using compressed air as the drilling fluid (or mist, unstable foam) [7]. This example is a

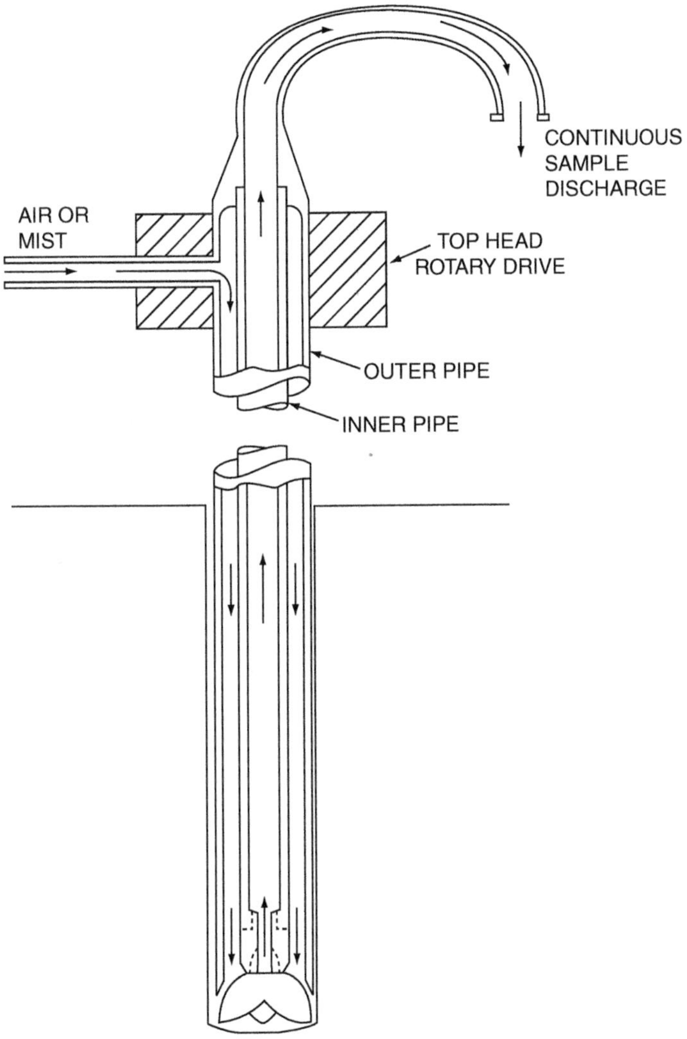

FIGURE 2-8. Dual tube (or dual drill pipe) closed reverse circulation operation.

dual tube (or dual drill pipe) closed reverse circulation system. The closed system is characterized by an annulus space bounded by the inside of the outer tube and the outside of the inner tube. This is a specialized type of reverse circulation and is usually limited to small single and double drilling rigs with top-head rotary drives (see Chapter 4 for drill pipe details).

Reverse circulation drilling operations require specially fabricated drill bits. Figure 2-9 shows a schematic of the interior flow channel of a tricone rotary drill

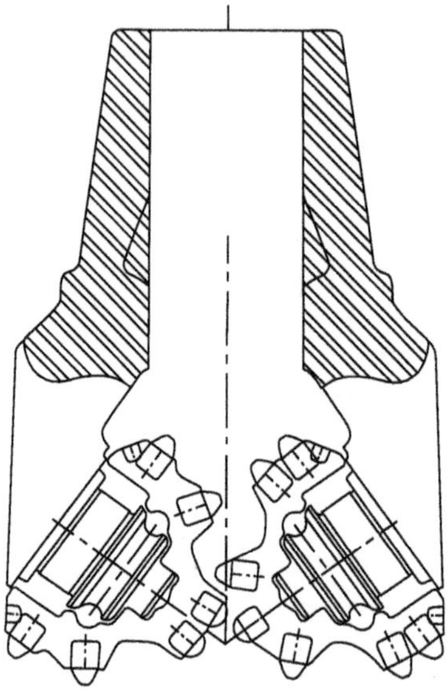

FIGURE 2-9. Schematic of the internal flow channel of a tricone roller cutter bit designed for reverse circulation operations (courtesy of Smith International Incorporated).

bit designed for reverse circulation. These drill bits utilize typical roller cutter cones exactly like those used in direct circulation drill bits (see Figure 2-1). These bits, however, have a large central channel opening that allows the circulation fluid flow with entrained rock cuttings to flow from the bottom of the borehole to the inside of the drill string and then to the surface.

Most tricone drill bits with a diameter of 5 3/4 inches (*146 mm*) or less are designed with the central flow channel as shown in Figure 2-9. Figure 2-1 showed the typical tricone drill bit for direct circulation operations. These direct circulation drill bits usually have three orifices that can be fitted with nozzles. Tricone roller cutter drill bits for reverse circulation operations are available in diameters from 4 1/2 inches (*114 mm*) to 31 inches (*787 mm*). The larger diameter bits for reverse circulation operations are usually custom designed and fabricated. Dual wall pipe reverse circulation operations require special skirted drill bits (see Chapter 4 for details). These skirted drill bits are specifically designed for the particular drilling operation. These specialized drill bits are usually manufactured by mining equipment companies.

2.3 COMPARISON OF MUD AND AIR DRILLING

The direct circulation model is used to make some important comparisons between mud drilling and air and gas drilling operations.

2.3.1 Advantages and Disadvantages

There are some very basic advantages and disadvantages to mud drilling and air drilling operations. The earliest recognized advantage of air and gas drilling technology was the increase in drilling penetration rate relative to mud drilling operations. Figure 2-10 shows a schematic of the various drilling fluids (the top four comprise air and gas drilling technology) and how these drilling fluids affect the drilling penetration rate. The drilling fluids in Figure 2-10 are arranged with the lightest at the top of the list and the heaviest at the bottom. The lighter the fluid column in the annulus (with entrained rock cuttings), the lower the confining pressure on the rock bit cutting face. This lower confining pressure allows the rock cuttings from the rock bit to be removed more easily from the cutting face (see Chapter 4 for more details).

Figure 2-11 shows a schematic of the various drilling fluids and their respective potential for avoiding formation damage. Formation damage is an important issue in

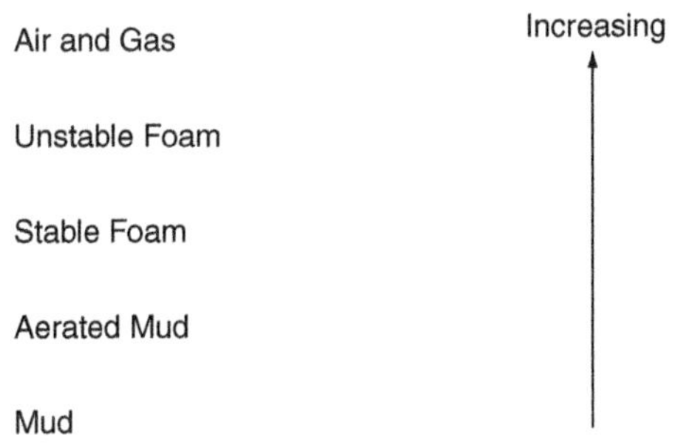

FIGURE 2-10. Improved penetration rate.

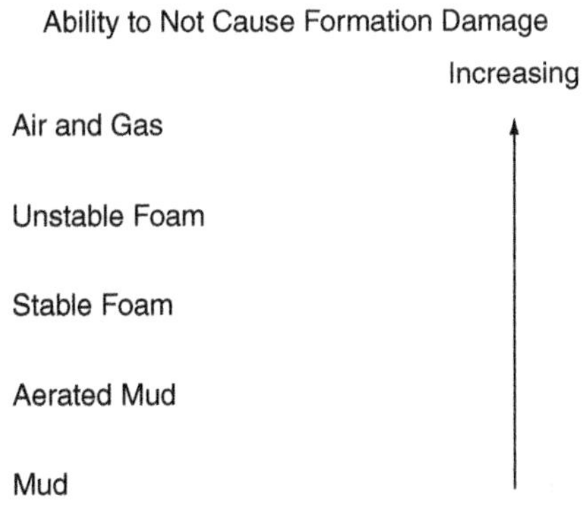

FIGURE 2-11. Formation damage avoidance.

fluid resource recovery (e.g., oil and natural gas, and geothermal fluids). The lighter the fluid column in the annulus (with entrained rock cuttings), the lower the potential for formation damage (arrow points upward to increasing avoidance of formation damage). Formation damage occurs when the fluid column pressure at the bottom of the borehole is higher than the pore pressure of the resource fluid (oil, gas, or water) in the potential producing rock formations. This higher bottom hole pressure forces the drilling fluid (with entrained rock cutting fines) into the exposed fractures and pore passages in the producing rock formations. These fines plug these features in the immediate region around the borehole. This damage is often called a "skin effect." This skin effect damage restricts later formation fluid flows to the borehole, thus reducing the productivity of the well.

Figure 2-12 shows a schematic of the various drilling fluids and their respective potential for avoiding loss of circulation. Loss of circulation occurs when drilling with drilling muds or treated water through rock formations that have fractures or large interconnected pores or vugs. If these features are sufficiently large and are not already filled with formation fluids, then as drilling progresses the drilling fluid that had been flowing to the surface in the annulus can be diverted into these fractures or pore structures. This diversion can result in no drilling fluid (with entrained rock cuttings) returning to the surface. The rock cuttings are left in the borehole and consolidate around the lower portion of the drill string and the drill bit. If this situation is not identified quickly, the drill string will begin to torque up in the borehole and mechanical damage to the drill string will occur. Such damage can sever the drill string and result in a fishing job to retrieve the portion of the drill string remaining in the borehole.

For deep oil and natural gas recovery wells, loss of circulation can result in even more catastrophic situations. If drilling fluids are lost to thief formations, the fluid column in the annulus can be reduced, resulting in a lower bottom hole pressure. This low bottom hole pressure can cause a high pressure oil and/or natural gas "kick" or geothermal fluid "kick" (a slug of formation fluid) to enter the

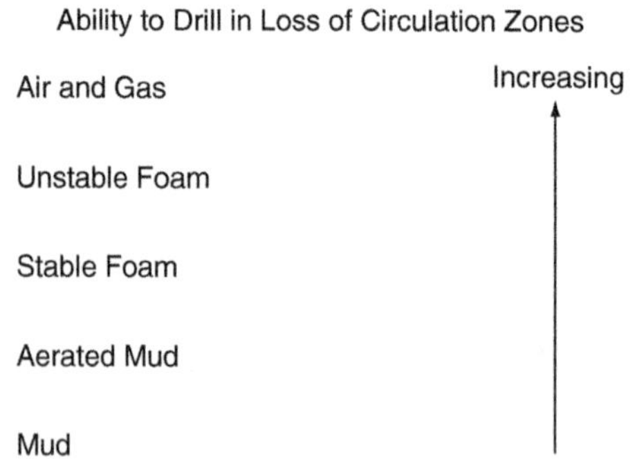

FIGURE 2-12. Loss of circulation avoidance.

annulus. Such kicks must be immediately and carefully circulated out of the annulus (to the surface), as an uncontrolled blowout of the well could occur. Here again heavier drilling fluids are generally more prone to loss of circulation (arrow points upward to increasing loss of circulation avoidance).

Figure 2-13 shows a schematic of the various drilling fluids and their respective potential for use in geologic provinces with high pore pressures. High pore pressures are encountered in oil, natural gas, and geothermal drilling operations. New discoveries of oil, natural gas, or geothermal fluid deposits are usually highly pressured. In order to safely drill boreholes to these deposits, heavily weighted drilling muds are utilized. The heavy fluid column in the annulus provides the high bottom hole pressure needed to balance (or overbalance) the high pore pressure of the deposit.

Figure 2-13 also shows that the heavier the drilling fluid column in the annulus, the more useful the drilling fluid is for controlling high pore pressure (the arrow points downward to increasing capability to control high pore pressure). There are limits to how heavy a drilling mud can be. As discussed earlier, too heavy a drilling mud results in overbalanced drilling, which can result in formation damage. However, there is a greater risk to overbalanced drilling. If the drilling mud is too heavy, the rock formations in the open hole section can fracture. These fractures will result in a loss of the circulating mud in the annulus that could result in a blowout.

In the past decade it has been observed that drilling with a circulation fluid that has a bottom hole pressure slightly below that of the pore pressure of the fluid deposit gives near optimum results. This type of drilling is denoted as underbalanced drilling. Underbalanced drilling allows the formation to produce fluid as the drilling progresses. This lowers or eliminates the risk of formation damage and eliminates the possibility of formation fracture and loss of circulation. In general, if the pore pressure of a deposit is high, an engineered adjustment to the

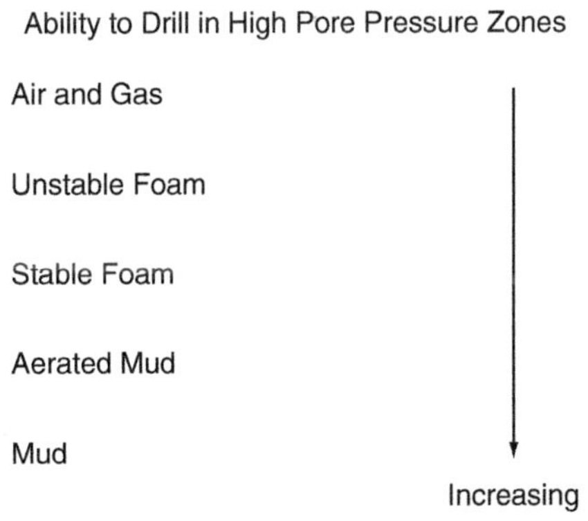

FIGURE 2-13. Controlling high pore pressure.

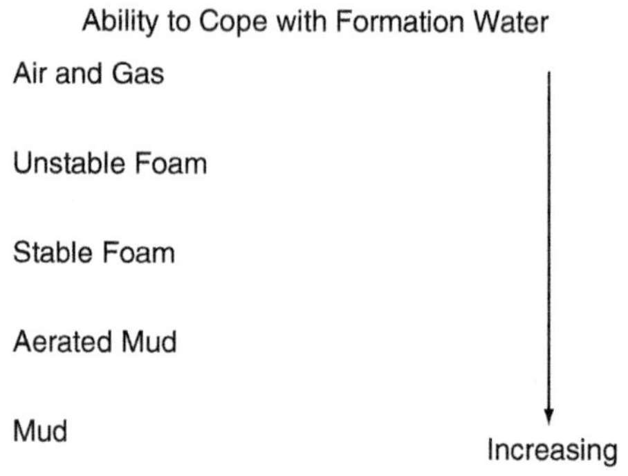

FIGURE 2-14. Control of the inflow of formation water.

drilling mud weight (with additives) can yield the appropriate drilling fluid to assure underbalanced drilling. However, if the pore pressure is not unusually high, then air and gas drilling techniques are required to lighten the drilling fluid column in the annulus.

Figure 2-14 shows a schematic of the various drilling fluids and their respective potential for keeping formation water out of the drilled borehole. Formation water is often encountered when drilling to a subsurface target depth.

This water can be in fractures and pore structures of the rock formations above the target depth. If drilling mud is used as the circulating fluid, the hydrostatic pressure of the mud column in the annulus is designed to be sufficient to keep formation water from flowing out of the exposed rock formations in the borehole. Lighter drilling fluids, such as compressed air or other gases and foam, have lower bottom hole pressure and, therefore, lower the pressure on any formation water in the exposed fractures or pore structures in the drilled rock formations.

Figure 2-14 shows that the heavier drilling fluids have a greater ability to cope with formation water flow into the borehole (the arrow points downward to increasing control of formation water). This has always been a distinct advantage for deep targets in young immature sedimentary rock formations. In some drilling situations, a foam drilling fluid system can be designed to take on formation water when the foam system has extra foamer added to it. Aerated drilling fluids are capable of handling influxes of formation fluids. However, all these air and gas based drilling fluids are restricted to older mature sedimentary geologic provinces.

2.3.2 Flow Characteristics

A comparison is made of the flow characteristics of mud drilling and air drilling in a deep well example. A schematic of this example well is shown in Figure 2-15.

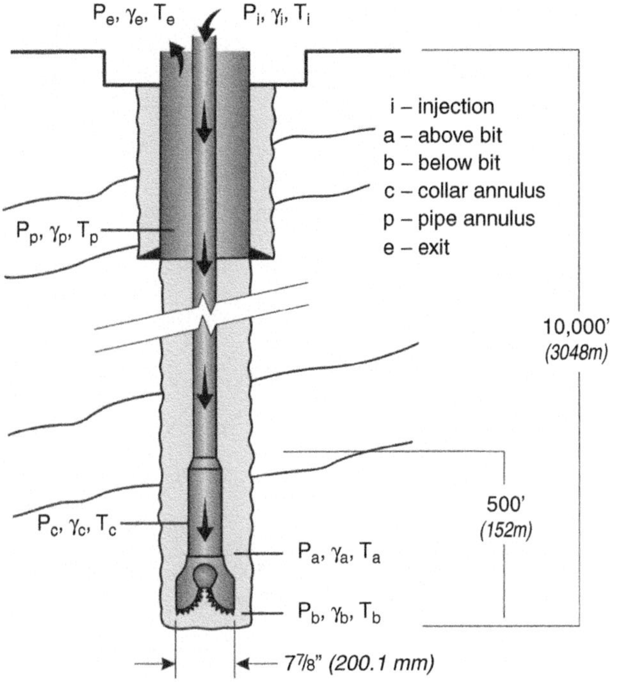

FIGURE 2-15. Comparison example well and drill string.

The well is cased from the surface to 7000 ft (*2133 m*) with API 8 5/8-in (*219 mm*) diameter, 32.00 lb/ft (*14.60 kg/m*) nominal casing. The well has been drilled out of the casing shoe with a 7 7/8-in (*200 mm*)-diameter drill bit. The comparison is made for drilling at 10,000 ft (*3048 m*). The drill string in the example well is made up of (bottom to top) a 7 7/8-in (*200 mm*)-diameter drill bit, ~500 ft (*152 m*) of 6 1/4-in (*159 mm*) outside diameter by 2 13/16-in (*71 mm*) inside diameter drill collars, and ~9500 ft (*2895 m*) of API 4 1/2-in (*114 mm*) diameter, 16.60-lb/ft (*7.52 kg/m*) nominal, IEU-S135, NC 46, drill pipe.

The mud drilling hydraulics calculations are carried out assuming that the drilling mud weight (i.e., specific weight) is 10 lb/gal (*density of 1201 kg/m^3 or 1.2 kg/liter*), the Bingham mud yield is 10 lb/100 ft^2 (*48 N/10 m^2*), and the plastic viscosity is 30 centipose. The drill bit is assumed to have three 12/32-in (*9.5 mm*)-diameter nozzles with a drilling mud circulation flow rate of 200 gal/min (*757 liters/min*). Figure 2-16 shows plots of the pressures in the incompressible drilling mud as a function of depth and shows a plot of the pressures inside the drill string. The injection pressure is approximately 1072 psig (*739 N/cm^2 gauge*) and 5877 psig (*4049 N/cm^2 gauge*) at the bottom of the inside of the drill string just above the bit nozzles. Also shown is a plot of the pressure in the annulus. The

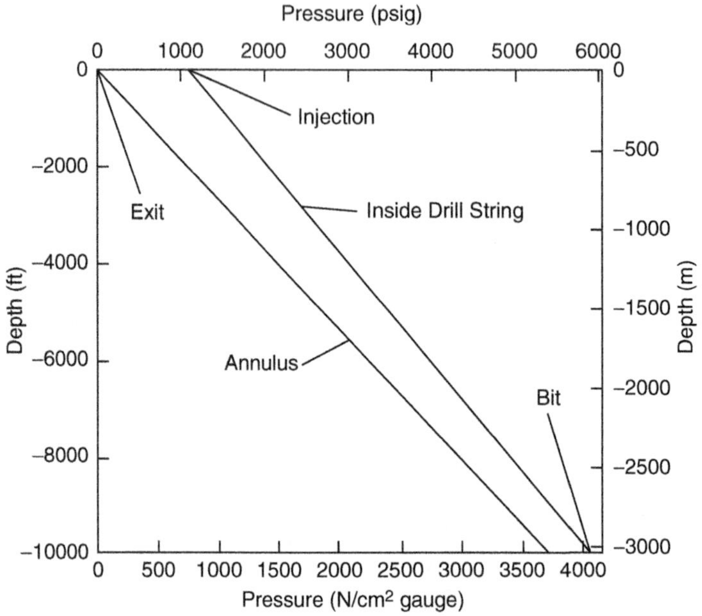

FIGURE 2-16. Mud drilling pressure versus depth.

pressure is approximately 5401 psig (*3721 N/cm² gauge*) at the bottom of the annulus just below the bit nozzles and 0 psig (*0 N/cm² gauge*) at the top of the annulus at the surface.

The pressures in Figure 2-16 reflect the hydrostatic weight of the column of drilling mud and the resistance to fluid flow from the inside and outside surfaces of the drill string, the surfaces of the borehole wall, and the bit nozzle orifices.

This resistance to flow is the result of friction losses of energy in the fluid. The total losses due to friction are the sum of all the aforementioned losses. This mud drilling example shows a total loss through the system of approximately 1071 psi (*739 N/cm²*). This includes the approximate 476 psi (*328 N/cm²*) loss through the drill bit. Smaller diameter nozzles would yield higher losses across the drill bit and higher injection pressures at the surface.

The air drilling calculations are carried out assuming the drilling operation is at sea level. There are two compressors capable of 1200 scfm (*566.3 standard liters/sec*) each, so the total volumetric flow rate to the drill string is 2400 scfm (*1133 standard liters/sec*). The drill bit is assumed to have three open orifices at 0.75 in (*19 mm*) diameter. Figure 2-17 shows the plots of the pressures in the compressible air as a function of depth and shows a plot of the pressure inside the drill string. The pressure is approximately 245 psia (*169 N/cm² absolute*) at injection and 201 psia (*139 N/cm² absolute*) at the bottom of the inside

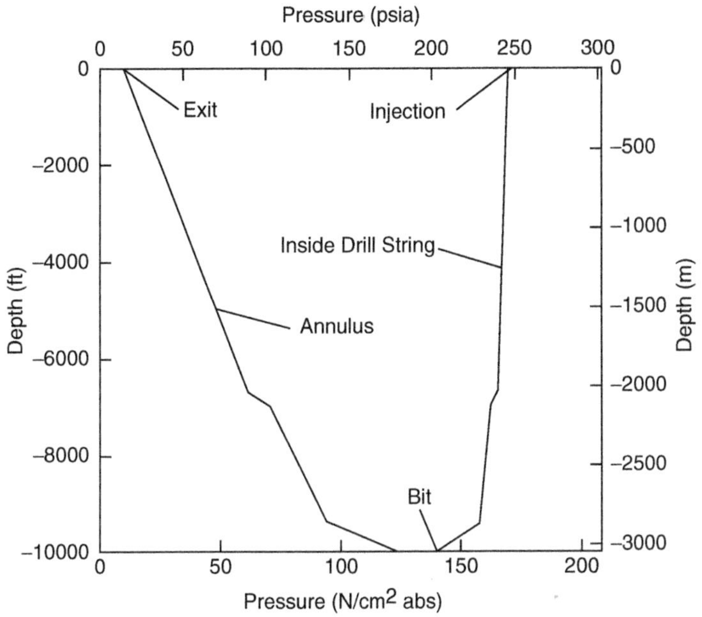

FIGURE 2-17. Air drilling pressure versus depth.

of the drill string just above the bit orifices. Also shown is a plot of the pressure in the annulus. The pressure is approximately 180 psia (*124 N/cm² absolute*) at the bottom of the annulus just below the bit orifices and 14.7 psia (*10.1 N/cm² absolute*) at the end of the blooey line at the surface (top of the annulus).

As in the mud drilling example, the pressures in Figure 2-17 reflect the hydrostatic weight of the column of compressed air and the resistance to air flow from the inside surfaces of the drill string and the surfaces of the annulus. This resistance to flow is due to friction losses. In this example the fluid is compressible. Considering the flow inside the drill string, the friction losses component dominates the hydrostatic weight component in the column, as the injection pressure into the inside of the drill string is higher than the pressure at the bottom of the drill string (inside the drill string just above the bit open orifices). This phenomenon is a strong function of the inside diameter of the drill string, and the type of choke or restriction is at the bottom of the inside of the drill string. For example, if the drill pipe diameter was larger and a smaller choke was present in the bottom of the drill string, the hydrostatic component would dominate.

Figure 2-18 shows the plots of the temperature in the incompressible drilling mud as a function of depth. The geothermal gradient for this example is 0.01°F/ft (*0.018°C/m*). Subsurface earth is nearly an infinite heat source. The drilling mud in a mud drilling circulation system is significantly denser than compressed

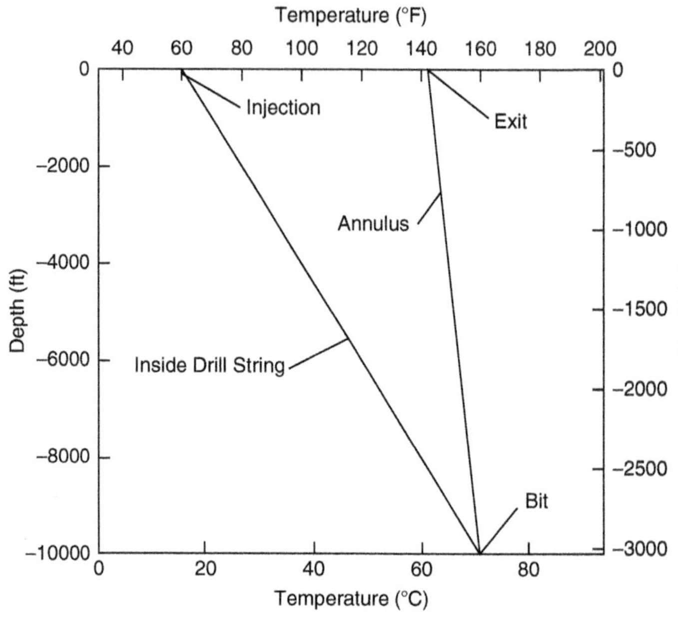

FIGURE 2-18. Mud drilling temperature versus depth.

air or other gases. Thus, as the drilling mud flows down the drill string and up through the annulus to the surface, heat is transferred from the rock formations through the surfaces of the borehole, through the drilling mud in the annulus, and through the steel drill string to the drilling mud inside. It is assumed that the drilling mud is circulated into the top of the inside of the drill string at 60°F (*15.6°C*).

As the drilling mud flows down the inside of the drill string, the drilling mud, because of its high specific heat, heats up as heat flows from the higher temperature rock formations and drilling mud in the annulus. At the bottom of the well the drilling mud temperature reaches the bottom hole temperature of 160°F (*71.1°C*). The drilling mud flowing up the annulus is usually laminar flow and is heated by the geothermal heat in the rock formation. The heated drilling mud flowing in the annulus heats the outside of the drilling string, which in turn heats the drilling mud flowing down the drill string. Because of its good heat storage capabilities, the drilling mud exits the annulus with a temperature greater than the injection temperature but less than the bottom hole temperature. In this example, the temperature of the drilling mud exiting the annulus is approximated to be the average of the injection and bottom hole temperatures, i.e., 130°F (*54.4°C*).

Figure 2-19 shows plots of the temperature in compressible air drilling fluid as a function of depth. The compressed air drilling fluid is significantly less dense

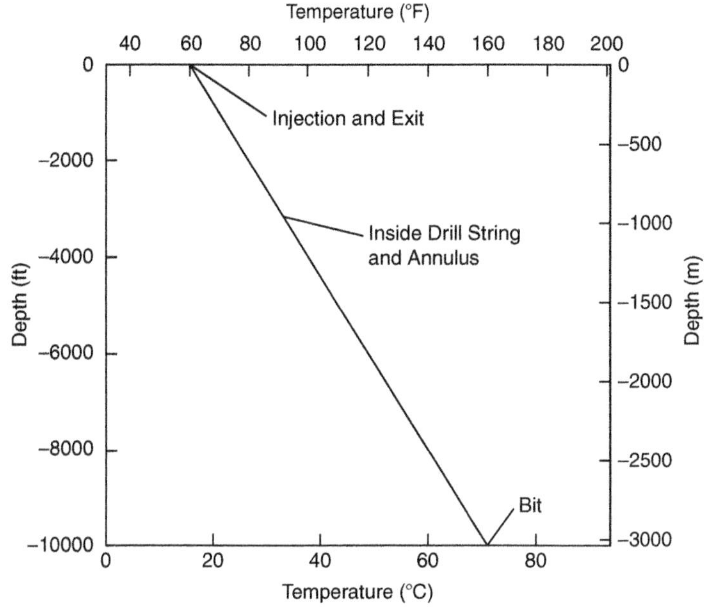

FIGURE 2-19. Air drilling temperature versus depth.

than drilling mud. Thus, compressed air has poor heat storage qualities relative to drilling mud. Also, compressed air flowing in the drilling circulation system is flowing rapidly and, therefore, the flow is turbulent inside the drill string and in the annulus. Turbulent flow is very efficient in transferring heat from the surface of the borehole to the flowing air in the annulus and in the inside of the drill string. It is assumed that the compressed air entering the top of the drill string at 60°F (*15.6°C*) will heat up rapidly and transfer heat in the rock formations to and from the steel piping and casing in the well. Under these conditions the compressed air exiting the annulus has approximately the same temperature as the air entering the top of the drill string.

Figure 2-19 shows that the temperature of the compressed air at any position in the borehole is approximately the same as the geothermal temperature at that depth. Thus, the temperature of the flowing air at the bottom of the hole is the bottom hole temperature of 160°F (*71.1°C*). There is some local cooling of the air as it exits the open orifices of the drill bit at the bottom of the hole. This cooling effect is more pronounced if nozzles are used in the drill bit (when using a down hole motor). This cooling effect is known as the Joule-Thomson effect and can be estimated [8]. However, it is assumed that this effect is small and that the air flow returns very quickly to the bottom hole geothermal temperature.

Figure 2-20 shows the plot of the specific weight of drilling mud for this example calculation. The drilling mud is incompressible and, therefore, the specific

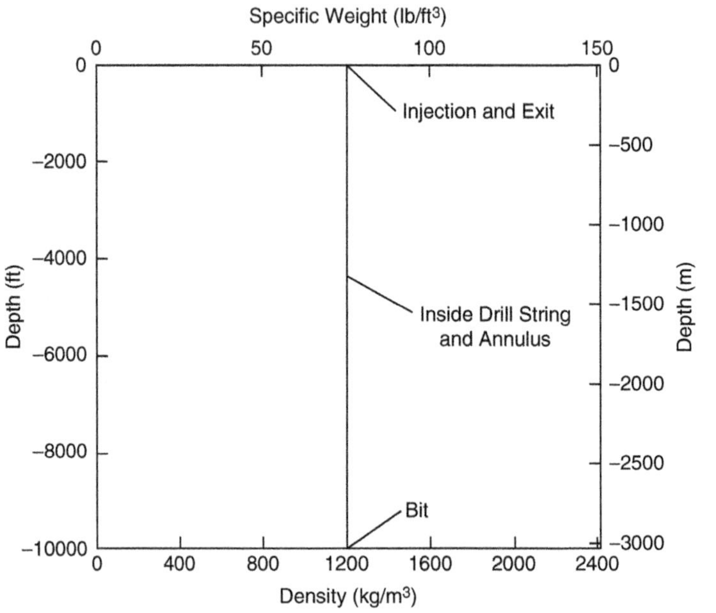

FIGURE 2-20. Mud drilling specific weight versus depth.

weight is 75 lb/ft^3 or 10 lb/gal (*density of 1.2 kg/liter*) at any position in the circulation system. There is some slight expansion of the drilling mud as a consequence of the increase in temperature as the drilling mud flows to the bottom of the well. This effect is quite small and is neglected in these engineering calculations.

Figure 2-21 shows the plot of the specific weight of the compressed air in this example. The compressed air is injected into the top of the drill string at a specific weight of 1.3 lb/ft^3 (*density of 20.8 kg/m^3*), a pressure of 250 psia (*172 N/cm^2 absolute*), and a temperature of 60°F (*15.6°C*). As the air flows down the drill string the pressure decreases. At the bottom of the inside of the drill string the specific weight is 0.84 lb/ft^3 (*density of 13.5 kg/m^3*), at the pressure of 190 psia (*131 N/cm^2 absolute*), and a temperature of 160°F (*71.1°C*). The compressed air exits the drill bit orifices into the bottom of the annulus (bottom of the well) with a specific weight of 0.77 lb/ft^3 (*density of 12.3 kg/m^3*), at a pressure of 180 psia (*124 N/cm^2 absolute*), and a temperature of 160°F (*71.1°C*).

As the compressed air flows to the surface through the annulus, it decompresses as it flows toward the atmospheric pressure at the exit to the blooey line. At the surface, the air exits the annulus (via the blooey line) with a specific weight of 0.0763 lb/ft^3 (*density of 1.22 kg/m^3*). The surface atmosphere for this example is assumed to be API Mechanical Equipment Standards standard atmospheric conditions, dry air, pressure of 14.696 psia (*10.1 N/cm^2 absolute*), and a temperature

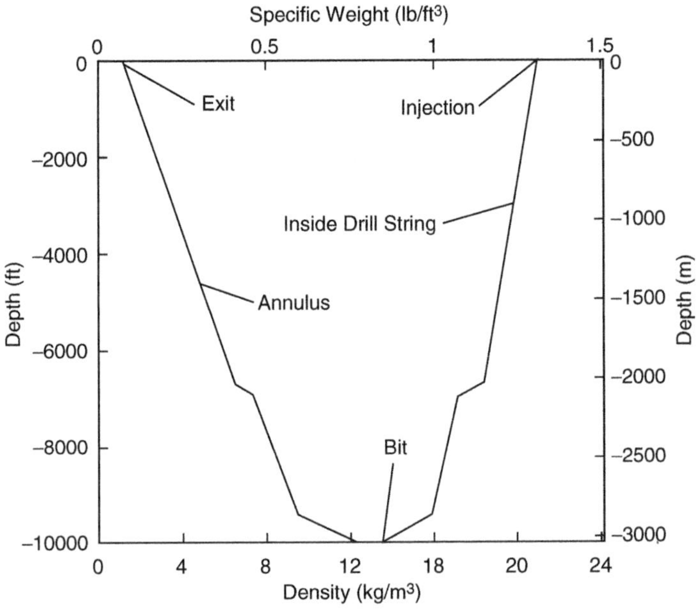

FIGURE 2-21. Air drilling specific weight versus depth.

of 60°F (*15.6°C*) [9]. Figure 2-21 shows a typical friction resistance-dominated drill string flow (as opposed to hydrostatic column weight dominated). This type of flow has a drill string injection pressure at the top that is higher than the pressure above the drill bit inside the drill string at the bottom of the string. Friction-dominated flow usually results when the drill bit is run with no nozzles.

2.3.3 Borehole Cleaning

Figure 2-22 is the concluding plot of these comparison calculations and shows a side-by-side comparison of the annulus velocities of the drilling mud and the compressed air as they flow to the surface. It is the power of these return flows up the annulus that keeps the rock cuttings entrained and moving to the surface at a rate that allows the drill bit to be safely advanced.

The drilling mud flows in the annulus around the drill collars with an average velocity of approximately 5.0 ft/sec (*1.5 m/sec*). The drilling mud slows to an average velocity of approximately 2.0 ft/sec (*0.61 m/sec*) in the annulus around the drill pipe (this is the critical velocity for the transportation of bit rock cuttings).

For the air drilling case, the lowest velocity of the compressed air flow in the annulus is just above the drill collars with a velocity of approximately 21.1 ft/sec

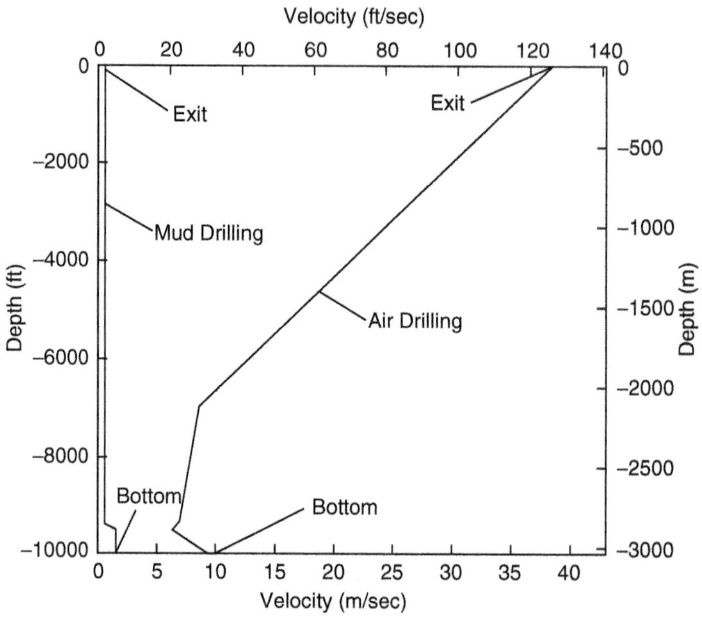

FIGURE 2-22. Annulus velocities for drilling mud and compressed air.

(*6.4 m/sec*). Here again, this is the critical velocity for the transportation of bit rock cuttings. In the air example, however, the air velocity increases up the annulus to approximately 126 ft/sec (*38.4 m/sec*) at the exit to the annulus.

It is instructive to compare the power (per unit volume) of these example flows at the positions in the annulus where the power is likely the lowest. For both of these examples the lowest power is just above the drill collars in the annulus around the bottom of the drill pipe. The kinetic energy per unit volume, *KE*, is [1, 10]

$$KE = \frac{1}{2}\rho V^2 \tag{2-1}$$

where KE is the kinetic energy per unit volume (lb-ft/ft^3, N-m/m^3), ρ is the density of the fluid (lb-sec^2/ft^4 = slug/ft^3, kg/m^3), and V is the average velocity of the fluid (ft/sec, m/sec).

The density of the fluid, ρ, is

$$\rho = \frac{\gamma}{g} \tag{2-2}$$

where γ is the specific weight of the fluid (lb/ft^3, N/m^3) and g is the acceleration of gravity (32.2 ft/sec^2, 9.81 m/sec^2).

In USCS units, in the mud drilling example, the specific weight of the drilling mud in the annulus of the drill collars is 75 lb/ft^3. Using this value in Equation (2-2), the density of the drilling mud is

$$\rho_m = \frac{75.0}{32.2}$$

$$\rho_m = 2.33\, lb - \sec^2/ft^4$$

or

$$\rho_m = 2.33\ \ slugs/ft^3.$$

The average velocity of the drilling mud at this position is approximately 2.0 ft/sec. Substituting these values into Equation (2-1), the kinetic energy per unit volume for the drilling mud example is

$$KE_m = \frac{1}{2}(2.33)(2.0)^2$$

$$KE_m = 4.7 lb - ft/ft^3.$$

In SI units, in the mud drilling example, the density of the drilling mud in the annulus of the drill collars is 12,001 kg/m^3. Also, the approximate velocity just above the drill collars is 0.61 m/sec. Using these values in Equation (2-1), the kinetic energy per unit volume for the mud example is

$$KE_m = \frac{1}{2}(1201)(0.61)^2$$

$$KE_m = 233.4 N - m/m^3.$$

In USCS units, in the air drilling example, the specific weight of the compressed air in the annulus just above the drill collars is 0.632 lb/ft^3. Using this value in Equation (2-2), the density of the compressed air is

$$\rho_a = \frac{0.632}{32.2}$$

$$\rho_a = 0.0196 lb - \sec^2/ft^4$$

or

$$\rho_a = 0.0196\ \ slugs/ft^3.$$

The average velocity of the compressed air at this position is approximately 21.1 ft/sec. Substituting these values into Equation (2-1), the kinetic energy per unit volume for the drilling mud example is

$$KE_a = \frac{1}{2}(0.0196)(21.1)^2$$

$$KE_a = 4.4 lb - ft/ft^3.$$

In SI units, in the compressed air example, the density of the air in the annulus of the drill collars is 10.1 kg/m^3. Also, the approximate velocity of the air flow just above the drill collars is 6.43 m/sec. Using these values in Equation (2-1), the kinetic energy per unit volume for the mud example is

$$KE_a = \frac{1}{2}(10.1)(6.43)^2$$

$$KE_a = 209N - m/m^3.$$

The kinetic energy per unit volume values in the two flow examples (drilling mud and compressed air) are similar in magnitude at the critical position in the annulus where it would be expected that the rock cutting carrying capacity of the drilling fluids is minimum. The flow kinetic energy per unit volume of the mud drilling fluid does not change as the drilling mud flows to the surface in the annulus (assuming a uniform cross-sectional area in the annulus). The flow kinetic energy per unit volume of the compressed air, however, increases dramatically as it seeks atmospheric conditions at the exit to the annulus. This is because the compressed air has stored internal energy (from the compression process) and as it starts up the annulus and resistance to flow decreases (i.e., lower hydrostatic head), this internal energy is converted to velocity. This kinetic energy per unit volume is a critical factor in assuring proper borehole cleaning drill bit-generated rock cuttings of both a mud-drilled and an air-drilled borehole.

REFERENCES

1. Marcus, R. D., Leung, L. S., Klinzing, G. E., and Rizk, F., *Pneumatic Conveying of Solids*, Chapman and Hall, 1990.
2. Singer, C., Holmyard, E. J., Hall, A. R., and Williams, T. I., *A History of Technology*, Vol. 4, Oxford Press, 1958.
3. Singer, C., Holmyard, E. J., Hall, A. R., and Williams, T. I., *A History of Technology*, Vol. 5, Oxford Press, 1958.
4. Singer, C., Holmyard, E. J., Hall, A. R., and Williams, T. I., *A History of Technology*, Vol. 6, Oxford Press, 1958.
5. H. J. Gruy, Personal communication, February 5, 1997.
6. API Recommended Practice for Drill Stem Design and Operating Limits, *API RP 7G*, 16th Edition, August 1998.
7. Roscoe Moss Company, *Handbook of Ground Water Development*, Wiley, 1990.
8. Burghardt, M. D., *Engineering Thermodynamics with Applications*, Harper and Row, 1982.
9. API Specifications for the Internal-Combustion Reciprocating Engines for Oil-Field Service, *API Std. 7B-11C*, Ninth Edition, 1994.
10. Angel, R. R., "Volumetric Requirements for Air or Gas Drilling," *Petroleum Transactions, AIME*, Vol. 210, 1957.

CHAPTER

Surface Equipment

3

Air and gas drilling operations require some special surface equipment not normally used in rotary mud drilling operations. Shallow drilling operations usually have this specialized equipment incorporated into the single rotary drilling rig design. For the deeper drilling operations that use double and triple rotary drilling rigs, this specialized surface equipment is usually provided by an air and gas drilling equipment contractor. These contractors supply the rotary drilling contractor (the drilling rig) with the necessary surface equipment to convert the mud drilling rig to an air and gas drilling rig. The rotary drilling contractor and the air and gas drilling contractor are usually contracted by an operating company.

3.1 DRILLING LOCATION

Nearly all air and gas drilling operations are land operations. Figure 3-1 shows a typical air drilling location plan for the drilling rig and the other important surface equipment [1]. The plan in Figure 3-1 shows the location of the drilling rig (borehole directly below the rotary table). This is a typical triple drilling rig configuration. The drilling rig floor is larger and, therefore, it is easier to show the important features of an air drilling operation with this type of rig. This rig is a typical mud rotary drilling rig that has been set up to drill with compressed air as the circulating fluid. The rig is powered by two prime movers on the rig floor. These prime movers provide their power to the rig equipment through the compound (a chain drive transmission system). The prime mover on a triple rotary drilling rig like that shown in Figure 3-1 is limited to operating either the rotary table or the draw works (hoist system), but not both simultaneously [2]. The development of the hydraulic top head rotary drive, which replaces the rotary table on most single and some double drilling rigs, allows the prime mover to simultaneously operate the rotary action and the hoist system. These smaller hydraulic top head rotary drive rigs use rig weight (via pull-down systems) to put axial force on the bit.

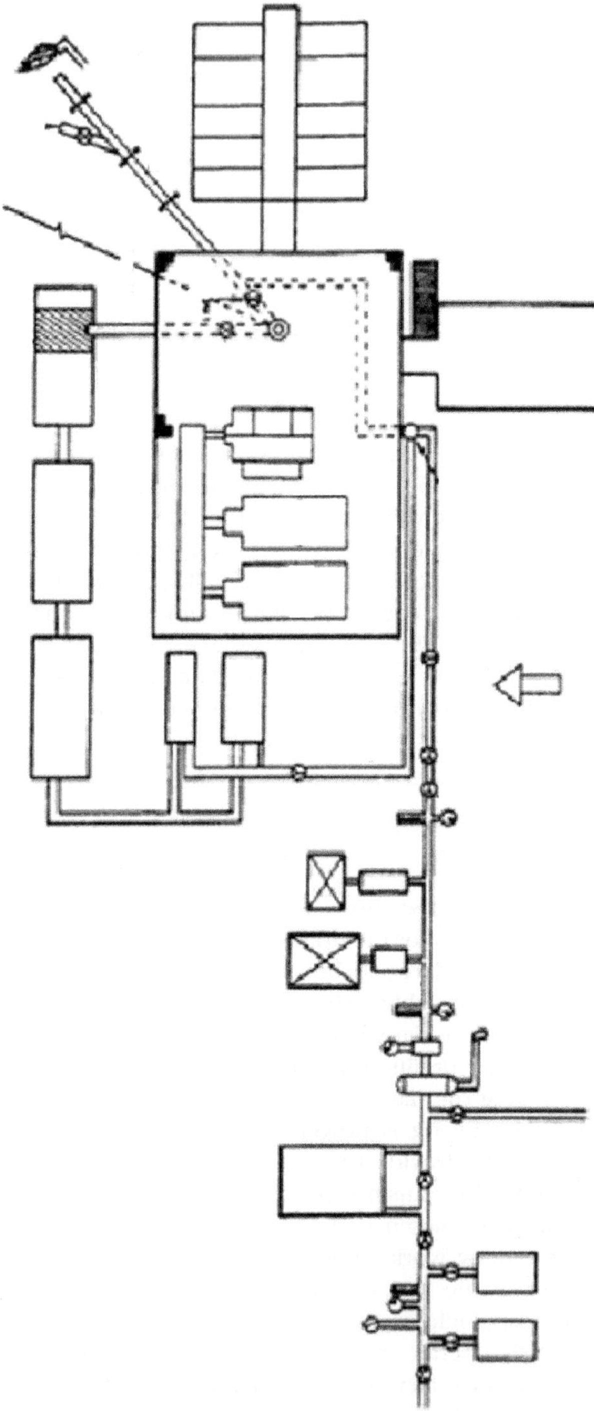

FIGURE 3-1. Drilling rig location plan for air drilling operations [1].

Figure 3-1 shows the primary compressors (low pressure) that supply compressed air to a flow line between the compressors and the rig standpipe. In this example there are two primary compressors supplying the rig. These compressors intake air from the atmosphere and compress the air in several stages of mechanical compression. These primary compressors are positive displacement fluid flow machines, either reciprocating piston, or rotary compressors (see Chapter 5 for more details). These primary compressors are usually capable of an intake rate of 1200 acfm (actual cubic feet per minute) (*566.3 actual liters/sec*) of atmospheric air and an output flow of compressed air at pressures up to approximately 300 psig (*207 N/cm^2 gauge*). These primary compressors expel their compressed air into the flow line to the standpipe of the drilling rig. This flow line is usually an API 2 7/8-in (*73 mm*)(OD) line pipe (or an ASME equivalent) or larger [3]. Downstream along this flow line from the primary compressors is the booster compressor. This booster compressor is a reciprocating piston compressor. The booster compressor is used to increase the flow pressure from the primary compressors to pressures up to approximately 1000 psig (*690 N/cm^2 gauge*). In most drilling operations the injection pressure is less than 300 psig (*207 N/cm^2 gauge*) and, therefore, the booster compressor is commonly used only for special drilling operations such as directional drilling with a downhole motor.

Downstream from the booster compressor are liquid pump systems that allow water to be injected into the compressed air flow to the rig. Also, solids can be injected into the compressed air flow. This is accomplished by injecting the solids into a small water tank and then the water with the entrained solids is injected into the air flow.

Along the flow line leading from the compressors to the drilling rig standpipe is an assembly of pressure gauges, temperature gauges, valves, and a volumetric flow rate meter [3]. This instrumentation is critical in successfully controlling air drilling operations. Also along this flow line is a safety valve. This flow line safety valve acts in a similar manner as the safety valves on each of the compressors in releasing pressure in the event the pressure exceeds safe limits. Also on the flow line is a valve allowing the compressed air flow to be diverted either to the atmosphere or to primary and secondary jets in the blooey line.

The blooey line runs from the top of the annulus to the burn pit and allows the compressed air with the entrained rock cuttings to exit the circulating system to the atmosphere. The blooey line is about 100 to 200 ft (*30.5 to 61.0 m*) in length. Usually the blooey line is an API 8 5/8-in (*219 mm*)(OD) casing or larger [4].

However, some blooey line systems are fabricated with two smaller diameter parallel lines. As shown in Figure 3-1, the exit (to the atmosphere) of the blooey line expels the air with the rock cuttings into a burn pit. For oil and natural gas drilling operations, a pilot flame is placed at the exit of the blooey line. This ignites any hydrocarbons produced exiting the blooey line with the circulating air.

Figure 3-1 shows how the drilling location is oriented so that the blooey line exit is downwind of the prevailing wind over the site. This keeps dust or smoke from blowing across the location.

3.2 FLOW LINE TO THE RIG

The flow line to the standpipe of the drilling rig acts as a manifold collecting the compressed air outputs from the primaries. These flow lines are API 2 7/8-in (*73 mm*)(OD) or 3 1/2 (*89 mm*)(OD) steel line pipe (or ASME equivalent)[5]. The valves in the flow line at the booster compressor allow the air flow from the primaries to be diverted to the booster when high compression of the air is needed. When higher compression is not needed, the booster compressor is isolated with the check valves in the flow line to the rig.

3.2.1 Bleed-Off Line

The bleed-off line allows pressure to be released throughout the flow line to the rig and inside the standpipe, rotary hose, kelly, and the drill pipe to the depth of the first float valve (see Figure 3-1). The bleed-off line is usually run to the blooey line and exits into that line. The bleed-off line is generally used when drill pipe connections are made, replacing the drill bit (making round trips), and for other operations where the well is opened to the atmosphere.

3.2.2 Scrubber

The scrubber removes excess water in the compressed air flow in the flow line. If the humidity of the atmospheric air is high, then as the air is compressed in the compressors much of the water will return to the liquid state. Dry air drilling operations require the removal of this water before the compressed air is injected into the well. The scrubber is incorporated into a surge tank. The water in the compressed air flow is collected in the bottom of the surge tank as the air flows through this tank and is vented from time to time to eliminate the water from the line.

3.2.3 Water Injection Pump

Unstable foam (mist) drilling operations require the injection of water into the compressed air flow before the air is injected into the well. The water injection pump injects water, chemical corrosion inhibitors, and liquid foamers into the compressed air flow line. Figure 3-2 shows a skid-mounted water injection pump for air and gas drilling operations. These skid-mounted water injection pumps are used for deep drilling operations. These pumps are capable of injecting up to 20 bbl/hr (*75.7 liters/hr*) into the air or gas flow to the well.

The smaller drilling rigs have onboard water injection pumps. These smaller rig water injection pumps have capabilities from 10 to 25 gal/min (*37.9 to 94.6 liters/min*). The small water injection pump carries out the same objective on these smaller rigs as the skid-mounted water pump for the larger double and triple drilling rigs. The injection of water and appropriate chemicals and foamer is a vital option for air and gas drilling operations. Very few air and gas drilling

FIGURE 3-2. Water injection pump (courtesy of Mountain Air Drilling Services).

operations are carried out without some water, chemical additives, and foam producing additives being injected.

3.2.4 Solids Injector

The solids injector is used to inject hole drying and hole stabilizing powders into the well to dry water seeping into the well from water-bearing rock formations. Other solids are often injected to reduce torque between the drill string and the borehole. These torque reduction solids are often necessary when drilling highly deviated and horizontal boreholes.

3.2.5 Valves

Both manually and remotely operated valves are located along the flow line to the rig. These valves are usually the gate or ball type. These valves cannot be operated in a partially open position. The abrasive nature of the compressed air flow in the flow line would erode the gate or ball of the valve and render the valve ineffective in the closed position. At strategic locations along the flow line are check valves. These special mechanical valves allow compressed air flow in only one direction (toward the standpipe). These check valves are spring loaded and the force of the flow allows the mechanism (flapper-gate or ball) to open in the correct direction of flow. If the flow is reversed, the mechanism is forced closed by the spring and the force of the reverse flow.

3.2.6 Gauges

Each of the compressors is equipped with independent gauges to assess its operating performance. In addition to the compressor gauges are those placed along the flow line. A low pressure gauge is placed downstream of the primary compressors but upstream of the booster compressor. This gauge allows assessment of the performance of the primaries. A high pressure gauge is placed downstream of the booster compressor to assess the performance of the primaries and booster when high pressure compressed air is required. Pressure gauges are also placed upstream and downstream of the water injection pump and the solids injector. These gauges allow assessment of the performance of these injection systems. All these gauges must be high-quality gas gauges. Most drilling rig floors are equipped with a mud pressure gauge. For air drilling operations this mud gauge must be replaced with a high-quality gas gauge having the appropriate pressure range.

3.2.7 Volumetric Flow Rate Meters

No driller would carry out a mud drilling operation without knowing the volumetric flow rate of mud being circulated to the well. The volumetric flow rate from a mud pump can be assessed easily by either counting strokes per minute of the mud pump (and knowing the capacity of the pump in gallons per stroke and then calculating the output of the pump in gallons per minute) or by providing the rig floor with an accurate volumetric flow rate gauge.

The volumetric flow rate of air (or other gases) to the well is vital knowledge for a successful drilling operation and its knowledge must also be made available to the rig personnel. The volumetric flow rate of air (or other gases) is referenced to the atmospheric conditions of the air entering the primary compressor. At sea level locations the volumetric flow rate is given as standard cubic feet per minute (scfm) or *standard liters per second* and sometimes *standard cubic meters per second*. At locations above sea level the volumetric flow rate is given as actual cubic feet per minute (acfm) or as *actual liters per second.*

There are other very practical reasons why the compressed air flow to the drilling operation must be metered. As will be seen in Chapter 5, there is a great deal of difference in primary compressors and their respective effectiveness to produce the volume rate of air required by the operation. The screw compressor is notorious for wearing and thus supplying less volumetric flow rate than advertised. In addition, all compressors must be derated when the units are operated at surface locations that are above sea level. This is one of the most important calculations that must be carried out by the drilling engineer and must be verified by on-site measurements.

There are two techniques for determining the air volumetric flow rate from the primary compressors (or natural gas from a pipeline). A gas production orifice plate with an associated recording system can be used in the flow line downstream of the compressors and scrubber, but upstream from the water injection pump. Figure 3-3 shows a simple schematic of an orifice plate with a differential pressure gauge to measure the difference between pressure upstream and downstream of the plate.

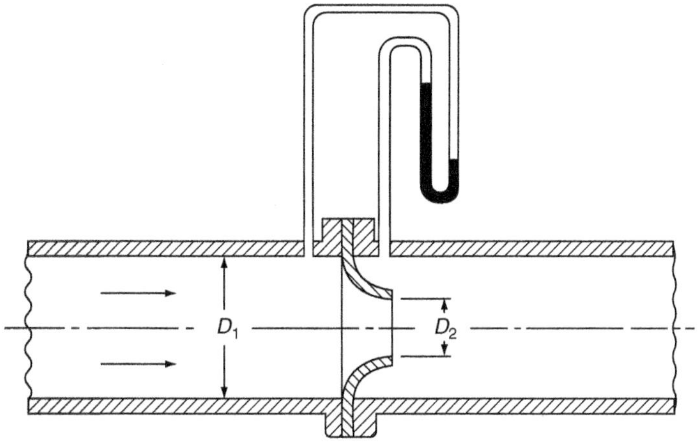

FIGURE 3-3. Schematic of orifice plate and manometer differential pressure gauge.

The other volumetric flow rate metering device is the gas turbine flowmeter. Figure 3-4 shows this type of flowmeter. Figure 3-5 shows the placement of this type of flowmeter in a ASME 2-in nominal steel pipe or a API 2 3/8-in (*73 mm*) (OD) line pipe. Figure 3-6 shows the digital readout that accompanies the turbine flowmeter. The turbine and readout need to be correlated for the flow gas specific gravity and the location atmospheric conditions. The digital readout can also be wired to the rig floor to allow the driller and other rig personnel to assess the operation of the compressors.

FIGURE 3-4. Turbine flowmeter (courtesy of Halliburton Energy Services, Incorporated).

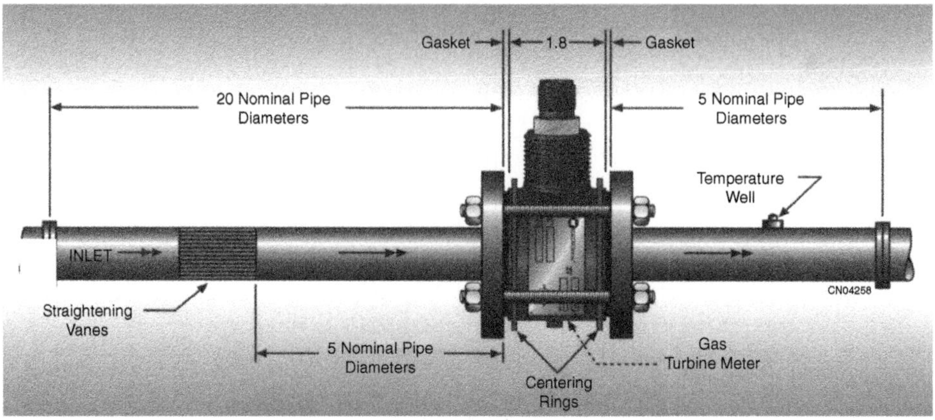

FIGURE 3-5. Installation of the turbine flowmeter in ASME 2-in nominal diameter pipe (courtesy of Halliburton Energy Services, Incorporated).

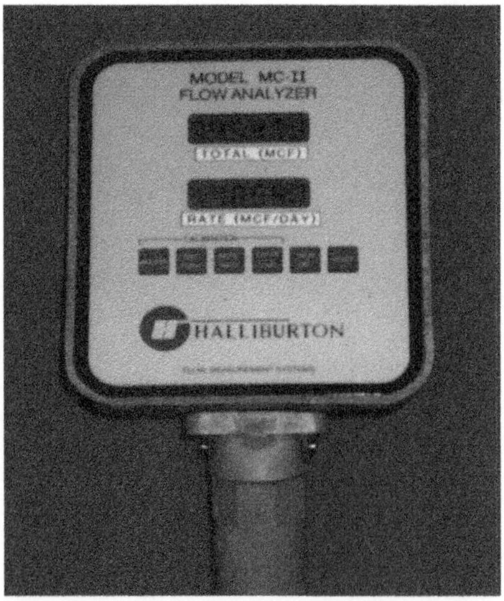

FIGURE 3-6. Digital display for the turbine flowmeter (courtesy of Halliburton Energy Services, Incorporated).

3.3 WELLHEAD EQUIPMENT

All air and gas drilling operations require the use of a rotating head (or similar air or gas flow diverter), which is installed below the rotary table. The blowout preventer (BOP) stack is always used when subsurface overpressured dangerous gases or fluids might be encountered while drilling (i.e., oil and natural gas drilling operations, and geothermal drilling operations). Figure 3-7 shows schematics of three typical wellhead assemblies for double and triple drilling rigs set up for air and gas drilling operations to recover oil and natural gas, or geothermal fluids [6].

3.3.1 Rotating Head

The rotating head or a similar air flow diverter was developed for use in air and gas drilling operations to keep air or gas with entrained rock cuttings from flowing to the drilling rig floor through the rotary table kelly bushings. Diverting this drilling fluid flow from the drilling rig floor is mandatory for all air and gas drilling operations. Even on small drilling rigs the air exiting the annulus (direct circulation) must be diverted in order to provide a safe work space for the rig operators. These diverter devices were developed with the introduction of air and gas drilling operations in the early 1930s.

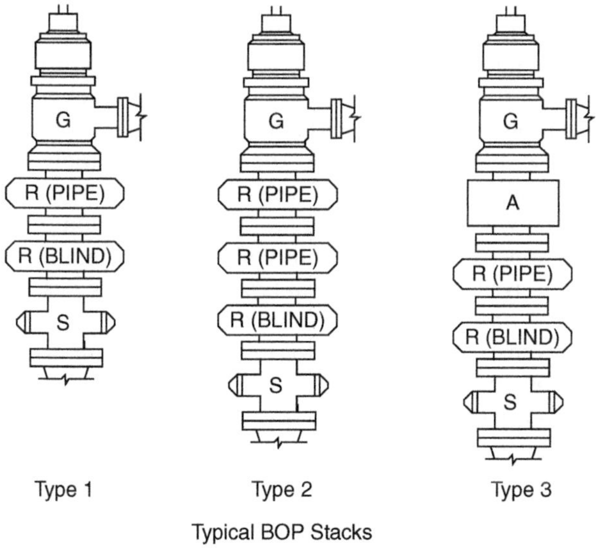

FIGURE 3-7. Typical wellhead assemblies for an air and gas drilling operation (G denotes a rotating head, A denotes an annular blowout preventer, R denotes pipe or blind rams, and S denotes the drilling spool).

FIGURE 3-8. Low pressure rotating head (courtesy of Weatherford International Ltd).

The right side of the rotating heads shown in Figure 3-7 shows the vent to the blooey line. Figure 3-8 shows a low pressure rotating head. This rotating head is capable of diverting a 500 psig (*345 N/cm²*) gauge air or gas flow while rotating at 100 rpm and 1000 psig (*690 N/cm²*) with no rotation. This rotating head is made up (via the flange fitting on the bottom of the head) to the top of a BOP stack or the top of a casing spool and casing. The BOP stack is incorporated in the wellhead assembly when overpressured dangerous gases or mixtures of gas and other fluids may be encountered in the drilling operation (see Figure 3-7). Typically the BOP is used for all deep wells. The type of rotating head shown in Figure 3-8 is used with large drilling rigs. Direct circulation or reverse circulation drilling operations can be carried out with these rotating heads. This particular rotating head is available in an 8.25-in (*209.6 mm*) bore design (Model 8000) and a 9.00-in (*228.6 mm*) bore design (Model 9000).

Figure 3-9 shows an exploded view of the four major sections of the rotating head. The top three sections are the internal sections of the head and are easily removed in the field from the fourth (bottom) section (the bowl or main housing and quick-lock clamp assembly). The top section shown in Figure 3-9 is the kelly driver with lugs on its side that lock into the bearing assembly shown below it. The bearing assembly has bearings and bearing seals that allow the inside of this assembly to rotate with the drill string and its outside to seal inside the nonrotating housing (i.e., the bowl and quick-lock clamp assembly). Attached to the bottom of the bearing assembly is the stripper rubber (or flexible packer). The stripper rubber is designed to fit tightly around and rotate with the kelly, the drill pipe, the drill pipe tool joints, and any crossover subs in the drill string. Any air or gas pressure in the annulus of the well acts to force the stripper rubber to fit more tightly around the kelly and drill string.

FIGURE 3-9. Rotating internal parts of the rotating head (courtesy of Weatherford International Ltd).

In order to place the drill string and kelly into the well, the quick-lock clamp must be unlocked and the three rotating internal sections lifted to the rig floor. The drill bit with the drill collars are placed in the well through the open rotating head. The internal sections of the rotating head are fitted over the bottom tool joint of the drill pipe. The bottom drill pipe joint is lowered into the well and the internal sections are placed into the rotating head and the quick-lock clamp locked. This secures the rotating head for drilling operations. The drill pipe can be lowered into the well through the rotating head as the drill bit is advanced. The kelly drive (together with the kelly bushing) fits snugly around the kelly and allows the internal rotating sections of the head to rotate with the rotation of the drill string. If it is anticipated that a well will be making large volumes of natural gas, the bottom hole assembly of the drill string is designed to allow the stripper rubber to be stripped over the drill collars to the drill bit.

Procedures for placing the drill string into a well and removing it from the well must be followed carefully when a well is making natural gas or geothermal steam or hot water (or any other dangerous gases or fluids). There are other operations performed by a drilling rig that involve use of the rotating head that must be carried out carefully if a well is making gas or geothermal steam (e.g., placing a casing string or a liner string in a well).

Figure 3-10 shows a cutaway view of a typical rotating head and shows the rotating head in the drilling position with the kelly inside the head. The kelly drive section has a kelly bushing, which can be changed to accommodate various kelly designs. The spindle subassembly rotates with the drill string. The spindle housing subassembly (which does not rotate) is seated into the main body housing of the head and seals to prevent air or gas from passing to the rig floor. The stripper rubber (flexible packer) diverts the air or gas with entrained rock cuttings to the outlet (to the blooey line). The flexible stripper rubber is forced by air (or gas) flow pressure against the outer surface of the kelly or drill pipe. There are other seals between the rotating assembly and the spindle housing. Together all these sealing agents divert the air or gas flow (with entrained rock cuttings) to the blooey line, away from passage to the rig floor (see Figure 3-10).

Although the rotating head was originally developed for air and gas drilling operations, this device was adapted for use in geothermal drilling operations and later for use in underbalanced drilling operations to recover oil and natural gas. These recent applications have encouraged the development of rotating heads capable of operating at higher pressures and temperatures. These high pressure heads are used in underbalanced drilling operations where lightweight drilling mud (or other drilling fluids) is used to drill through pressured oil or natural gas rock formations. The lightweight drilling mud (or other drilling fluids) allows oil and gas to flow into the well bore as the drill bit advances into the rock formation. When these reservoir fluids are circulated to the surface they impose high pressures on the wellhead equipment. These high pressure rotating heads are capable of operating at pressures up to approximately 1500 psig *(1035 N/cm^2 gauge)*(while rotating the drill string at about 100 rpm) and up to approximately 3000 psig *(2070 N/cm^2 gauge)*(for the nonrotating drill string). The high temperature rotating heads are generally used in geothermal drilling operations. Most of these heads can operate with steam and hot water flows at temperatures up to about 500°F *(260°F)*.

Rotating heads are also used in air and gas drilling operations where subsurface high overpressured oil, natural gas, or geothermal fluids are not expected.

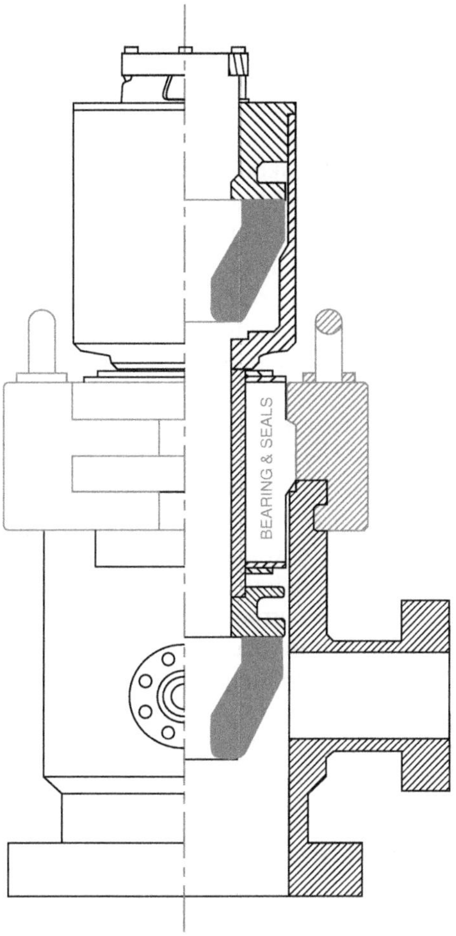

FIGURE 3-10. Cutaway view of a rotating head with dual stripper rubbers (courtesy of Weatherford International Ltd).

These are deep water well, deep monitoring well, deep mining borehole, and deep geotechnical borehole drilling operations where double and triple drilling rigs are required. These rotating heads are used to keep air or gas (in this case nitrogen) flow with entrained rock cuttings from flowing to the rig floor (for direct circulation). However, many of these non-oil, natural gas, or geothermal recovery drilling operations utilize reverse circulation. Reverse circulation requires that the compressed air or gas be injected into the "outlet" of the rotating head to the annulus space of the borehole (see Figure 3-10). In this situation the rotating head still keeps the air or gas from flowing to the rig floor.

Air and gas drilling operations using small single drilling rigs drill only shallow (usually less than 1000 ft in depth) water wells, monitoring wells, mining boreholes, and geotechnical boreholes (see Figure 2-3). Some air drilling operations on these small rigs use direct circulation, but can be converted for reverse circulation operations. These small single drilling rigs usually have hydraulic top head rotary drives. The rig "floor," or breakout platform, of these small rigs is protected from cutting returns by a rubber seal around the drill string and a flexible skirt around the edge of the floor (skirt not shown in figure). When direct circulating, the air returning up the annulus (with the entrained rock cuttings) is kept from coming through the rig floor by the rubber seal around the drill string. The drilling cuttings are allowed to accumulate on the surface of the ground around the top of the borehole where the skirt slows the air flow and allows the cuttings to be dropped out.

Reverse circulation provides a useful way for dealing with the return flow of compressed air and entrained rock cuttings from the borehole. The compressed air is injected into the annulus of the borehole via a sealed fitting at the top of the annulus, or a dual drill pipe annulus fitting. After circulating through the bit, the air with entrained rock cuttings exits the borehole through the inside of the drill string, then flows through the top head rotary drive, and then through the rotary hose. The air with the cuttings can be diverted to a pit away from the rig with a hose extension, or a hose extension run to a cyclone separator where cutting samples can be obtained for analysis. Air drilling operations using small single drilling rigs are generally safer for the operators and more environmentally clean when utilizing reverse circulation. More details on reverse circulation operations are given in Chapter 7, 9, and 10.

3.3.2 Blowout Prevention Stack

Blowout prevention equipment was developed for use in drilling deep wells for the recovery of oil and natural gas. Later this unique oil and gas industry equipment was adapted for use in drilling deep geothermal wells. Natural deposits of oil and gas exist in porous rock formations deep in the Earth's crust. These deposits were created by millions of years of sediment burial and confinement by geologic structures. Over time, increased sedimentary burial created high pressures and temperatures in these deposits. Most newly discovered oil and natural gas deposits have static pressures up to about 8000 psi (*5518* N/cm^2) and temperatures of about 300°F (149°F). A few abnormally pressured natural gas deposits have static pressures as high as 16,000 psi (*10,346* N/cm^2). These pressures, although found in deposits at depths of 10,000 ft (*3048 m*) or greater, are quite dangerous to drilling rig personnel and the environment. Blowout prevention equipment (or the BOP stack) was developed to provide protection of the surface from these high pressured deposits.

A blowout occurs when oil and/or natural gas deposits are allowed to flow uncontrollably to the atmosphere at the surface. The first line of defense against the dangers of these high pressure deposits is weighted drilling mud. Water- and oil-based drilling muds can be designed so that their specific weights are sufficiently high to provide bottom hole pressures that are slightly higher than the static pressure of the deposits when the drill bit penetrates the host rock formation. When drilling exploratory wells it is not possible to precisely know the static pressure in target oil or natural gas deposits. Therefore, geologic and engineering judgment must be used to estimate the static pressures that might be encountered. These estimates are used to design the weighted drilling mud. However, even after the first exploratory wells have been drilled successfully and the oil or gas field is being developed with follow-on development wells, surprises in deposit pressures can occur. When too light a drilling mud is used and a high pressure deposit is drilled, the well will receive a liquid or gas "kick." A kick is a slug of formation liquid and/or gas that has flowed from the formation into the annulus of the well bore. The kick is composed of fluids that have lower specific weights than the heavily weighted drilling muds. Therefore, the kick will "float" in the drilling mud and rise rapidly to the surface. If the kick is mostly natural gas, the gas will expand as it moves up the drill string annulus to the surface. The surface wellhead equipment is the second line of defense against a blowout. The wellhead equipment in the form of the BOP stack must be engineered so that it is capable of containing the high pressure of this gas when it reaches the top of the annulus. This BOP stack must contain this gas pressure while the slug is circulated under control to the surface and expelled from the annulus via a flow line to a remote burn area where the slug can be burned off safely.

The BOP stack can be composed of two types of preventers: (1) the ram-type blowout preventer and, (2) the annular-type preventer. The ram-type preventer can be a blind (shear) ram and/or a pipe ram. The blind ram is capable of sealing the well completely by compressing the drill pipe from two sides and failing the pipe steel structure in a manner to prevent the well fluids from escaping to the surface through either the inside of the drill pipe or around the outside of the drill pipe. This vise-like action of the two rams essentially forces the pipe to deform between the two rams. The pipe ram acts in a somewhat similar manner as the blind ram, except that the pipe ram has a geometric shape on the end of the ram that conforms to the outside surface of the drill pipe. Thus the pipe ram seals against the outside of the drill pipe and prevents well fluids from escaping to the surface around the outside of the drill pipe. The pipe ram does not fail the pipe structure; therefore, drilling mud can be circulated down the inside of the drill pipe to safely allow the kick to be circulated to the surface.

Figure 3-11 shows a cutaway view of a twin ram-type blowout preventer. A typical twin preventer will have a pipe ram on the top and a blind ram on the bottom. The cutaway shows the bottom blind ram. In the event of a blowout, the pipe ram would be used to seal the well and allow the slug in the annulus to be circulated to the surface safely. In the event that the pipe ram cannot seal the well for the safe circulation of the slug, the blind ram can be actuated to seal

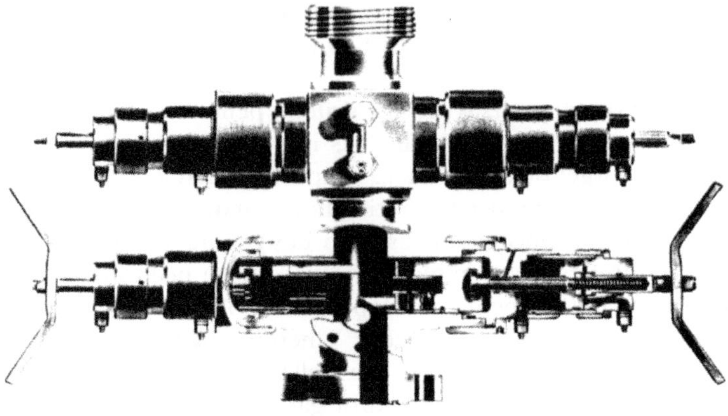

FIGURE 3-11. Typical twin ram-type blowout preventer (courtesy of Bowen Tools, Incorporated).

the well. These rams can be actuated manually or hydraulically. Figure 3-11 shows the blind rams on the bottom that are set up to be actuated manually.

This twin ram-type blowout preventer is flange connected (made up) to the top of the well casing. The bottom of the casing spool is threaded (or welded) to the top of the casing. The top of the spool is flange connected to the bottom flange fitting of the twin ram-type blowout preventer.

Figure 3-7 shows a Type 1 BOP stack that utilizes only the twin ram-type blowout preventer for well control. This BOP stack is fitted with a rotating head flange connected to the top of the twin ram blowout preventer. This is the standard well control setup for air or gas drilling operations directed toward the recovery of oil and natural gas deposits with static bottom hole pressures of the order of 3000 psi (*2069* N/cm^2) or less. Figure 3-7 shows a Type 2 BOP stack that utilizes three ram-type blowout preventers for well control. This BOP stack is configured with two pipe rams on the top and a blind ram on the bottom. The two pipe rams allow some flexibility in carrying out well control when drilling deep wells with a tapered drill string and when placing a liner string in an air- or gas-drilled well. This BOP stack is fitted with a rotating head flange connected to the top pipe ram blowout preventer. This stack can be configured for the recovery of oil and natural gas deposits with static bottom hole pressures of up to 5000 psi (*3449* N/cm^2).

The annular-type blowout preventer can also be used in a BOP stack (see Type 3 in Figure 3-7). An annular preventer is operated hydraulically. Figure 3-12 shows a cutaway view of a typical annular preventer. The closing of the preventer is actuated by hydraulic pressure. This hydraulic pressure forces the operating piston upward against a pusher plate (see Figure 3-12). The pusher plate in turn displaces (compresses) an elastomer donut inward to close and seal on the outer surface of drill pipe, drill collar, casing, or liner. Utilizing an annular preventer in conjunction with ram blowout preventers greatly increases well

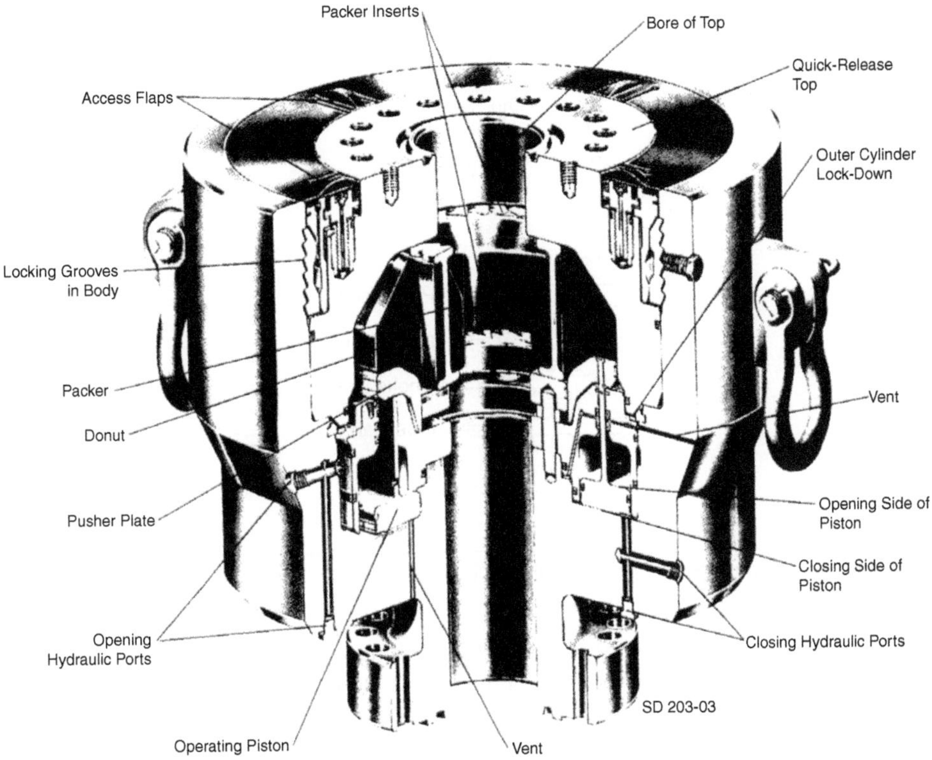

FIGURE 3-12. Typical annular blowout preventer (courtesy of Cooper Cameron Corporation).

control flexibility and general rig safety when drilling with air and gas drilling fluids. The Type 3 BOP stack in Figure 3-7 is configured with a twin ram-type blowout preventer on the bottom (pipe ram on top and blind ram on the bottom), an annular preventer flange connected to the top of the twin ram preventer, and a rotating head flange connected to the top of the annular preventer. This BOP stack can be configured for the recovery of oil and natural gas deposits with static bottom hole pressures of up to 10,000 psi (*6897 N/cm*^{2}).

Figure 3-13 shows a schematic of a more recent innovation in BOP stack design. This configuration is a variation of the standard Type 3 shown in Figure 3-7. Figure 3-13 shows the addition of a pipe ram below the drilling spool. This BOP configuration has evolved for use in underbalanced drilling and completion operations. Underbalanced drilling operations allow the oil and natural gas fluids to continue to be produced by the reservoir formation as the rock is penetrated by the advance of the drill bit. In order for underbalanced drilling operations to be successful, the oil and natural gas formations must be allowed to flow even when connections are being made, during liner operations, or

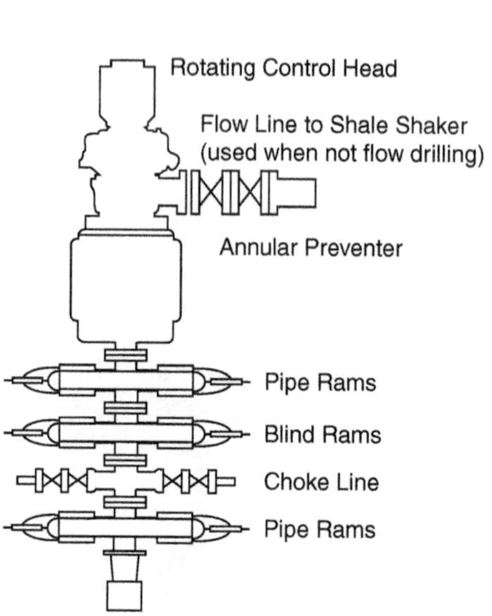

FIGURE 3-13. Schematic of recent BOP stack design for underbalanced drilling operations.

during well completion operations (after drilling operations). The addition of a pipe ram below the drilling spool increases BOP flexibility to accommodate these operations. With the drill string or tubing string in the well and with the upper pipe ram closed, drilling on completion fluids with entrained formation fluids can be circulated safely to the surface through the choke line (attached to the drilling spool). The bottom pipe ram provides a backup well control device during these operations [7, 8].

3.4 FLOW LINE FROM RIG

Air and gas drilling operations require a variety of flow line designs from the drilling rig. Drilling operations using compressed air or other compressed gases require the use of large inside diameter flow lines. These return flow lines should be designed not to choke the air or gas flow as it exits the circulating system. This line is known as the "blooey line," which derives its name from the sound made when a slug of formation water is ejected from the line with high velocity air or gas (see Figure 3-1). Aerated drilling operations require return flow lines similar to those of conventional mud drilling operations, as volumetric flow rates are very similar. These return lines are usually longer in length than conventional mud return flow lines. The air in the returning aerated fluid with entrained rock cuttings is released to the atmosphere as the fluid exits the flow line. Foam drilling return flow lines are large diameter pipelines and are unique in that they must be equipped with valves to allow choking of the return flows.

3.4.1 Blooey Line

Figure 3-1 shows a blooey line exiting from the drilling rig annulus for direct circulation operations. Blooey lines (or equivalent) are required for all air and gas drilling operations and are needed to keep drilling rock dust and cuttings away from the drilling rig and rig personnel. Blooey lines must be secured to the ground surface with tie-downs (see Figure 3-1). The high velocity of the air or gas flow from the well will interact with the flexible blooey line to set up an aerodynamic flutter situation, which is very similar to the motion of a water hose on the ground when the water valve is turned on. This flutter situation can result in high dynamic forces and resulting blooey line movement. This potential movement must be constrained along its length by tie-downs to the ground.

The blooey line should be designed with an inside cross-sectional area greater (by a factor of ≈1.1) than the annulus cross-sectional area at the top of the well. In general, this is not practical when drilling shallow larger diameter borehole sections. This requirement applies to the drilling of the deep smaller diameter borehole sections. Therefore, the inside diameter of the blooey line, d_b (inches, or *mm*), should be approximately

$$d_b \geq \left[1.1\left(d_c^2 - d_p^2\right)\right]^{0.5},$$

where d_c is the inside diameter of the casing at the top of the well (inches, or *mm*) and d_p is the outside diameter of the drill pipe at the top of the well (inches, or *mm*).

The typical length of the blooey line for large drilling rigs is from 100 to 300 ft (*30 to 90 m*). This line is run from the annulus to a burn pit (see Figure 3-1). The air or natural gas drilling fluid with the entrained rock cuttings flows from the annulus down the blooey line and exits at the burn pit. The rock cuttings are dropped in the burn pit as the air flow with the natural gas exits the blooey line.

In some operations the single blooey line is often replaced with two parallel smaller diameter lines. In this situation the inside cross-sectional area of the two lines should also be designed to be greater (by a factor of ≈1.1) than the annulus cross-sectional area at the top of the well.

All blooey lines should be equipped with two high pressure gate valves. These valves are located on the horizontal blooey line at its entrance (just downstream from the tee turn where the return flow from the annulus turns to horizontal flow in the blooey line). Figure 3-13 shows these two valves on the horizontal flow line (blooey line) just below the rotating (control) head. During drilling operations these valves are in the full open position to prevent erosion. These valves are an added safety feature allowing the well to be closed when the surface pressure in the well is low. However, the valves can also be used to carry out some rudimentary well testing operations (e.g., static wellhead pressure, wellhead flowing pressure, and volumetric flow rate).

3.4.2 Burn Pit

Figures 3-1 and 3-14 show the burn pit at the exit end of the blooey line. The burn pit should always be located away from the standard mud drilling reserve pit (water storage for an emergency mud drilling operation). This design of pit location prevents any hydrocarbon liquids from flowing into the reserve pit, thus preventing reserve pit fires near the rig. The burn pit is located downwind from the drilling rig. Such a location keeps the smoke and any dust from the drilling operation from blowing back over the drilling rig.

The burn pit must be lined with an impermeable layer of commercial clay to prevent the contamination of surrounding soil and ground water. Usually the burn pit is designed with a high berm (≈6 ft, or *2 m*) at one side of the pit (opposite the exit from the blooey line). This berm prevents high velocity rock particles and liquid slugs from passing over the burn pit. The burn pit is part of the drilling site location preparation.

3.4.3 Primary and Secondary Jets

Figure 3-14 shows the exit positions in the blooey line of the primary and secondary jet flow lines from the compressors. Figure 3-15 shows the high pressure vent lines from the compressor for the primary and secondary jet flow lines. These jet flow line installations in the blooey line are only required for drilling operations directed toward the recovery of oil, natural gas, or geothermal gas products.

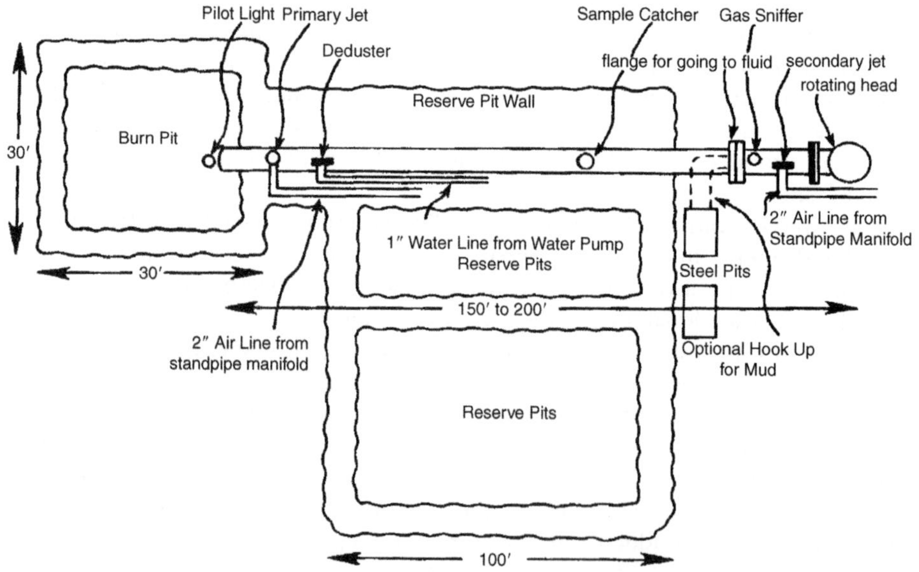

FIGURE 3-14. Schematic of burn pit, reserve pit, and blooey line plan [3].

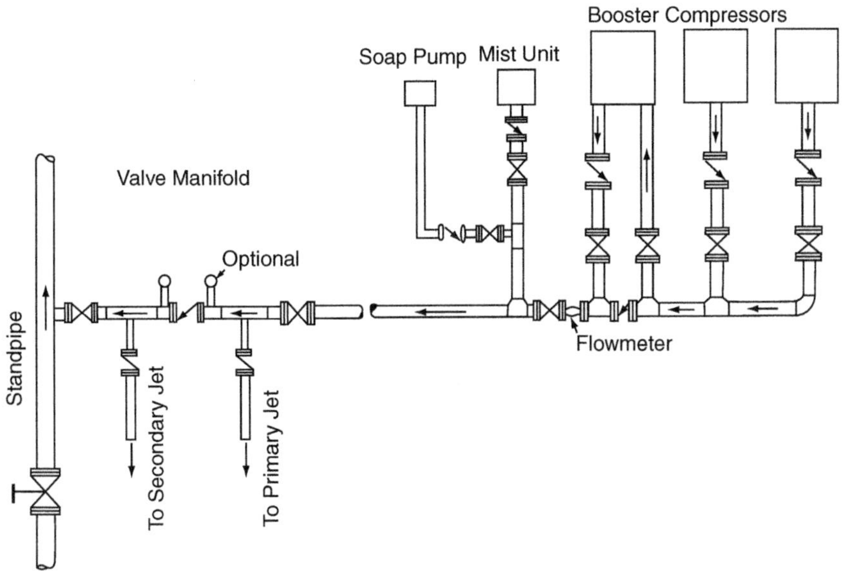

FIGURE 3-15. High pressure vent lines to the primary and secondary jets [3].

Primary and secondary jets are incorporated into the blooey line to allow safe venting of the top of the wellhead when the well is producing natural gas or other dangerous gas. These lines allow for the direct discharge of compressed air from the compressors into the blooey line. This discharge into the closed blooey provides jet pumping action, which forces any gas venting from an atmosphere exposed wellhead to flow to the blooey line and exit this line at the burn pit.

Figure 3-15 shows the surface layout of the flow line from the compressors to the standpipe and shows the high pressure vent flow lines to the primary and secondary jets. The primary jet is positioned near the exit end of the blooey line, and the secondary jet is near the entrance end of the blooey line just downstream of the tee from the annulus (see Figure 3-14).

3.4.4 Sample Catcher

Sample catchers are usually required for any air and gas drilling operation. The sample catcher allows small rock cutting samples to be obtained from a well during the drilling operation. The sample catcher is installed in the body of the blooey line usually near the entrance to the blooey line (see Figure 3-14). Figure 3-16 shows a typical sample catcher design. This design has a small diameter (ASME 2-in nominal diameter or smaller) transport pipe welded through the blooey line body. A short section of this small pipe protrudes into the flow stream

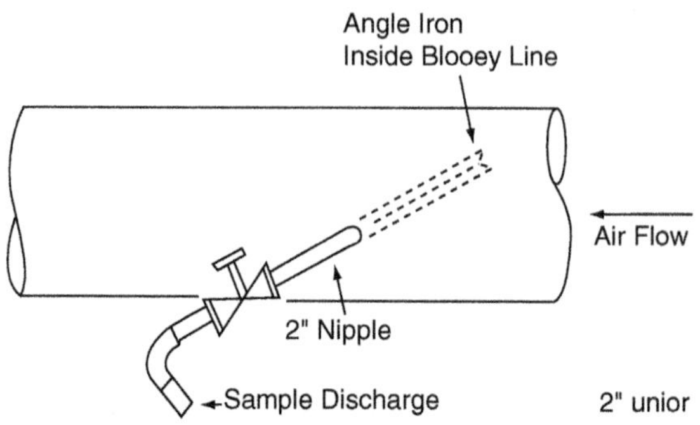

FIGURE 3-16. Typical sample catcher designs [3].

inside the blooey line. Inside the blooey line there is a short section of angle iron welded to the small transport pipe. This angle iron directs the cuttings into the small transport pipe. Outside the blooey line there is a gate valve on the small pipe to allow discharge of sample rock cuttings. Because the flow of air or gas up the annulus is at high velocities (of the order of 50 ft/sec to 80 ft/sec, *15 m/sec to 25 m/sec*), cuttings sampling can be correlated accurately to subsurface rock formations being drilled.

The securing of rock cuttings from the depths is an essential practice when drilling deep boreholes. At the drilling location, these rock cutting samples can be studied under a microscope and analyzed to ascertain chemical and physical properties. Knowledge of the rock characteristics and properties allows geologists and drilling engineers to identify the rock formations being penetrated as the drill bit is advanced. This information allows the drilling operation to accurately drill to a subsurface target area.

3.4.5 Deduster

Dedusters are usually required for any air or gas drilling operation. Figure 3-14 shows the location of the deduster near the exit end of the blooey line. Figure 3-17 shows a typical deduster design. The deduster is a small diameter pipe (ASME 2-in nominal diameter or smaller) water system located inside the blooey line. A pump supplies the system with water. The water is sprayed on the dry rock dust particles that exit the line. This reduces or eliminates the dust clouds typical of dry air or natural gas drilling operations.

3.4.6 Gas Detector

Gas detectors (gas sniffers) are used only in air drilling operations directed at the recovery of oil and natural gas. Figure 3-14 shows the position of the gas detector on the blooey line. This detector can detect very small quantities of hydrocarbons

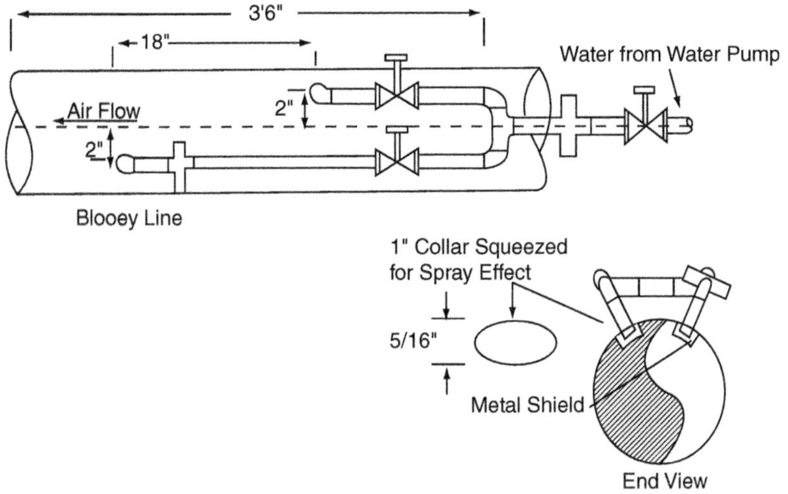

FIGURE 3-17. Typical deduster design [3].

that might enter the blooey line from the annulus. As the drill bit is advanced and hydrocarbon-producing formations are drilled, the hydrocarbons are entrained in the return air flow to the surface with entrained rock cuttings. The detector alerts the drilling rig crew that hydrocarbons are in the annulus. This alert allows rig personnel to take safety precautions against subsurface and surface fires or explosions.

3.4.7 Pilot Light

Figures 3-1 and 3-14 show a pilot light at the end of the blooey line. Pilot lights are used only in air or gas drilling operations directed at the recovery of oil and natural gas. The pilot light is a small open flame (propane or natural gas) maintained at the end of the blooey line to ignite and burn any hydrocarbons that might exit the line as the drilling operation progresses. Many new air and gas drilling operations are equipped with electric igniters instead of open flame pilot lights.

REFERENCES

1. B. J. Mitchell, Personal communication, November 1982.
2. Moore, W. W., *Fundamentals of Rotary Drilling*, Energy Publications, 1981.
3. Hook, R. A., Cooper, L. W., and Payne, B. R., "Air, Mist and Foam Drilling: A Look at Latest Techniques: Parts I and II," *World Oil*, April and May 1977.

4. Bulletin on Performance Properties of Casing, Tubing, and Drill Pipe, *API Bul 5C2*, Twentieth Edition, May 31, 1987.

5. API Specification for Line Pipe, *API Spec. 5L,* Thirty-Fourth Edition, May 31, 1984.

6. API Recommended Practices for Blowout Prevention Equipment Systems for Drilling Wells, *API RP 53,* First Edtion, February 1976.

7. Bourgoyne, A. T., "Rotating Control Head Applications Increasing," *Oil and Gas Journal*, October 9, 1995.

8. Hannegan, D. M., "RCHs Lower Cost, Boost Productivity," *The American Oil and Gas Reporter*, April 1996.

CHAPTER

4 Downhole Equipment

Air and gas drilling operations require some special subsurface equipment and drilling methods that are not normally used in rotary mud drilling operations. Deep direct circulation operations use rotary drill strings that are similar to those used in mud drilling. However, even these drill strings are equipped with downhole tools unique to air and gas drilling operations.

Larger diameter shallow and intermediate depth wells are usually drilled with reverse circulation techniques. These techniques and their associated equipment are virtually unknown to those who drill deep small diameter wells.

4.1 ROTARY DRILL STRING

Two general types of drill strings are used in air and gas drilling operations. The standard drill string discussed here is used almost exclusively for deep direct circulation operations. The dual wall pipe drill string is used exclusively for intermediate and shallow depth reverse circulation operations.

4.1.1 Standard Drill String

Figure 4-1 shows a schematic of a standard rotary drill string used to drill deep boreholes with direct circulation. Such a drill string would be used on large drilling rigs. At the bottom of the drill string is the drill bit. The drill bit is threaded (made up) to a bit sub. The drill bit has a threaded pin pointing up. The bit sub is a short thick wall pipe that has a threaded box on both ends. Above the bit sub are the drill collars. Each of the drill collars and most of the remainder of the components in the drill string are designed with a threaded pin down and a threaded box up. The bit sub is used to protect the bottom threads of the bottom drill collar from the wear caused by the frequent drill bit changes typical for all deep drilling operations. A drill collar is a thick wall pipe that provides the weight or vertical axial force on the drill bit, allowing the drill bit to be advanced as it is rotated (see Figure 2-2). Usually there are a number of drill collars in a drill string.

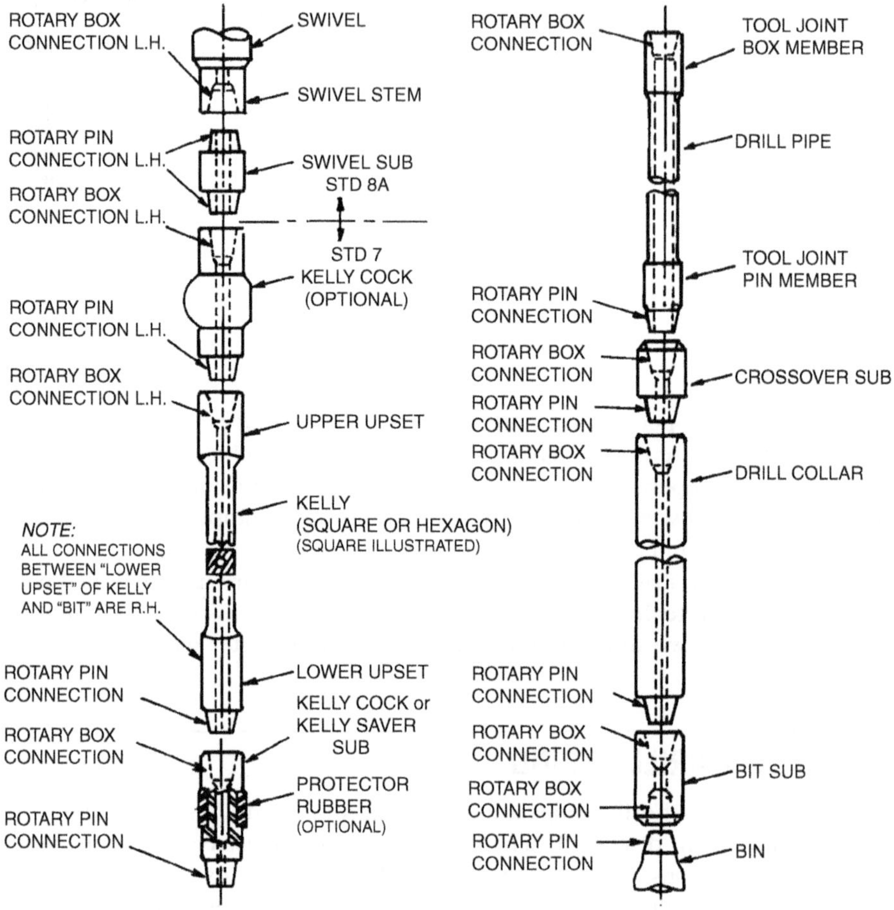

FIGURE 4-1. Standard rotary drill string for direct circulation.

The number of drill collars in a drill string depends on how much weight on bit (WOB) is required to allow the drill bit to be advanced efficiently (drill string design is discussed in Section 4.6). The drill pipe and collar lengths are in accordance with the range designations of Table 2-1 [1].

Generally the drill collars in a drill string have the same thread design. Above the drill collars are the drill pipe joints. The drill pipe joint lengths are also in accordance with the range designations of Table 2-1. The threads of the drill collar connections are usually not the same as the threads of the drill pipe joint connections (tool joints). Therefore, a special crossover sub must be used to mate the drill collars to the drill pipe. The crossover sub is a short thick-walled pipe with a threaded pin down (with the drill collar threads) and a threaded box up (with the drill pipe threads). The number of drill pipe joints is determined by the depth of the borehole to be drilled. Only the drill collars can be placed in compression (to place weight on the bit). The drill pipe joints are always kept in tension [1].

All of the threaded connections in drill strings are API-threaded shoulder connections. There are a variety of these connections and they are discussed in detail in Sections 4.3 and 4.4. Figure 4-2 shows a typical API-threaded shoulder connection for a drill pipe. As can be seen, the connection has matching flat shoulders on the pin and on the box. When a pin and box are made up, the flat surfaces of the shoulders mate against each other and seal to form a strong structure that is also leak proof. The shouldered connection protects the thinner walled body of the drill pipe and the threads inside the connection from damage when the drill string (and the connection) is flexed when bent in a deviated borehole [2, 3].

At the top of the drill pipe section is the kelly cock (or saver) sub. The kelly cock sub is another crossover sub. However, this sub is used to protect the bottom threads of the kelly. Even if the threads at the bottom of the kelly are the same as the drill pipe threads, this special crossover sub is usually used. As drilling progresses, additional elements of drill pipe are added to the top of the drill string. The kelly is a special type of drill pipe with a square or hexagon outer surface. The rotary table grips the outside of this pipe and provides the torque to the drill string to make it rotate. Thus, as additional drill pipes are added to the drill string as the bit advances in the borehole, the drill pipe must be disconnected and a new pipe joint added. The bottom-threaded box of the kelly takes the wear of these repeated connections of drill pipe. All of the threaded components below the top-threaded connection of the kelly are right-hand threads. The rotary table rotates to the right (clockwise from the top view of the table). This rotation tightens the right-hand threads below the table.

At the top of the kelly is a left-hand thread connection (threaded box). As drilling progresses, the rotary table, in addition to providing torque to rotate the drill string, also allows the kelly to slide through the table, allowing the borehole to be deepened. Since the torque is applied along the square or hexagon outer

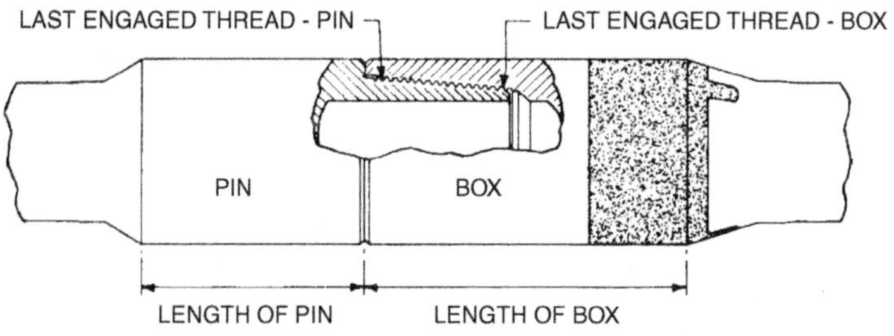

FIGURE 4-2. Cutaway of a made-up API shouldered connection.

surfaces of the kelly, the left-hand thread at the top of the kelly is tightened by the inertial drag of the nonrotating components above the kelly. All of the components above the kelly are left-hand thread connections. Above the kelly is a kelly cock sub (optional). The kelly cock is a special valve that allows sealing off of the inside of the drill string in a blowout event during oil or natural gas drilling operations. The kelly cock sub has a threaded pin connection down and a threaded box connection up. Above the kelly cock sub is a swivel sub. The swivel sub protects the swivel and has a threaded pin connection down and a threaded pin up. Above the swivel sub is the swivel itself. The bottom of the swivel has a threaded box connection pointed down. The swivel is split into two sections: a rotating section on the bottom and a nonrotating section on the top. The nonrotating section of the swivel is held in the mast by the traveling block and hoisting system. A sealed bearing allows the bottom section of the swivel to rotate while the top section can be held in position by the traveling block. The swivel allows the circulation fluid (drilling mud or compressed air or natural gas) to flow through the swivel to the rotating drill string.

For direct circulation, the circulation fluid flows down the inside of the drill string to the drill bit, flows through the drill bit orifices (or nozzles), entrains the rock cuttings from the drill bit, and flows up the annulus between the outside surface of the drill string and the inside surface of the borehole.

4.1.2 Dual Wall Pipe Drill String

Conductor and surface casing wellbores are drilled to shallow depth with large diameter boreholes. These boreholes can be drilled with direct circulation techniques. However, reverse circulation techniques are often more efficient. The drilling industry has developed some very unique downhole tools for reverse circulation air drilling operations. Figure 4-3 shows a schematic configuration of a rotary reverse circulation operation using dual wall drill pipe.

Reverse circulation techniques are not restricted to air drilling operations. Reverse circulation techniques can also use standard drill string like that shown in Figure 4-1. In the past two decades there has been a dramatic increase in the use of air drilling reverse circulation techniques for drilling water of deep wells, monitoring wells, geotechnical boreholes, and other shallow (i.e., less than 3000 ft) wells. The increased use of reverse circulation techniques has been encouraged by the development of new technologies. One of these innovations is the development of dual wall drill pipe.

Rotary dual wall pipe reverse circulation operations must be used on drilling rigs equipped with hydraulic rotary top drive systems (for single drilling rigs) or with hydraulic or electric power swivel systems (for double and triple drilling rigs) to rotate the drill string. Dual wall pipe is quite rigid and has a much higher weight per unit length than standard single wall drill pipe. Thus, dual wall pipe can be used like drill collars (the lower portion of the drill string can be placed in compression). The dual wall drill string in Figure 4-3 is shown rotating a

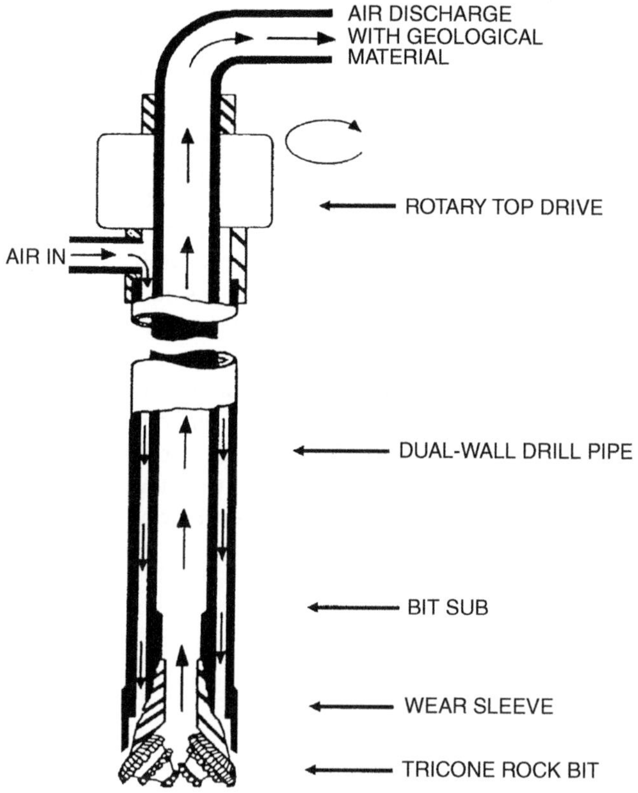

FIGURE 4-3. Dual wall pipe drill string for reverse circulation operations (courtesy of Foremost Industries Incorporated).

tricone drill bit. The top drive system rotates the entire drill string. The tricone drill bits used in reverse circulation operations have the same cutting structures as tricone bits used in direct circulation operations. However, the reverse circulation bits are fabricated to allow the compressed air to flow from the annulus between the two walls of the dual wall pipe to the bit rock cutting face. At the bottom of the well the air flow entrains the rock cuttings and flows to the surface through a large center orifice in the drill bit that leads to the inside of the inner pipe of the dual wall pipe. The drill bit used in a dual wall pipe reverse circulation operation is selected to have a diameter that is slightly larger than the outside diameter of the dual wall pipe. Thus, the outside annulus between the inside of the borehole and the outside of the dual wall pipe is kept quite small. This outer microannulus, together with a skirt structure fabricated on the drill bit like that shown in Figure 2-9, restricts the return flow of air to the surface and entrained rock cuttings to the inside of the inner pipe [4].

The drill string for a dual wall pipe reverse circulation air drilling operations is rather simple. These specially fabricated reverse circulation drill bits have a threaded box connection. The bit sub has a threaded pin connection down and a threaded box connection up. The bottom-threaded pin connection (of the bottom dual wall pipe joint in the drill string) is made up to the top connection of the bit sub. Additional dual wall pipe are made up to each other, and the top-threaded box connection of the top pipe is made up to a special side inlet sub (shown in Figure 4-4). This sub allows compressed air to be injected into the dual wall pipe annulus through the nonrotating outer section of the sub. The return flow of air and entrained rock cuttings from the inside of the inside pipe flows up through the rotating inner section of the sub to and through the rotary top drive.

The side inlet sub acts as the “first swivel” for the dual wall pipe drill string. This first swivel accommodates the injection of air into the circulation system while allowing the air and entrained rock cuttings to pass through. The “second swivel,” which is the rotary top drive, is above the side inlet sub. This second swivel provides torque and the rotation of the drill string while allowing the air and entrained rock cuttings to return to the surface.

FIGURE 4-4. Side inlet sub for dual wall pipe reverse circulation operations (courtesy of Holte Manufacturing Company).

The threaded pin and box connections on most dual wall pipe are shouldered connections. Most manufacturers of this type of pipe do not use API thread profiles. The threaded connections for dual wall pipe are discussed in Section 4.4.

4.2 DRILL BITS

There are three basic types of rotary drill bits used in air and gas drilling operations. These are drag bits, roller cutter bits, and air hammer bits.

4.2.1 Drag Bits

The original drag bits used in the early use of rotary drilling have fixed cutter blades or elements that are integral with the body of the bit. The earliest drag bits were simply steel cutter blades attached rigidly to a steel body that is made up to the bottom of the drill string. Later, natural diamonds were used as the cutter elements. A diamond drill bit has natural diamonds embedded in a tungsten carbide matrix body that is made up to the bottom of the drill string. The most recent development in drag bit technology is the polycrystalline diamond compact (PDC) bit. These drill bits have specially designed diamond cutter elements bonded to small tungsten carbide studs. These studs in turn are embedded in a steel body that is made up to the bottom of the drill string. Drag bits have no moving parts. Their cutting mechanism is a scrapping or machining action that is best used to drill rock formations that fail structurally in a plastic mode (e.g., soft, firm, and medium-hard, nonabrasive rock formations). Although PDC bits have been used extensively in past decades, they have been confined to drilling operations using water- and oil-based drilling muds. However, recent field experiments have shown that PDC bits can be used successfully in air operations using gas, aerated, and foam drilling fluids.

4.2.2 Roller Cutter Bits

Roller cutter bits use primarily a crushing action to remove rock from the cutting face and to advance the drill bit. The weight or axial force that is applied to the drill bit is transferred to the tooth or teeth on the bit. These teeth are pointed (mill tooth bit) or rounded (insert tooth bit), and the force applied is sufficient to fail the rock in shear and tension and cause particles of the rock to separate from the cutting face. The drill bits are designed to remove a layer of rock with each successive rotation of the bit. Figure 4-5a shows the tooth of a tricone bit being forced against the rock face. Figure 4-5b shows the rock particles created by the failure of the rock face due to the "crushing" action of the tooth. The circulation fluid entrains these rock cuttings and carries them away from the rock face.

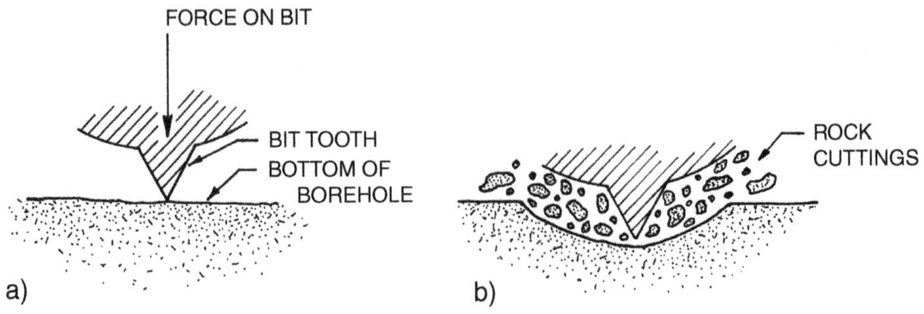

FIGURE 4-5. The rock crushing action of a tooth of a roller cutter drill bit (a) prior to failure of rock and (b) after failure of rock.

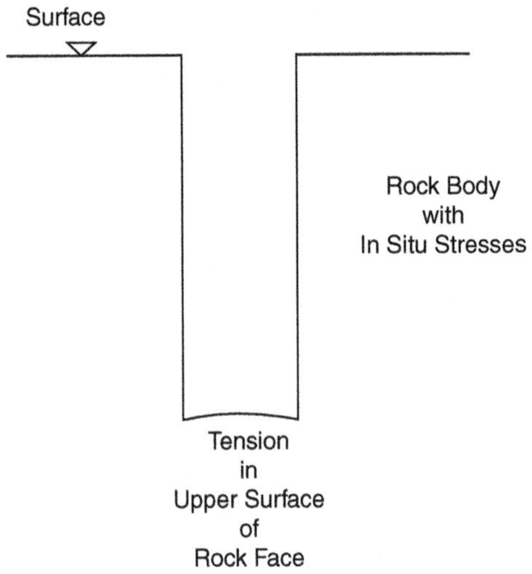

FIGURE 4-6. Schematic of tension rock face at bottom of air drilled borehole.

The roller cutter element(s) of a drill bit has a series of teeth that are designed to crush rock over the entire rock face after a single rotation of the drill bit. The repeated crushing by the teeth in conjunction with the flow of the circulation fluid allows rock particles at the rock face to be continuously removed and the drill bit advanced.

When this crushing action takes place at the bottom of a well filled with drilling mud, the hydrostatic pressure due to the fluid column compresses the rock face and places the rock face and the rock material in the immediate vicinity of the rock face in compression. This makes the crushing action less efficient and ultimately reduces the overall drilling rate of the drill bit (for a given WOB).

When this crushing action takes place at the bottom of a well filled only with compressed air, there is little hydrostatic pressure on the rock face. Further, the drilling process has removed a column of rock (above the rock face) from a semi-infinite block of prestressed rock. The in situ preexisting stresses that were in this block of rock prior to the drilling operation and the vertical cylindrical void of the new borehole create a thin tension stress field in the rock material just below the rock (see Figure 4-6).

The aforementioned argument explains why the drilling rate for an air and gas drilling operation is approximately two to four times greater than that of a similar mud drilling operation (given similar geology and drilling parameters).

There are four styles of roller cutter drill bits. These are quad-cone drill bits, tricone drill bits, dual-cone bits, and single cone bits. Quad-cone drill bits and dual-cone drill bits are used for special mud drilling operations and have little application in air and gas drilling. Tricone drill bits are used extensively in air and gas drilling operations.

Tricone Bits

The most widely used roller cutter bit is the tricone drill bit. The tricone drill bit has three roller cutter cones. Each of these cones has a series of teeth that crushes rock on the rock face as they roll over the face when the drill string (and thus the drill bit) is rotated. Figure 4-7 shows a schematic of the configuration of the tricone bit. Figure 4-7a shows a cross-section view of a cone (for a soft

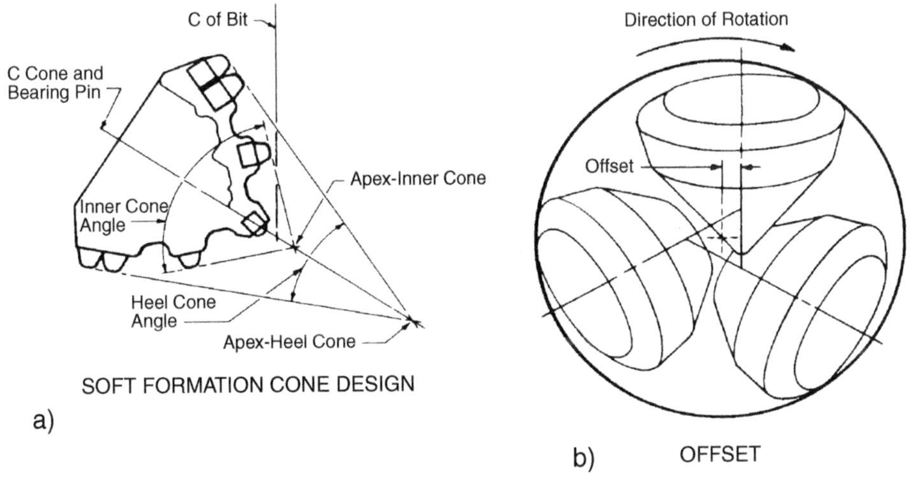

FIGURE 4-7. Schematic of the three cones of a tricone bit on the bottom of the borehole: (a) cross section of cone and (b) top view of action of three cones during rotation [5].

rock formation). Figure 4-7b shows the three roller cones at the bottom of a borehole. This latter schematic shows the offset of the cones. The offset is the degree the cones of the bit are designed to depart from a true rolling action on the rock face. Offset indicates that two or more cones of the bits do not have their centerlines of rotation passing through the center of bit rotation. Figure 4-7b shows a bit with no unique center point through which all three cone centerlines pass [5].

Tricone drill bits can be used to drill a wide variety of rock formations. Figure 4-8 shows a typical insert tricone drill bit. The "insert" term for this bit refers to the fact that the "teeth" or "buttons" are tungsten-carbide studs that are inserted (shrink-fit) into holes drilled in the cone material. The insert tricone drill bits are used to drill medium-to-hard rock formations.

Most tricone drill bits are fabricated for use with drilling muds. However, most manufacturers produce a few of their drill bit styles for air drilling operations. These tricone drill bits are designed with special internal air passages to provide the bit bearings with the appropriate cooling from the less dense compressed air or natural gas. Figure 4-9 shows a cutaway of a tricone drill bit used for air operations.

Tricone drill bits used for air and gas drilling are usually designed with increased gauge protection (relative to their mud drilling counterparts). This

FIGURE 4-8. Insert tooth 7 7/8-in (200 mm)-diameter tricone roller cutter bit IADC Code 517 (courtesy of Reed Rock Bit Company).

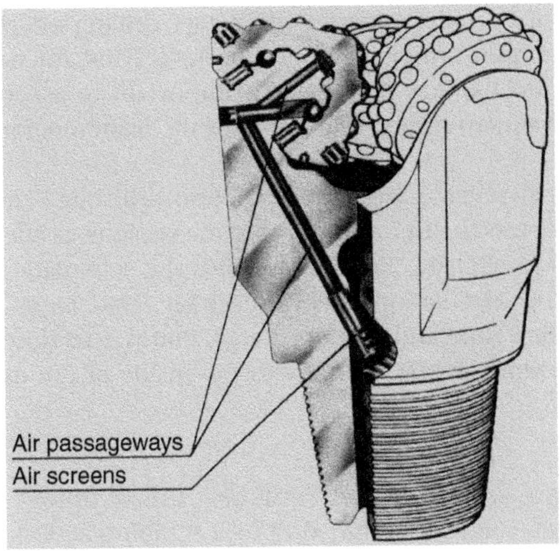

FIGURE 4-9. Cutaway of the interior of a tricone drill bit (courtesy of Reed Rock Bit Company).

gauge protection allows air bits to drill long abrasive sections without appreciable loss of gauge. However, it should be noted that some gauge loss will always occur in hard abrasive formations. It is good practice to design the well profile (i.e., borehole diameters and associated casing diameters) in such a manner that long sections having hard abrasive formations can be drilled with the drill bit diameter sequence of 6 in (*152 mm*), $6^1/_8$ in (*156 mm*), $6^1/_4$ in (*159 mm*), $6^1/_2$ in (*165 mm*), and $6^3/_4$ in (*171 mm*), the sequence of $7^1/_8$ in (*181 mm*), $7^3/_8$ in (*187 mm*), $7^5/_8$ in (*194 mm*), and $7^7/_8$ in (*200 mm*), or the sequence of $8^3/_8$ in (*213 mm*), $8^1/_2$ in (*216 mm*), $8^3/_4$ in (*222 mm*), and 9 in (*229 mm*). Using one of these drill bit diameter sequences allows anticipation of loss of gauge. The top of a long hard abrasive section can be drilled with a $6^1/_2$-in (*165 mm*)-diameter drill bit and, when there is a bit change, followed by a $6^1/_4$-in (*159 mm*)-diameter drill bit, and then near the bottom of the section followed by a $6^1/_8$-in (*156 mm*)-diameter drill bit for the last bit change.

Most air or natural gas drilling operations use insert tricone drill bits. Even though previous drilling operations with mud may have shown that a mill tooth bit had been successful in drilling a particular rock formation, the mechanics of the rock cuttings creation process at the bottom of the air borehole require that an insert bit be used in order to generate smaller rock cuttings. The smaller the rock cuttings generated by the drill bit, the more efficient the rock cutting creation and transport of cuttings particles from the bottom of the borehole.

Nearly all tricone drill bits are equipped to accept nozzle inserts in three open orifice flow channels in the drill bit body. Nozzles of various sizes (in 32nds of an inch) are used extensively in mud drilling operations. Standard practice for vertical direct circulation air or natural gas rotary drilling operations is to use tricone drill bits with open orifices (i.e., no nozzles). Thus, for well planning calculations (discussed in Chapters 8 to 12) it is important to ascertain from the drill bit manufacturer the open orifice minimum inside diameters for the drill bits used in the operation.

There are special reverse circulation tricone drill bits. These are fabricated using the same mill tooth and insert tooth cone designs as the direct circulation drill bits discussed previously. Figure 2-9 shows the schematic of the inside flow channel of a reverse circulation tricone drill bit. This large orifice allows the return flow of drilling fluid and entrained rock cuttings to flow from the annulus through the large orifice in the bit body to the inside of the drill string.

Single Cone Bits

There are single cone or "monocone" drill bits. Unlike tricone drill bits that drill by a crushing action, single cone bits drill by a scraping action. Thus, single cone drill bits utilize wear-resistant tungsten-carbide inserts in the cutting structure. These drill bits are most effective in smaller diameters: $2^3/_4$ in (*70 mm*) to $6^1/_8$ in (*156 mm*). These drill bits, with the appropriate cutting structure, are suitable for drilling soft, as well as medium-and-hard, rock formations.

The principal advantage of single cone drill bits is the large size of the support bearing of the cone and the tungsten-carbide inserts in the small drill bit diameters. Small diameter tricone drill bits are very fragile and subject to pinching and bearing damage if forced into an out-of-gauge borehole or used to ream an out-of-gauge borehole. These single cone drill bits are not subject to pinching and other damage when used to ream out-of-gauge boreholes. It is therefore good drilling practice to use single cone drill bits to drill small diameter sections in deep wells.

Figure 4-10 shows a typical single cone drill bit. Single cone drill bits are also equipped to accept nozzle inserts in three open orifice flow channels in the drill bit body. However, like the tricone drill bits used for air and gas drilling operations, it is standard practice to use single cone drill bits with open orifices (i.e., no nozzles). Thus, for well planning calculations (discussed in Chapters 8 to 10) it is important to ascertain from the drill bit manufacturer the open orifice minimum inside diameters for the drill bits to be used in the operation.

4.2.3 Air Hammer Bits

Percussion air hammers have been used for decades in shallow air drilling operations. These shallow operations have been directed at the drilling of water wells, monitoring wells, geotechnical boreholes, and mining boreholes. In the past decade, however, percussion air hammers have seen increasing use in drilling deep oil and natural gas wells. Percussion air hammers have a distinct advantage over roller cutter bits in drilling abrasive, hard rock formations.

FIGURE 4-10. Single cone roller cone drill bit (courtesy of Rock Bit International Incorporated).

The use of percussion air hammers (or down-the-hole air hammers) is an acceptable option to using rotating tricone or single cone drill bits for air and gas drilling operations. The air hammer utilizes an internal piston (or hammer) that is actuated by the compressed air (or other gas) flow inside the drill string.

The internal piston moves up and down in a chamber under the action of air pressure applied either below or above the piston through ports in the inside of the air hammer. In the downward stroke, the hammer strikes the bottom of the upper end of the drill bit shaft (via a coupling shaft) and imparts an impact load to the drill bit. The drill bit in turn transfers this impact load to the rock face of the bit. This impact load creates a crushing action on the rock face very similar to that discussed earlier at the beginning of Section 4.2.2. However, in this situation, the crushing action is dynamic and is more effective than the quasi-static crushing action of tricone and single cone drill bits. Therefore, air hammer drilling operations require far less WOB than comparable drilling operations using tricone or single cone drill bits.

FIGURE 4-11. Two typical air hammer bits with a concave face (courtesy of Rock Bit International Incorporated).

Figure 4-11 shows two typical air hammer bits. The air hammer is made up to the bottom of the drill string and at the bottom of the air hammer is the air hammer bit. In Figure 4-11, the larger bit (standing on its shank end) is an $8^5/_8$-in (*219 mm*)-diameter concave bit and the smaller bit (on its side) is a 6-in (*152 mm*)-diameter concave bit.

The drill string with an air hammer must be rotated just like a drill string that utilizes tricone or single cone drill bits. The rotation of the drill string allows the inserts (i.e., tungsten-carbide studs) on the bit face to move to a different location on the rock face surface. This rotation allows a different position on the rock face to receive the impact load transferred from the piston through the hammer shank. In direct circulation operations, air flow passes through the hammer section, through the drill bit channel and orifices to the annulus. As the air passes into the annulus, the flow entrains the rock cuttings and carries the cuttings to the surface in the annulus. Direct circulation air hammers are available in a wide variety of outside housing diameters from 3 in (*76 mm*) to 16 in (*406 mm*). These air hammers drill boreholes with diameters from $3^5/_8$ in (*92 mm*) to $17^1/_2$ in (*445 mm*).

There are also reverse circulation air hammers. These unique air hammers allow air pressure in the annulus to actuate the hammer via ports in the outside housing of the hammer. The reverse circulation air hammer bits are designed with two large orifices in the bit face that allow the return air flow with entrained rock cuttings to flow to the inside of the drill string and then to the surface. Reverse circulation air hammers are available in larger outside housing diameters, 6 in (*152 mm*) to 24 in (*610 mm*). These air hammers drill boreholes with diameters from $7^7/_8$ in (*200 mm*) to 33 in (*838 mm*).

There are five air hammer bit cutting face designs. Figure 4-12a shows the profile of the drop center bit, and Figure 4-12b shows the profile of the concave bit. Figure 4-13a shows the profile of the step gauge bit, and Figure 4-13b shows the profile of the double gauge bit. Figure 4-14 shows the profile of the flat face bit.

In the past, air hammer manufacturers have provided the air hammer bits for their specific air hammers. This practice ensured compatibility of bit with hammer housing. Increased air hammer use in drilling deep oil and natural gas recovery wells has attracted traditional oil field drill bit manufacturers to fabricate air hammer bits. Although the air hammer bit faces are somewhat uniform in design, the shafts are different for each air hammer manufacturer. The air hammer face and shafts are integral to the bit; manufacturing air hammer bits is complicated. Fortunately, the air hammer has proven in the past decade to be very effective in drilling deep boreholes. This has given rise to competition among traditional

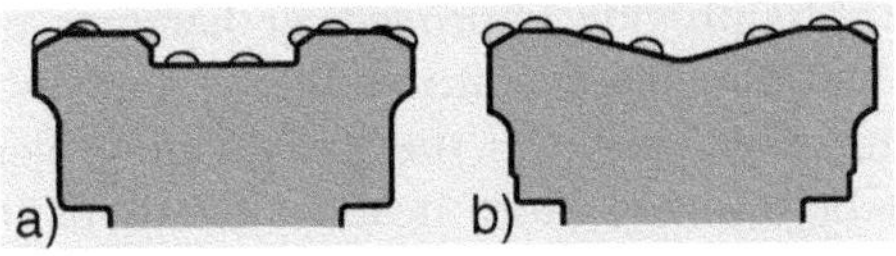

FIGURE 4-12. Air hammer bit face profile designs: (a) drop center bit and (b) concave bit (courtesy of AB Sandvik Rock Tools).

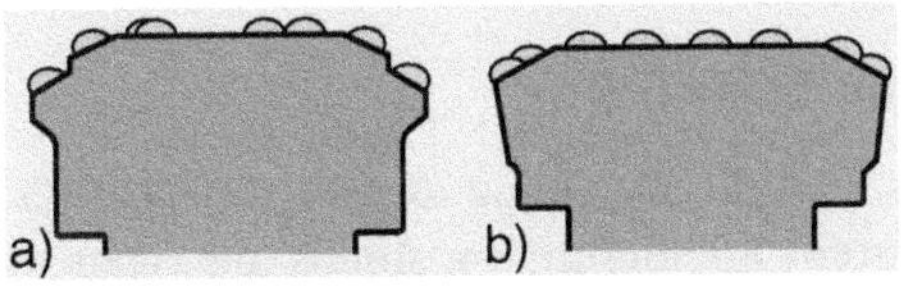

FIGURE 4-13. Air hammer bit face profile designs: (a) step gauge bit and (b) double gauge bit (courtesy of AB Sandvik Rock Tools).

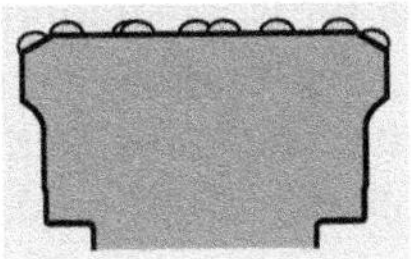

FIGURE 4-14. Air hammer bit face profile design, flat face bit (courtesy of AB Sandvik Rock Tools).

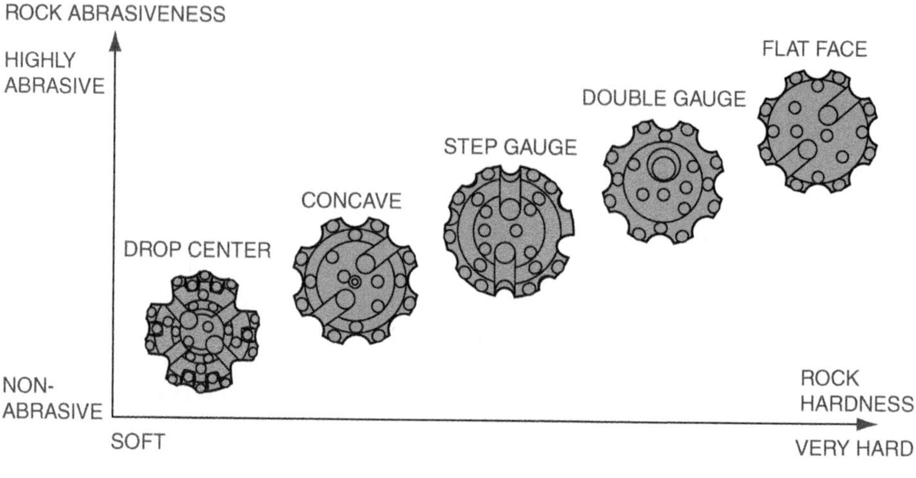

FIGURE 4-15. Air hammer bit face profile designs and application to rock formation abrasiveness and hardness (courtesy of AB Sandvik Rock Tools).

drill bit manufacturers to provide improved air hammer bits for deep drilling operations. This competition has in turn resulted in an increase in the quality and durability of air hammer bits (over the traditional air hammer manufacturer-supplied air hammer bits) in the more hostile environments of the deep boreholes. Operational use of the air hammer is discussed in detail in Chapter 11.

These five bit cutting face designs are applicable for a variety of drilling applications from nonabrasive, soft rock formations to highly abrasive, very hard rock formations. The applications of these five face designs are shown in Figure 4-15.

4.2.4 Classification of Drill Bits

The International Association of Drilling Contractors (IADC) has approved a standard classification system for identifying similar bit types available from various manufacturers [6]. Table 4-1 gives an example IADC numerical classification chart for insert drill bits (the three digits in the first column).

In general, the classification system adopted is a three-digit code. The first digit in the bit classification code is the rock formation series number. The letter "D" precedes the first digit if the bit is a diamond or PDC drag type bit. The first digit can be the numbers 1 to 3. These are reserved for milled tooth bits for soft, medium, and hard formation categories, respectively. If the first digit is 5 to 8, then these are insert bits for soft, medium, hard, and extremely hard formation categories, respectively.

The second digit is called the type number. Type 0 is reserved for PDC drag bits. Types 1 to 4 designate formation hardness subclassifications (ranging from the softest to the hardest formations with each series category).

Table 4-1. Comparison IADC Chart for Four Manufacturers of Insert Tricone Drill Bits (Courtesy of Reed Rock Bit Company)

Insert Bit Comparison by Type

	IADC CODE	REED	HUGHES	SECURITY	SMITH
	SOFT FORMATIONS				
	435	—	X11	—	—
	437	—	J11	—	—
HS-51	515	—	X22	S84	2JS
	517	FP51A	J22	S84F	F2
	525	S52	—	—	—
	527	FP52	—	—	—
	SOFT TO MEDIUM FORMATIONS				
HP-SM	532	—	HH33	—	—
	535	S53	X33	—	3JS
	537	FP53/FP53A	J33	S86F	F3
	542	—	—	S8JA	—
	545	—	—	S88	—
	547	—	—	S88F	—
HP-M	612	—	HH44	—	4JA
	615	—	X44	—	4JS
	617	FP62	J44	M84F	F4/F45
	622	Y62 JA/Y62B JA	—	M8JA	5JA
	625	S62	—	—	4JS
	627	FP62B/FP62X	J44C/J55R	M88F/M89TF	F5
	MEDIUM TO HARD FORMATIONS				
HP-MH	632	Y63 JA	HH55	—	—
	635	S63	—	—	—
	637	FP63	J55	M89F	F47/F57
	HARD FORMATIONS				
	717	—	—	—	F6
HP-H	722	—	—	—	7JA
	725	S72	—	—	—
	727	FP72	—	—	F7
	732	Y73 JA	HH77	—	—
	737	FP73	J77	H84F	—
	739	Y73 RAP	—	—	—
	742	—	—	H8JA	—
	745	S74	—	H88	—
	747	FP74	—	H88F	—
	812	—	HH88	H9JA	—
	815	—	—	H99	—
	817	—	—	H99F	—
	832	Y83 JA	HH99	H10JA	9JA
	835	S83	—	H100	—
	837	FP83	J99	H100F	F9

The third digit feature numbers for roller cutter bits are 1 to 7 and refer, respectively, to standard roller bearing, standard roller bearing for air applications, standard roller bearing with gauge protection, sealed roller bearing, sealed roller bearing with gauge protection, sealed journal bearing, and sealed journal bearing with gauge protection.

The third digit is the feature number. These numbers are interpreted differently for drag bits and roller cutter bits. For diamond and PDC drag bits, the feature numbers are 1 to 8 and refer, respectively, to step type, long taper, short taper, nontaper, downhole motor, side-track, oil base, and core ejector.

Table 4-1 also shows an insert rock bit comparison chart for four manufacturers (second through fifth columns). Feature 2 on all the charts of the bit manufacturers shows the insert bits designed for air drilling operations. Although the comparison chart shows insert tricone drill bits, the sealed roller bearing and sealed journal bearing bits are also often selected for air and gas drilling operations.

It should be noted that single cone drill bits and air hammer bits are not presently classified in accordance to the IADC code system.

4.3 BOTTOM HOLE ASSEMBLY

Figure 4-16 shows a typical bottom hole assembly (BHA) for a direct circulation rotary drilling operation. The BHA is the section of the drill string below the drill pipe (see Figure 4-1). This section of the drill string is the most rigid length of the string. It is this section of the drill string that determines how much weight can be placed on the drill bit and how "straight" a vertical borehole will be drilled with the drill string.

The assembly in Figure 4-16 is composed of a drill bit at the bottom, drill collar tubulars, a near bit stabilizer directly above the bit, a stabilizer at the middle of the assembly, and a stabilizer at the top of the assembly. The addition of stabilizers to the drill collar string generally improves the straight drilling capability of the drill string. Highly stabilized drill strings are necessary when drilling in "crooked hole country." Crooked hole country usually refers to rock formations that tend to deflect the bit and, thus, the drill string as the drill bit is advanced.

Hard-to-medium rock formations that are tilted to a high angle from horizontal are one of the main causes of severe borehole deviations from vertical. All deep rotary drilled boreholes will tend to have some deviation and tend to have a corkscrew three-dimensional shape (usually to the right). The deviation from vertical can usually be kept below 3° to 5° with good drilling practices. In general, air drilled boreholes can have more deviation than mud drilled boreholes (assuming same rock formations). However, most of the increased deviation from vertical is due to the fact that air drilling penetration rates are significantly higher than a mud drilling operation and drillers tend to take advantage of that increased drilling rate and let the deviation get away from them. To correct this tendency, it is good practice to utilize a more stabilized BHA when drilling an air drilled borehole than would be used in a comparable mud drilled

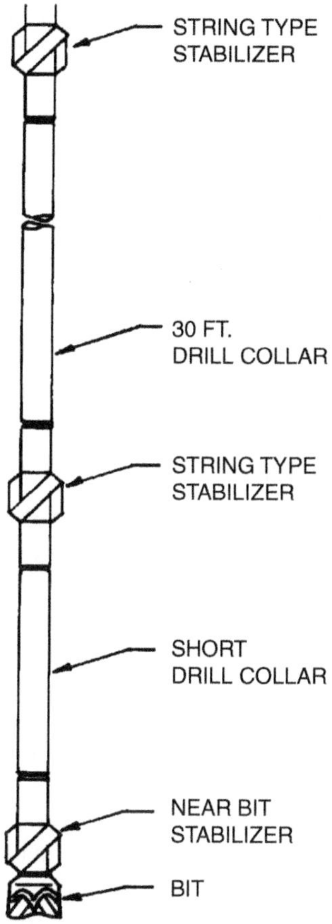

FIGURE 4-16. Typical bottom hole assembly for direct circulation rotary drilling operations.

borehole. For more details regarding the design of stabilized BHA and how to apply such assemblies, readers are referred elsewhere [1, 4] or to service company literature.

4.3.1 Drill Collars

Drill collars are thick-walled tubulars used at the bottom of the drill string (see Figure 4-1). Their principal purpose in the drill string is to provide the axial force needed to advance the drill bit (see Figure 2-2). When drilling a vertical borehole, the axial force is the weight of the drill collars. Drill collars are available in the API range lengths given in Table 2-1. Figure 4-16 shows a BHA with Range 2 (~30 ft or *9 m* long) drill collars. Range 2 lengths are typical for double and triple land rotary drilling rigs. Also shown in Figure 4-16 are short drill collar lengths used to adjust positions of the stabilizers in the BHA. These shorter drill collar lengths are selected from the Range 1 stock of drill collars. Dimensions and mechanical properties for API drill collars are given elsewhere [1].

Drill collars are usually fabricated from American Iron and Steel Institute (AISI) 4140 or 4145 heat-treated steel or equivalent. These are chrome-molybdenum steel alloys and have yield stresses in excess of 100,000 psi. In addition to drill collars fabricated of steel are special drill collars fabricated of nonmagnetic nickel alloys (e.g., usually Monel K-500). These nickel alloy drill collars (usually three) are used at the bottom of the drill string to allow magnetic compass-like equipment to be used to survey the borehole as the well is drilled. These nickel alloy drill collars have material properties that are almost identical to that of AISI 4140 heat-treated steel of the standard drill collars.

4.3.2 Stabilizers and Reamers

Stabilizers and rolling cutter reamers are special thick-walled drill collar subs that are placed in the BHA to force the drill collars to rotate at or near the center of the borehole. By keeping the drill collars at or near the center of the borehole the drill bit will drill on a nearly straight course projected by the center axis of the rigid BHA. Stabilizers and rolling cutter reamers have blades or rolling cutters that protrude from the sub wall into the annulus to near the borehole diameter. The space between blades or rolling cutters allows the air or natural gas flow with entrained rock cuttings to return to the surface nearly unobstructed.

Figure 4-17 shows three rotating blade stabilizers. These three stabilizers are, respectively, the integral blade (usually a spiral blade configuration) stabilizer, the big bear stabilizer (a larger type integral blade stabilizer), and the welded blade (spiral blade) stabilizer. The blades on these three stabilizers are machined into (integral) the stabilizer body or are rigidly attached to the stabilizer body and, therefore, rotate with the body of the stabilizer and, thus, with the drill string itself.

Figure 4-18 shows two sleeve types of blade stabilizers. These stabilizers have replaceable sleeves (with blades). These two stabilizers are, respectively, the sleeve type stabilizer and the rubber sleeve stabilizer. The sleeve type stabilizer has a metal sleeve with attached metal blades (sleeve rotates) and can be replaced on the stabilizer body when the blades wear. The rubber sleeve stabilizer has a sleeve that has a rubber sheath over a metal substructure (sleeve does not rotate). The rubber sleeve can be replaced on the stabilizer body when the blades wear.

In general, the rotating blade stabilizers are shop repairable. The integral blade stabilizers have gauge protection in the form of tungsten-carbide inserts or replaceable wear pads. Integral blade stabilizers can be used in abrasive, hard rock formations. When the blades are worn, the stabilizers can be returned to the machine shop and the inserts or wear pads replaced. Welded blade stabilizers are not recommended for use in abrasive, hard rock formations. When their

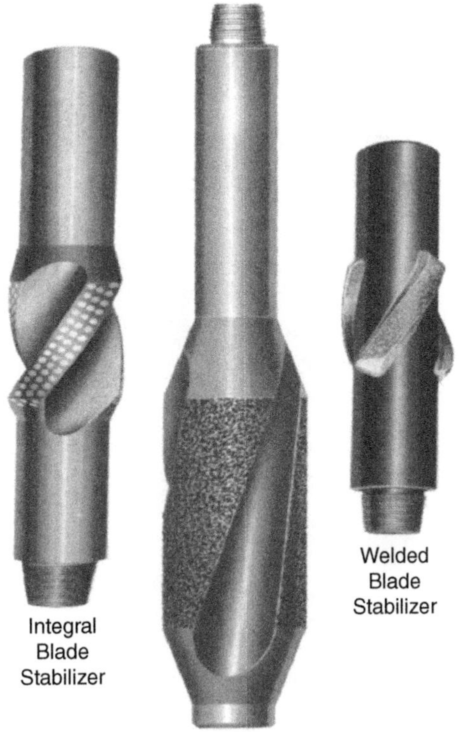

FIGURE 4-17. Rotating blade stabilizers (courtesy of Smith International Incorporated).

blades become worn or damaged, they can be returned to the machine shop for repairs.

Nonrotating blade stabilizers can be repaired at the drilling rig location. Worn sleeves can be removed and new ones placed on the stabilizer body. This is an important advantage over the rotating stabilizer. The nonrotating stabilizer is most effective in abrasive, hard rock formations, as the sleeve is stationary and acts like a drilling bushing. This action decreases wear on the metal sleeve blades.

Stabilizers are used extensively to improve the straight hole drill capability of a BHA for both mud drilling operations and air drilling operations. However, care must be exercised in using stabilizers in air drilling operations. The wear rate on stabilizer blades in air drilling operations will be greater than in a mud drilling operation (assuming similar geologic conditions).

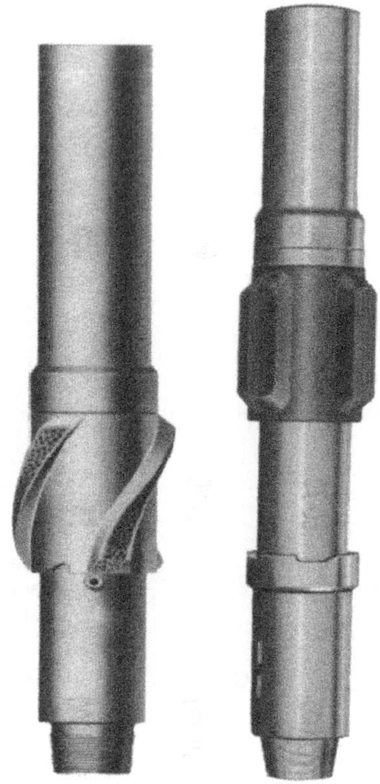

FIGURE 4-18. Nonrotating blade stabilizers (courtesy of Smith International).

Figure 4-19 shows a three-point rolling cutter reamer. These reamers have the roller cutters 120° apart on the circumference. The rolling cutter reamer is a special type of stabilizer tool that provides "blades," which are cylindrical roller cutter elements that can crush and remove rock from the borehole wall as the drill bit is advanced. Often the reamer is placed just above the drill bit (replacing the near bit stabilizer, see Figure 4-16). Reamers are also available in a four-point rolling cutter reamer. These reamers have the rolling cutters 90° apart on the circumference.

Such rolling cutter reamers are used when drilling in abrasive, hard rock formations. The gauge of the rolling cutter reamers can be adjusted by replacing the rolling cutter elements on the stabilizer body with different outside diameter elements. Also, damaged rolling cutters can be replaced. These replacements can be accomplished at the drilling rig location. When drilling abrasive, hard rock formations, the gauge of the rolling cutter reamers is usually adjusted to be slightly

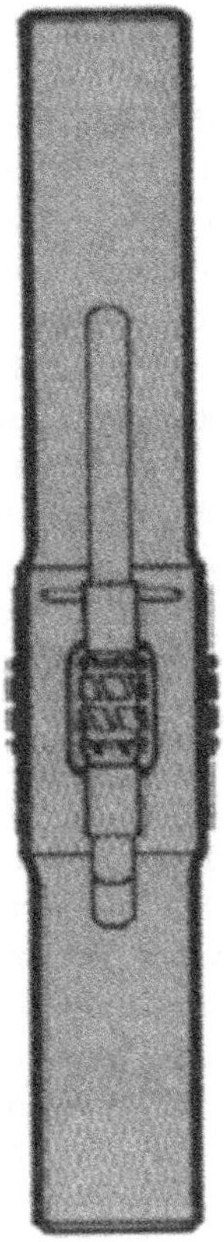

FIGURE 4-19. Three-point rolling cutter reamer (courtesy of Smith International).

under the drill bit gauge or at the drill bit gauge. The reamers provide the near-bit stabilization needed for straight drilling in abrasive formations.

4.3.3 Downhole Surveying Equipment

Bottom hole assemblies for drill strings used to drill deep vertical wells are usually fitted with several nonmagnetic drill collars. These drill collars are usually near the bottom of the BHA (just above the drill bit). These nonmagnetic collars are needed to carry out the downhole surveys required by most natural resource regulatory agencies. Downhole surveys are used to describe the deviation of a drilled borehole from the ideal vertical centerline of the intended well. Knowledge of where an actual well has been drilled is needed to ensure economic and environmentally safe recovery of mineral resources.

The survey is usually accomplished by using a magnetic single-shot instrument. This instrument is usually part of the equipment inventory of a typical double and triple rotary drilling rig. The single-shot survey is usually carried out by the rig crew. The single-shot instrument contains a small compass that floats in a liquid and gives borehole compass direction information. The floating compass is also designed with a half sphere top and an extended pendulum bottom. The spherical top of the compass is etched with a traditional compass rose allowing direction determination when viewing the compass from above and down the axis of the instrument. Also etched on the spherical top are concentric circles that represent different angles of inclination from the vertical. When viewing the compass from above and down the axis of the instrument, a set of crosshairs shows the concentric circles of angles of inclination. A small single-shot camera is installed in the instrument above the compass. The camera shutter mechanism, exposure light, and timer are battery operated. The instrument timer is set at the surface to give sufficient time for the instrument to be lowered to the bottom of the inside of the drill string. The instrument is lowered on a slick wire line (a simple wire line not having electrical transmission capability). When the instrument is in place at the bottom of the inside of the drill string, the timer actuates the light exposing the small circular film cartridge. Figure 4-20 shows a typical single-shot exposure. This exposed single-shot picture shows a direction of magnetic north (or an azimuth of 0°) and an inclination of 1.8° from vertical. As a well is drilled, single-shot survey pictures can be taken every few hundred feet. Calculations can be made using these survey pictures and the measured distance to each survey point to give a three-dimensional plot of the drilling course of the well.

Since the magnetic single-shot instrument utilizes a simple compass for directional information, the instrument must be placed in a nonmagnetic portion of

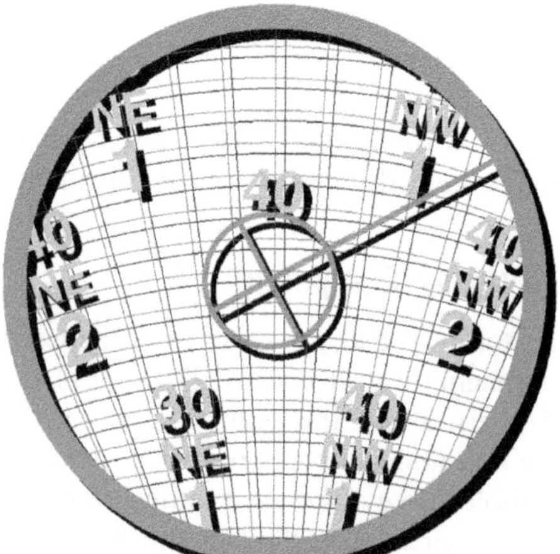

FIGURE 4-20. Typical single-shot exposure that reads north and 1.8° inclination from vertical (courtesy of Sperry-Sun Drilling Services, a Halliburton Company).

the drill string in order for the compass to give accurate azimuth readings. This is why nonmagnetic drill collars are placed at the bottom of the drill string. When running the single-shot survey, care must be taken to make sure that the single-shot instrument is located approximately midway along the nonmagnetic drill collar length section before the camera film is allowed to be exposed.

More details regarding directional drilling operations and surveys are discussed in Chapter 12.

4.4 DRILL PIPE

There are four types of drill pipe used in air and gas drilling operations. These are standard API drill pipe, heavyweight drill pipe, drill rod, and dual wall pipe.

4.4.1 Standard API Drill Pipe

Standard API drill pipe is used in most rotary drilling operations (shallow and deep). API drill pipes are fabricated by various API-certified manufacturers around the world, principally for use in the drilling of deep wells for the recovery of oil and natural gas. The nominal weight per unit length defines the wall thickness and the inside diameter of the pipe body. When tool joints (box and pin) are added to a pipe body, the average actual weight per unit length of the drill pipe element is increased above the pipe nominal weight per unit length. API drill pipe is also classified by API material (steel) grade. Tables 4-2a and 4-2b give

Table 4-2a. API Drill Pipe Steel Grade Minimum and Maximum Mechanical Properties in USCS Units [1, 7]

API Grade	E75	X95	G105	S135
Minimum yield (psi)	75,000	95,000	105,000	135,000
Maximum yield (psi)	105,000	125,000	135,000	165,000
Minimum tensile (psi)	100,000	105,000	115,000	145,000

Table 4-2b. API Drill Pipe Steel Grade Minimum and Maximum Mechanical Properties in SI Units

API Grade	E75	X95	G105	S135
Minimum yield (MPa)	517	655	724	931
Maximum yield (MPa)	724	862	931	1138
Minimum ultimate tensile (MPa)	690	724	793	1000

the API material grade designations and respective standards for minimum yield, maximum yield, and minimum tensile strengths of the steel grade. For an AISI grade of steel to be used to fabricate the drill pipe of a particular API grade, the AISI grade must satisfy the minimum and maximum specifications given in Tables 4-2a and 4-2b.

Tables 4-2a and 4-2b show that for API E75 grade the minimum yield is 75,000 psi (*517 MPa*), the maximum yield is 105,000 psi (*724 MPa*), and the minimum tensile (ultimate) strength is 100,000 psi (*690 MPa*). Similar data are shown for API X95, API G105, and API S135 steel grades.

4.4.2 Heavyweight Drill Pipe

Heavyweight drill pipe is an intermediate weight per unit length drill string element. This type of drill pipe has a heavy wall pipe body with attached extra length tool joints (see Figure 4-21). Heavyweight drill pipe has the approximate outside dimensions of standard drill pipe to allow easy handling on the drill rig [7].

The unique characteristic of this type of drill pipe is that it can be run in compression in the same manner as drill collars. Most heavyweight drill pipe is fabricated in Range 2 and 3 API lengths. It is also available in custom lengths shorter than Range 2. Heavyweight drill pipe is available in $3^1/_2$-in (*89 mm*), 4-in (*102 mm*), $4^1/_2$-in (*114 mm*), 5-in (*127 mm*), $5^1/_2$-in (*140 mm*), and $6^5/_8$-in (*168 mm*) nominal outside diameters. Figure 4-21 shows two typical heavyweight drill pipe elements standing in a regular drill pipe rack in a drill rig. One unique feature of heavyweight drill pipe is the wear pad in the center of the element. The wear pad acts as a stabilizer and improves the stiffness of the heavyweight stand in the drill string and thus reduces the deviation of boreholes.

Tables 4-3a and 4-3b give the dimensional and mechanical properties for Range 2 heavyweight drill pipe (tube body and tool joints). For heavyweight drill pipe sizes from $3^1/_2$ in (*89 mm*) and greater, the pipe tube body is fabricated of steel alloys with a minimum yield stress of 55,000 psi (*797 MPa*). The tool joints for these sizes are fabricated of steel alloys with a minimum yield stress of 120,000 psi (*828 MPa*). The $2^7/_8$-in drill pipe is fabricated entirely of 110,000 psi (*759 MPa*) steel alloy.

Figure 4-22 shows a "tapered" drill string with a drill bit at the bottom, drill collars above the drill bit, heavyweight drill pipe above the drill collars, and standard drill pipe above the heavyweight drill pipe.

Heavyweight drill pipe elements are used in a number of applications in rotary drilling. Because this drill pipe can be used in compression, this drill pipe can be used in place of drill collars in shallow wells with small single or double rotary drilling rigs. This drill pipe is also used in conventional drill string for vertical drilling operations as transitional stiffness elements between the stiff drill collars and the very limber drill pipe. Their use as transitional stiffness elements reduces the mechanical failures in the bottom drill pipe elements of the drill string. The practice is to run from 6 to 30 heavyweight drill pipe on top of a conventional BHA.

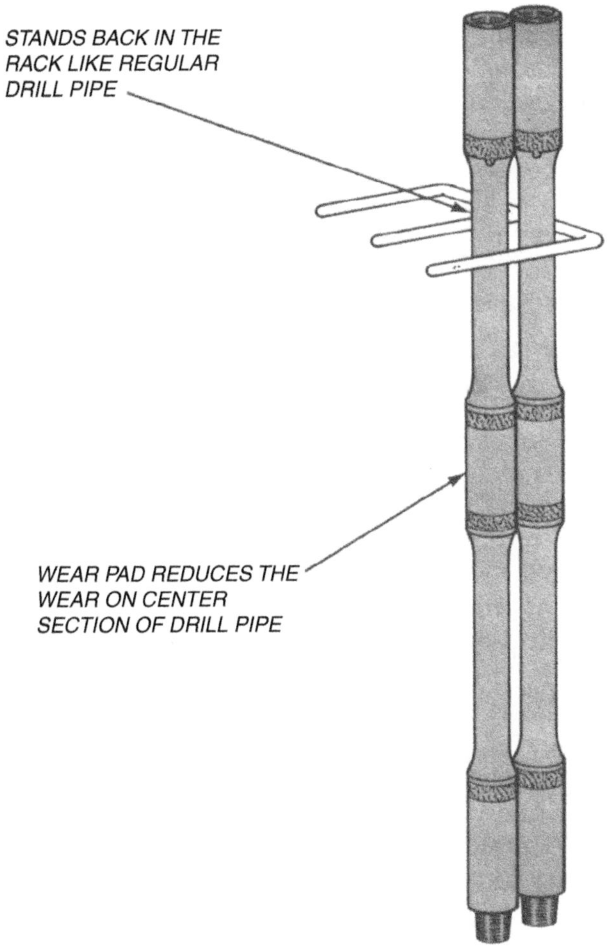
STANDS BACK IN THE
RACK LIKE REGULAR
DRILL PIPE
WEAR PAD REDUCES THE
WEAR ON CENTER
SECTION OF DRILL PIPE

FIGURE 4-21. Heavyweight drill pipe standing in rig rack (courtesy of Smith International).

In off-shore drilling operations, heavyweight drill pipe has become essential. Heavyweight drill pipe is used in directional drilling operations where drill collars can be replaced by the heavyweight pipe. Using heavyweight drill pipe in place of drill collars reduces the rotary torque and drag and increases directional control.

In Tables 4-3a and 4-3b there is an unusual drill pipe size. This is the $5^7/_8$-in (*149 mm*) heavyweight drill pipe, and its associated standard drill pipe companion is used for drilling long reach highly deviated directional boreholes (particularly in offshore operations). This drill pipe allows higher annulus drilling fluid velocities for improved cuttings removal (particularly in horizontal boreholes) and lower drilling fluid velocities inside the drill string to reduce pipe friction losses.

Table 4-3a. Heavyweight Drill Pipe Range 2 Dimensions and Mechanical Properties in USCS [7]

Nominal OD (in)	Approx Wt (lb/ft)	Tube ID (in)	Wall Thick (in)	Conn Type	Took Jt OD (in)	Took Jt ID (in)	Tensible Yield (lb)
2 7/8	17.26	1.5000	0.688	NC26	3 3/8	1.5000	519,750
3 1/2	23.70	2.2500	0.625	NC38	4 3/4	2.2500	310,475
4	29.90	2.5625	0.719	NC40	5 1/4	2.5625	407,500
4 1/2	40.80	2.8125	0.844	NC46	6 1/4	2.8125	533,060
5	50.38	3.0000	1.000	NC50	6 5/8	3.0000	691,130
5 1/2	61.60	3.2500	1.125	5 1/2 FH	7 1/4	3.2500	850,465
5 7/8	57.42	4.0000	0.938	XT57	7	4.0000	799,810
6 5/8	71.43	4.5000	1.063	6 5/8 FH	8	4.5000	1,021,185

Table 4-3b. Heavyweight Drill Pipe Range 2 Dimensions and Mechanical Properties in SI Units

Nominal OD (mm)	Approx M (kg/m)	TUbe ID (mm)	Wall Thick (mm)	Conn Type	Tool Jt OD (mm)	Tool Jt ID (mm)	Tensible Yield (kN)
73	25.68	38	17	NC26	86	38	2,312
89	35.26	57	16	NC38	121	57	1,381
102	44.48	65	18	NC40	133	65	1,813
114	60.70	71	21	NC46	159	71	2,371
127	74.95	76	25	NC50	168	76	3,074
140	91.64	83	29	5 1/2 FH	184	83	3,783
149	86.42	102	24	XT57	178	102	3,558
168	106.26	114	27	6 5/8 FH	203	114	4,542

4.5 SAFETY EQUIPMENT

Drill strings used in direct circulation drilling operations for the recovery of oil, natural gas, or geothermal fluids are usually fitted with several safety valves.

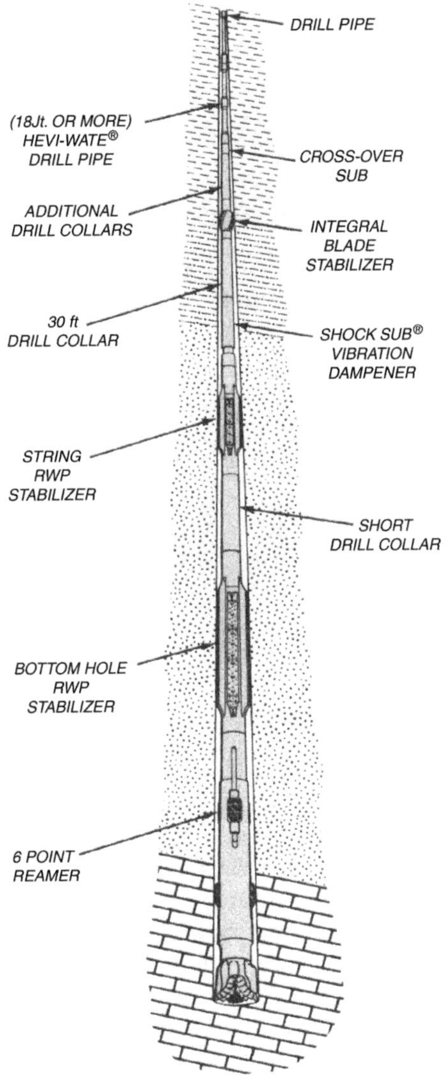

FIGURE 4-22. Typical tapered drill string using heavyweight drill pipe (courtesy of Smith International).

4.5.1 Float Valves

Figures 4-23 and 4-24 show the two typical drill string float valves. These are also known as nonreturn valves. These two types are the dart type and the flapper type. These are safety valve devices that prevent the back flow or U-tube flow of liquids or gases from the annulus into the inside of the drill string.

FIGURE 4-23. Typical dart type float valve for direct circulation operations (courtesy of Baker Oil Tools).

The dart type valve is spring activated, which opens to allow the direct circulation flow to pass around the dart. This type of valve provides a more secure shut-off against high and low pressure back flows. The flapper type valve opens fully during circulation to provide an unrestricted bore through the valve and closes when back flow pressure is applied. These valves are used in nearly all deep rotary air and gas drilling operations. The dart valve is used in the bit sub just above the drill bit. In practice, at least one flapper valve is placed just above the drill collars or above a downhole motor. A second is often placed about 1000 ft below the surface. It is not unusual for a long drill string to have three or four float valves.

The fire stop is a special type of flapper valve. It is essentially an upside-down float valve. It is usually placed just above the drill bit. These valves have a zinc ring that holds back a spring-loaded flapper mechanism to allow the compressed air or gas to be circulated directly from the surface through the inside of the drill string. Wire line equipment can be run through these valves when the fire stop is in the normal open position. In the event of a downhole fire, the zinc element melts, releasing the spring-loaded flapper. This shuts down the flow of air or gas into the bottom of the well, thus shutting off the source of fuel for the fire. Fire stop valves are rarely used in present drilling operations.

FIGURE 4-24. Typical flapper type float valve for direct circulation operations (courtesy of Baker Oil Tools).

4.5.2 Kelly Sub Valves

At the top of the drill string (just above the kelly) is a kelly cock sub, which is fitted with a ball valve (see Figure 4-1). In the event of a subsurface blowout, the kelly cock's ball valve can be closed and the sub left made up to the top of the kelly. With the ball valve closed, a pressure gauge can be made up to the top of the sub. Using this pressure gauge, the ball valve can be opened and vital pressure information obtained for the pressure inside the drill string (together with casing head annulus pressure). This information is needed to design the well control procedure.

4.6 DRILL STRING DESIGN

One of the initial planning steps for a rotary drilling operation is the design of the drill string. The drill string must have the strength to drill to the intended target depth and be light enough so that the hoisting system can extract the string from the well when the target depth has been reached. This section describes the API procedure for drill string design [1, 7].

The axial tension force (or load), F (lb, N), at the top of the drill string in a vertical well is

$$F = \left[(L_p\ w_p) + (L_c\ w_c)\right] K_b, \tag{4-1}$$

where L_p is the length of the drill pipe (ft, m), w_p is the weight per unit length of drill pipe (lb/ft, N/m), L_c is the length of the drill collars (ft, m), w_c is the weight per unit length of drill collars (lb/ft, N/m), and K_b is the buoyancy factor.

The buoyancy factor is

$$K_b = \left(1 - \frac{\gamma_m}{\gamma_s}\right), \tag{4-2}$$

where γ_m is the specific weight of the drill fluid (lb/ft^3, N/m^3 or lb/gal, kg/liter) and γ_s is the specific weight of steel (lb/ft^3, N/m^3 or lb/gal, kg/liter).

The maximum allowable design axial tension force, F_a (lb, N), is

$$F_a = 0.9\ F_y, \tag{4-3}$$

where F_y is the drill pipe tension force to produce material (steel) yield (incipient failure of the drill pipe)(lb, N).

For most drill string designs, a factor of safety is used to ensure that there is a margin of overpull (*MOP*) to allow for a stuck drill string. The drill string design factor of safety, *FS*, is given by

$$FS = \frac{F_a}{F}. \tag{4-4}$$

The *MOP* is determined by the rotary drilling rig hoisting capacity. Thus, *MOP* is

$$MOP = F_c - F, \tag{4-5}$$

where F_c is the hoisting capacity of the rotary drill rig (lb, N). However, the total hoisting axial force cannot exceed F_a.

Illustrative Example 4.1 A section of the example vertical well (see Figure 2-15) is to be rotary air drilled from 7,000 ft (*2133.5 m*) to 10,000 ft (*3047.9 m*) with a $7^7/_8$-in (*200 mm*) tricone roller cutter drill bit. This is to be a direct circulation drilling operation. Above the drill bit the BHA is made up of 500 ft (*152.4 m*) of $6^1/_4$-in (*158 mm*) by $2^{13}/_{16}$-in (*71.4 mm*) drill collars and similar diameter survey subs and nonmagnetic drill collars. The drill pipe available from the drilling contractor is API $4^1/_2$-in (*114 mm*), 16.60 lb/ft (*24.70 kg/m*) nominal, IEU-S135, NC46, new drill pipe. Determine the *FS* and the effective *MOP* associated with using this drill string to drill this section of the well. The drilling rig has a hook load capability of 300,000 lb (*1334.4 kN*).

USCS units solution: Reference 1 gives 83 lb/ft (*123.48 kg/m*) for the unit weight of the $6^1/_4$-in (*158 mm*) by $2^{13}/_{16}$-in (*71.4 mm*) drill collar, and an actual unit weight of the API $4^1/_2$-in (*114 mm*) drill pipe is 19.00 lb/ft (*28.27 kg/m*) for the drill pipe. Because the drilling fluid is air, then $\gamma_m = 0$, and $K_b = 1.0$. The maximum axial tension force in the top drill pipe element of the drill string is when the depth is 10,000 ft *(3047.9 m)* and when the drill bit is lifted off the bottom of the well (after the target depth has been reached). The maximum axial tension force is determined from Equation (4-1). This is

$$F = \left[(9500)(19.00) + (500)(83.00)\right](1.0)$$

$$F = 222000 \text{ lb.}$$

This tension force to yield is 595,004 lb *(2646.6 kN)*[1]. Equation (4-3) can be used to determine the maximum allowable axial tension force for the drill pipe. Thus, Equation (4-3) is

$$F_a = 0.9(595004)$$

$$F_a = 535504 \text{ lb.}$$

The factor of safety is determined from Equation (4-4). Equation (4-4) gives

$$FS = \frac{535504}{222000}$$

$$FS = 2.41.$$

The selected rotary drilling rig has a maximum hoisting capacity of 300,000 lb, thus using Equation (4-5) the effective *MOP* is

$$MOP = 300000 - 222000$$

$$MOP = 78000 \text{ lb.}$$

The calculations just given show that an additional 78,000 lb (above the weight of the drill string) can be pulled by the hoist system or with the assurance that the drill string will not fail structurally.

SI units solution: The unit mass of the 158-mm by 71.4-mm drill collar is 123.48 kg/m, and the actual unit mass of the API 114 mm drill pipe is 28.27 kg/m. Because the drilling fluid is air, then $\gamma_m = 0$ and $K_b = 1.0$. The maximum axial tension force in the top drill pipe element of the drill string is when the depth is 3047.9 m and when the drill bit is lifted off the bottom of the well (after the target depth has been reached). The maximum axial tension force is determined from Equation (4-1). This is

$$F = \left[(2895.5)(28.27)(9.81) + (152.4)(123.48)(9.81)\right](1.0)$$

$$F = 987613 \text{ N.}$$

This tension force to yield is 2646.6 kN [1]. Equation (4-3) can be used to determine the maximum allowable axial tension force for the drill pipe. Thus, Equation (4-3) is

$$F_a = 0.9\ (2646600)$$

$$F_a = 2381940 \text{ N.}$$

The factor of safety is determined from Equation (4-4). Equation (4-4) gives

$$FS = \frac{2381940}{987613}$$

$$FS = 2.41$$

The selected rotary drilling rig has a maximum hoisting capacity of 1334.4 kN lb, thus using Equation (4-5) the effective *MOP* is

$$MOP = 1334400 - 987613$$

$$MOP = 346787 \text{ N}$$

The calculations just given show that an additional 349,787 N (above the weight of the drill string) can be pulled by the hoist system or with the assurance that the drill string will not fail structurally.

REFERENCES

1. API Recommended Practices for Drill Stem Design and Operating Limits, *API RP7G,* Sixteenth Edition, August1998.
2. *Drilling Manual*, International Association of Drilling Contractors (IADC), Eleventh Edition, 1992.
3. *Drilling Assembly Handbook*, Smith Services, Division of Smith International, 2006.
4. Roscoe Moss Company, *Handbook of Ground Water Development*, Wiley, 1990.
5. Bourgoyne, A. T., Millheim, K. K., Chenevert, M. E., and Young, F. S., *Applied Drilling Engineering*, SPE, First Printing, 1986.
6. Durrett, E., "Rock Bit Identification Simplified by IADC Action," *Oil and Gas Journal*, Vol. 76, May 22, 1972.
7. *Drill Stem Design and Operation*, Standard DS-1, Vol. 2, Third Edition, T. H. Hill Associates, 2004.

CHAPTER

Compressors and Nitrogen Generators

5

There are a variety of air and gas compressor designs in use throughout industry. These designs vary greatly in the volume amounts of air or gas moved and the pressures attained. The largest usage of compressors is in the oil and gas production and transportation industries and in the chemical industry. Information regarding this technology will be used to develop an understanding of how compressors can be used in air and gas drilling operations.

Air or gas compressors are very similar in basic design and operation to liquid pumps. The basic difference is that compressors are movers of compressible fluids; pumps are movers of incompressible fluids (i.e., liquids).

5.1 COMPRESSOR CLASSIFICATION

Similar to the classification of pumps, compressors are grouped in one of two general classes: continuous flow (i.e., dynamic) and intermittent flow (i.e., positive displacement)(see Figure 5-1)[1, 2]. Intermittent flow or positive displacement compressors move the compressible fluid through the compressor in separate volume packages of compressed fluid (these volume packages are separated by moving internal structures in the machine). The most important subclass examples of positive displacement compressors are reciprocating and rotary compressors. Continuous flow or dynamic compressors utilize the kinetic energy of the continuously moving compressible fluid in combination with the internal geometry of the compressor to compress the fluid as it moves through the device. The most important subclass examples of dynamic compressors are centrifugal and axial-flow compressors.

Each of the two general classes of compressors and their subclasses have certain advantages and disadvantages regarding their respective volumetric flow rate capabilities and overall compression pressure ratios. Figure 5-2 shows the typical application range in volumetric flow rates (actual cfm) and compression pressure ratios for most important compressor classes and subclasses [2].

In general, positive displacement compressors are best suited for handling high pressure ratios (i.e., up to approximately 200), but this can be accomplished with

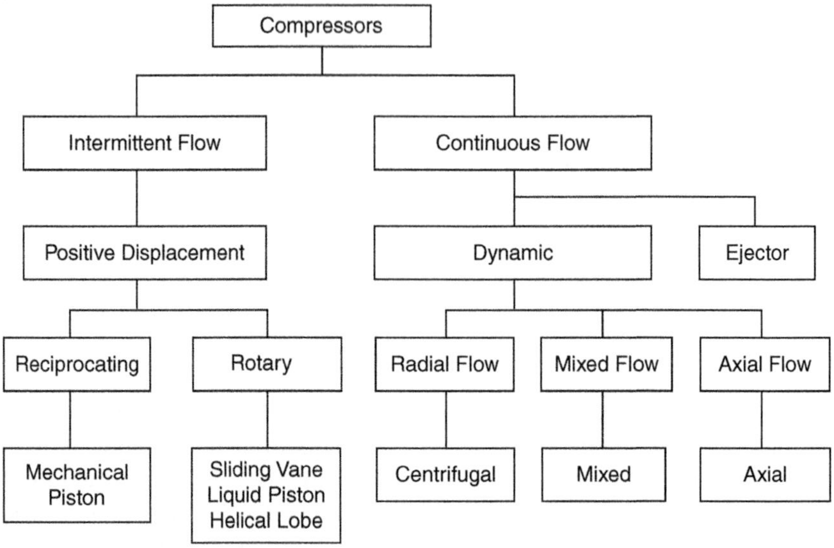

FIGURE 5-1. Compressor classification [2].

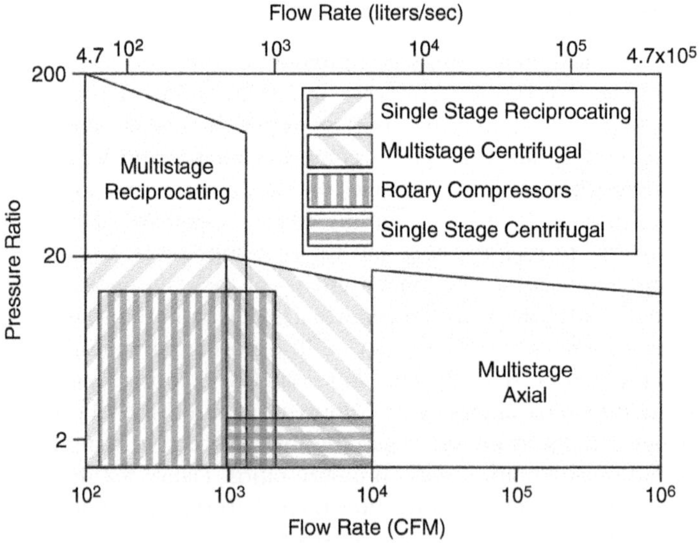

FIGURE 5-2. Typical application ranges of compressor types [2].

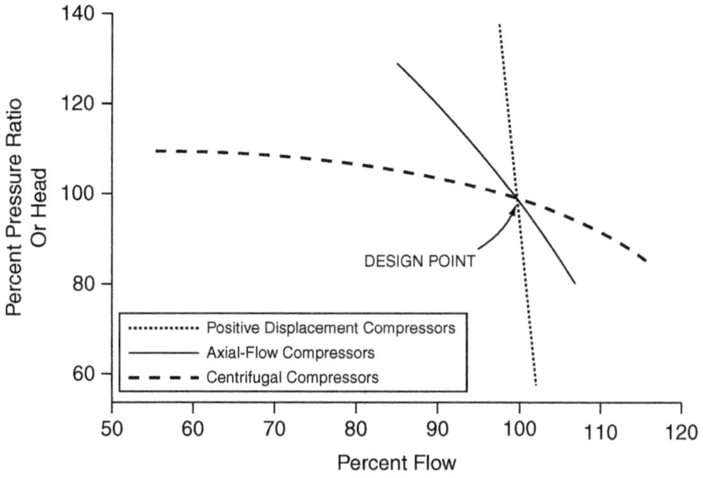

FIGURE 5-3. General performance curves for various compressor types [2].

only moderate volumetric flow rate magnitudes (i.e., up to about 2×10^3 actual cfm or 10^3 actual liters/sec). Dynamic compressors are best suited for handling large volumetric flow rates (i.e., up to about 10^6 actual cfm or 5×10^5 actual liters/sec), but with only moderate pressure ratios (i.e., up to approximately 20).

Figure 5-3 gives general performance curves for various positive displacement and dynamic compressors [2]. Positive displacement compressors, particularly the multistage reciprocating compressors, are very insensitive to downstream back pressure changes. These compressors will produce their rated volumetric flow rate even when the pressure ratio approaches the design limit of the machine. Rotary compressors are fixed pressure ratio machines and are generally insensitive to downstream back pressure ratio changes as long as the output pressures required are below the maximum design pressures of the machines (i.e., problems with slippage). Dynamic compressors are quite sensitive to pressure ratio changes. The volumetric flow rates will change drastically with rather small changes in the downstream back pressure (relative to the pressure ratio around which the machine has been designed). Thus, positive displacement compressors are normally applied to industrial operations where volumetric flow rates are critical and pressure ratios are variable.

Dynamic compressors are generally applied to industrial operations where the volumetric flow rate and pressure ratio requirements are relatively constant.

5.2 STANDARD UNITS

In the United States, a unit of air or any gas is referenced to the standard cubic foot of dry air. The API Mechanical Equipment Standards standard atmospheric conditions for dry air are a temperature of 60°F, which is 459.67 + 60 = 519.67°R (*15.6°C, which is 273.15 + 15.6 = 288.7 K*) and a pressure of 14.696 psia (*10.134 N/cm² abs*). The equation of state for the perfect gas can be written as

$$\frac{P}{\gamma} = \frac{\mathbf{R}_u T}{m_w}, \tag{5-1}$$

where P is the pressure (lb/ft² abs, N/m² abs), γ is the specific weight (lb/ft³, N/m³), $\mathbf{R}_u$ is the universal gas constant (1545.4 ft-lb/lb-mole-°R or 848.9 N-m/N-mole-K), T is the temperature (°R, K), and m_w is the mole weight of the gas (lb/lb-mole, N/N-mole).

USCS units: Using Equation (5-1), the specific weight γ or the weight of 1 ft³ of dry air is

$$\gamma = \frac{14.696(144)(28.96)}{1545.4(519.67)}$$

$$\gamma = 0.0763 \text{ lb/ft}^3,$$

where $m_w = 28.96$ lb/lb-mole (for dry air). Thus, a dry cubic foot of air at the API Mechanical Equipment Standards standard atmospheric condition weighs 0.0763 pounds (or a specific weight of 0.0763 lb/ft³)[2–4].

SI units: Using Equation (5-1), the specific weight γ or the weight of 1 m³ of dry air is

$$\gamma = \frac{101342(28.96)}{848.9(288.70)}$$

$$\gamma = 12.01 \text{ N/m}^3,$$

where $m_w = 28.96$ N/N-mole (for dry air). Thus, a dry cubic foot of air at the API Mechanical Equipment Standards standard atmospheric condition weighs 12.01 newtons (or a specific weight of 12.01 N/m³)[2–4].

Other organizations within the United States and regions around the world have established slightly different standard atmospheric conditions. For example, in the Untied States, another industrial standard is that of the American Society of Mechanical Engineering (ASME), which uses a standard atmosphere with a temperature of 68°F (*20°C*), a pressure of 14.7 psia (*10.137 N/cm² abs*), and a relative humidity of 36%. Most of the industries in the United Kingdom use a standard atmosphere with a temperature of 60°F (*15.6°C*) and a pressure of 30.00 in of Hg abs (*762 mm of Hg abs*). Most industries in continental Europe use a standard atmosphere with a temperature of *15°C* (59°F) and pressure of a bar of *750 mm of Hg abs* (29.53 inches of Hg abs)[5]. This later pressure of a bar converts to 14.5 psia.

When selecting and sizing compressors for a particular application, it is necessary to determine which standard condition has been used to rate a compressor under consideration. This is very important if the compressor has been manufactured in a foreign country. All further discussions in this section will utilize only API Mechanical Equipment Standards standard atmospheric conditions.

Table 5-1. Average Annual Atmosphere at Elevation (Mid altitudes North America) above Mean Sea Level [6]

Elevation (ft, m)	Pressure (psia, N/cm^2 abs)	Temperature (°F, °C)
0	14.696 (10.136)	59.00 (15.00)
2,000 (609.6)	13.662 (9.423)	51.87 (11.04)
4,000 (1219.1)	12.685 (8.749)	44.74 (7.08)
6,000 (1828.7)	11.769 (8,117)	37.60 (3.11)
8,000 (2438.9)	10.911 (7.525)	30.47 (–0.85)
10,000 (3047.9)	10.108 (6.972)	23.36 (–4.80)

Compressors are rated by their maximum volumetric flow rate input and their associated maximum pressure output. Volumetric flow rate input ratings are usually specified in units of standard cubic feet per minute (scfm) or standard liters per second. The maximum pressure output is in psig or N/m^2 gauge (at a specified standard atmospheric condition, e.g., API, ASME). The scfm and standard liters per second volumetric flow rate refer to the compressor intake. The pressure rating refers to the output pressure capability referenced to an atmospheric standard gauge pressure, e.g., API and ASME.

When a compressor is operated at surface location elevations above sea level, the volumetric flow rate intake is referred to as actual cubic feet per minute (actual cfm or actual liters/sec). Table 5-1 gives the average annual atmospheric pressure and annual average atmospheric temperature for middle northern latitudes (applicable to latitudes 30° North to 60° North)[6]. Appendix B gives additional graphic data for surface elevation atmospheric pressures and temperatures. These data will be used in follow-on examples.

5.3 CONTINUOUS FLOW (DYNAMIC) COMPRESSORS

The most widely used continuous flow compressors in industry are centrifugal and axial flow (or compressors that combine the two designs).

5.3.1 Centrifugal Compressors

The centrifugal compressor was the earliest developed dynamic compressor. This type of compressor allows for the continuous flow of the gas through the machine. There is no distinct closed boundary enclosure in which compression takes place. Compression of the gas results from the speed of the flow through a specified geometry within the compressor. The basic concept of the centrifugal

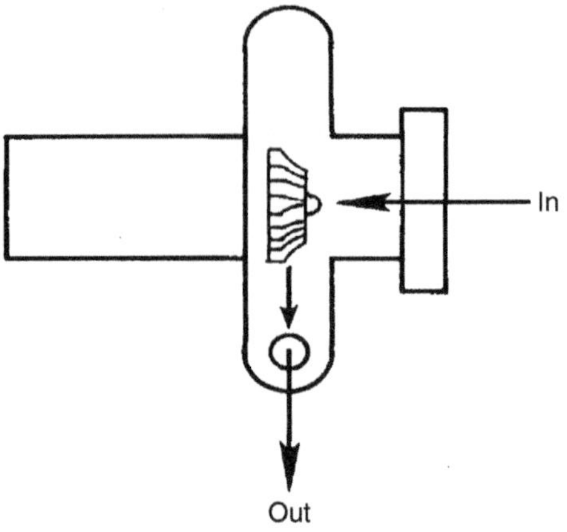

FIGURE 5-4. Single-stage centrifugal compressor [2].

compressor is the use of centrifugal forces on the gas created by high velocity flow of the gas in the cylindrical housing. Figure 5-4 shows a diagram of a single-stage centrifugal compressor [2]. The gas to be compressed flows into the center of the rotating impeller. The impeller throws the gas out to the periphery by means of its radial blades rotating at high speed. The gas is then guided through the diffuser where the high velocity gas is slowed, which results in a higher pressure in the gas. In multistage centrifugal compressors, the gas is passed to the next impeller from the diffuser of the previous impeller (or stage). In this manner, the compressor may be staged to increase the pressure of the ultimate discharge (see Figure 5-5)[2]. Because the compression pressure ratio at each stage is usually

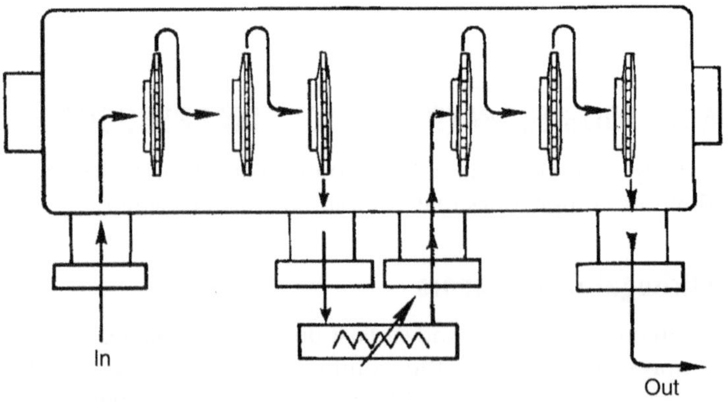

FIGURE 5-5. Multistage centrifugal compressor with intercooler [2].

rather low, of the order of 2, there is little need for intercooling between each stage (Figure 5-5 shows an intercooler after first three stages).

The centrifugal compressor must operate at rather high rotation speeds to be efficient. Most commercial centrifugal compressors operate at speeds on the order of 20,000 to 30,000 rpm. With such rotation speeds, very large volumes of gas can be compressed with equipment having rather modest external dimensions. Commercial centrifugal compressors can operate with volumetric flow rates up to approximately 10^4 actual cfm (*4.7 × 10^3 actual liters/sec*) and with overall multistage compression ratios up to about 20.

Centrifugal compressors are usually used in large processing plants and in some pipeline applications. They can be operated with some small percentage of liquid in the gas flow.

These machines are used principally to compress large volumetric flow rates to rather modest pressures. Thus, their use is more applicable to the petroleum refining and chemical processing industries.

More details regarding the centrifugal compressor can be found elsewhere [2, 7–9].

5.3.2 Axial-Flow Compressors

Axial-flow compressors are very high-speed, large volumetric flow rate machines. This type of compressor flows gas into the intake ports and propels the gas axially through the compression space via a series of radial arranged rotating rotor blades and stationary stators (or diffuser) blades (see Figure 5-6)[1]. As in the centrifugal compressor, the kinetic energy of the high-velocity flow exiting each rotor stage is converted to pressure energy in the follow-on stator (diffuser) stage.

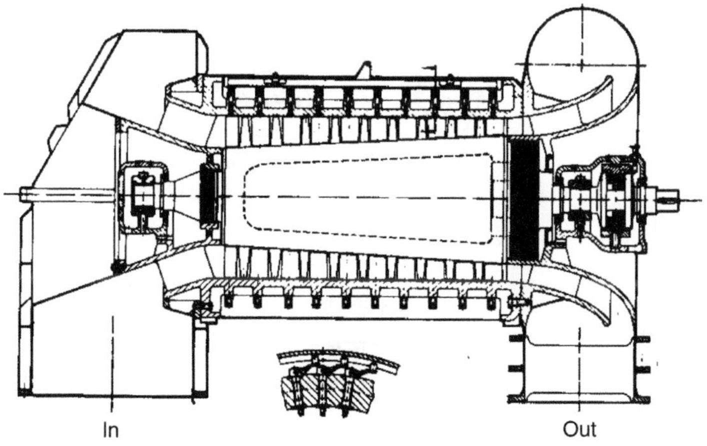

FIGURE 5-6. Multistage axial-flow compressor [1].

Axial-flow compressors have a volumetric flow rate range of approximately 10^4 to 10^6 actual cfm (*4.7 to 4.7 × 10^5 actual liters/sec*). Their compression ratios are typically around 10 to 20. Because of their small diameter, these machines are the principal compressor design for jet engine applications. There are also applications for axial-flow compressors in large process plant operations where very large constant volumetric flow rates at low compression ratios are needed.

More detail regarding axial-flow compressors can be found elsewhere [2, 7-9].

5.4 INTERMITTENT (POSITIVE DISPLACEMENT) COMPRESSORS

In general, only the reciprocating compressor allows for rather complete reliable flexibility in applications requiring variable volumetric flow rates and variable pressure ratios. The rotary compressor (which has a fixed pressure ratio built into the compressor design) does not allow for much variation in either.

5.4.1 Reciprocating Compressors

The reciprocating compressor is the simplest example of the positive displacement class of compressors. This type of compressor was also the earliest designed. Like reciprocating incompressible fluid pumps, reciprocating compressors can also be either single acting or double acting. Single-acting compressors are usually of the trunk type (see Figure 5-7)[1]. Double-acting compressors are usually of the crosshead type (see Figure 5-8)[1].

Reciprocating compressors are available in both lubricated and nonlubricated versions. Lubricated versions provide lubrication for the moving pistons (in the

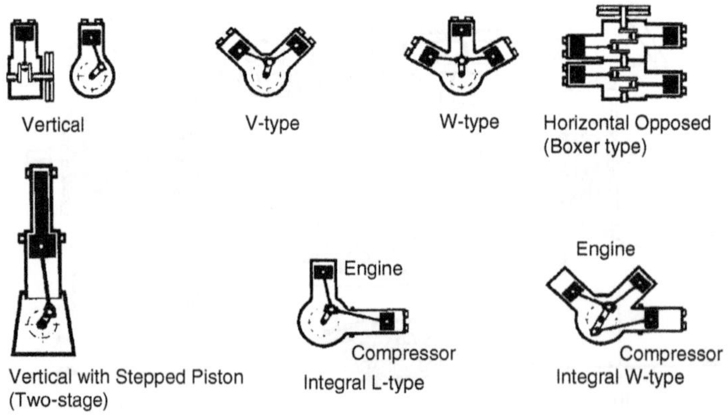

FIGURE 5-7. Single-acting (trunk type) reciprocating piston compressor [1].

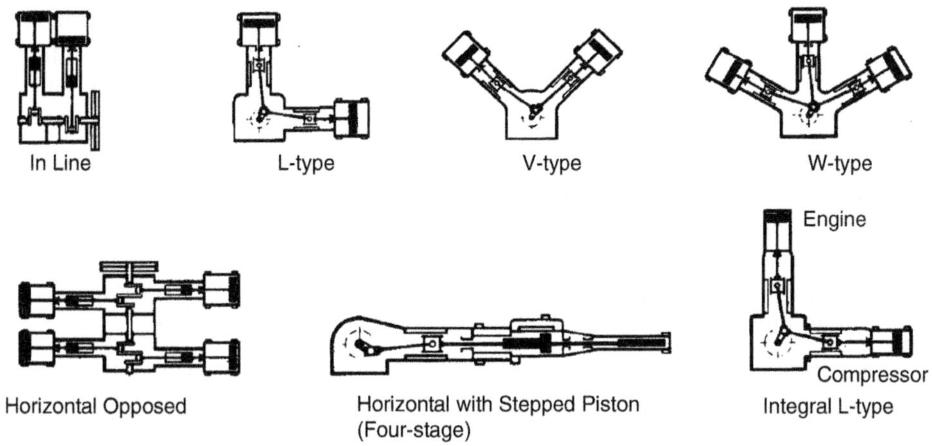

FIGURE 5-8. Double-acting (crosshead type) reciprocating piston compressor [1].

cylinder) either through an oil-lubricated intake gas stream or via an oil pump and injection of oil to the piston sleeve. There are some applications where oil must be omitted completely from the compressed air or gas exiting the machine. For such applications where a reciprocating piston type of compressor is required, there are nonliquid lubricated compressors. These compressors have piston rings and wear bands around the periphery of each piston. These wear bands are made of special wear-resistant dry lubricating materials such as polytetrafluorethylene. Trunk type nonlubricated compressors have dry crankcases with permanently lubricated bearings. Crosshead type compressors usually have lengthened piston rods to ensure that no oil wet parts enter the compression space [1, 7].

Most reciprocating compressors have inlet and outlet valves (on the piston heads) and are actuated by a pressure differential. These are called *self-acting valves*.

The main advantage of the multistage reciprocating piston compressor is the positive control of both the volumetric flow rate, which can be put through the machine, and the pressure of the output. Many reciprocating piston compressors allow for the rotation to be adjusted, thus changing the throughput of air or gas. Also, provided that there is adequate input power from the prime mover, reciprocating piston compressors can adjust to any back pressure changes and maintain proper rotation speed (which in turn maintains a given volumetric flow rate).

The main advantage of this subclass of compressor is extremely high pressure output capability and reliable volumetric flow rates (see Figures 5-2 and 5-3). The main disadvantage to multistage reciprocating piston compressors is that they cannot be practically constructed in machines capable of volumetric flow rates much beyond 1400 actual cfm (*660 actual liters/sec*). Also, the higher capacity compressors are rather large and bulky and generally require more maintenance than similar capacity rotary compressors. In any positive displacement compressor, such as a liquid positive displacement pump, the real volume flow rate is slightly smaller than the mechanical displacement volume. This is due to the following factors:

- Pressure drop on the suction side
- Heating up of the intake air
- Internal and external leakage
- Expansion of the gas trapped in the clearance volume (reciprocating piston compressors only)

Reciprocating compressors can be designed with multiple stages. Such multistage compressors are designed with nearly equal compression ratios for each stage (it can be shown that equal stages of compression lead to minimum input power requirements). Thus, because the volumetric flow rate (in actual cfm) is reduced from one stage to the next, the volume displacement of each stage (its geometric size) is progressively smaller.

These compressors can be used as either primary or booster compressors in drilling operations.

5.4.2 Rotary Compressors

Another important positive displacement compressor is the rotary compressor. This type of compressor is usually of rather simple construction, having no valves and being lightweight. These compressors are constructed to handle volumetric flow rates up to around 2000 actual cfm (*actual 944 liters/sec*) and pressure ratios up to around 15 (see Figure 5-2). Rotary compressors are available in a variety of designs. The most widely used rotary compressors are the sliding vane, helical lobe (screw), and liquid piston.

The most important characteristic of this type of compressor is that all have a fixed, built-in, compression ratio for each stage of compression (as well as a fixed, built-in volume displacement)[1]. Thus, at a given rotational speed (provided by the prime mover), there will be a predetermined volumetric flow rate through the compressor (the geometry of the compressor inlet is fixed), and the pressure at the outlet will be equal to the built-in design pressure ratio of the machine multiplied by the inlet pressure.

The top pressure versus volumetric flow rate plot in Figure 5-9 shows the typical situation when the back pressure on the outlet side of the compressor is equal to the built-in design output pressure. Under these conditions, there is no expansion of the output gas as it exits the compressor, passes through the expansion tank and continues into the initial portion of the pipeline [1].

The middle pressure versus volumetric flow rate plot in Figure 5-9 shows the typical situation when the back pressure on the outlet side of the compressor is above the built-in design output pressure. Under these conditions, the compressor

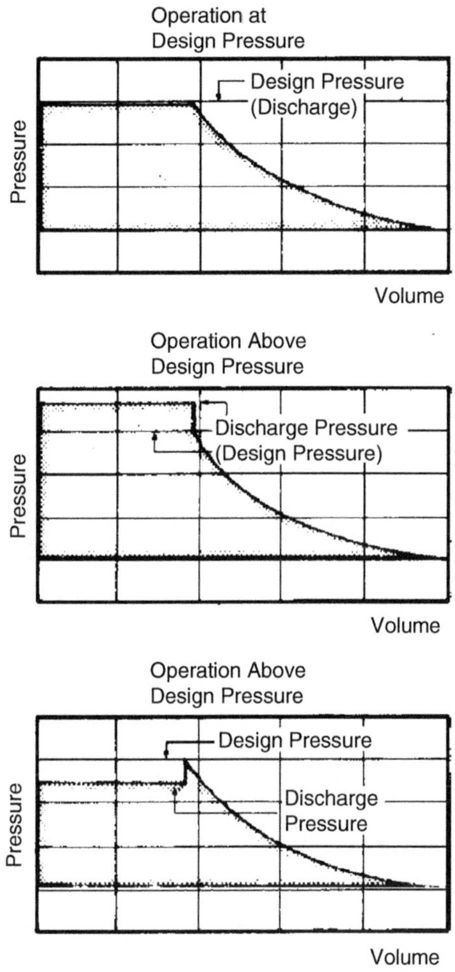

FIGURE 5-9. Rotary compressors operating with back pressure at the built-in design pressure (upper), with back pressure above the built-in design pressure (middle), and with back pressure below the built-in design pressure (lower)[1].

cannot efficiently expel the gas volume within it. Thus, the fixed volumetric flow rate (at a given rotation speed) will be reduced by slippage within the compressor from the volumetric flow rate when the back pressures are equal to or less than the built-in design output pressure [1]. This slippage can be the result of a seal structural failure. This is why a safety valve is usually stalled on these compressors so that the compressor will automatically shut down (or be slowed down) when the back pressure to the compressor exceeds the design fixed pressure.

The bottom pressure versus volumetric flow rate plot in Figure 5-9 shows the typical situation when the back pressure on the outlet side of the compressor is less than the built-in design output pressure. Under these conditions, the gas exiting the compressor expands in the expansion tank and the initial portion of the pipeline until the pressure is equal to the pipeline back pressure [1].

Rotary compressors can also be designed with two stages. Such compressors are designed with nearly equal compression ratios for each stage (i.e., minimum input power requirements). Thus, because the volumetric flow rate (in actual cfm) is reduced from one stage to the next, the volume displacement of each stage (its geometric size) is progressively smaller.

Rotary compressors are only used as primary compressors in drilling operations.

Sliding Vane Compressors

The typical sliding vane compressor stage is a rotating cylinder located eccentric to the center line of a cylindrical housing (see Figure 5-10)[1, 2]. The vanes are in slots in the rotating cylinder and are allowed to move in and out in these slots to adjust to the changing clearance between the outside surface of the rotating cylinder and the inside bore surface of the housing. The vanes are always in contact with the inside bore due to either pressured gas under the vane (in the slots) or spring forces under the vane. The top of the vanes slides over the inside surface of the bore of the housing as the inside cylinder rotates. Gas is brought into the compression stage through the inlet suction port. The gas is then trapped between the vanes, and as the inside cylinder rotates, the gas is compressed to a smaller volume as the clearance is reduced. When the clearance is the smallest, the gas has rotated to the outlet port. At the outlet port, the compressed gas is discharged to a surge tank or pipeline system connected to the outlet side of

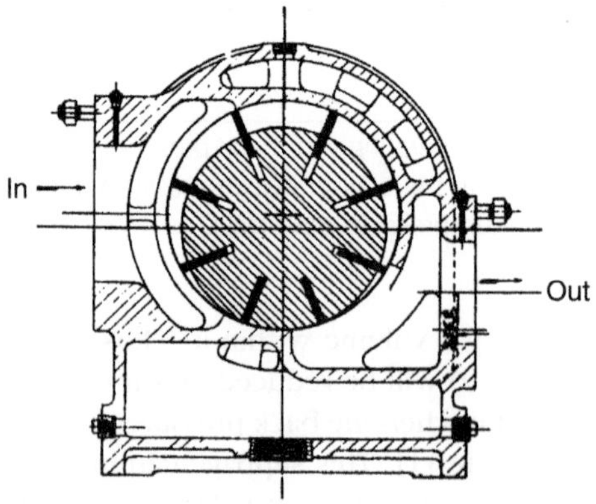

FIGURE 5-10. Sliding vane rotary compressor.

the compressor. As each set of vanes reaches the outlet port, the gas trapped between the vanes is discharged. The clearance between the rotating cylinder and the stationary cylindrical housing is fixed, and thus the pressure ratio of compression for the stage is fixed. The geometry (e.g., cylinder length, diameter, the inside housing diameter, the inlet area, and the outlet area) of each compressor stage determines the stage displacement volume and compression ratio.

The principal seals within the sliding vane compressor are provided by the interface force between the end of the vane and the inside surface of the cylindrical housing. The sliding vanes must be made of a material that will not damage the inside surface of the housing and slide easily on that surface. Therefore, most vane materials are composites such as phenolic resin-impregnated laminated fabrics. Usually vane compressors require oil lubricants to be injected into the gas entering the compression cavity. This lubricant allows smooth action of the sliding vanes against the inside of the housing. There are, however, some sliding vane compressors that may be operated nearly oil free. These utilize bronze or carbon/graphite vanes [7].

The volumetric flow rate for a sliding vane compression stage, q_s, is approximately

$$q_s = 2.0\, al\, (d_2 - mt)\, RPM \tag{5-2}$$

and

$$a = \frac{d_2 - d_1}{2}, \tag{5-3}$$

where q_s is the volumetric flow rate (cfm, m^3/minute), a is the eccentricity (ft, m), l is the length of the cylinder (ft, m), d_1 is the outer diameter of the rotary cylinder (ft, m), d_2 is the inside diameter of the cylindrical housing (ft, m), t is the vane thickness (ft, m), m is the number of vanes, and RPM is the speed of the rotating cylinder (rpm).

Some typical values of a vane compressor stage geometry are $d_1/d_2 = 0.88$, $a = 0.06\, d_2$, and $l/d_2 = 2.00$ to 3.00. Typical vane tip speed usually should not exceed 50 ft/sec (*15 m/sec*).

Helical Lobe (Screw) Compressors

A typical helical lobe (screw) rotary compressor stage is made up of two rotating helical-shaped shafts, or screws. One is a female rotor and the other is a male rotor. These two rotating components turn counter to one another (counterrotating)(see Figure 5-11)[1]. As with all rotary compressors, there are no valves. The gas flows (due to negative pressure conditions at the inlet) into the inlet port and is squeezed between the male and the female portions of the rotating intermeshing screw elements and the housing. The compression ratio of the stage and its volumetric flow rate are determined by the fixed geometry of the two rotating screw elements and the rotation speed. Thus, the rotary screw compressor is a fixed ratio machine.

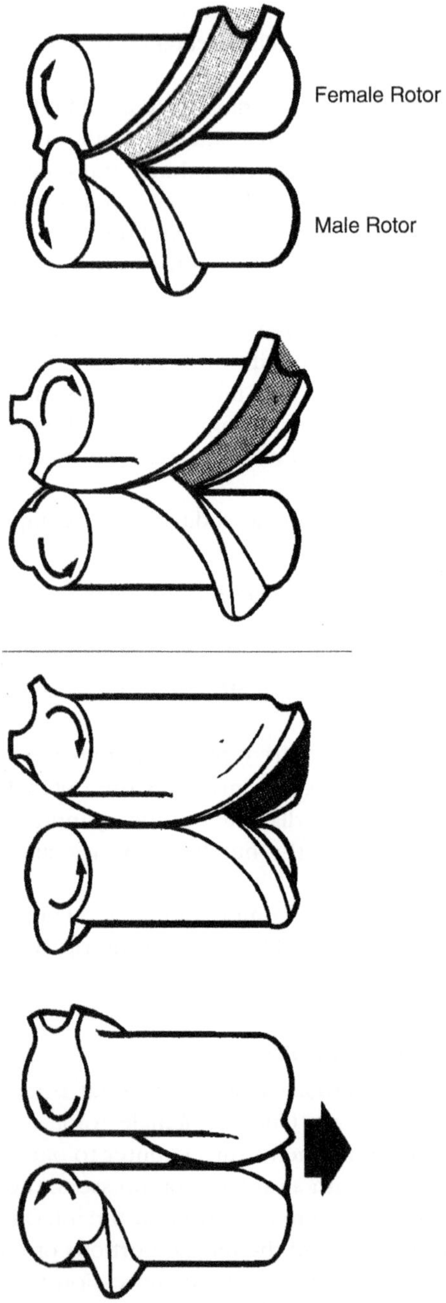

FIGURE 5-11. Helical lobe (screw) rotary compressor [1].

Screw compressors operate at rather high speeds. Thus, they are rather high volumetric flow rate compressors with relatively small exterior dimensions (i.e., small location footprint).

Most helical lobe rotary compressors use lubricating oil within the compression space. This oil is injected into the compression space and is recovered, cooled, and recirculated. The lubricating oil has several functions:

- Seal the internal clearances
- Cool the gas (usually air) during compression
- Lubricate the rotors
- Eliminate the need for timing gears

There are versions of the helical lobe rotary compressor that utilize water injection (rather than oil). The water accomplishes the same purposes as the oil, but the air delivered by these machines can be oil free.

Some helical lobe rotary compressors have been designed to operate with an entirely liquid-free compression space. Because the rotating elements of the compressor need not touch each other or the housing, lubrication can be eliminated. However, such helical lobe rotary compressor designs require timing gears. These machines can deliver totally oil-free, water-free dry gas.

The helical lobe rotary compressor can be staged very much like the sliding vane compressor. However, these compressors are restricted to two-stage systems. In general, this is due to the fact that the helical lobe rotary compressor has some characteristics for the dynamic compressor type. A helical lobe rotary compressor stage can be altered from their original fixed design by adjusting the cross-sectional area of the flow at the exit from that stage. This is essentially converting the kinetic energy of the flow to pressure.

The helical lobe rotary compressor has one serious disadvantage. It is very sensitive to dust particles in the input air. This dust degrades the seals on the helical surfaces. Once the seals are degraded, the stage begins to take on more characteristics of a dynamic compressor. This problem is evidenced by the fact that nearly all drilling operation primary helical lobe compressors will have a measured volumetric flow rate output that is approximately 7 to 8% less than the volumetric flow rate specified by manufacturers (even when accounting for altitude).

Detailed calculations regarding the design of the helical lobe rotary compressor are beyond the scope of this book. Additional details can be found elsewhere [1, 7].

Liquid Piston Compressors

The liquid piston (or liquid ring) rotary compressor utilizes a liquid ring as a piston to perform gas compression within the compression space. The liquid piston compressor stage uses a single rotating element that is located eccentric to the center of the housing (see Figure 5-12)[2]. The rotor has a series of vanes extending radial from it with a slight curvature toward the direction of rotation. A liquid, such as oil, partially fills the compression space between the rotor and the

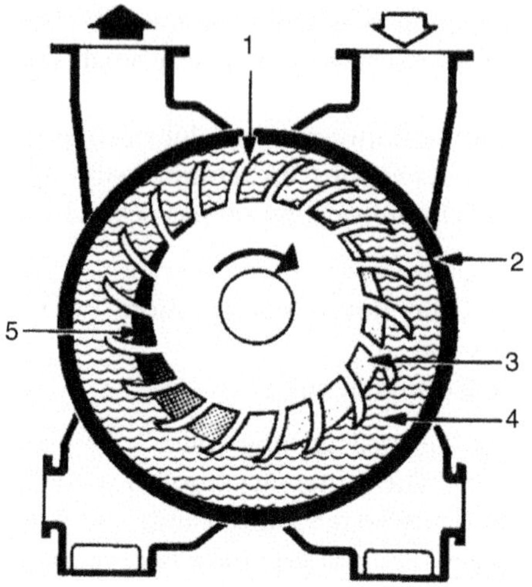

FIGURE 5-12. Liquid piston rotary compressor: (1) impeller, (2) housing, (3) intake port, (4) working liquid, and (5) discharge port [1].

housing walls. As rotation takes place, the liquid forms a ring as centrifugal forces and the vane geometry force the liquid to the outer boundary of the housing. Since the element is located eccentric to the center of the cylindrical housing, the liquid ring (or piston) moves in an oscillatory manner. The compression space in the center of the stage communicates with the gas inlet and outlet parts and allows a gas pocket. The liquid ring alternately uncovers the inlet part and the outlet part. As the system rotates, gas is brought into the pocket, compressed, and released to the outlet port.

The liquid piston compressor has rather low overall efficiency, about 50%. The main advantage to this type of compressor is that it can be used to compress gases with significant liquid content in the stream.

5.4.3 Summary of Positive Displacement Compressors

The main advantages of reciprocating piston compressors are as follow: (1) a dependable near constant volumetric flow rate, (2) a variable pressure capability (up to the maximum pressure capability of the compressor), and (3) can be used as primary and booster compressors. The disadvantages are (1) bulky, (2) high initial capital costs (relative to the rotary compressor of similar capabilities), and (3) relatively high maintenance costs due to a greater number of moving parts (relative to most rotary compressors).

The main advantages of rotary compressors are (1) initial low capital cost (relative to reciprocating compressors), (2) less bulky (relative to the reciprocating compressors of similar capabilities), and (3) general ease of maintenance, as these compressors have few moving parts. The main disadvantages are as follow: (1) cannot adjust to flow line back pressures (fixed compression ratios), (2) need frequent specific maintenance of rotating wear surfaces to prevent slippage, (3) most rotary compressors operate with oil lubrication in the compression chambers, and (4) can only be used as primary compressors [1, 2, 7].

5.5 COMPRESSOR SHAFT POWER REQUIREMENTS

The most important single factor affecting the successful outcome of air and gas drilling operations is the availability of constant, reliable volumetric flow rates of air or gas to the well. This must be the case even when significant (and frequent) changes in back pressure occur during these operations. The only two compressor subclasses that can meet these flexibility requirements are the reciprocating compressor and the rotary compressor. In what follows, the important calculation techniques that allow for the proper evaluation and selection of the appropriate compressors for air and gas drilling operations are reviewed [1, 7, 10]. This section derives the theoretical power required at the compressor shaft to compress the gas in the compressor.

5.5.1 Basic Single-Stage Shaft Power Requirement

Figure 5-13 shows a pressure-volume (P-v) diagram for a simple compression cycle process (where P is pressure and v is specific volume in any set of consistent units). In Figure 5-13, point c represents the final state, or state 2, of the gas leaving the compressor.

The area *odcm* measures the product $P_2 v_2$, which is the flow work (lb-ft/lb or N-m/N) required for delivery of the gas from the compressor. Point b represents

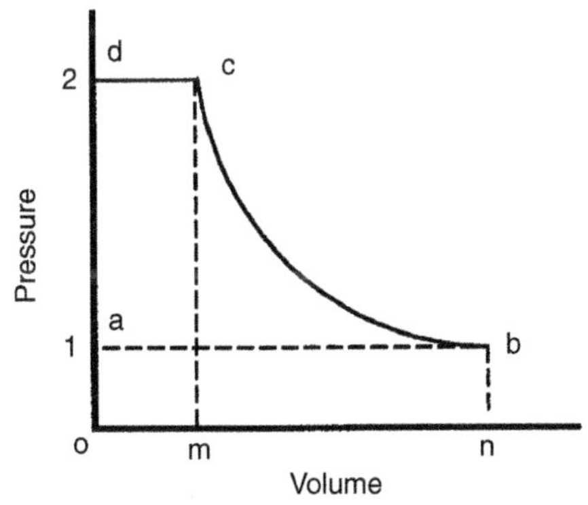

FIGURE 5-13. Basic pressure–volume diagram.

the initial state 1 of the gas, and area *oabn* measures the product $P_1 v_1$, which is the flow-work (ft-lb/lb or N-m/N) supplied in the passage of the fluid to the compressor. The line *bc* represents the state change of the gas during compression. The area *mcbn* measures

$$-\int_1^2 P\,dv \tag{5-4}$$

or

$$+\int_2^1 P\,dv, \tag{5-5}$$

which is the work ideally required for effecting the actual compression within the compressor. Thus, aside from the work required for increasing the kinetic energy, the net area *abcd* measures the net shaft work required for the induction, compression, and delivery of the gas under conditions assumed to be ideal (absence of fluid or mechanical friction, and mechanically reversible).

The total shaft work W_s required for compression can be written as

$$W_s = \frac{V_2^2}{2g} - \frac{V_1^2}{2g} + \left(P_2\,V_2 - \int_1^2 P\,dv - P_1\,v_1\right) \tag{5-6}$$

or

$$W_s = \frac{V_2^2}{2g} - \frac{V_1^2}{2g} + \left(P_2\,v_2 + \int_2^1 P\,dv - P_1\,v_1\right), \tag{5-7}$$

where W_s is the total shaft work (lb-ft/lb, N-m/N), V_1 is the velocity of the gas entering the compressor (ft/sec, m/sec), V_2 is the velocity of the gas exiting the compressor (ft/sec, m/sec), and g is the acceleration of gravity (32.2 ft/sec^2, 9.81 m/sec^2).

Equation (5-6) can be rewritten as

$$W_s = \frac{V_1^2}{2g} - \frac{V_2^2}{2g} + P_1\,v_1 - P_2\,v_2 = -\int_1^2 P\,dv. \tag{5-8}$$

Assuming a polytropic process, where $P_1 v_1^n = P_2 v_2^n = Pv^n =$ constant, and the simplest polytropic process where the exponent term $n \approx k$, where k is the ratio of specific heat for the gas involved in the process (e.g., for air, $k = 1.4$), Equation (5-8) can be rewritten as

$$-\int_1^2 P\,dv = -P_1 v_1^k \int_1^2 \frac{dv}{v^k}. \tag{5-9}$$

The right-hand side of the aforementioned equation can be integrated to yield

$$-\int_1^2 P\,dv = -\frac{P_1\,v_1}{k-1}\left[\left(\frac{v_1}{v_2}\right)^{k-1} - 1\right]. \tag{5-10}$$

For engineering calculations, Equation (5-1) is often simplified to the form

$$\frac{P}{\gamma} = \frac{\mathbf{R_e}\, T}{S_g} \tag{5-11}$$

$$\mathbf{R_e} = \frac{\mathbf{R_u}}{m_w}, \tag{5-12}$$

where $\mathbf{R}_e$ is the engineering gas constant for API standard condition air (53.36 lb-ft/lb-°R, 29.31 N-m/N-K) and S_g is the specific gravity of the particular gas used (for API standard condition air $S_g = 1.0$).

Equation (5-11) can be simplified further by defining **R** (lb-ft/lb-°R, N-m/N-K) as the gas constant for any specified gas (e.g., air, natural gas, nitrogen). Therefore, the gas constant for any gas is approximately

$$\mathbf{R_g} = \frac{\mathbf{R_e}}{S_g}. \tag{5-13}$$

Substituting Equation (5-13) into Equation (5-11) yields

$$\frac{P}{\gamma} = \mathbf{R_g}\, T. \tag{5-14}$$

The specific volume and specific weight of a gas are related by

$$v = \frac{1}{\gamma}. \tag{5-15}$$

Substituting Equation (5-15) into Equation (5-14) yields

$$P\, v = \mathbf{R_g}\, T \tag{5-16}$$

or, specifically,

$$P\, v_1 = \mathbf{R_g}\, T_1. \tag{5-17}$$

Using the definitions for a polytropic process given earlier and Equation (5-17), then Equation (5-10) can be reduced to

$$-\int_1^2 P\, dv = -\frac{\mathbf{R_g}\, T_1}{k-1}\left[\left(\frac{P_2}{P_1}\right)^{\frac{k-1}{k}} - 1\right]. \tag{5-18}$$

Again, using the definitions for a polytropic process and Equation (5-16), then a general relationship between P and T can be obtained. This is

$$\left(\frac{P_2}{P_1}\right)^{\frac{k-1}{k}} = \frac{T_2}{T_1}. \tag{5-19}$$

Substituting Equation (5-19) into Equation (5-18) yields

$$-\int_1^2 P\, dv = \frac{\mathbf{R_g}\, T_1}{k-1}\left[\frac{T_2}{T_1} - 1\right] = \frac{\mathbf{R_g}}{k-1}(T_2 - T_1). \tag{5-20}$$

Substituting Equation (5-20) into Equation (5-8) gives

$$W_s + \frac{V_2^2}{2g} - \frac{V_1^2}{2g} + P_1\, v_1 - P_2\, v_2 = \frac{\mathbf{R_g}}{k-1}(T_2 - T_1). \tag{5-21}$$

Equation (5-17) defined $P\,v$ at state 1. Similarly, the definition of $P\,v$ at state 2 is

$$P_2\, v_2 = \mathbf{R_g}\, T_2. \tag{5-22}$$

Substituting Equations (5-17) and (5-22) into Equation (5-21) and rearranging gives

$$W_s = \frac{V_2^2}{2g} - \frac{V_1^2}{2g} + \mathbf{R_g}\,(T_2 - T_1) + \frac{\mathbf{R_g}}{k-1}(T_2 - T_1). \tag{5-23}$$

The aforementioned can be rearranged to give

$$W_s = \frac{k}{k-1}\,\mathbf{R_g}\, T_1\left(\frac{T_2}{T_1} - 1\right) + \frac{V_2^2 - V_1^2}{2g}. \tag{5-24}$$

It is more useful to have Equation (5-24) written in terms of pressure instead of temperature. This can be accomplished by substituting Equations (5-17) and (5-19) into Equation (5-24). This gives

$$W_s = \frac{k}{k-1}\, P_1\, v_1\left[\left(\frac{P_2}{P_1}\right)^{\frac{k-1}{k}} - 1\right] + \frac{V_2^2 - V_1^2}{2g}. \tag{5-25}$$

The last term (the kinetic energy) in Equation (5-25) can be shown in practical applications to be quite small relative to the first term. Thus, this latter term is usually neglected. Therefore, the shaft work for compression of gas in a compressor is reduced to

$$W_s = \frac{k}{k-1}\, P_1\, v_1\left[\left(\frac{P_2}{P_1}\right)^{\frac{k-1}{k}} - 1\right]. \tag{5-26}$$

Compressors can actually be considered steady-state flow mechanical devices (even intermittent flow machines). If the weight rate of flow through the compressor is $\dot{w}$ (lb/sec, N/sec), then the time rate of shaft work done $\dot{W}_s$ (lb-ft/sec, N-m/sec) to compress gas in a compressor can be obtained by multiplying Equation (5-26) by $\dot{w}$. This gives

$$\dot{W}_s = \frac{k}{k-1}\, P_1\, \dot{w}\, v_1\left[\left(\frac{P_2}{P_1}\right)^{\frac{k-1}{k}} - 1\right]. \tag{5-27}$$

The term $\dot{W}_s$ (lb-ft/sec, N-m/sec) is the theoretical power required by the compressor to compress the gas. Using the general relationship between specific volume and specific weight given in Equation (5-15), the volumetric flow rate at state 1 (entering the compressor) Q_1 is

$$Q_1 = \frac{\dot{w}}{\gamma_1} = \dot{w}\, v_1 \tag{5-28}$$

Therefore, Equation (5-27) can be rewritten as

$$\dot{W}_s = \frac{k}{k-1} P_1\, Q_1 \left[\left(\frac{P_2}{P_1} \right)^{\frac{k-1}{k}} - 1 \right]. \tag{5-29}$$

This expression is valid for a single-stage compressor and for any set of consistent units. With Equation (5-29), the shaft power to compress a continuous flow rate of gas can be determined knowing properties of the gas (specifically k), the initial pressure and volumetric flow rate of the gas entering the compressor (state 1), and the exit pressure of gas exiting the compressor (state 2).

5.5.2 Multistage Shaft Power Requirements

Using Equation (5-24), the minimum shaft power required for a multistage compressor can be derived [10]. Equation (5-24) can be used for each stage of a multistage compressor and added together for the total shaft work required by the compressor. Such an expression can be minimized to obtain the conditions for minimum shaft work for a multistage compressor. Minimum shaft work is attained when a multistage compressor is designed with equal compression ratios for each stage and with intercoolers that cool the gas entering each stage to a temperature that is nearly the same as the temperature entering the first stage of the compressor [10]. Once these conditions are obtained for the multistage compressor, the expression for the shaft power can be obtained in the same manner as given in Equations (5-25) to (5-29).

Figure 5-14 shows an example schematic of a two-stage compressor with an intercooler between the compression stages. The temperature exiting stage 1 (at position 2) will be governed by Equation (5-19). The intercooler cools the gas between positions 2 and 3 under constant volume conditions. Thus, the temperature of the gas entering stage 2 (at position 3) is the same as the temperature entering stage 1 (i.e., position 1). Using these conditions for the minimum shaft power, an expression for the shaft power of a multistage compressor can be derived. This expression is

$$\dot{W}_s = \frac{n_s\, k}{k-1} P_1\, Q_1 \left[\left(\frac{P_2}{P_1} \right)^{\frac{k-1}{n_s k}} - 1 \right], \tag{5-30}$$

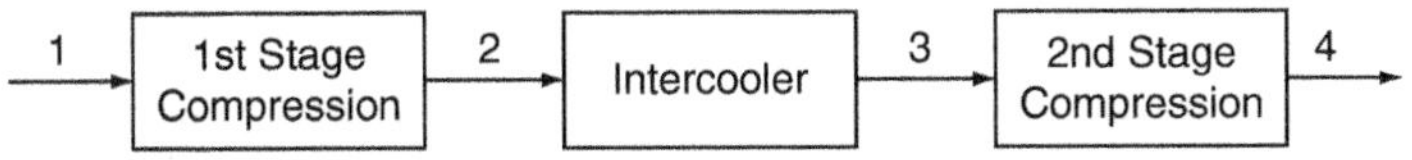

FIGURE 5-14. Schematic of a two-stage compressor [10].

where n_s is the number of compression stages in the multistage compressor, P_i is the pressure entering the first stage of the compressor (lb/ft^2 abs, N/m^2 abs), P_o is the pressure exiting the last stage of the compressor (lb/ft^2 abs, N/m^2 abs), and Q_i is the volumetric flow rate entering the first stage of the compressor (ft^3/sec, m^3/sec).

The compression ratio for each stage r_s is given by

$$r_s = \left(\frac{P_o}{P_i}\right)^{\frac{1}{n_s}}. \tag{5-31}$$

Again, Equations (5-30) and (5-31) are valid for any set of consistent units.

In what follows, two sets for field equations are derived for determining the theoretical power to compress the gas passing through the compressor. One set is derived for use with the USCS unit and the other set for use with the SI unit.

Equation (5-30) is obtained in USCS units by substituting the following for P_i, P_o, and Q_i (and dividing by 550 ft-lb/hp):

$$P_i = p_i\, 144 \tag{5-32}$$

$$P_o = p_o\, 144 \tag{5-33}$$

$$Q_i = \frac{q_i}{60}, \tag{5-34}$$

where p_i is input pressure (psia), p_o is output pressure (psia), and q_i is the input volumetric flow rate (ft^3/min).

The expression for theoretical shaft power $\dot{W}_s$ in USCS units (horsepower) is

$$\dot{W}_s = \frac{n_s\, k}{k-1}\,\frac{p_i\, q_i}{229.17}\left[\left(\frac{p_o}{p_i}\right)^{\frac{k-1}{n_s\, k}} - 1\right]. \tag{5-35a}$$

Using Equation (5-13), an alternate expression of Equation (5-35a) can be obtained. Equation (5-35b) is convenient for booster compressor calculations

$$\dot{W}_s = \frac{n_s\, k}{k-1}\,\frac{\dot{w}_g}{550}\left(\frac{\mathrm{R_c}\, T_i}{S_g}\right)\left[\left(\frac{p_o}{p_i}\right)^{\frac{k-1}{n_s k}} - 1\right], \tag{5-35b}$$

where T_i is the input gas temperature (°R) and $\dot{w}_g$ is the weight of flow of gas through the compressor (lb/sec).

The expression for theoretical shaft power $\dot{W}_s$ in SI units (watts) is

$$\dot{W}_s = \frac{n_s\, k}{k-1} P_i\, Q_i\left[\left(\frac{P_o}{P_i}\right)^{\frac{k-1}{n_s k}} - 1\right], \tag{5-36a}$$

where Q_i is the input volumetric flow rate (m^3/sec), P_i is the input pressure (N/m^2 abs), and P_o is the output pressure (N/m^2 abs).

Using Equation (5-13), an alternate expression of Equation (5-36a) can be obtained. Equation (5-36b) is convenient for booster compressor calculations

$$\dot{W}_s = \frac{n_s\, k}{k-1}\dot{w}_g\left(\frac{\mathbf{R}_e\, T_i}{S_g}\right)\left[\left(\frac{P_o}{P_i}\right)^{\frac{k-1}{n_s\, k}} - 1\right], \tag{5-36b}$$

where T_i is the input gas temperature (K) and $\dot{w}_g$ is the gas weight rate of flow (N/sec).

The compression ratio for each stage r_s is given by

$$r_s = \left(\frac{P_o}{P_i}\right)^{\frac{1}{n_s}} = \left(\frac{p_o}{p_i}\right)^{\frac{1}{n_s}}. \tag{5-37}$$

Equations (5-35) through (5-37) are valid for both reciprocating and rotary compressors (together with the other pertinent equations in this chapter).

5.6 PRIME MOVER INPUT POWER REQUIREMENTS

The previous section discussed the theoretical shaft power required to compress the gas in a compressor. In order to obtain the complete picture of compressors, it is necessary to ascertain the prime mover input power requirement to operate the compressor shaft and the actual power available from the prime mover. The application of the equations described earlier will be slightly different depending on whether reciprocation piston compressors or rotary compressors are being analyzed.

The actual power available from a prime mover is a function of elevation above sea level and whether the prime mover is naturally aspirated or turbocharged. Figure 5-15 gives the power percentage reduction in power necessary for naturally aspirated and turbocharged prime movers as a function of elevation above sea level [1].

5.6.1 Compressor System Units

Compressors used for air and gas drilling operations are driven by stationary prime movers. For small drilling rigs, the compressor is often integrated into the rig design and the compressor is driven by a shared stationary prime mover that also drives the rig draw works and rotating table (or hydraulic pump and motor rotating system). Larger drilling operations have separate compressor system units that are fabricated with their own dedicated stationary prime movers. These separate compressor system units are usually provided to the operator by a contractor other than the drilling rig contractor.

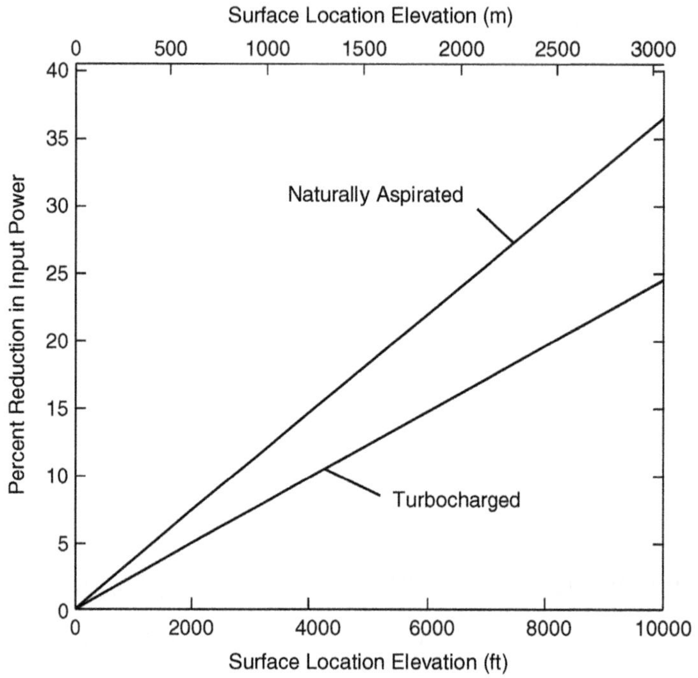

FIGURE 5-15. Prime mover percentage reduction in power as a function of elevation above sea level [1].

Primary Compressor System Unit

Primary compressor system units take air directly from the atmosphere. These compressor systems can be rotary or reciprocating compressors. The compressors in these units can be single stage or multistage. The prime movers for these units can be fueled by gasoline, propane/butane, diesel, or natural gas. These units can be fabricated as skid mounted, as wheeled trailers, or semitrailer mounted.

Booster Compressor System Unit

Booster compressor system units are operated downstream from a primary compressor system (sometimes these units are denoted as secondary units). They accept compressed air from the primary compressor system and compress the air to a higher pressure before sending the air to the drilling rig. Also, a booster compressor system can accept compressed natural gas (or other gases) from a pipeline and compress that gas to a higher pressure before sending the gas to the drilling rig. The compressors in these units are all reciprocating compressors. The compressors in these units can be single stage or multistage. The prime movers for these units can be fueled by gasoline, propane/butane, diesel, or natural gas. These units can be fabricated as skid mounted, semitrailer mounted, or as wheeled trailers.

5.6.2 Reciprocating Compressor Unit

A reciprocating compressor can adjust its output pressure to match the back pressure on the machine. Thus, the reciprocating compressor is somewhat more flexible than the rotary compressor and will tend to use less fuel for a given application than a similarly configured rotary compressor [1, 11].

The intake volumetric flow rate of a real reciprocating compressor is slightly smaller than the theoretical sweep volume (i.e., the calculated intake volumetric flow rate). This is due to the fact that the piston compressor cannot be fabricated without a clearance volume. This clearance volume at the top of the piston cylinder is necessary in order to have space for the valves and to keep the piston from striking the top of the cylinder. This clearance volume results in a volumetric efficiency term ε_v that is unique to the reciprocating compressor. This volumetric efficiency is only applicable to the first stage of the reciprocating compressor. The expression for the volumetric efficiency ε_v of a reciprocating compressor can be approximated as

$$\varepsilon_v = 0.96\left\{1 - c\left[r_s^{\frac{1}{k}} - 1\right]\right\}, \tag{5-38}$$

where c is the clearance volume ratio for the compressor model. The clearance volume ratio is the clearance volume divided by the sweep volume of the first-stage piston. The range of values for the clearance volume term c is from 0.02 to 0.08 [1].

The mechanical efficiency term ε_m is used with the volumetric efficiency term to determine the input power requirement. This mechanical efficiency term is a measure of the friction losses in the mechanical application of the prime mover power to the compressor. These losses are due to the friction in the bearings and linkages in the compressor system. The values of mechanical efficiency for typical reciprocating compressor systems can vary from about 0.84 to 0.99.

Illustrative Example 5.1 A two-stage reciprocating primary air compressor system unit is rated to have a volumetric flow rate of 950 scfm (*448.3 standard liters/sec*) and a maximum pressure capability of 300 psig (*206.9 N/cm² gauge*) at API Mechanical Equipment Standards atmospheric conditions. The compressor has a diesel prime mover that is rated to have a maximum power of 350 hp (*261 kW*) at a prime mover output shaft speed of 1800 rpm (at API Standard conditions). The prime mover is turbocharged. This reciprocating compressor system has a clearance ratio of 0.02 and a mechanical efficiency of 0.95.

Determine the horsepower required by the prime mover to operate the compressor against a flow-line back pressure of 150 psig (*103.5 N/cm² gauge*) for (a) a surface location at sea level (use API Mechanical Equipment Standards atmospheric conditions for mean sea level) and (b) a surface location elevation of 6000 ft (*1829 m*) (use average midlatitudes data in Table 5-1).

(a) Surface location at sea level (USCS units)

This is a reciprocating piston compressor and, thus, has a volumetric efficiency term. The volumetric efficiency ε_v can be determined using Equation (5-38).

$$n_s = 2$$

$$p_i = 14.696 \text{ psia}$$

$$p_o = 150 + 14.696 = 164.696 \text{ psia}.$$

Equation (5-31) becomes

$$r_s = \left(\frac{164.696}{14.696}\right)^{\frac{1}{2}} = 3.348.$$

The specific heat ratio for air is

$$k = 1.4.$$

The clearance volume ratio for this compressor is

$$c = 0.02\cdot$$

With these values, Equation (5-38) becomes

$$\varepsilon_v = 0.96\left\{1 - 0.02\left[(3.348)^{\frac{1}{1.4}} - 1\right]\right\}$$

$$\varepsilon_v = 0.934.$$

The rated volumetric flow rate into the compressor is 950 ft^3/min. For this example, the compressor is located at mean sea level (API Standard conditions), thus

$$q_i = 950 \text{ scfm}.$$

With these terms, the theoretical shaft horsepower required to compress the air moving through the machine is given by Equation (5-35a). Thus, the theoretical shaft horsepower is

$$\dot{W}_s = \frac{(2)\,(1.4)}{(0.4)}\frac{(14.696)\,(950)}{229.17}\left[\left(\frac{164.696}{14.696}\right)^{\frac{(0.4)}{(2)\,(1.4)}} - 1\right]$$

$$\dot{W}_s = 175.9\ hp.$$

The actual shaft power $\dot{W}_{as}$ is actual power needed to compress the air to a pressure of 150 psig. Actual shaft power is given by

$$\dot{W}_{as} = \frac{\dot{W}_s}{\varepsilon_m\,\varepsilon_v}. \qquad (5\text{-}39)$$

For this example,

$$\varepsilon_m = 0.95$$

and Equation (5-39) becomes

$$\dot{W}_{as} = \frac{175.9}{(0.95)(0.934)} = 198.3\ hp.$$

The above determined 198.3 hp is the actual shaft power needed by the compressor to match the back pressure of 150 psig. Because this power level is less than the prime mover's rated input power of 350 hp, this compressor system is capable of operating at a sea level surface location.

(a) Surface location at sea level (SI units)

This is a reciprocating piston compressor and, thus, has a volumetric efficiency term. The volumetric efficiency ε_v *can be determined using Equation (5-38).*

$$n_s = 2$$

$$P_i = 101360\ N/m^2\ abs$$

$$P_o = 1034550.0 + 101360.0 = 1135910.0\ N/\mathrm{m}^2\ abs.$$

Equation (5-31) becomes

$$r_s = \left(\frac{1135910.0}{101360.0}\right)^{\frac{1}{2}} = 3.348.$$

The specific heat ratio for air is

$$k = 1.4.$$

The clearance volume ratio for this compressor is

$$c = 0.02.$$

With these values, Equation (5-38) becomes

$$\varepsilon_v = 0.96\left\{1 - 0.02\left[(3.348)^{\frac{1}{1.4}} - 1\right]\right\}$$

$$\varepsilon_v = 0.934.$$

The rated volumetric flow rate into the compressor is 448.3 liters/sec. For this example, the compressor is located at mean sea level (API Standard conditions), thus

$$Q_i = 0.4483\ \text{standard}\ m^3/\text{sec}.$$

With the aforementioned terms, the theoretical shaft horsepower required to compress the air moving through the machine is given by Equation (5-36a). Thus, the theoretical shaft horsepower is

$$\dot{W}_s = \frac{(2)\,(1.4)}{(0.4)}(101360.0)\,(0.4483)\left[\left(\frac{1135910.0}{101360.0}\right)^{\frac{(0.4)}{(2)\,(1.4)}} - 1\right]$$

$$\dot{W}_s = 131200\ watts.$$

The actual shaft power $\dot{W}_{as}$ is actual power needed to compress the air to a pressure of 150 psig. Equation (5-39) becomes

$$\dot{W}_{as} = \frac{131200}{(0.95)\,(0.934)} = 147864\ watts.$$

The above determined 147.9 kW is the actual shaft power needed by the compressor to match the back pressure of 103.5 N/cm^2 abs. Because this power level is less than the prime mover's rated input power of 261 kW, this compressor system is capable of operating at a sea level surface location.

(b) Surface location at 6000 ft above sea level (USCS units)

The volumetric efficiency ε_v *can be determined using Equation (5-39).*

$$n_s = 2$$

$$p_i = 11.769 \text{ psia}$$

$$p_o = 150 + 11.769 = 161.769 \text{ psia}.$$

Equation (5-37) becomes

$$r_s = \left(\frac{161.769}{11.769}\right)^{\frac{1}{2}} = 3.708.$$

The specific heat ratio for air is

$$k = 1.4.$$

The clearance volume ratio for this compressor is

$$c = 0.02.$$

With these values, Equation (5-38) becomes

$$\varepsilon_v = 0.96\left\{1 - 0.02\left[(3.708)^{\frac{1}{1.4}} - 1\right]\right\}$$

$$\varepsilon_v = 0.930.$$

The rated volumetric flow rate into the compressor is 950 ft^3/min. For this example, the compressor is located at 6000 ft above sea level, thus

$$q_i = 950 \text{ acfm}.$$

With these terms the theoretical shaft horsepower required to compress the air moving through the machine is given by Equation (5-35a). Thus, the theoretical shaft horsepower is

$$\dot{W}_s = \frac{(2)\,(1.4)\,(11.769)\,(950)}{(0.4) \qquad 229.17}\left[\left(\frac{161.769}{11.769}\right)^{\frac{(0.4)}{(2)\,(1.4)}} - 1\right]$$

$$\dot{W}_s = 155.1\ hp.$$

For this example,

$$\varepsilon_m = 0.95$$

and from Equation (5-39), the actual shaft power is

$$\dot{W}_{as} = \frac{155.1}{(0.95)\,(0.930)} = 175.6\ hp.$$

The above determined 175.6 hp is the actual shaft power needed by the compressor to match the back pressure of 150 psig (at the surface location elevation of 6000 ft above sea level). At this surface location, the input power available from the prime mover is a derated value (derated from the rated 350 hp available at 1800 rpm). In order for the compressor system to operate at this 6000-ft surface location elevation, the derated input power available must be greater than the actual shaft power needed. Figure 5-15 shows that for a 6000-ft elevation the power of a turbocharged prime mover must be derated by 15%. The derated input horsepower available from the prime mover $\dot{W}_i$ is

$$\dot{W}_i = 350\,(1 - 0.15) = 297.5\ hp.$$

For this example, the prime mover derated input power is greater than the actual shaft power needed, thus the compressor system can be operated at this 6000-ft surface location elevation.

(b) Surface location at 1829 m above sea level (SI units)

The volumetric efficiency ε_v *can be determined using Equation (5-39).*

$$n_s = 2$$

$$P_i = 81170\ \mathrm{N/m^2}\ abs$$

$$P_o = 1034550.0 + 81170.0 = 1115720.0\ N/m^2\ abs.$$

Equation (5-37) becomes

$$r_s = \left(\frac{1115720.0}{81170.0}\right)^{\frac{1}{2}} = 3.708.$$

The specific heat ratio for air is

$$k = 1.4.$$

The clearance volume ratio for this compressor is

$$c = 0.02.$$

With the values just given, Equation (5-38) becomes

$$\varepsilon_v = 0.96\left\{1 - 0.02\left[(3.708)^{\frac{1}{1.4}} - 1\right]\right\}$$

$$\varepsilon_v = 0.930.$$

The rated volumetric flow rate into the compressor is 448.3 liters/sec. For this example, the compressor is located at 1829 m above sea level, thus

$$Q_i = 0.448\ m^3/\text{sec}.$$

With the aforementioned terms, the theoretical shaft horsepower required to compress the air moving through the machine is given by Equation (5-36a). Thus, the theoretical shaft horsepower is

$$\dot{W}_s = \frac{(2)(1.4)}{(0.4)}(81170.0)(0.4483)\left[\left(\frac{1115720.0}{81170.0}\right)^{\frac{(0.4)}{(2)(1.4)}} - 1\right]$$

$$\dot{W}_s = 115660\ watts.$$

For this example,

$$\varepsilon_m = 0.95$$

and from Equation (5-39), the actual shaft power is

$$\dot{W}_{as} = \frac{115660}{(0.95)\ (0.930)} = 130947\ watts.$$

The above determined 131.0 kW is the actual shaft power needed by the compressor to match the back pressure of 103.5 N/cm^2 gauge (at the surface location elevation of 1829 m above sea level). At this surface location, the input power available from the prime mover is a derated value (derated from the rated 261 kW available at 1800 rpm). In order for the compressor system to operate at this 1829-m surface location elevation, the derated input power available must be greater than the actual shaft power needed. Figure 5-15 shows that for a 1829-m elevation the power of a turbocharged prime mover must be derated by 15%. The derated input horsepower available from the prime mover $\dot{W}_i$ is

$$\dot{W}_i = 261(1 - 0.15) = 221.9\ kW.$$

For this example, the prime mover derated input power is greater than the actual shaft power needed, thus the compressor system can be operated at this 1829-m surface location elevation.

5.6.3 Rotary Compressor System Unit

Rotary compressors have fixed compressor ratios for each stage. These machines have a constant pressure output. These compressors cannot adjust output pressures to match the back pressure in the exit flow line. The rotary compressors will inject compressed gas into the flow line at the compressor's rated pressure output. If the back pressure in the exit flow line is less than the rated pressure output, the gas will expand in the flow line (or surge tank upstream of the flow line) to match the back pressure [1, 7].

There is no volumetric efficiency term for the rotary compressor. There is a mechanical efficiency term for these compressors (denoted by the term ε_m).

Values of the mechanical efficiency for typical rotary compressor systems can vary from about 0.84 to 0.99.

Illustrative Example 5.2 A two-stage helical lobe (screw) primary air compressor system unit is rated to have a volumetric flow rate of 950 scfm *(448.3 standard liters/sec)* and a fixed design pressure output of 300 psig *(206.9 N/cm² gauge)* at API Mechanical Equipment Standards atmospheric conditions. The compressor has a diesel prime mover that is rated to have a maximum power of 350 hp *(261 kW)* at a compressor shaft speed of 1800 rpm (at API Standard conditions). The prime mover is turbocharged. This rotary compressor system has a mechanical efficiency of 0.95.

Determine the horsepower required by the prime mover to operate the compressor against a flow line back pressure of 150 psig *(103.5 N/cm² gauge)* for (a) a surface location at sea level (use API Mechanical Equipment Standards atmospheric conditions for mean sea level) and (b) a surface location elevation of 6000 ft *(1829 m)*(use average midlatitudes data in Table 5-1).

(a) Surface location at sea level (USCS units)

The rotary compressor is a fixed ratio machine. Therefore, the output of the compressor will be 300 psig.

$$n_s = 2$$

$$p_i = 14.696 \text{ psia}$$

$$p_o = 300 + 14.696 = 314.696 \text{ psia.}$$

The total fixed compression ratio across the two stages of the compressor r_{tf} is

$$r_{tf} = \frac{314.696}{14.696}$$

$$r_{tf} = 21.414.$$

The rotary compressor volumetric efficiency ε_v *is*

$$\varepsilon_v = 1.0.$$

The rated volumetric flow rate into the compressor is 950 ft³/min. For this example, the compressor is located at mean sea level (API Standard conditions), thus,

$$q_i = 950 \text{ scfm.}$$

With these terms the theoretical shaft horsepower required to compress the air moving through the machine is given by Equation (5-35a). Thus, the theoretical shaft horsepower is

$$\dot{W}_s = \frac{(2)(1.4)}{(0.4)} \frac{(14.696)(950)}{229.17} \left[\left(\frac{314.696}{14.696} \right)^{\frac{(0.4)}{(2)(1.4)}} - 1 \right]$$

$$\dot{W}_s = 234.3 \text{ } hp.$$

The actual shaft power $\dot{W}_{as}$ is actual power needed to compress the air to a pressure of 300 psig. Actual shaft power from Equation (5-39) is

$$\dot{W}_{as} = \frac{\dot{W}_s}{\varepsilon_m\ \varepsilon_v}.$$

For this example,

$$\varepsilon_m = 0.95$$

and Equation (5-39) becomes

$$\dot{W}_{as} = \frac{234.3}{(0.95)(1.0)} = 246.6\ hp.$$

The above determined 246.6 horsepower is the actual shaft power needed by the compressor to give a fixed design pressure output of 300 psig. This power level is less than the prime mover's rated input power of 350 hp, thus this compressor system is capable of operating at a sea level surface location.

(a) Surface location at sea level (SI units)

The rotary compressor is a fixed ratio machine. Therefore, the output of the compressor will be 300 psig.

$$n_s = 2$$

$$p_i = 101360\ N/m^2\ abs$$

$$p_o = 2069000.0 + 101360.0 = 2170360.0\ \mathrm{N/m^2}\ abs.$$

The total fixed compression ratio across the two stages of the compressor r_{tf} is

$$r_{tf} = \frac{2170360.0}{101360.0}$$

$$r_{tf} = 21.414.$$

The rotary compressor volumetric efficiency ε_v *is*

$$\varepsilon_v = 1.0.$$

The rated volumetric flow rate into the compressor is 448.3 liters/sec. For this example, the compressor is located at mean sea level (API Standard conditions), thus

$$Q_i = 0.4483\ \text{standard}\ m^3/\text{sec}.$$

With the aforementioned terms, the theoretical shaft horsepower required to compress the air moving through the machine is given by Equation (5-36a). Thus, the theoretical shaft horsepower is

$$\dot{W}_s = \frac{(2)(1.4)}{(0.4)}(101360.0)(0.4483)\left[\left(\frac{2170360.0}{101360.0}\right)^{\frac{(0.4)}{(2)(1.4)}} - 1\right]$$

$$\dot{W}_s = 174721 \; watts.$$

The actual shaft power $\dot{W}_{as}$ is actual power needed to compress the air to a pressure of 206.9 N/cm^2 gauge. Equation (5-39) becomes

$$\dot{W}_{as} = \frac{174721}{(0.95)(1.0)} = 183917 \; watts.$$

The above determined 183.9 kW is the actual shaft power needed by the compressor to develop the fixed design pressure output of 206.9 N/cm^2 gauge. This power level is less than the prime mover's rated input power of 261 kW, thus this compressor system is capable of operating at a sea level surface location.

(b) Surface location at 6000 ft above sea level (USCS units)

From (a) given earlier, the total fixed r_{tf} pressure ratio of the rotary compressor when operating at the assumed design API Standard conditions was found to be

$$r_{tf} = 21.414.$$

This fixed ratio cannot be changed. From Table 5-1, the approximate atmospheric pressure at 6000 ft above sea level can be assumed to be 11.769 psia. Therefore, the new fixed output pressure of this compressor at 6000 ft above sea level will be

$$p_o = p_i \, r_{tf}$$

$$p_o = (11.769)(21.414)$$

$$p_o = 252.0 \; psia$$

or

$$p_o = 240.2 \; psig.$$

The rotary compressor volumetric efficiency ε_v is

$$\varepsilon_v = 1.0.$$

The rated volumetric flow rate into the compressor is 950 ft^3/min. For this example, the compressor is located at mean sea level (API Standard conditions), thus

$$q_i = 950 \text{ acfm}.$$

With these terms the theoretical shaft horsepower required to compress the air moving through the machine is given by Equation (5-35a). Thus, the theoretical shaft horsepower is

$$\dot{W}_s = \frac{(2)(1.4)}{(0.4)} \frac{(11.769)(950)}{229.17} \left[\left(\frac{252.021}{11.769} \right)^{\frac{(0.4)}{(2)(1.4)}} - 1 \right]$$

$$\dot{W}_s = 187.6 \ hp.$$

The actual shaft power $\dot{W}_{as}$ is actual power needed to compress the air to the fixed design pressure of 240.2 psig. Actual shaft power from Equation (5-39) is

$$\dot{W}_{as} = \frac{\dot{W}_s}{\varepsilon_m \ \varepsilon_v}.$$

For this example,

$$\varepsilon_m = 0.95$$

and Equation (5-39) becomes

$$\dot{W}_{as} = \frac{187.6}{(0.95)(1.0)} = 197.5 \ hp.$$

The above determined 197.5 hp is the actual shaft power needed by the compressor to produce the 240.2-psig fixed design pressure output (at the surface location elevation of 6000 ft above sea level). At this surface location, the input power available from the prime mover is a derated value (derated from the rated 350 hp available at 1800 rpm). In order for the compressor system to operate at this 6000-ft surface location elevation, the derated input power available must be greater than the actual shaft power needed. Figure 5-15 shows that for a 6000-ft elevation the input power of a turbocharged prime mover must be derated by 15%. The derated input horsepower available from the prime mover $\dot{W}_i$ is

$$\dot{W}_i = 350(1 - 0.15) = 297.5 \ hp.$$

For this example, the prime mover's derated input power is greater than the actual shaft horsepower needed, thus the compressor system can be operated at this 6000-ft surface location elevation.

(b) Surface location at 1829 m above sea level (SI units)

From (a) given earlier, the total fixed r_{tf} pressure ratio of the rotary compressor when operating at the assumed design API Standard conditions was found to be

$$r_{tf} = 21.414.$$

This fixed ratio cannot be changed. From Table 5-1, the approximate atmospheric pressure at 1829 m above sea level can be assumed to be 81,170 N/m^2 abs. Therefore, the new fixed output pressure of this compressor at 1829 m above sea level will be

$$P_o = P_i\, r_{tf}$$

$$P_o = (81170.0)(21.414)$$

$$P_o = 1738174.4\ N/m^2\ abs$$

or

$$P_o = 1657004.4\ N/m^2\ gauge.$$

The rotary compressor volumetric efficiency ε_v *is*

$$\varepsilon_v = 1.0.$$

The rated volumetric flow rate into the compressor is 448.3 liters/sec. For this example, the compressor is located at mean sea level (API Standard conditions), thus

$$Q_i = 0.4483 m^3/\text{sec}.$$

With the terms just described, the theoretical shaft horsepower required to compress the air moving through the machine is given by Equation (5-36a). Thus, the theoretical shaft horsepower is

$$\dot{W}_s = \frac{(2)(1.4)}{(0.4)}(81170.0)(0.4483)\left[\left(\frac{1738174.4}{81170.0}\right)^{\frac{(0.4)}{(2)(1.4)}} - 1\right]$$

$$\dot{W}_s = 139937\ watts.$$

The actual shaft power $\dot{W}_{as}$ is actual power needed to compress the air to the fixed design pressure of 165.7 N/cm^2 gauge. Equation (5-39) becomes

$$\dot{W}_{as} = \frac{139937}{(0.95)(1.0)} = 147302\ watts.$$

The above determined 147.3 kW is the actual shaft power needed by the compressor to produce the 165.7 N/cm^2 gauge fixed design pressure output (at the surface location elevation of 1829 m above sea level). At this surface location, the input power available from the prime mover is a derated value (derated from the rated 261 kW available at 1800 rpm). In order for the compressor system to operate at this 1829-m surface location elevation, the derated input power available must be greater than the actual shaft power needed. Figure 5-15 shows that for an 1829-m elevation the input power of a turbocharged prime mover must be derated by 15%. The derated input horsepower available from the prime mover $\dot{W}_i$ is

$$\dot{W}_i = 261(1 - 0.15) = 221.9\ kW.$$

For this example, the prime mover's derated input power is greater than the actual shaft horsepower needed, thus the compressor system can be operated at this 6000-ft surface location elevation.

5.6.4 Fuel Consumption

In order to plan air or gas drilling operations effectively, it is important to be able to predict the fuel consumption by the prime movers of compressor system units. The consumption of either liquid or gaseous fuels is initially calculated in pounds of fuel used per horsepower-hour (lb/hp-hr, N/kW-hr). Fuel consumption is very dependent upon the prime mover's power ratio, or percent power utilization of total power available.

Figure 5-16 gives the approximate fuel consumption for prime movers fueled with gasoline, propane/butane, or diesel [3, 12]. Most stationary prime movers can be field converted to operate on propane/butane or natural gas. Figure 5-17 gives the approximate fuel consumption for prime movers fueled by natural gas [3, 12].

Figures 5-16 and 5-17 are approximate values covering most prime mover designs. However, more accurate estimates for prime mover fuel consumptions can be obtained from the manufacturer specifications.

Illustrative Example 5.3 Determine the diesel fuel consumption rate (in gallons per hour) for the primary reciprocating piston compressor system and conditions given in Illustrative Example 5.1 for (a) a surface location at sea level and (b) a surface location elevation of 6000 ft (*1829 m*).

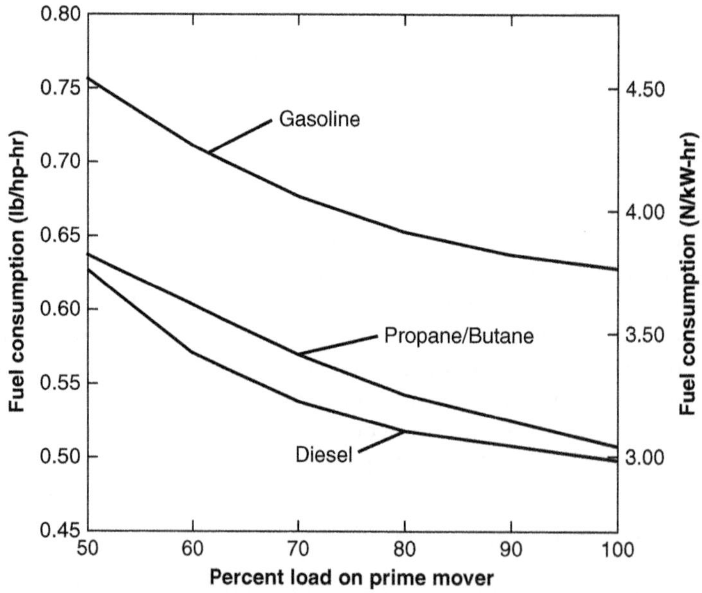

FIGURE 5-16. Fuel consumption for gasoline, propane/butane, and diesel [3, 12].

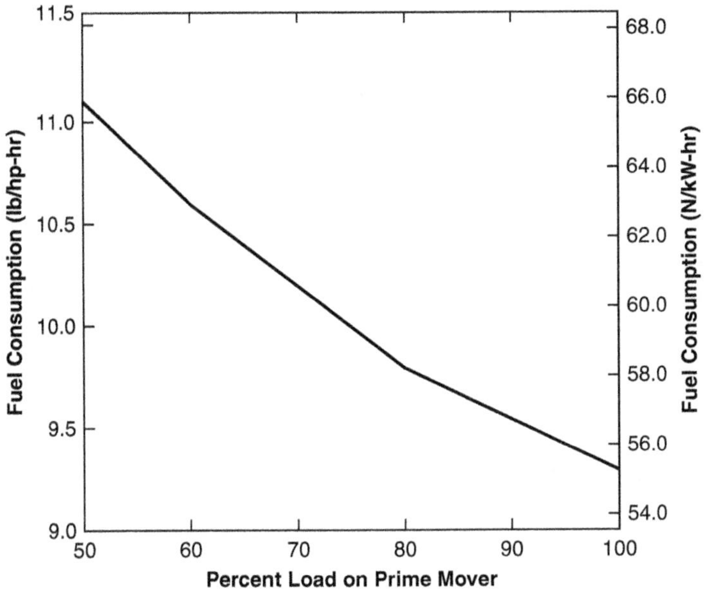

FIGURE 5-17. Fuel consumption for natural gas [3, 12].

(a) Surface location at sea level (USCS units)

The prime mover has a maximum 350 hp available at 1800 rpm at sea level conditions. The actual shaft power needed by the two-stage reciprocating compressor to match the back pressure of 150 psig is 198.3 hp. Thus, the percent load on the prime mover, or power ratio *PR* is

$$PR = \frac{198.3}{350}(100) = 56.7.$$

Entering the abscissa of Figure 5-16 with the power ratio percent, the approximate fuel consumption rate can be read on the ordinate using the diesel fuel curve. The approximate fuel consumption rate at this power level is 0.585 lb/hp-hr. The total weight of diesel fuel consumption per hour $\dot{w}_f$ is

$$\dot{w}_f = 0.585(198.3) = 116.0 \text{ lb/hr}.$$

The specific gravity of diesel fuel is 0.8156 at 60°F [5]. The total volume (in U.S. gallons) of diesel fuel consumption per hour q_f is

$$q_f = \frac{116.0}{(0.8156)(8.33)} = 17.1 \text{ gal/hr},$$

where the specific weight of fresh water is 8.33 lb/gal.

(a) Surface location at sea level (SI units)

The prime mover has a maximum 261 kW available at 1800 rpm at sea level conditions. The actual shaft power needed by the two-stage reciprocating compressor to match the back pressure of 103.5 N/cm^2 gauge is 147.9 kW. Thus, the percent load on the prime mover, or power ratio *PR* is

$$PR = \frac{147.9}{261}(100) = 56.7.$$

Entering the abscissa of Figure 5-16 with the power ratio percent, the approximate fuel consumption rate can be read on the ordinate using the diesel fuel curve. The approximate fuel consumption rate at this power level is 3.490 N/kW-hr. The total weight of diesel fuel consumption per hour $\dot{w}_f$ is

$$\dot{w}_f = 3.490(147.9) = 516.2\ \text{N/hr}.$$

The specific gravity of diesel fuel is 0.8156 at 15.6°C [5]. The total volume (in U.S. gallons) of diesel fuel consumption per hour q_f is

$$q_f = \frac{516.2}{(0.8156)(9.81)} = 64.5\ \text{liters/hr},$$

where the specific weight of freshwater is 9.81 N/liter.

(b) Surface location at 6000 ft above sea level (USCS units)

The prime mover has a derated maximum 297.5 hp available at 1800 rpm at 6000 ft elevation. The actual shaft power needed by the two-stage reciprocating compressor to match the back pressure of 150 psig is 175.6 hp. Thus, the percent load on the prime mover, or power ratio *PR* is

$$PR = \frac{175.6}{297.5}(100) = 59.0.$$

Entering the abscissa of Figure 5-16 with the power ratio percent, the approximate fuel consumption rate can be read on the ordinate using the diesel fuel curve. The approximate fuel consumption rate at this power level is 0.580 lb/hp-hr. The total weight of diesel fuel consumption per hour is

$$\dot{w}_f = 0.580(175.6) = 101.9\ \text{lb/hr}.$$

The specific gravity of diesel fuel is 0.8156, thus the total volume (in U.S. gallons) of diesel fuel consumption per hour is

$$q_f = \frac{101.9}{(0.8156)(8.33)} = 15.0\ \text{gal/hr}.$$

(b) Surface location at 1829 m above sea level (SI units)

The prime mover has a derated maximum 221.9 kW available at 1800 rpm at 1829 m elevation. The actual shaft power needed by the two-stage reciprocating compressor to match the back pressure of 103.4 N/cm^2 gauge is 131.0 kW. Thus, the percent load on the prime mover, or power ratio *PR* is

$$PR = \frac{131.0}{221.9}(100) = 59.0.$$

Entering the abscissa of Figure 5-16 with the power ratio percent, the approximate fuel consumption rate can be read on the ordinate using the diesel fuel curve. The approximate fuel consumption rate at this power level is 3.460 N/kW-hr. The total weight of diesel fuel consumption per hour is

$$\dot{w}_f = 3.460(131.0) = 453.3 \text{ N/hr}.$$

The specific gravity of diesel fuel is 0.8156, thus the total volume (in U.S. gallons) of diesel fuel consumption per hour is

$$q_f = \frac{453.3}{(0.8156)(9.81)} = 56.7 \text{ liters/hr}.$$

Illustrative Example 5.4 Determine the diesel fuel consumption rate (in gallons per hour) for the helical lobe (screw) air compressor system and conditions given in Illustrative Example 5.2 for (a) a surface location at sea level and (b) a surface location elevation of 6000 ft (*1829 m*).

(a) Surface location at sea level (USCS units)

The prime mover has a maximum 350 hp available at 1800 rpm at sea level conditions. The actual shaft power needed by the two-stage helical lobe (screw) compressor to produce the fixed design pressure of 300 psig is 246.6 hp. Thus, the percent load on the prime mover, or power ratio *PR* is

$$PR = \frac{246.6}{350}(100) = 70.5.$$

Entering the abscissa of Figure 5-16 with the power ratio percent, the approximate fuel consumption rate can be read on the ordinate using the diesel fuel curve. The approximate fuel consumption rate at this power level is 0.535 lb/hp-hr. The total weight of diesel fuel consumption per hour $\dot{w}_f$ is

$$\dot{w}_f = 0.535(246.6) = 129.5 \text{ lb/hr}.$$

The specific gravity of diesel fuel is 0.8156 at 60°F [5]. The total volume (in U.S. gallons) of diesel fuel consumption per hour q_f is

$$q_f = \frac{129.5}{(0.8156)(8.33)} = 19.1 \text{ gal/hr},$$

where the specific weight of fresh water is 8.33 lb/gal.

(a) Surface location at sea level (SI units)

The prime mover has a maximum 261 hp available at 1800 rpm at sea level conditions. The actual shaft power needed by the two-stage helical lobe (screw) compressor to produce the fixed design pressure of 206.9 N/cm^2 gauge is 183.9 kW. Thus, the percent load on the prime mover, or power ratio PR is

$$PR = \frac{183.9}{261}(100) = 70.5.$$

Entering the abscissa of Figure 5-16 with the power ratio percent, the approximate fuel consumption rate can be read on the ordinate using the diesel fuel curve. The approximate fuel consumption rate at this power level is 3.191 N/kW-hr. The total weight of diesel fuel consumption per hour $\dot{w}_f$ is

$$\dot{w}_f = 3.191(183.9) = 587.0 \text{ N/hr}.$$

The specific gravity of diesel fuel is 0.8156 at 15.6°C [5]. The total volume (in U.S. gallons) of diesel fuel consumption per hour q_f is

$$q_f = \frac{587.0}{(0.8156)(9.81)} = 73.4 \text{ liters/hr},$$

where the specific weight of fresh water is 9.81 N/liter.

(b) Surface location at 6000 ft above sea level (USCS units)

The prime mover has a maximum 297.5 hp available at 1800 rpm at sea level conditions. The actual shaft power needed by the two-stage helical lobe (screw) compressor to produce the fixed design pressure of 240.2 psig is 197.5 hp. Thus, the percent load on the prime mover, or power ratio PR is

$$PR = \frac{197.5}{297.5}(100) = 66.4.$$

Entering the abscissa of Figure 5-16 with the power ratio percent, the approximate fuel consumption rate can be read on the ordinate using the diesel fuel curve. The approximate fuel consumption rate at this power level is 0.550 lb/hp-hr. The total weight of diesel fuel consumption per hour is

$$\dot{w}_f = 0.550(197.5) = 108.6 \text{ lb/hr}.$$

The specific gravity of diesel fuel is 0.8156, thus the total volume (in U.S. gallons) of diesel fuel consumption per hour is

$$q_f = \frac{108.6}{(0.8156)(8.33)} = 16.0 \text{ gal/hr}.$$

(b) Surface location at 1829 m above sea level (SI units)

The prime mover has a derated maximum 221.9 kW available at 1800 rpm at 1829 m elevation. The actual shaft power needed by the two-stage helical lobe (screw) compressor to produce the fixed design pressure of 173.8 N/cm^2 gauge is 147.3 kW. Thus, the percent load on the prime mover, or power ratio PR is

$$PR = \frac{147.3}{221.9}(100) = 66.4.$$

Entering the abscissa of Figure 5-16 with the power ratio percent, the approximate fuel consumption rate can be read on the ordinate using the diesel fuel curve. The approximate fuel consumption rate at this power level is 3.281 N/kW-hr. The total weight of diesel fuel consumption per hour is

$$\dot{w}_f = 3.281(147.3) = 483.3\ \text{N/hr}.$$

The specific gravity of diesel fuel is 0.8156, thus the total volume (in U.S. gallons) of diesel fuel consumption per hour is

$$q_f = \frac{483.3}{(0.8156)(9.81)} = 60.4\ \text{liters/hr}.$$

Summary

Table 5-2 gives a summary of the aforementioned fuel consumption examples. It is clear that with a reciprocating piston primary and a helical lobe (screw) primary with similar specifications, the screw compressor will have higher fuel consumption rates (independent of elevation above sea level). Also shown with Table 5-2 is a clear reduction in fuel consumption by both types of primaries as they are operated at higher elevations above sea level. This is because the compressors are working on fewer molecules of the atmospheric air as elevation above sea level is increased.

The main reason that the screw compressor has poor fuel consumption performance is because of the fixed pressure ratio design of this class of compressor. This poor performance characteristic is shared by all rotary compressors.

Table 5-2. Summary of Fuel Consumption for Reciprocating Piston and Helical Lobe (Screw) Compressors

	USCS Units (gal/hr)	SI Units (liters/hr)
Reciprocating piston		
Sea level	17.1	64.5
6000 ft	15.0	56.7
Helical lobe (screw)		
Sea level	19.1	73.4
1829 m	16.0	60.4

5.7 EXAMPLE COMPRESSOR SYSTEM UNITS

In this section, several operational example compressor system units are presented. These units represent the typical variety of units in operation around the United States and the world. In follow-on illustrative examples for air and gas drilling operations, stable foam drilling operations, and aerated fluid drilling operations, one or more of these compressor system units will be used as the provider of the compressed air or gas.

It must be understood that all primary compressors are specified by (1) the volumetric flow rate of atmospheric air that they intake and (2) the maximum output pressure that they can exhaust.

The atmospheric air volumetric flow rate intake capability is given as *acfm* (actual cubic feet per minute) or as *actual liters/second*. This primary compressor specification recognizes that the actual internal design of a positive displacement compressor cannot be altered as the machine is moved from one surface location above sea level to another (i.e., the internal volume chambers cannot be altered). Primary compressors intake their specified volumetric flow rate of the atmospheric air regardless where that location is above sea level. The maximum output pressure specification pertains to a structural design limitation on the housing and moving parts of the positive displacement compressor.

Booster compressors are specified by (1) the minimum pressure limit of the compressed air the booster will accept from the primary compressors, (2) the intake volumetric flow rate (at that minimum pressure limit) in *scfm* (standard cubic feet per minute) or *standard liters/second*, and (3) the maximum pressure output capability of the booster.

The minimum intake pressure limit and its volumetric flow rate are related to the restriction on available booster suction power. The output pressure limit is related to the structural design limitations of the booster.

5.7.1 Small Reciprocating Primary and Booster Compressor System

Most of the compressor systems for air and gas drilling operations are provided as a separate primary compressor system unit (either with a reciprocating compressor or with a rotary compressor) and a separate booster compressor system unit (always a reciprocating compressor). These units are usually skid mounted and each has its own prime mover.

Figure 5-18 shows a skid-mounted primary compressor system unit that has a Gardner Denver Model WEN, two-stage reciprocating piston compressor. The Gardner Denver Model WEN is operated at 1000 rpm. At this speed, the compressor has a volumetric flow rate of 700 acfm (*330 actual liters/sec*) and a maximum pressure capability of 350 psig (*241 N/cm^2 gauge*) at API Mechanical Equipment Standards. The compressor and its cooling subsystem are shown at the right end of the skid in Figure 5-18. On the left end of the skid in Figure 5-18 is the Caterpillar Model D353 prime mover. The Caterpillar Model D353 is an in-line six cylinder, diesel-fueled prime mover. This prime mover is turbocharged and

FIGURE 5-18. Skid-mounted, reciprocating piston primary compressor system unit (courtesy of Mountain Air Drilling Service).

aftercooled and is rated to produce a peak of 270 hp (*201 kW*) at a speed of 1000 rpm (at API Mechanical Equipment Standard).

Figure 5-19 shows a skid-mounted booster compressor system unit that has a Gardner Denver Model MDY, two-stage reciprocating piston compressor. The compressor is seen at the front end (right end) of the skid in Figure 5-19. The compressor is driven by a Caterpillar Model D353 prime mover. This is the same prime mover used for the primary compressor system unit discussed earlier. The prime mover is rated to

FIGURE 5-19. Skid-mounted, reciprocating piston booster compressor system unit (courtesy of Mountain Air Drilling Service).

produce a peak 270 hp (*201 kW*) at a speed of 1000 rpm (at API Mechanical Equipment Standards). However, unlike the primary compressor system discussed earlier, the booster compressor system may be operated at speeds different from the 1000 rpm. The booster prime mover would produce an approximate peak of 230 hp (*172 kW*) at 900 rpm and a peak of 300 hp (*224 kW*) at 1100 rpm.

This booster compressor system unit is usually operated in series with the primary compressor system. When the drilling rig requires compressed air at pressures above the 350 psig (*241 N/cm^2 gauge*) maximum capability of the primary, the air flow from the primary is conducted to the inlet of the booster. In this series configuration, the booster is engaged and the volumetric flow rate from the primary is further compressed to pressures above the 350 psig (*241 N/cm^2 gauge*) level. However, the booster can also be used to compress pipeline natural gas to pressures above typical pipeline pressures for use on a drilling rig.

5.7.2 Four-Stage Reciprocating Compressor System

The compressor system unit shown in Figure 5-20 is another unusual design. This is a semitrailer-mounted unit. The compressor is a large Dresser Clark CFB-4, four-stage reciprocating piston compressor. The compressor is operated at a rotating speed of 900 rpm. At this speed the compressor produces a volumetric flow rate of 1200 acfm (*566.3 actual liters/sec*) and a continuous maximum pressure of 1000 psig (*690 N/cm^2 gauge*) at API Mechanical Equipment Standards. The compressor can produce a maximum pressure of 1200 psig (*828 N/cm^2 gauge*) in intermittent service. Technically, this compressor system is a primary compressor system unit, but in application the unit performs as a combined primary booster all-in-one compressor system.

FIGURE 5-20. Semitrailer-mounted, four-stage reciprocating piston compressor system unit (courtesy of Air Comp Drilling).

This compressor system is probably one of the most reliable systems in the drilling industry. These compressor systems were introduced in the early 1980s to provide compressed air for deep steam production wells in the Geysers geothermal fields of northern California. This application required the drilling of 11-in (*280 mm*)-diameter wells to depths in excess of 10,000 ft (*3048 m*). During the drilling of these wells, steam inflows to the wells were nearly always encountered, which increased the injection pressure requirements greatly. No similar air drilling requirements have been encountered in the drilling of deep oil and natural gas wells. Not only do these compressor systems produce a very useful volumetric flow rate, but they can produce this volumetric flow rate at a great variety of pressures (i.e., from 100 to 1200 psig, or 69 to 828 N/cm^2 gauge). Because these units are reciprocating piston compressors, the volumetric flow rate is not affected by the pressure output. This compressor system is presently the highest quality unit on the market and should be considered for any critical or risky air drilling operation.

The compressor is directly coupled to a Caterpillar Model D398 prime mover. The Caterpillar Model D398 is a V-12 piston configuration. The prime mover is diesel fueled, turbocharged, and aftercooled. At the rotation speed of 900 rpm, the prime mover can produce a peak of 760 hp (*567 kW*).

5.7.3 Rotary Primary and Reciprocating Piston Booster Compressor System

Figure 5-21 shows a skid-mounted primary compressor system unit with an Ingersoll Rand Model XHP 1170, two-stage helical lobe (screw) compressor. The compressor is operated at 1800 rpm. At this speed the compressor has a volumetric flow rate of 1170 acfm (*552.1 actual liters/sec*) and a fixed pressure output of

FIGURE 5-21. Skid-mounted, two-stage rotary primary compressor system unit (courtesy of Weatherford International).

350 psig (*241 N/cm² gauge*) at API Mechanical Equipment Standards. The after-cooling and innercooling subsystems are at the right end of the skid in Figure 5-21. The prime mover is in the middle of the skid, and the compressor is mounted at the left end (or aft end) of the skid. The prime mover is a diesel-fueled, turbo-charged Caterpillar Model ACERT C15 with an in-line 6-cylinder configuration. This prime mover produces a peak of 540 hp (*403 kW*) at a speed of 1800 rpm at API Mechanical Equipment Standards.

Figure 5-22 shows a skid-mounted, low pressure booster compressor system unit that has a Joy Model WB-12, single- or two-stage, double-acting, reciprocating piston compressor. The primer mover is a Detroit Model 12V71T DDEC diesel fueled and turbocharged. The prime mover is rated at 585 hp (*436 kW*) at 2100 rpm. On the right end of the skid is the aftercooler, at the left end of the skid is the prime mover, and in the center is the Joy compressor.

Operating in a single-stage mode, the booster is capable of an equivalent input volumetric flow rate of 3000 scfm (*1416 standard liters/sec*)(at a minimum input pressure of 165 psig, or 114 N/cm² gauge) and increasing the flow pressure to 650 psig (*448 N/cm² gauge*). Operating in a two-stage mode, the booster is capable of an input volumetric flow rate of 2150 scfm (*1014 standard liters/sec*) (at a minimum input pressure of 165 psig, or 114 N/cm² gauge) and increasing flow pressure to 1400 psig (*966 N/cm² gauge*).

Figure 5-23 shows a skid-mounted, high pressure booster compressor system unit that has a Joy Model WB-11, single-stage, double-acting, reciprocating piston compressor. The primer mover is a Caterpillar Model 3306 diesel fueled and turbocharged. The prime mover is rated at 225 hp (*155 KW*) at 1800 rpm. On the left end of the skid is the aftercooler, at the right end of the skid is the prime mover, and in the center is the Joy compressor.

FIGURE 5-22. Skid-mounted, reciprocating piston low pressure booster compressor system unit (courtesy of Weatherford International).

FIGURE 5-23. Skid-mounted, reciprocating piston high pressure booster compressor system unit (courtesy of Weatherford International).

The unit is capable of an equivalent input volumetric flow rate of 800 scfm (*378 standard liters/sec*)(at a minimum input pressure of 1400 psig, or 966 N/cm^2 gauge) and increasing flow pressure to 5000 psig (*3449 N/cm^2 gauge*).

5.8 MEMBRANE FIELD NITROGEN GENERATOR

Using compressed atmospheric air in drilling or production operations exposes the well and operating personnel to fire and explosion risks. Cryogenic nitrogen has been used in drilling and production operation since the 1940s to negate this risk in certain high risk operations. Using cryogenic technology is very expensive and logistically cumbersome when the locations are remote. The very recent development of membrane filter units to reduce the oxygen percentage in atmospheric air has been driven by the need to find a more cost-effective method for eliminating the risk of downhole fires and explosions when drilling boreholes in rock formations containing hydrocarbons. This problem was recognized in the early years of the development of air and gas drilling technology. In those early years, the solution was to use natural gas as the drilling fluid. However, using natural gas as a drilling fluid increased the risk of surface fire or explosions in and around the rig location. Also, although natural gas was inexpensive in early years, natural gas today has a sizable share of the energy market and the cost of using natural gas for drilling operations has become prohibitive.

The risk of downhole fires and explosions exists for both vertical and horizontal drilling operations. However, this risk is far more acute for horizontal drilling operations. This is due to the fact that during a typical horizontal drilling operation, the horizontal interval drilled in the hydrocarbon bearing rock formations is several times longer than in a typical vertical interval drilled in a vertical drilling operation (assuming similar hydrocarbon bearing rock formations). Further, the drilling rate of penetration for a horizontal drilling operation will be about half that of vertical drilling (assuming the similar rock type). This further increases the drilling time in the hydrocarbon bearing formation.

5.8.1 Allowable Oxygen Concentrations

For the past two decades, membrane technologies have been used to separate oxygen (and some other molecules) from gas mixtures, particularly atmospheric air. Because of the high cost of horizontal wells, membrane technology has been developed that can provide high volumetric flow rates of inert atmospheric air (i.e., field-generated nitrogen) for the drilling operation. Figure 5-24 shows a typical membrane filter unit used in drilling operations. This unit is basically housed in a structure similar to a freight container. This particular unit can be placed on a skid mount, semitrailer, or moved by barge to an offshore platform.

Membrane filter units are operated so that the oxygen content in the compressed air flow-through units will be below the level needed to support fire or explosion. Membrane filter technology for drilling operations is incorporated in portable skid-mounted units or semitrailer-mounted units that can be placed in series in the gas flow line between the primary compressor(s) and the drill rig (or between the primary compressors and the booster compressors).

Drilling operation membrane units are available in input flow rate capacities that are rated as 750 scfm (*354 standard liters/sec*), 1500 scfm (*708 standard liters/sec*), and 3000 scfm (*1416 standard liters/sec*). These membrane units

FIGURE 5-24. Typical membrane filter field unit for drilling operations (courtesy of Weatherford International).

must be matched with a variety of primary compressors that feed the units with compressed atmospheric air.

Figure 5-25 shows a schematic of the basic components inside a membrane filter unit. This figure represents the various processes that take place within the unit (with their associated equipment).

Because of the variety of membrane unit capacities and the variety of primary compressors and booster compressors, the flow rate capacities of each must be carefully matched. Figure 5-26 shows a schematic of the complexity of the field location that utilizes these units and shows the layout for a typical drilling operation that provides high pressure field-generated nitrogen to the drilling operation (requiring primaries, low pressure boosters, and a high pressure booster).

5.8.2 Membrane Unit Efficiencies

It is not necessary for these membrane filter units to filter all the oxygen from the atmospheric air in order to render the injected drilling gas inert to downhole hydrocarbon ignition. Filtering oxygen down to approximately 5% (by volume) of the resulting gas will make the gas inert [14, 15]. When the membrane filter units were introduced, Burlington Resources Incorporated (San Juan Division) carried out field tests to evaluate the efficiency of the units. Figure 5-27 gives the approximate results of these tests. Thus, if the drilling operation requires that the oxygen content in atmospheric air be reduced to a level of approximately 5% (by volume), Figure 5-27 shows that the efficiency of the unit will be approximately 50%. Therefore, if a primary compressor unit rated at 1500 scfm (*708 standard liters/sec*) is used to supply a membrane filter unit also rated at 1500 scfm (*708 standard liters/sec*), then the output from the filter unit will only be 750 scfm (*354 standard liters/sec*) of inert atmospheric air (i.e., going to the drill rig).

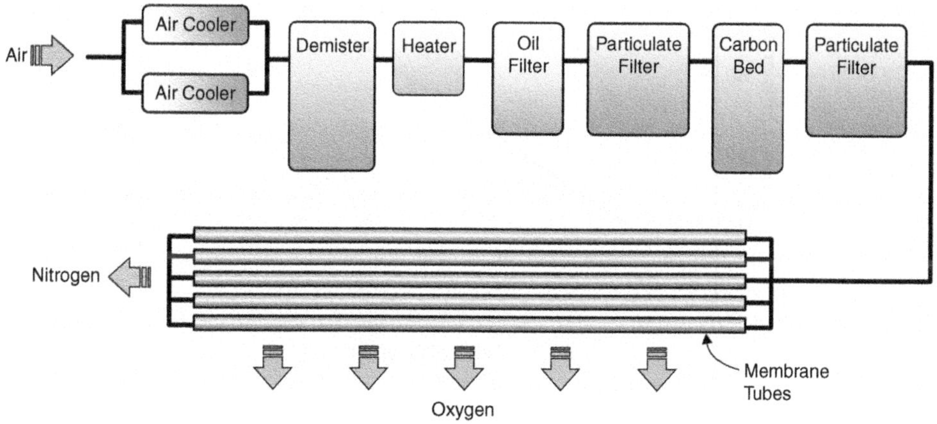

FIGURE 5-25. Schematic membrane filter unit internal components [13].

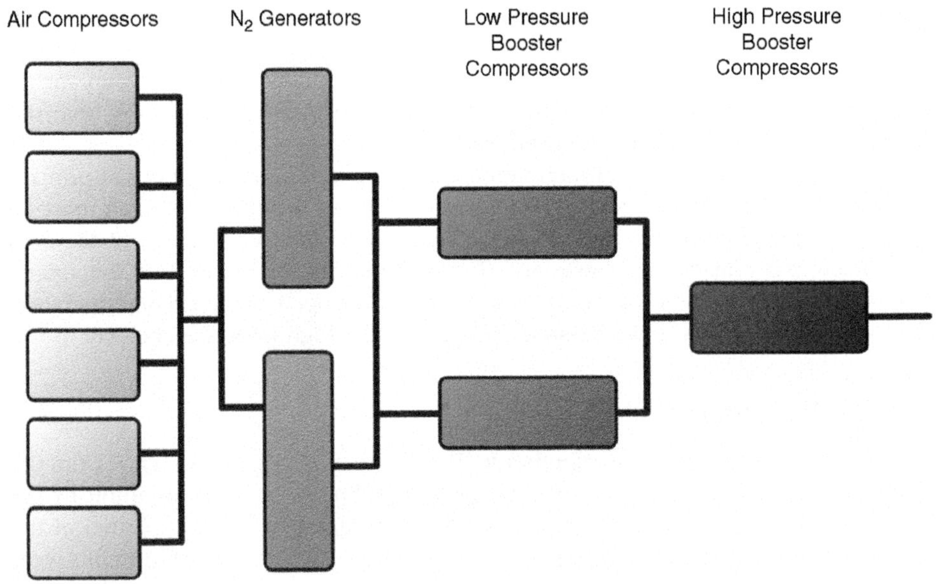

FIGURE 5-26. Schematic of field location to provide high pressure field-generated nitrogen for drilling operation.

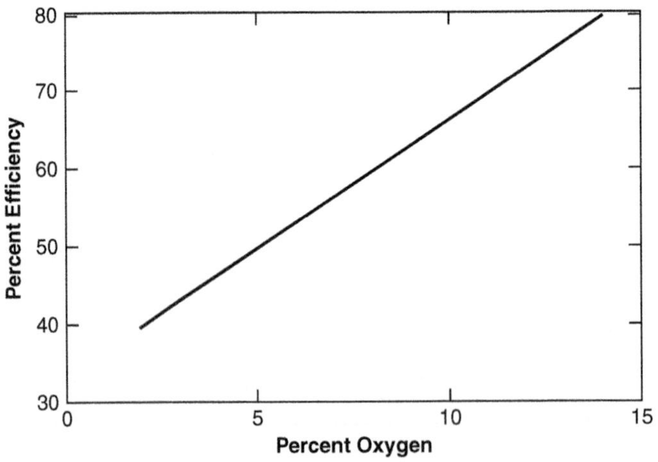

FIGURE 5-27. Membrane filter unit volumetric flow rate efficiency versus percent of oxygen (by volume) remaining in output [13].

REFERENCES

1. *Atlas Copco Manual*, Atlas Copco Company, Fourth Edition, 1982.
2. Brown, R. N., *Compressors: Selection and Sizing*, Gulf Publishing Company, 1986.
3. *API* Specification for the Internal-Combustion Reciprocating Engines for Oil-Field Service, *API Std. 7B-11C*, Ninth Edition, 1994.
4. API Recommended Practice for Installation, Maintenance, and Operation of Internal-Combustion Engines, *API RP 7C-11F*, Fifth Edition, 1994.
5. Baumeister, T., *Marks' Standard Handbook for Mechanical Engineers*, Seventh Edition, McGraw-Hill Book Company, 1979.
6. *U. S. Standard Atmosphere Supplement, 1966*. U. S. Committee on Extension to the Standard Atmosphere, United States Government Publication, 1966.
7. Loomis, A. W., *Compressed Air and Gas Data*, Third Edition, Ingersoll-Rand Company, 1980.
8. Pichot, P., *Compressor Application Engineering, Vol. 1: Compression Equipment*, Gulf Publishing Company, 1986.
9. Pichot, P., *Compressor Application Engineering, Vol. 2: Drivers for Rotating Equipment*, Gulf Publishing Company, 1986.
10. Burghardt, M. D., *Engineering Thermodynamics with Applications*, Second Edition, Harper and Row Publishers, 1982.
11. Block, H. E., and Hoefner, J. J., *Reciprocating Compressors Operations and Maintenance*, Gulf Publishing Company, 1996.
12. Lyons, W. C., *Standard Handbook of Petroleum and Natural Gas Engineering*, Vol. 1, Gulf Publishing Company, 1996.
13. Allan, P. D., "Nitrogen Drilling System for Gas Drilling Applications," *SPE 28320*, Presented at the SPE 69th Annual Technical Conference and Exhibition, New Orleans, Louisiana, September 25–28, 1994.
14. Zabetakis, M. G., "Flammability Characteristics of Combustible Gases and Vapors," *Bureau of Mines Bulletin 627*, Washington, DC, 1964.
15. Coward, H. F., and Jones, G. W., "Limits of Flammability of Gases and Vapors," *Bureau of Mines Bulletin 503*, Washington, DC, 1952.

CHAPTER

Direct Circulation Models

6

In order to make reasonable predictions of the flow characteristics for direct circulation air and gas drilling operations, aerated fluids drilling operations, and stable foam drilling operations, it is necessary to derive a consistent theory that can be used, with certain simplifying limitations, to develop specific equations to model each of the aforementioned operations. All three basic drilling fluid circulation models, air and gas, aerated, and stable foam, must utilize a combination of mathematical theory and empirical correlations to develop a complete calculation model for each.

6.1 BASIC ASSUMPTIONS

Direct circulation is defined as the injection of drilling fluid into the inside of the top of the drill string, the flow of the fluid down the inside of the drill string, through the bit orifices or nozzles, the entraining of the rock cuttings into the drilling fluid at the bottom of the borehole, and then the flow of the drilling fluid with the entrained cuttings up the annulus between the outside of the drill string and the inside of the borehole.

Figure 6-1 shows a simplified U-tube schematic representation of direct circulation flow. In general, in air and gas drilling operations, two-phase flow occurs in the inside of the drill string and through the orifices or nozzles in the drill bit. Three-phase flow occurs when the fluids with entrained rock cuttings move up the annulus from the bottom of the well to the surface. The three phases are a compressible gas, an incompressible fluid, and the solid rock cuttings from the advance of the drill bit. The compressible gases that are used most in drilling are air, natural gas, and nitrogen (or air stripped of oxygen). The incompressible fluids used are treated freshwater, treated salt water (formation water), and water-based drilling muds. Diesel oil, oil-based drilling muds, and crude oil (formation oil) are somewhat compressible.

It is assumed that compressible gases can be approximated by the perfect gas law. Further, it is assumed that the mixture of compressed gas and incompressible

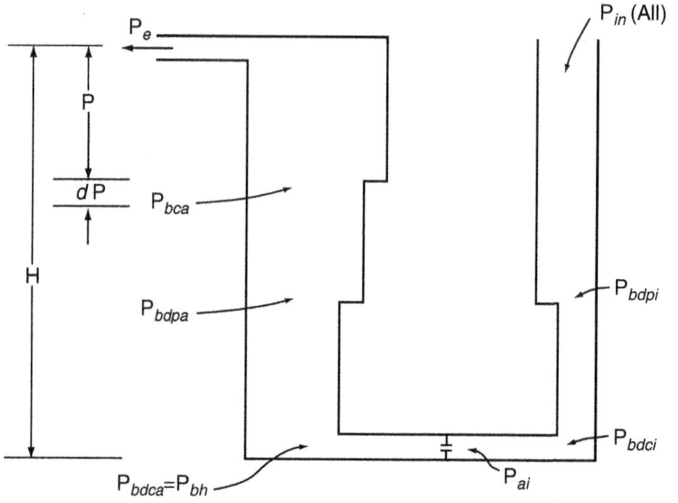

FIGURE 6-1. Schematic of direct circulation. P_{in} is the injection pressure into the top of the drill string, P_{bdpi} is pressure at the bottom of the drill pipe inside the drill string, P_{bdci} is the pressure at the bottom of drill collars inside the drill string, P_{ai} is the pressure above the drill bit inside the drill string, P_{bdca} is the pressure at the bottom of drill collars in the annulus, P_{bh} is the bottom hole pressure in the annulus, P_{bdpa} is the pressure at bottom of drill pipe in the annulus, P_{bca} is the pressure at the bottom of casing in the annulus, and P_e is the pressure at the top of the annulus.

fluid will be uniform and homogeneous. When solid rock cuttings are added to the mixture of compressible gas and incompressible fluid, the solid rock particles are assumed to be uniform in size and density and will be distributed uniformly in the mixture of gas and fluid [1].

The assumption of uniformity of the two or three phases in the mixtures is an important issue in light of the technology developed for gas lift-assisted oil production [2, 3]. The aeration of oil (or other formation produced fluids) from the bottom of a well with the flow of gas from the surface (down the annulus between the casing and the production tubing) is similar to the aeration of fluid and rock cuttings from the bottom of a well with a flow of gas and fluid from the surface (down the inside of the drill string). However, in most oil production situations, the two-phase flow takes place inside of the tubing. In the drilling situation, the gas and fluid are injected together into the top of the drill string and move together down the inside of the drill string, through the bit orifices or nozzles, and then the resulting three-phase flow (gas, fluid, and rock cuttings) moves up the annulus to the surface. Thus, the geometry of flow for the two operations is quite different and probably not comparable [4].

6.2 GENERAL DERIVATION

The term P_{in} represents the pressure of the injected drilling fluids into the top of the drill string. The U-tube representation in Figure 6-1 shows the larger inside diameter of the drill pipe at the top of the drill string where the drilling fluids are injected. Below the drill pipe is shown the smaller inside diameter of the drill collars, and below the drill collars is shown a schematic of the drill bit orifices (or nozzles). The schematic shows the smaller annulus space between the outside of the drill collars and the inside of the open borehole. Above is the annulus space between the outside of the drill pipe and the inside of the open borehole. Then at the top (in the annulus space) is the largest annulus space between the outside of the drill pipe and the inside of the casing. At the top of the annulus the drilling fluids with the entrained cuttings exit the circulation system at pressure P_e.

As in all compressible flow problems, the process of solution must commence with a known pressure and temperature and, in this case, the pressure and temperature at the exit from the circulation system. Therefore, the derivation will begin with the analysis of the flow of the gas and incompressible fluids in the annulus and will continue through the circulation system in the upstream direction. Thus, this derivation will start with the annulus, continue through the drill bit orifices, and then continue up the inside of the drill string to the surface. Figure 6-1 shows pressure P at any position in the annulus, which is referenced from the surface to a depth by the term h. The total depth of the well is H. The differential pressure dP in the upward flowing three-phase flow occurs over an incremental distance of dh. This differential pressure can be approximated as [1]

$$dP = \gamma_{mix}\left[1 + \frac{fV^2}{2g(D_h - D_p)}\right]dh, \tag{6-1}$$

where P is fluid pressure (lb/ft^2 abs, N/m^2 abs); h is the reference depth (ft, m); H is the total depth (ft, m); γ_{mix} is the specific weight of the mixture of air (or other gas), incompressible fluid, and rock cuttings (lb/ft^3, N/m^3), f is the Darcy-Weisbach friction factor; V is the average velocity in the annulus (ft/sec, m/sec); D_h is the inside diameter of the borehole (ft, m); D_p is the outside diameter of the drill pipe (ft, m); and g is the acceleration of gravity (32.2 ft/sec^2, 9.81 m/sec^2).

The first term on the right side of Equation (6-1) represents the incremental pressure change due to the hydrostatic weight of the column of fluid (with entrained rock cuttings) in the annulus. The second term on the right side of Equation (6-1) represents the incremental increase pressure change due to the friction loss of the flowing fluid mixture.

6.2.1 Weight Rate of Flow of the Gas

In order to carry out derivation of the governing equations for direct circulation, the weight rate of flow of air (or gas) to the well must be determined. Assuming that the compressed air is provided by a compressor(s), the weight rate of flow through the circulating system is determined from the atmospheric pressure and temperature of the air at the compressor location on the surface of the earth, and the characteristics of the compressor(s). For air, the atmospheric pressure and average annual temperatures for sea level and various elevations above sea level can be approximated for most of North America by midlatitudes data given in Tables 5-1a and 5-1b. These reference pressures are denoted as p_r and t_r. For engineering calculations, the actual atmospheric pressure and temperature of the air entering a primary compressor(s) are p_{at} and t_{at} and can be approximated as

$$p_{at} \approx p_r \tag{6-2}$$

$$t_{at} \approx t_r, \tag{6-3}$$

where p_{at} is the atmospheric pressure (psia, N/cm^2 abs), p_r is the reference atmospheric pressure (psia, N/cm^2 abs), t_{at} is the atmospheric pressure (psia, N/cm^2 abs), and t_r is the reference atmospheric temperature (°F, °C). Also see Appendix B for more detailed plots for atmospheric conditions for North America midlatitudes. Similar data as that given in Tables 5-1a and 5-1b and Appendix B for North America midlatitudes can be obtained for most other continents and latitudes around the world.

The aforementioned pressures and temperatures in field units must be converted to units consistent with Equations (5-11) and (6-1). These are

$$P_g = P_r \tag{6-4}$$

$$T_g = T_r, \tag{6-5}$$

where P_r is the reference atmospheric pressure (lb/ft^2 abs, N/m^2 abs), and T_r is the absolute atmospheric reference temperature (°R, K).

Substituting Equations (6-2) and (6-3) into Equation (5-11), the specific weight of the gas (air) γ_g entering the compressor(s) is

$$\gamma_g = \frac{P_g\, S_g}{\mathbf{R_e}\, T_g} = \frac{P_{rl}\, S_g}{\mathbf{R_e}\, T_{rl}}, \tag{6-6}$$

where γ_g is specific weight (lb/ft^3, N/m^3), $\mathbf{R_e}$ is the engineering gas constant (53.36 ft-lb/lb-°R, 29.31 N-m/N-K), and S_g is the specific gravity of the gas ($S_g = 1.0$ for air at standard conditions). The weight rate of flow of the gas $\dot{w}_g$ through the compressor is

$$\dot{w}_g = \gamma_g Q_g, \tag{6-7}$$

where $\dot{w}_g$ is the weight rate of flow of gas (lb/sec, N/sec) and Q_g is the volumetric flow rate of air into the circulation system (actual ft^3/sec, actual m^3/sec).

If the gas for the drilling operation is natural gas from a pipeline and the pressure and temperature of the gas in the pipeline are p_{pl} and t_{pl}, then the pressure and temperature of the gas entering either directly into the circulation system or to the booster compressor are

$$P_g = P_{pl} \tag{6-8}$$

$$T_g = T_{pl}, \tag{6-9}$$

where p_{pl} is the pipeline pressure (psia, N/cm^2 abs), P_{pl} is the pipeline pressure (lb/ft^2 abs, N/m^2 abs), t_{pl} is the pipeline temperature (°F, °C), and T_{pl} is the absolute pipeline temperature (°R, K).

Substituting Equations (6-8) and (6-9) into Equation (5-11), the specific weight of the gas from a pipeline can be obtained. This is

$$\gamma_g = \frac{P_g\, S_g}{\mathbf{R_e}\, T_g} = \frac{P_{pl}\, S_g}{\mathbf{R_e}\, T_{pl}}. \tag{6-10}$$

In this equation, S_g would be the specific gravity of the pipeline natural gas (e.g., usually between 0.65 and 0.85).

Substituting the result from Equation (6-10) into Equation (6-7) gives the weight rate of flow of gas from a pipeline, where Q_g is the volumetric flow rate of natural gas from the pipeline at pressure p_{pl} and temperature t_{pl}. Note that the volumetric flow rate in a pipeline is usually given by flow meters in either scfm or standard m^3/sec regardless of surface elevation location. This value must be converted to obtain the actual volumetric flow rate at p_{pl} and t_{pl} (see Appendix A). As discussed in Chapter 5, care must be taken to determine which set of standard conditions are being used to define the "scfm or standard m^3/sec." Chapter 5 gives a summary of the most common standard condition specifications used throughout the industrialized world.

6.2.2 Three-Phase Flow in the Annulus

This general solution for three-phase flow is valid for aerated (gasified) drilling fluids where the three phases in the annulus are gas, incompressible fluid, and solids (cuttings).

The weight rate of flow of incompressible drilling fluid (usually drilling mud), $\dot{w}_m$, into the well is

$$\dot{w}_m = \gamma_m\, Q_m, \tag{6-11}$$

where $\dot{w}_m$ is the weight rate of flow of drilling mud (lb/sec, N/sec), γ_m is the specific weight of the drilling mud (lb/ft^3, N/m^3), and Q_m is the volumetric flow rate of drilling mud (ft^3/sec, m^3/sec).

The weight rates of flow $\dot{w}_g$ and $\dot{w}_m$ entering the well through the top of the drill string and flow to the bottom of the string and exit into the annulus through the openings in the drill bit (open orifices or the bit nozzles). After passing through the drill bit, the fluids entrain the rock cuttings generated by the drill bit as the bit is advanced. The entrained weight rate of flow of the solids, $\dot{w}_s$, is

$$\dot{w}_s = \frac{\pi}{4}\, D_h^2\, \gamma_w\, (2.7)\, \kappa, \tag{6-12}$$

where $\dot{w}_s$ is the weight rate of flow of solid rock cuttings (lb/sec, N/sec), D_h is the diameter of the drilled hole (i.e., the bit diameter) (ft, m), γ_w is the specific weight of the fresh water (lb/ft^3, N/m^3), and κ is the penetration rate (ft/sec, m/sec).

For USCS units, in Equation (6-12), the specific weight of fresh water is 62.4 lb/ft^3. For SI units, in Equation (6-12), the specific weight of fresh water is the density of fresh water, 1000 kg/m^3, multiplied by the acceleration of gravity in the SI units 9.81 m/sec^2. The average specific gravity of sedimentary rocks is approximately 2.7. If igneous or metamorphic rocks are to be drilled, average values of 2.80 and 3.00, respectively, can be used [5].

The total weight rate of flow $\dot{w}_t$ in the annulus from the bottom of the well to the surface is

$$\dot{w}_t = \dot{w}_g + \dot{w}_m + \dot{w}_s. \tag{6-13}$$

The drilling mud and the rock cutting solids are assumed to not change volume when pressure is changed (note that if the liquid phase fluid is oil, the volume changes with pressure can be taken into account). However, the air (or gas) does change volume as a function of pressure change and, therefore, as a function of depth. Thus, the specific weight of the gas at any position in the annulus is

$$\gamma = \frac{P S_g}{\mathbf{R_e} T_{av}}, \tag{6-14}$$

where T_{av} is the average temperature of the gas over a depth interval (°R, K). This average temperature term is determined by taking the average of the sum of the geothermal temperatures at the top and bottom of the depth interval. The geothermal temperature at depth, t_b, is determined from the approximate expression

$$t_b = t_r + \beta H, \tag{6-15}$$

where t_r is the average annual atmospheric reference temperature (°F, °C), t_b is the geothermal temperature at depth (°F, °C), and β is the geothermal temperature gradient (°F/ft, °C/m).

The reference surface geothermal temperature t_r is assumed to be the temperatures given in Tables 5-1a and 5-1b for sea level and various elevations above sea level. These temperatures represent North American midlatitude year-round averages. It is assumed that these temperatures also represent an average constant deep soil or rock temperatures near the surface of the earth at the elevations given in the table. The value of the geothermal gradient constant is determined from temperature logs of offset wells and other geophysical data. An average value of the geothermal gradient that can be used when the actual gradient has not been determined is 0.01°F/ft, or 0.018°C/m. The temperature at depth can be expressed as absolute temperatures using the following:

$$T_b = T_r + \beta H, \tag{6-16}$$

where T_r is the reference atmospheric temperature (°R, K). Once the reference temperature is changed to absolute, no other changes need to be made in Equation (6-16).

The absolute average temperature T_{av} over the first depth interval below the surface is

$$T_{av1} = \frac{T_r + T_{b1}}{2}. \tag{6-17}$$

The T_{av} for follow-on intervals will be the average of the absolute temperature at the top and the absolute temperature at the bottom of the interval. Follow-on average temperatures will be

$$T_{av2} = \frac{T_{b1} + T_{b2}}{2}\ldots, \tag{6-18}$$

where T_{b1} is the temperature at the bottom of the first interval (°R, K) and T_{b2} is the temperature at the bottom of the second interval (°R, K). Follow-on T_{av} interval temperatures are determined in sequence in a similar method as above.

The relationship between the weight rate of flow of the gas and the specific weight and volumetric flow rate of gas at any position in the annulus is given by

$$\dot{w}_g = \gamma_g\, Q_g = \gamma\, Q. \tag{6-19}$$

Substituting Equations (6-6) and (6-14) into the two terms on the right side of Equation (6-19) gives a relationship between the specific weight and volumetric flow rate at the surface and the specific weight and volumetric flow rate at any position in the annulus. This is

$$\frac{P_g\, S_g}{\mathbf{R_e}\, T_g}\, Q_g = \frac{P\, S_g}{\mathbf{R_e}\, T_{av}}\, Q. \tag{6-20}$$

Solving Equation (6-20) for Q yields

$$Q = \left(\frac{P_g}{P}\right)\left(\frac{T_{av}}{T_g}\right) Q_g. \tag{6-21}$$

The three-phase flow of gas, incompressible fluid, and rock cuttings up the annulus can be described by a mixed specific weight term, which is a function of its position in the annulus. This mixed specific weight γ_{mix} is

$$\gamma_{mix} = \frac{\dot{w}_t}{\left(\frac{P_g}{P}\right)\left(\frac{T_{av}}{T_g}\right) Q_g + Q_m}. \tag{6-22}$$

In the derivation of Equation (6-22), the volume contribution of the solids (the rock cuttings) is assumed to be small and negligible relative to the volumes of the gas and the incompressible fluid in the mixture (i.e., contributes only to the $\dot{w}_t$ term).

The velocity of this mixture changes as a function of its position in the annulus. The velocity V of the three-phase flow in the annulus is

$$V = \frac{Q + Q_m}{\frac{\pi}{4}(D_h^2 - D_p^2)}, \tag{6-23}$$

where D_h is the inside diameter of the annulus (ft, m) and D_p is the outside diameter of the drill string (drill pipe or collars) (ft, m).

Substituting Equation (6-21) into Equation (6-23) yields

$$V = \frac{\left(\frac{P_g}{P}\right)\left(\frac{T_{av}}{T_g}\right) Q_g + Q_m}{\frac{\pi}{4}(D_h^2 - D_p^2)}. \tag{6-24}$$

Substituting Equations (6-22) and (6-24) into Equation (6-1) yields

$$dP = \left[\frac{\dot{w}_t}{\left(\frac{P_g}{P}\right)\left(\frac{T_{av}}{T_g}\right) Q_g + Q_m}\right]\left\{1 + \frac{f}{2g(D_h - D_p)}\left[\frac{\left(\frac{P_g}{P}\right)\left(\frac{T_{av}}{T_g}\right) Q_g + Q_m}{\frac{\pi}{4}(D_h^2 - D_p^2)}\right]^2\right\} dh. \tag{6-25}$$

Equation (6-25) contains only two independent variables: P and h. All of the other terms in the equation are known constants. Separating variables in Equation (6-25) and integrating from the surface to the bottom of the well yields

$$\int_{P_e}^{P_{bh}} \frac{dP}{B_a(P)} = \int_0^H dh, \tag{6-26}$$

where P_e is the exit pressure at the top annulus (lb/ft^2 abs, N/m^2 abs), P_{bh} is the bottom hole pressure in the annulus (lb/ft^2 abs, N/m^2 abs), and

$$B_a(P) = \left[\frac{\dot{w}_t}{\left(\frac{P_g}{P}\right)\left(\frac{T_{av}}{T_g}\right) Q_g + Q_m}\right]\left\{1 + \frac{f}{2g(D_h - D_p)}\left[\frac{\left(\frac{P_g}{P}\right)\left(\frac{T_{av}}{T_g}\right) Q_g + Q_m}{\frac{\pi}{4}(D_h^2 - D_p^2)}\right]^2\right\}.$$

For this general derivation, exit pressure P_e is the pressure at the entrance to the blooey line (in the case of air or gas drilling), the pressure at the entrance to the return flow line (in the case of aerated fluid drilling), or the back pressure upstream of the control valve in the exit flow line (in the case of stable foam drilling).

The Darcy-Weisbach friction factor f given in the aforementioned equation is determined by the standard fluid mechanics empirical correlations relating the friction factor to the Reynolds number, diameter (or hydraulic diameter), and absolute pipe roughness. In general, the values for Reynolds number, diameter, and absolute pipe roughness are known. The classic correlation for the Reynolds number is

$$\mathbf{N_R} = \frac{(D_h - D_p)V}{\nu}, \tag{6-27}$$

where $D_h - D_p$ is the hydraulic diameter for the annulus (ft, m), V is the velocity (ft/sec, m/sec), and ν is the kinematic viscosity of the drilling fluid (ft^2/sec, m^2/sec).

Three flow conditions can exist in the annulus. These are laminar, transitional, and turbulent [1].

The empirical correlation for the friction factor for laminar flow conditions is

$$f = \frac{64}{\mathbf{N_R}}. \tag{6-28}$$

This equation can be solved directly once the Reynolds number is known. In general, Equation (6-28) is valid for values of Reynolds numbers from 0 to 2000.

Up until recently it was necessary to use the Colebrook empirical correlation for transitional flow conditions and the von Karman empirical correlation for the wholly turbulent flow conditions to obtain an analytic value for the friction factor. These empirical correlations were difficult trial and error solutions. A new empirical correlation for the friction factor can be used for both transitional flow conditions and wholly turbulent flow conditions (for Reynolds numbers greater than 2000). This empirical correlation is the Haaland correlation [6]. This empirical expression is

$$f = \left[\frac{1}{-1.8\log\left[\left(\dfrac{\left(\frac{e_{av}}{D_h - D_p}\right)}{3.7}\right)^{1.11} + \dfrac{6.9}{N_R}\right]}\right]^2, \tag{6-29}$$

where e_{av} is the absolute roughness of the annulus surfaces (ft, m). Note that the logarithm in the aforementioned equation is to the base 10. Equation (6-30) gives the approximation for e_{av} for the open hole section of the annulus. This approximation is

$$e_{av} = \frac{e_r\, D_{ob}^2 + e_p\, D_p^2}{D_{ob}^2 + D_p^2}. \tag{6-30}$$

For follow-on calculations for flow in the annulus, the absolute roughness for steel pipe, $e_p = 0.0005$ ft or $e_p = 0.0002$ m, will be used for the outside surfaces of the drill pipe and drill collars, and the inside surface of the casing. The open hole surfaces of boreholes can be approximated with an absolute roughness, $e_r = 0.01$ ft or $e_r = 0.003$ m (i.e., this example value is the same as concrete pipe, which approximates borehole surfaces in limestone and dolomite sedimentary rocks, or in similar competent igneous and metamorphic rocks, see Table 8-1).

Equations (6-26) through (6-30) can be used in sequential integration steps starting at the top of the annulus (with the known exit pressure) and continuing for each subsequent change in the annulus cross-sectional area until the bottom hole pressure is determined. These sequential calculation steps require trial and error solutions. The trial and error process requires selection of the upper limit of pressure in each integral on the right side of Equation (6-26). This upper limit pressure selection must give a left side integral solution equal to the right side integral solution.

6.2.3 Two-Phase Flow Through the Bit

There are two basic calculation techniques for determining the pressure change through the constrictions of the drill bit orifices or nozzles.

The first technique assumes that the mixture of incompressible fluid and the gas passing through the orifices has a high fraction of incompressible fluid volume. Under these conditions the mixture is assumed to act as an incompressible fluid. Thus, borrowing from mud drilling technology, the pressure change through the drill bit ΔP_b can be approximated by [8, 9]

$$\Delta P_b = \frac{(\dot{w}_g + \dot{w}_m)^2}{2g\ \gamma_{mixbb}\ C^2 \left(\frac{\pi}{4}\right)^2 D_e^4}, \tag{6-31}$$

where ΔP_b is pressure change (lb/ft^2, N/m^2), γ_{mixbb} is the specific weight of the mixture at the bottom of the annulus (lb/ft^3, N/m^3), C is the fluid flow loss coefficient for drill bit orifices or nozzles (the value of this constant is dependent flow conditions), and D_e is the equivalent single orifice inside diameter (ft, m). For drill bits with n equal diameter orifices (or nozzles), D_e is

$$D_e = \sqrt{n\ D_n^2}, \tag{6-32}$$

where n is the number of equal diameter orifices (or nozzles) and D_n is the orifice (or nozzle) inside diameter (ft, m).

The pressure change obtained from Equation (6-31) is added to the bottom hole annulus pressure P_{bb} obtained from Equation (6-26). The pressure above the drill bit inside the drill string P_{ai} is

$$P_{ai} = P_{bb} + \Delta P_b, \tag{6-33}$$

where P_{ai} is pressure above the drill bit inside the drill string (lb/ft^2 abs, N/m^2 abs) and P_{bb} is bottom hole pressure (lb/ft^2 abs, N/m^2 abs).

For fluid mixtures that are nearly all gas (with little incompressible fluid), the pressure above the drill bit inside the drill string will depend on whether the critical flow conditions exist in the orifices or nozzle throats. The critical pressure through the bit orifices or nozzles is [1]

$$\left(\frac{P_{bb}}{P_{ai}}\right)_c = \left(\frac{2}{k+1}\right)^{\frac{k}{k-1}}, \tag{6-34}$$

where k is the ratio of specific heats for the gas.

The right-hand side of Equation (6-34) is determined only by the value of the specific heat ratio constant of a gas (e.g., for air $k = 1.4$ and for natural gas $k = 1.28$). Thus, for air the critical pressure ratio is 0.528 and for natural gas the critical ratio is 0.549. Therefore, if P_{ai} is determined to be

$$\text{for air} \qquad P_{ai} \geq \frac{P_{bb}}{0.528} \tag{6-35}$$

$$\text{for natural gas} \qquad P_{ai} \geq \frac{P_{bb}}{0.549}, \tag{6-36}$$

then the flow through the orifice or nozzle throat is sonic. Under these sonic flow conditions, upstream pressure P_{ai} does not depend on downstream pressure P_{bh}. For these sonic flow conditions, the upstream pressure P_{ai} is

$$P_{ai} = \frac{\dot{w}_g \, T_{bb}^{0.5}}{A_n \left[\left(\frac{g\, k\, S_g}{\mathbf{R_e}} \right) \left(\frac{2}{k+1} \right)^{\left(\frac{k+1}{k-1}\right)} \right]^{0.5}}, \tag{6-37}$$

where T_{bb} is the temperature of the gas at the bottom of the well (°R, K) and A_n is the total cross-sectional area of drill bit orifices or nozzles (ft^2, m^2).

If the upstream pressure is less than the right-hand side of either Equation (6-35) or Equation (6-36), the flow through the orifices or nozzles is subsonic and the upstream pressure will be dependent on the pressure and temperature at the bottom of the borehole annulus. The subsonic flow condition is a more complicated calculation situation than the sonic flow condition. In this calculation situation and knowing the bottom hole pressure and bottom hole temperature, the bottom hole specific weight must be determined using Equation (5-11). Knowing the bottom hole pressure, temperature, and specific weight, the upstream pressure can be obtained for the subsonic condition. This equation is

$$P_{ai} = P_{bb} \left[\frac{\left(\frac{\dot{w}_g}{A_n} \right)^2}{2g \left(\frac{k}{k-1} \right) P_{bb}\, \gamma_{bb}} + 1 \right]^{\frac{k}{k-1}}, \tag{6-38}$$

where γ_{bb} is the specific weight of the gas at the bottom of the annulus (lb/ft^3, N/m^3).

6.2.4 Two-Phase Flow in the Drill String

The combined gas and incompressible fluid are injected into the top of the inside drill string. This is two-phase flow. The differential pressure dP in the downward flowing two-phase flow occurs over the incremental distance of dh at the depth h for a total depth of well of H.

Using Figure 6-1, the differential pressure at a depth h can be approximated as

$$dP = \gamma_{mix} \left[1 - \frac{f V^2}{2g\, D_i} \right] dh, \tag{6-39}$$

where D_i is the inside diameter of the drill pipe, or drill collars (ft, m).

The two-phase flow of gas and incompressible fluid down the inside of the drill string can be described by a mixed specific weight term, which is a function of position in the drill string. This mixed specific weight term is

$$\gamma_{mix} = \frac{\dot{w}_g + \dot{w}_m}{\left(\frac{P_g}{P} \right) \left(\frac{T_{av}}{T_g} \right) Q_g + Q_m}. \tag{6-40}$$

The velocity of this mixture changes as a function of its position in the drill string. The velocity of the two-phase flow in the drill string is

$$V = \frac{\left(\frac{P_g}{P}\right)\left(\frac{T_{av}}{T_g}\right)Q_g + Q_m}{\frac{\pi}{4}D_i^2}. \quad (6\text{-}41)$$

Equations (6-40) and (6-41) are functions of the pressure P at a depth of h. Substituting Equations (6-40) and (6-41) into Equation (6-39) yields

$$dP = \left[\frac{\dot{w}_g + \dot{w}_m}{\left(\frac{P_g}{P}\right)\left(\frac{T_{av}}{T_g}\right) Q_g + Q_m}\right]\left\{1 - \frac{f}{2gD_i}\left[\frac{\left(\frac{P_g}{P}\right)\left(\frac{T_{av}}{T_g}\right) Q_g + Q_m}{\frac{\pi}{4} D_i^2}\right]^2\right\} dh. \quad (6\text{-}42)$$

Equation (6-42) contains only two independent variables: P and h. Separating variables in Equation (6-42) and integrating from the bottom of the inside of the drill string to the surface of the well yields

$$\int_{P_{in}}^{P_{at}} \frac{dP}{B_i(P)} = \int_0^H dh, \quad (6\text{-}43)$$

where

$$B_i(P) = \left[\frac{\dot{w}_g + \dot{w}_m}{\left(\frac{P_g}{P}\right)\left(\frac{T_{av}}{T_g}\right) Q_g + Q_m}\right]\left\{1 - \frac{f}{2g\ D_i}\left[\frac{\left(\frac{P_g}{P}\right)\left(\frac{T_{av}}{T_g}\right) Q_g + Q_m}{\frac{\pi}{4}\ D_i^2}\right]^2\right\}$$

and P_{in} is the injection pressure into the inside of the drill string (lb/ft^2 abs, N/m^2 abs).

The friction factor f given in the aforementioned equation is determined by the standard fluid mechanics empirical correlations relating the friction factor to the Reynolds number, diameter, and absolute pipe roughness. In general, values for Reynolds number, diameter, and absolute pipe roughness are known. The classic correlation for the Reynolds number is

$$\mathbf{N_R} = \frac{D_i V}{\nu}, \quad (6\text{-}44)$$

where D_i is the inside diameter for the drill string (ft, m).

Three flow conditions can exist in the inside of the drill string. These are laminar, transitional, and turbulent.

The empirical correlation for the friction factor for laminar flow conditions is

$$f = \frac{64}{\mathbf{N_R}}. \quad (6\text{-}45)$$

This equation can be solved directly once the Reynolds number is known. In general, Equation (6-45) is valid for values for Reynolds numbers from 0 to 2000.

The empirical correlation for the friction factor for both transitional flow conditions and wholly turbulent flow conditions (for Reynolds numbers greater than 2000) can be determined from the Haaland correlation [6]. This empirical expression is

$$f = \left[\frac{1}{-1.8 \log \left[\left(\frac{\left(\frac{e}{D_i} \right)}{3.7} \right)^{1.11} + \frac{6.9}{N_R} \right]} \right]^2 \tag{6-46}$$

where e is the absolute roughness of the inside of the surface of the steel drill string (ft, m).

Equations (6-43) through (6-46) can be used in sequential integration steps starting at the bottom of the inside of the drill string (with the known pressure above the drill bit inside the drill string) and continuing for each subsequent change in the cross-sectional area inside the drill string until the injection pressure is determined. These sequential calculation steps require trial and error solutions. The trial and error process requires the selection of the upper limit of the pressure in each integral on the right side of Equation (6-43). This upper limit pressure selection must give a left side integral solution equal to the right side integral solution.

6.3 AERATED FLUID DRILLING MODEL

Aerated (or gasified) drilling fluid governing equations are changed very little from the direct circulation general derivation given in Section 6.2. The gases used in aerated fluid drilling are usually either air or membrane generated nitrogen (air stripped of oxygen). The fluids used are usually drilling mud, diesel oil, or formation oil.

The basic mathematical model described in Section 6.2 must be augmented with specialized empirically derived correlation models that take into account changes in flow viscosity and liquid holdup experienced in actual aerated drilling operations [2–4]. In particular, these empirical additions to the mathematical model demonstrate the origins of the increased injection and bottom hole pressures experienced in field operations. Chapter 9 will show how these correlations can be incorporated into the basic mathematical model. The resulting mathematical and empirical models will be demonstrated with illustrative examples.

Equations (6-11) through (6-30) describe the flow of aerated drilling fluids in the annulus.

Equations (6-31) through (6-38) describe the flow of aerated drilling fluids through the drill bit orifices or nozzles.

Equations (6-39) through (6-46) describe the flow of aerated drilling fluids through the inside of the drill string.

6.4 STABLE FOAM DRILLING MODEL

Stable foam drilling is a special case of the general derivation given in Section 6.2. In essence, it is a special case of aerated drilling fluids. In stable foam drilling operations, the mixture of gas (usually air or membrane generated nitrogen) and water (with a surfactant) combine under the action of high shear flow through the drill bit nozzles to form a foam.

Referring to Figure 6-1, foam gas volume fraction (or foam quality) Γ at any position in the annulus is defined as

$$\Gamma = \frac{Q_g}{Q_g + Q_f}, \tag{6-47}$$

where Q_g is the volumetric flow rate of gas (ft^3/sec, m^3/sec) and Q_f is the volumetric flow rate of the incompressible fluid (ft^3/sec, m^3/sec).

The objective of the foam operation is to control the foam column inside the annulus in such a manner as to maintain continuous unbroken (or stable) foam throughout the annulus. In order to accomplish this, normal operational practice is to place a back pressure valve on the return flow line at the surface. The back pressure valve and appropriate instrumentation just upstream of the valve allow monitoring of the foam gas volume fraction at that position (i.e., Γ_{bp}). Knowing the foam gas volume fraction at this position, the foam gas volume fraction at the bottom of the annulus Γ_{bb} can be determined via modeling and operational experience. The gas volume fraction at the bottom of the annulus must be maintained at approximately 0.60 or greater [10]. Thus, as the drilling operation progresses and the drill bit is advanced to greater depths, the foam gas volume fraction upstream of the back pressure valve must be adjusted to allow the bottom hole gas volume fraction to be maintained at 0.60 or greater.

If the bottom hole foam gas volume fraction drops much below 0.60, the foam will collapse to the three separate phases. To maintain the bottom hole foam gas volume fraction in the annulus at 0.60 or greater, the foam gas volume fraction upstream of the back pressure valve is usually maintained in the range of 0.90 to 0.99. This return line foam gas volume fraction is dependent on the geometry of the well and of course on the quality of the additives (mainly the surfactant).

The flow up the annulus is a stable foam carrying the entraining cutting solids. The exit pressure for stable foam drilling operations is the back pressure P_{bp} upstream of the valve in the return flow line from the annulus. Equation (6-26) becomes

$$\int_{P_{bp}}^{P_{bb}} \frac{dP}{B_a(P)} = \int_0^H dh, \tag{6-48}$$

where P_{bp} is the back pressure on the annulus (lb/ft^2 abs, N/m^2 abs) and

$$B_a(P) = \left[\frac{\dot{w}_t}{\left(\frac{P_g}{P}\right)\left(\frac{T_{av}}{T_g}\right) Q_g + Q_f}\right]\left\{1 + \frac{f}{2g(D_b - D_p)}\left[\frac{\left(\frac{P_g}{P}\right)\left(\frac{T_{av}}{T_g}\right) Q_g + Q_f}{\frac{\pi}{4}\left(D_b^2 - D_p^2\right)}\right]^2\right\}.$$

Somewhat like the aerated drilling fluid model, the stable foam model requires empirically derived correlations to complete the model. Field operational data show that stable foam drilling operations on large diameter surface using a factor of less compressed air volumetric flow rates as would be required for an air drilling operation. This is because the stable foam bubble structure surface tension acts with an effective high viscous when flowing and with pseudo gel strength when flowing and when static.

To emulate these stable foam characteristics, two basic empirical correlation methodologies are presently used in the drilling industry. Both of these methodologies depend on simple laboratory tests that utilize the actual water, surfactant, and other additives that will be used in a future stable foam drilling operation. These tests are principally conduced to obtain the stable foam *half-life*. To a lesser extent, the tests can also give some information on the effective viscosity and the gel strength of the foam.

1. *Viscosity Correlation*: This correlation utilizes results of the laboratory tests to modify the viscosity of the fluid mixture to give an effective viscosity that will allow the overall model to match field results. This correlation utilizes the traditional Reynolds number calculation and the laminar, transition, and turbulent correlations [Equations (6-49), (6-50), and (6-51)] to obtain the annulus and pipe friction factor [Equation (6-48)][11].

2. *Friction Factor Correlation*: This correlation utilizes the laboratory test results to directly obtain the value for the friction factor in Equation (6-48).

Chapter 10 will present the typical laboratory test procedures used in the industry and the correlations that utilize the test results. These correlations will be incorporated into the mathematical model and illustrative examples shown.

The classic correlation for the Reynolds number is

$$\mathbf{N_R} = \frac{(D_b - D_p)V}{\nu}. \tag{6-49}$$

Three flow conditions can exist in the annulus. These are laminar, transitional, and turbulent.

The empirical correlation for the friction factor for laminar flow conditions is

$$f = \frac{64}{\mathbf{N_R}}. \tag{6-50}$$

This equation can be solved directly once the Reynolds number is known. In general, Equation (6-50) is valid for values for Reynolds numbers from 0 to 2000.

The Haaland empirical correlation can be used to determine the friction factor in Equation (6-48). This empirical expression is

$$f = \left[\frac{1}{-1.8 \log \left[\left(\frac{\frac{\varepsilon_{av}}{D_b - D_p}}{3.7} \right)^{1.11} + \frac{6.9}{N_R} \right]} \right]^2 . \tag{6-51}$$

Equations (6-30) and (6-47) through (6-51) can be used in sequential trial and error integration steps starting at the top of the annulus (with the specified foam gas volume fraction and, thus, pressure upstream of the return line valve) and continuing for each subsequent geometry change in the annulus cross-sectional area until the bottom hole pressure and foam gas volume fraction values are determined.

The flow through the drill bit nozzles is assumed to be an aerated drilling fluid. Therefore, Equations (6-31) through (6-38) are used to model this flow.

The flow through the inside of the drill string is also assumed to be an aerated drilling fluid. Therefore, Equations (6-39) through (6-46) are used to model this flow.

6.5 AIR AND GAS DRILLING MODEL

Unlike the aerated and stable foam drilling fluid models, the air and gas drilling model requires no special empirical correlations to adjust the results to provide results that agree more closely to field data. Chapter 8 will give illustrative examples for this model.

Air (or gas) drilling is a special case of the theory derived in Section 6.2. The governing equations for air (or gas) drilling operations can be obtained by setting $Q_m = 0$ in the equations derived in Section 6.2. The aforementioned assumption restricts the flow in the annulus to two-phase flow (gas and rock cuttings). Setting $Q_m = 0$ in Equation (6-25) yields

$$dP = \left[\frac{\dot{w}_t}{\left(\frac{P_g}{P}\right)\left(\frac{T_{av}}{T_g}\right) Q_g} \right] \left\{ 1 + \frac{f}{2g\,(D_h - D_p)} \left[\frac{\left(\frac{P_g}{P}\right)\left(\frac{T_{av}}{T_g}\right) Q_g}{\frac{\pi}{4}\left(D_h^2 - D_p^2\right)} \right]^2 \right\} dh, \tag{6-52}$$

where [see Equation (6-13)]

$$\dot{w}_t = \dot{w}_g + \dot{w}_s .$$

The exit pressure in direct circulation air (or gas) drilling operations is atmospheric pressure P_{at} at the top of the annulus. Separating variables in Equation (6-52) yields

$$\int_{P_{at}}^{P_{bh}} \frac{dP}{B_a(P)} = \int_0^H dh \tag{6-53}$$

where

$$B_a(P) = \left[\frac{\dot{w}_t}{\left(\frac{P_g}{P}\right)\left(\frac{T_{av}}{T_g}\right)Q_g}\right]\left\{1 + \frac{f}{2g(D_b - D_p)}\left[\frac{\left(\frac{P_g}{P}\right)\left(\frac{T_{av}}{T_g}\right)Q_g}{\frac{\pi}{4}(D_b^2 - D_p^2)}\right]^2\right\}.$$

Using Equations (6-6), (6-7), and (6-13), Equation (6-53) can be rearranged to give

$$\int_{P_{at}}^{P_{bb}} \frac{PdP}{(P^2 + b_a\, T_{av}^2)} = \frac{a_a}{T_{av}} \int_0^H dh, \tag{6-54}$$

where

$$a_a = \left(\frac{S_g}{\mathbf{R_e}}\right)\left[1 + \left(\frac{\dot{w}_s}{\dot{w}_g}\right)\right] \tag{6-55}$$

$$b_a = \frac{f}{2g\,(D_b - D_p)}\left(\frac{\mathbf{R_e}}{S_g}\right)^2 \frac{\dot{w}_g^2}{\left(\frac{\pi}{4}\right)^2 (D_b^2 - D_p^2)^2}. \tag{6-56}$$

In the form just given, both sides of Equation (6-54) can be integrated. Using the constants in Equations (6-55) and (6-56), the solution to Equation (6-54) is

$$\left|\frac{1}{2}\ln(P^2 + b_a\, T_{av}^2)\right|_{P_{at}}^{P_{bb}} = \frac{a_a}{T_{av}} |h|_0^H . \tag{6-57}$$

Evaluating Equation (6-57) at the limits and rearranging the results give

$$\ln\left[\frac{P_{bb}^2 + b_a T_{av}^2}{P_{at}^2 + b_a T_{av}^2}\right] = \frac{2a_a}{T_{av}} H. \tag{6-58}$$

Raising both sides of Equation (6-58) to the natural exponent e gives

$$\frac{P_{bb}^2 + b_a T_{av}^2}{P_{at}^2 + b_a T_{av}^2} = e^{\frac{2a_a H}{T_{av}}}. \tag{6-59}$$

Equation (6-59) can be rearranged and a solution obtained for P_{bb}. This is

$$P_{bb} = \left[(P_{at}^2 + b_a\, T_{av}^2)e^{\frac{2a_a H}{T_{av}}} - b_a\, T_{av}^2\right]^{0.5}. \tag{6-60}$$

The von Karman empirical correlation for wholly turbulent flow conditions can be used to determine the friction factor in Equation (6-56)[1]. This empirical expression is

$$f = \left[\frac{1}{2\log\left(\frac{D_b - D_p}{e}\right) + 1.14}\right]^2. \tag{6-61}$$

Equations (6-30), (6-55), (6-56), (6-60), and (6-61) can be used in sequential calculation steps starting at the top of the annulus and continuing for each subsequent change in cross-sectional area in the annulus until the bottom hole pressure is determined.

The flow condition through the drill bit orifices or nozzles is single phase (air or gas) flow. The character (sonic or subsonic) of the gas flow through the drill bit orifices or nozzles is determined by the critical pressure ratio equation. The critical pressure ratio equation for bottom hole conditions is

$$\left(\frac{P_{bb}}{P_{ai}}\right)_c = \left(\frac{2}{k+1}\right)^{\frac{k}{k-1}}, \tag{6-62}$$

where k is the ratio of specific heats for the gas.

If P_{ai} is determined to be

$$\text{for air } Pai \geq \frac{P_{bb}}{0.528} \tag{6-63}$$

$$\text{for natural gas } P_{ai} \geq \frac{P_{bb}}{0.549} \tag{6-64}$$

then the flow through the orifice or nozzle throat is sonic. Under these sonic flow conditions, the upstream pressure P_{ai} does not depend on downstream pressure P_{bb}. For sonic flow conditions through the nozzles the upstream pressure is

$$P_{ai} = \frac{\dot{w}_g T_{bb}^{0.5}}{A_n \left[\left(\frac{gkS_g}{\mathbf{R_e}}\right)\left(\frac{2}{k+1}\right)^{\left(\frac{k+1}{k-1}\right)}\right]^{0.5}}. \tag{6-65}$$

If the upstream pressure is less than the right-hand side of either Equation (6-63) or (6-64), the flow through the orifices or nozzles is subsonic and the upstream pressure will be dependent on the pressure and temperature at the bottom of the well (downstream). For these subsonic conditions the upstream pressure is

$$P_{ai} = P_{bb}\left[\frac{\left(\frac{\dot{w}_g}{A_n}\right)^2}{2g\left(\frac{k}{k-1}\right)P_{bb}\,\gamma_{bb}} + 1\right]^{\frac{k}{k-1}}. \tag{6-66}$$

The flow in the inside of the drill string is single phase flow (air or another gas). Setting $Q_m = 0$ in Equation (6-42) yields

$$dP = \left[\frac{\dot{w}_g}{\left(\frac{P_g}{P}\right)\left(\frac{T_{av}}{T_g}\right)Q_g}\right]\left\{1 - \frac{f}{2g\,D_i}\left[\frac{\left(\frac{P_g}{P}\right)\left(\frac{T_{av}}{T_g}\right)Q_g}{\frac{\pi}{4}D_i^2}\right]^2\right\}dh. \tag{6-67}$$

Separating variables in Equation (6-67) yields

$$\int_{P_{in}}^{P_{ai}} \frac{dP}{B_i(P)} = \int_0^H dh, \tag{6-68}$$

where

$$B_i(P) = \left[\frac{\dot{w}_g}{\left(\frac{P_g}{P}\right)\left(\frac{T_{av}}{T_g}\right)Q_g}\right]\left\{1 - \frac{f}{2g\,D_i}\left[\frac{\left(\frac{P_g}{P}\right)\left(\frac{T_{av}}{T_g}\right)Q_g}{\frac{\pi}{4}D_i^2}\right]^2\right\}.$$

Using Equations (6-6), (6-7), and (6-13), Equation (6-68) can be rearranged to give

$$\int_{P_{in}}^{P_{ai}} \frac{PdP}{(P^2 - b_i\,T_{av}^2)} = \frac{a_i}{T_{av}}\int_0^H dh, \tag{6-69}$$

where

$$a_i = \frac{S_g}{\mathbf{R_e}} \tag{6-70}$$

$$b_i = \frac{f}{2gD_i}\left(\frac{\mathbf{R_e}}{S_g}\right)^2 \frac{\dot{w}_g^2}{\left(\frac{\pi}{4}\right)^2 D_i^4}. \tag{6-71}$$

In the form just described, both sides of Equation (6-69) can be integrated. Using the constants in Equations (6-70) and (6-71), the solution to Equation (6-69) is

$$\left|\frac{1}{2}\ln(P^2 - b_iT_{av}^2)\right|_{P_{in}}^{P_{ai}} = \frac{a_i}{T_{av}}\,|\,h\,|_0^H\,. \tag{6-72}$$

Evaluating Equation (6-72) at the limits and rearranging the results give

$$\ln\left[\frac{P_{ai}^2 - b_iT_{av}^2}{P_{in}^2 - b_iT_{av}^2}\right] = \frac{2a_i}{T_{av}}H. \tag{6-73}$$

Raising both sides of Equation (6-73) to the natural exponent gives

$$\frac{P_{ai}^2 - b_iT_{av}^2}{P_{in}^2 - b_iT_{av}^2} = e^{\frac{2a_iH}{T_{av}}}. \tag{6-74}$$

Equation (6-74) can be rearranged and a solution obtained for P_{in}. This is

$$P_{in} = \left[\frac{P_{ai}^2 + b_iT_{av}^2\left(e^{\frac{2a_iH}{T_{av}}} - 1\right)}{e^{\frac{2a_iH}{T_{av}}}}\right]^{0.5}. \tag{6-75}$$

The von Karman empirical correlation can be used to determine the friction factor in Equation (6-71)[1]. This empirical expression is

$$f = \left[\frac{1}{2\log\left(\frac{D_i}{e}\right) + 1.14}\right]^2. \tag{6-76}$$

Equations (6-70), (6-71), (6-75), and (6-76) can be used in sequential calculation steps starting at the bottom of the inside of the drill string and continuing for each subsequent change in the cross-sectional area in the drill string until the surface injection pressure is determined.

REFERENCES

1. Daugherty, R. L., Franzini, J. B., and Finnemore, E. J., *Fluid Mechanics with Engineering Applications*, Eighth Edition, McGraw-Hill, 1985.
2. Brown, K. E., and Beggs, H. D., *The Technology of Artificial Lift Methods*, Vol. 1, PennWell Books, 1977.
3. Brown, K. E., et al, *The Technology of Artificial Lift Methods*, Vol. 2a, PennWell Books, 1980.
4. Personal communications with Stefan Miska, Department of Petroleum Engineering, University of Tulsa, January 1999.
5. Lapedes, D. H., *McGraw-Hill Encyclopedia of the Geological Sciences*, McGraw-Hill, 1978.
6. Kaminski, D. A., and Jensen, M. K., *Introduction to Thermal and Fluid Engineering*, Wiley and Sons, 2005.
7. Guo, B., Hareland, G., and Rajtar, J., "Computer Simulation Predicts Unfavorable Mud Rate and Optimum Air Injection Rate for Aerated Mud Drilling," *SPE Paper 26892*, Presented at the SPE Eastern Regional Conference and Exhibition, Pittsburgh, Pennsylvania, November 2–4, 1993.
8. Gatlin, C., *Petroleum Engineering: Drilling and Well Completions*, Prentice-Hall, 1960.
9. Bourgoyne, A. T., Millheim, K. K., Chenevert, M. E., and Young, F. S., *Applied Drilling Engineering*, SPE, First Printing, 1986.
10. Beyer, A. H., Millhone, R. S., and Foote, R. W., "Flow Behavior of Foam as a Well Circulating Fluid," *SPE Paper 3986*, Presented at the SPE 47th Annual Fall Meeting, San Antonio, Texas, October 8–11, 1972.
11. *Underbalanced Drilling Manual*, Gas Research Institute, GRI Reference No. GRI-97/0236, 1997.

CHAPTER

Reverse Circulation Models 7

In order to make reasonable predictions of the flow characteristics for direct circulation air and gas drilling operations, aerated fluids drilling operations, and stable foam drilling operations, it is necessary to derive a consistent theory that can be used, with certain simplifying limitations, to develop specific equations to model each of the aforementioned operations. All three basic drilling fluid circulation models, air and gas, aerated, and stable foam, must utilize a combination of mathematical theory and empirical correlations to develop a complete calculation model for each.

7.1 BASIC ASSUMPTIONS

Reverse circulation is defined as the injection of the drilling fluid into the top of the annulus space, the flow of the fluid down the inside of the annulus (between the inside of the casing or open hole), entrain the rock cuttings as the drilling fluid flows into the large opening in the drill bit, and then flow with the cuttings to the surface through the inside of the drill string.

Figure 7-1 shows a simplified U-tube schematic representation of reverse circulation flow. In general, in air and gas drilling operations, two-phase flow occurs in the inside of the annulus. Three-phase flow occurs when fluids with entrained rock cuttings pass through the large opening drill bit and then move up the inside of the drill string from the bottom of the well to the surface. The three phases are a compressible gas, an incompressible fluid, and the solid rock cuttings from the advance of the drill bit. The compressible gases used most in drilling are air, membrane generated nitrogen, nitrogen, or natural gas. The incompressible fluids used are treated fresh water, treated salt water (formation water), and water-based drilling muds. Diesel oil, oil-based drilling muds, and crude oil (formation oil) are somewhat compressible.

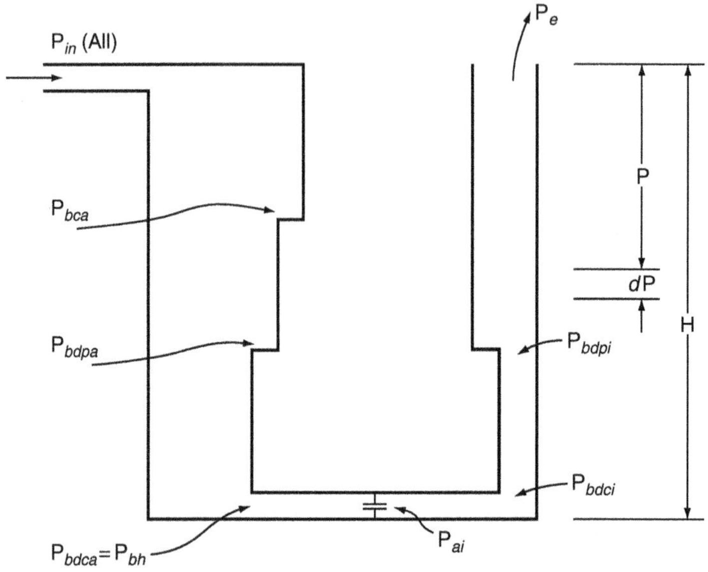

FIGURE 7-1. Schematic of reverse circulation. P_{in} is the injection pressure into the top of the annulus, P_{bca} is the pressure at the bottom of the casing in the annulus, P_{bdpa} is the pressure at the bottom of drill pipe in the annulus, P_{bdca} is the pressure at the bottom of drill collars in the annulus, P_{bh} is the bottom hole pressure in the annulus, P_{ai} is the pressure above the drill bit inside the drill string, P_{bdci} is the pressure at the bottom of drill collars inside the drill string, P_{bdpi} is the pressure at the bottom of the drill pipe inside the drill string, and P_e is the pressure at the exit at the top of the inside of the drill string.

It is assumed that compressible gases can be approximated by the perfect gas law. Further, it is assumed that the mixture of compressed gas and incompressible fluid will be uniform and homogeneous. When solid rock cuttings are added to the mixture of compressible gas and incompressible fluid, the solid rock particles are assumed to be uniform in size and density and will be distributed uniformly in the mixture of gas and fluid [1].

The assumption of uniformity of the two or three phases in the mixtures is an important issue in light of the technology developed for gas lift-assisted oil production [2, 3]. The aeration (or gasification) of oil (or other formation produced fluids) from the bottom of a well with the flow of gas from the surface (down the annulus between the casing and the production tubing) is basically the same as the aeration of drilling fluid and rock cuttings from the bottom of a well to the surface through the inside of the drill string using a flow of gas and fluid from the surface (down the annulus). However, in most oil production situations the two-phase flow takes place inside of the tubing [4].

7.2 GENERAL DERIVATION

The term P_{in} represents the pressure of the injected drilling fluid into the top of the annulus. The U-tube representation in Figure 7-1 shows the largest annulus space between the outside of the drill pipe and the inside of the casing. Next is the annulus space between the outside of the drill pipe and the inside of the open hole. Then at the bottom of the annulus is the space between the outside of the drill collars and the inside of the open hole. At the bottom of the drill string is the single large opening in the drill bit that allows the drilling fluids with entrained rock cuttings to pass into the inside of the drill string. The schematic shows the smaller inside diameter of the drill collars. Above the drill collars is the larger inside diameter of the drill pipe. At the top of the drill pipe the drilling fluid with the entrained cuttings exit the circulation system at pressure P_e.

As in all compressible flow problems, the process of solution must commence with a known pressure and temperature and, in this case, the pressure and temperature at the exit. Therefore, the derivation will begin with the analysis of the inside of the drill string. Figure 7-1 shows pressure P at any position in the inside of the drill string, which is referenced from the surface to a depth h. The total depth of the well is H. The differential pressure dP in the upward flowing three-phase flow occurs over an incremental distance of dh. This differential pressure can be approximated as [1]

$$dP = \gamma_{mix}\left[1 + \frac{f\,V^2}{2g\,D_i}\right]dh, \tag{7-1}$$

where P is fluid pressure (lb/ft^2 abs, N/cm^2 abs); h is the reference depth (ft, m); H is the total depth (ft, m); γ_{mix} is the specific weight of the mixture of air (or other gas), incompressible fluid, and rock cuttings (lb/ft^3, N/m^3); f is the Darcy–Weisbach friction factor; V is the average velocity in the annulus (ft/sec, m/sec); D_i is the inside diameter of the drill string (ft, m); and g is the acceleration of gravity (32.2 ft/sec^2, 9.81 m/sec^2).

The first term on the right side of Equation (7-1) represents the incremental pressure change due to the hydrostatic weight of the column of fluids (with entrained rock cuttings) inside the drill string. The second term on the right side of Equation (7-1) represents the incremental pressure change due to the friction resistance to the flowing fluids in the drill string.

7.2.1 Weight Rate of Flow of the Gas

In order to carry out derivation of the governing equations for reverse circulation, the weight rate of flow of air (or gas) to the well must be determined. Assuming that the compressed air is provided by a compressor(s), the weight rate of flow through the circulating system is determined from the atmospheric pressure and temperature of the air at the compressor location on the surface of the earth, and the characteristics of the compressor(s). For air, the atmospheric pressure and average annual temperatures for sea level and various elevations above sea level can be approximated for most of North America by midlatitudes data given in Tables 5-1a and 5-1b. These reference pressures are denoted as p_r and t_r. For engineering calculations, the actual atmospheric pressure and temperature of the air entering a primary compressor(s) are p_{at} and t_{at} and can be approximated as

$$p_{at} \approx p_r \tag{7-2}$$

$$t_{at} \approx t_r, \tag{7-3}$$

where p_{at} is atmospheric pressure (psia, N/cm^2 abs), p_r is the reference atmospheric pressure (psia, N/cm^2 abs), t_{at} is atmospheric pressure (psia, N/cm^2 abs), and t_r is the reference atmospheric temperature (°F, °C). Also see Appendix B for more detailed plots for atmospheric conditions for North America midlatitudes. Similar data as that given in Tables 5-1a and 5-1b and Appendix B for North America midlatitudes can be obtained for most other continents and latitudes around the world.

The aforementioned pressures and temperatures in field units must be converted to units consistent with Equations (5-11) and (7-1). These are

$$P_g = P_r \tag{7-4}$$

$$T_g = T_r, \tag{7-5}$$

where P_r is the reference atmospheric pressure (lb/ft^2 abs, N/m^2 abs) and T_r is the absolute atmospheric reference temperature (°R, K).

Substituting Equations (7-2) and (7-3) into Equation (5-11), the specific weight of the gas (air) γ_g entering the compressor(s) is

$$\gamma_g = \frac{P_g\, S_g}{\mathbf{R_e}\, T_g} = \frac{P_r\, S_g}{\mathbf{R_e}\, T_r}, \tag{7-6}$$

where γ_g is specific weight (lb/ft^3, N/m^3), $\mathbf{R_e}$ is the engineering gas constant (53.36 ft-lb/lb-°R, 29.31 N-m/N-K), and S_g is the specific gravity of the gas (S_g = 1.0 for air at standard conditions). The weight rate of flow of the gas, $\dot{w}_g$ through the compressor is

$$\dot{w}_g = \gamma_g Q_g, \tag{7-7}$$

where $\dot{w}_g$ is the weight rate of flow of gas (lb/sec, N/sec) and Q_g is the volumetric flow rate of air into the circulation system (actual ft^3/sec, actual m^3/sec).

If the gas for the drilling operation is natural gas from a pipeline and the pressure and temperature of the gas in the pipeline are p_{pl} and t_{pl}, then the pressure and temperature of the gas entering either directly into the circulation system or to the booster compressor are

$$P_g = P_{pl} \tag{7-8}$$

$$T_g = T_{pl}, \tag{7-9}$$

where p_{pl} is the pipeline pressure (psia, N/cm^2 abs), P_{pl} is the pipeline pressure (lb/ft^2 abs, N/m^2 abs), t_{pl} is the pipeline temperature (°F, °C), and T_{pl} is the absolute pipeline temperature (°R, K).

Substituting Equations (7-8) and (7-9) into Equation (5-11), the specific weight of the gas from a pipeline can be obtained. This is

$$\gamma_g = \frac{P_g S_g}{\mathbf{R_e}\, T_g} = \frac{P_{pl} S_g}{\mathbf{R_e}\, T_{pl}}. \tag{7-10}$$

In this equation, S_g would be the specific gravity of the pipeline natural gas (e.g., usually between 0.65 and 0.85).

Substituting the result from Equation (7-10) into Equation (7-7) gives the weight rate of flow of gas from a pipeline, where Q_g is the volumetric flow rate of natural gas from the pipeline at pressure p_{pl} and temperature t_{pl}. Note that the volumetric flow rate in a pipeline is usually given by flow meters in either scfm or standard m^3/sec regardless of surface elevation location. This value must be converted to obtain the actual volumetric flow rate at p_{pl} and t_{pl} (see Appendix A). As discussed in Chapter 5, care must be taken to determine which set of standard conditions are being used to define the "scfm or standard m^3/sec." Chapter 5 gives a summary of the most common standard condition specifications used throughout the industrialized world.

7.2.2 Three-Phase Flow in the Drill String

This general solution for three-phase flow is valid for aerated (gasified) drilling fluids where the three phases in the annulus are gas, incompressible fluid, and solids (cuttings).

The weight rate of flow of incompressible drilling fluid (usually drilling mud), $\dot{w}_m$, into the well is

$$\dot{w}_m = \gamma_m\, Q_m, \tag{7-11}$$

where $\dot{w}_m$ is the weight rate of flow of drilling mud (lb/sec, N/sec), γ_m is the specific weight of the drilling mud (lb/ft^3, N/m^3), and Q_m is the volumetric flow rate of drilling mud (ft^3/sec, m^3/sec).

The weight rates of flow $\dot{w}_g$ and $\dot{w}_m$ enter the well through the top of the annulus and flow to the bottom of the annulus and exit the bottom hole through the large opening in the drill bit into the inside of the drill bottom of the drill string just above the drill bit. As the flow passes through the drill bit, the fluids entrain the rock cuttings generated by the drill bit as the bit is advanced. The entrained weight rate of flow of the solids, $\dot{w}_s$, is

$$\dot{w}_s = \frac{\pi}{4}\, D_b^2 \gamma_w (2.7) \kappa, \tag{7-12}$$

where $\dot{w}_s$ is the weight rate of flow of solid rock cuttings (lb/sec, N/sec), D_b is the diameter of the drilled hole (i.e., the bit diameter) (ft, m), γ_w is the specific weight of the fresh water (lb/ft^3, N/m^3), and κ is the penetration rate (ft/sec, m/sec).

For USCS units, in Equation (7-12), the specific weight of fresh water is 62.4 lb/ft^3. For SI units, in Equation (7-12), the specific weight of fresh water is the density of fresh water, 1000 kg/m^3, multiplied by the acceleration of gravity in the SI units 9.81 m/sec^2. The average specific gravity of sedimentary rocks is approximately 2.7. If igneous or metamorphic rocks are to be drilled, average values of 2.80 and 3.00, respectively, can be used [5].

The total weight rate of flow, $\dot{w}_t$, inside the drill string and flowing from the bottom of the well to the surface is

$$\dot{w}_t = \dot{w}_g + \dot{w}_m + \dot{w}_s. \tag{7-13}$$

The drilling mud and the rock cutting solids are assumed to not change volume when pressure is changed (note that if the liquid phase fluid is oil, the volume changes with pressure can be taken into account). However, the air (or gas) does change volume as a function of pressure change and, therefore, as a function of depth. Thus, the specific weight of the gas at any position inside the drill string is

$$\gamma_g = \frac{PS_g}{\mathbf{R_e}\, T_{av}}, \tag{7-14}$$

where T_{av} is the average temperature of the gas over a depth interval (°R, K). This average temperature term is determined by taking the average of the sum of the geothermal temperatures at the top and bottom of the depth interval. The geothermal temperature at depth t_b is determined from the approximate expression

$$t_b = t_r + \beta H, \tag{7-15}$$

where t_r is the average annual atmospheric reference temperature (°F, °C), t_b is the geothermal temperature at depth (°F, °C), and β is the geothermal temperature gradient (°F/ft, °C/m).

The reference surface geothermal temperature, t_r, is assumed to be the temperatures given in Tables 5-1a and 5-1b for sea level and various elevations above sea level. These temperatures represent North American midlatitude year-round averages. It is assumed that these temperatures also represent an average constant of deep soil or rock temperatures near the surface of the earth at the elevations given in the table. The value of the geothermal gradient constant is determined from temperature logs of offset wells and other geophysical data. An average value of the geothermal gradient that can be used when the actual gradient has not been determined is 0.01°F/ft, or 0.018°C/m. The temperature at depth can be expressed as absolute temperatures using the following:

$$T_b = T_r + \beta H, \tag{7-16}$$

where T_r is the reference atmospheric temperature (°R, K). Once the reference temperature is changed to absolute, no other changes need to be made in Equation (7-16).

The absolute average temperature T_{av} over the first depth interval below the surface is

$$T_{av1} = \frac{T_r + T_{b1}}{2}. \tag{7-17}$$

The T_{av} for follow-on intervals will be the average of the absolute temperature at the top and the absolute temperature at the bottom of the interval. Follow-on average temperatures will be

$$T_{av2} = \frac{T_{b1} + T_{b2}}{2} \ldots, \tag{7-18}$$

where T_{b1} is the temperature at the bottom of the first interval (°R, K) and T_{b2} is the temperature at the bottom of the second interval (°R, K). Follow-on T_{av} interval temperatures are determined in sequence in a similar method as shown above.

The relationship between the weight rate of flow of the gas and the specific weight and volumetric flow rate of gas at any position inside the drill string is given by

$$\dot{w}_g = \gamma_g \, Q_g = \gamma Q. \tag{7-19}$$

Substituting Equations (7-6) and (7-14) into the two terms on the right side of Equation (7-19) gives a relationship between the specific weight and volumetric flow rate at the surface and the specific weight of volumetric flow rate at any position inside the drill string. This is

$$\frac{P_g \, S_g}{\mathbf{R_e} \, T_g} Q_g = \frac{P S_g}{\mathbf{R_e} \, T_{av}} Q. \tag{7-20}$$

Solving Equation (7-20) for Q yields

$$Q = \left(\frac{P_g}{P}\right)\left(\frac{T_{av}}{T_g}\right) Q_g. \tag{7-21}$$

The three-phase flow of gas, incompressible fluid, and rock cuttings up the inside of the drill string can be described by a mixed specific weight term, which is a function of its position in the drill string. This mixed specific weight, γ_{mix}, is

$$\gamma_{mix} = \frac{\dot{w}_t}{\left(\frac{P_g}{P}\right)\left(\frac{T_{av}}{T_g}\right) Q_g + Q_m}. \tag{7-22}$$

In the derivation of Equation (7-22), the volume of the solids (the rock cuttings) is assumed to be small and negligible relative to the volumes of the gas and the incompressible fluid in the mixture (i.e., contributes only to the $\dot{w}_t$ term).

The velocity of this mixture changes as a function of its position inside the drill string. The velocity, V, of the three-phase flow inside the drill string is

$$V = \frac{Q + Q_m}{\frac{\pi}{4} D_i^2}, \tag{7-23}$$

where D_i is the inside diameter of the drill string (ft).

Substituting Equation (7-21) into Equation (7-23) yields

$$V = \frac{\left(\frac{P_g}{P}\right)\left(\frac{T_{av}}{T_g}\right)Q_g + Q_m}{\frac{\pi}{4}D_i^2}. \tag{7-24}$$

Substituting Equations (7-22) and (7-24) into Equation (7-1) yields

$$dP = \left[\frac{\dot{w}_t}{\left(\frac{P_g}{P}\right)\left(\frac{T_{av}}{T_g}\right)Q_g + Q_m}\right] \left\{1 + \frac{f}{2gD_i}\left[\frac{\left(\frac{P_g}{P}\right)\left(\frac{T_{av}}{T_g}\right)Q_g + Q_m}{\frac{\pi}{4}D_i^2}\right]^2\right\} dh. \tag{7-25}$$

Equation (7-25) contains only two independent variables: P and h. All of the other terms in the equation are known constants. Separating variables in Equation (7-25) and integrating from the exit (at the surface) to the bottom of the inside of the drill string yields

$$\int_{P_e}^{P_{at}} \frac{dP}{B_i(P)} = \int_0^H dh, \tag{7-26}$$

where P_e is the exit pressure at the top of the inside of the drill string (lb/ft^2 abs, N/m^2 abs), P_{at} is the pressure above the bit inside the bottom of the drill string (lb/ft^2 abs, N/m^2 abs), and

$$B_i(P) = \left[\frac{\dot{w}_t}{\left(\frac{P_g}{P}\right)\left(\frac{T_{av}}{T_g}\right)Q_g + Q_m}\right] \left\{1 + \frac{f}{2gD_i}\left[\frac{\left(\frac{P_g}{P}\right)\left(\frac{T_{av}}{T_g}\right)Q_g + Q_m}{\frac{\pi}{4}D_i^2}\right]^2\right\}.$$

For this general derivation, exit pressure, P_e, is atmospheric pressure at the end of the blooey line from the top of the inside of the drill string (in the case of air or gas drilling) and at the end of the return flow line from the top of the inside of the drill string (in the case of aerated fluid drilling or stable foam drilling).

The Darcy-Weisbach friction factor f given in the aforementioned equation is usually determined by standard fluid mechanics empirical correlations relating the friction factor to the Reynolds number, diameter, and absolute pipe roughness.

In general, the values for Reynolds number, diameter, and absolute pipe roughness are known. The classic correlation for the Reynolds number is

$$\mathbf{N_R} = \frac{D_i V}{\nu}, \tag{7-27}$$

where D_i is the inside diameter for the drill string (ft, m), V is the velocity (ft/sec, m/sec), and ν is the kinematic viscosity of the drilling fluid (ft^2/sec, m^2/sec).

Three flow conditions can exist inside the drill string. These are laminar, transitional, and turbulent [1].

The empirical correlation for the friction factor for laminar flow conditions is

$$f = \frac{64}{\mathbf{N_R}}. \tag{7-28}$$

This equation can be solved directly once the Reynolds number is known. In general, Equation (7-28) is valid for values for Reynolds numbers from 0 to 2000.

Up until recently it was necessary to use the Colebrook empirical correlation for transitional flow conditions and the von Karman empirical correlation for the wholly turbulent flow conditions to obtain an analytic value for the friction factor. These empirical correlations required difficult trial and error calculations. A new empirical correlation for the friction factor can be used for both transitional flow conditions and wholly turbulent flow conditions (for Reynolds numbers greater than 2000). This empirical correlation is the Haaland correlation [6]. This empirical expression is

$$f = \left[\frac{1}{-1.8 \log \left[\left(\frac{\left(\frac{e_p}{D_p} \right)}{3.7} \right)^{1.11} + \frac{6.9}{N_R} \right]} \right]^2. \tag{7-29}$$

For follow-on calculations for the flow inside the steel drill string, the absolute roughness for steel pipe is approximately $e_p = 0.0005$ ft or $e_p = 0.0002$ m [1].

Equations (7-26) through (7-29) can be used in sequential trial and error integration steps starting at the top of the inside of the drill string (with the known exit pressure) and continuing for each subsequent geometry change in the drill string inside the cross-sectional area until the pressure at the bottom of the inside of the drill string is determined. These sequential calculation steps require trial and error solutions. The trial and error process requires selection of the upper limit of pressure in each integral on the right side of Equation (7-26). This upper limit pressure selection must give a left side integral solution that is equal to the right side integral solution.

7.2.3 Three-Phase Flow Through the Bit

There are three basic calculation techniques for determining the pressure change through the constriction of the single drill bit (or water course) orifice.

The first technique assumes that the mixture of incompressible fluid, gas, and rock cuttings passes through the single orifice. Under these conditions, the mixture is assumed to act as an incompressible fluid. Thus, borrowing from mud drilling technology, the pressure change through the drill bit, ΔP_b, can be approximated by [7, 8]

$$\Delta P_b = \frac{\dot{w}_t^2}{2g\, \gamma_{mixai} C^2 \left(\frac{\pi}{4}\right)^2 D_{bi}^4}, \qquad (7\text{-}30)$$

where ΔP_b is pressure change (lb/ft^2, N/cm^2), γ_{mixai} is the mixture-specific weight above the drill bit inside the drill string (lb/ft^3, N/m^3), C is the fluid flow loss coefficient for drill bit orifices or nozzles (the value of this constant is dependent on the flow components), and D_{bi} is the drill bit single orifice inside diameter (ft, m). The pressure change obtained from Equation (7-30) is subtracted from the pressure above the drill bit inside the drill string P_{ai} obtained from Equation (7-26). The annulus bottom hole pressure P_{bb} is

$$P_{bb} = P_{ai} - \Delta P_b, \qquad (7\text{-}31)$$

where P_{bb} is bottom hole pressure (lb/ft^2 abs, N/cm^2 abs) and P_{ai} is pressure above the drill bit inside the drill string (lb/ft^2 abs, N/cm^2 abs).

For fluid mixtures that are nearly all gas (with little incompressible fluid) and subsonic flow conditions, the pressure above the drill bit inside the drill string P_{ai} can be determined from [1]

$$2g\left(\frac{k}{k-1}\right)P_{bb}\,\gamma_{bb}\left[\left(\frac{P_{ai}}{P_{bb}}\right)^{\frac{2}{k}} - \left(\frac{P_{ai}}{P_{bb}}\right)^{\frac{k+1}{k}}\right] = \left(\frac{\dot{w}_g}{A_{bi}}\right)^2, \qquad (7\text{-}32)$$

where k is the ratio of specific heats for the gas (e.g., for air $k = 1.4$ and for natural gas $k = 1.28$), γ_{bb} is the specific weight of the gas at the bottom of the annulus (lb/ft^3, N/m^3), and A_{bi} is the cross-sectional area of the single drill bit orifice (ft^2, m^2). Equation (7-32) must be solved by trial and error for P_{bb}.

The equations just given will generally yield results that show that the annulus bottom hole pressure P_{bb} differs very little from the pressure above the drill bit inside the drill string P_{ai} for most practical parameters. Therefore, it can usually be assumed that

$$P_{bb} \approx P_{ai}. \qquad (7\text{-}33)$$

7.2.4 Two-Phase Flow in the Annulus

The downward flow condition in the annulus is two phase. Differential pressure, dP, occurs over the incremental distance of dh. This differential pressure can be approximated as

$$dP = \gamma_{mix}\left[1 - \frac{fV^2}{2g(D_h - D_p)}\right]dh, \tag{7-34}$$

where D_h is the inside diameter of the annulus (ft, m)(the borehole) and D_p is the outside diameter of the drill string (pipe and collars) (ft, m).

The two-phase flow of gas and incompressible fluid down the annulus space to the bottom of the well can be described by a mixed specific weight term, which is a function of position in the annulus. This mixed specific weight term is

$$\gamma_{mix} = \frac{\dot{w}_g + \dot{w}_m}{\left(\frac{P_g}{P}\right)\left(\frac{T_{av}}{T_g}\right)Q_g + Q_m}. \tag{7-35}$$

The velocity of this mixture changes as a function of its position in the annulus. The velocity of the two phase flow in the annulus is

$$V = \frac{\left(\frac{P_g}{P}\right)\left(\frac{T_{av}}{T_g}\right)Q_g + Q_m}{\frac{\pi}{4}\left(D_h^2 - D_p^2\right)}. \tag{7-36}$$

Substituting Equations (7-35) and (7-36) into Equation (7-34) yields

$$dP = \left[\frac{\dot{w}_g + \dot{w}_m}{\left(\frac{P_g}{P}\right)\left(\frac{T_{av}}{T_g}\right)Q_g + Q_m}\right]$$
$$\left\{1 - \frac{f}{2g(D_h - D_p)}\left[\frac{\left(\frac{P_g}{P}\right)\left(\frac{T_{av}}{T_g}\right)Q_g + Q_m}{\frac{\pi}{4}\left(D_h^2 - D_p^2\right)}\right]^2\right\}dh. \tag{7-37}$$

Equation (7-37) contains only two independent variables: P and h. Separating variables in Equation (7-37) and integrating from the injection (at the surface) to the bottom of the annulus yields

$$\int_{P_{in}}^{P_{bh}} \frac{dP}{B_a(P)} = \int_0^H dh, \tag{7-38}$$

where

$$B_a(P) = \left[\frac{\dot{w}_g + \dot{w}_m}{\left(\frac{P_g}{P}\right)\left(\frac{T_{av}}{T_g}\right)Q_g + Q_m} \right]$$

$$\left\{ 1 - \frac{f}{2g(D_b - D_p)} \left[\frac{\left(\frac{P_g}{P}\right)\left(\frac{T_{av}}{T_g}\right)Q_g + Q_m}{\frac{\pi}{4}\left(D_b^2 - D_p^2\right)} \right]^2 \right\},$$

where P_{in} is the injection pressure into the top of the annulus space (lb/ft^2 abs, N/m^2 abs).

The friction factor f given in the equation just shown is determined by the standard fluid mechanics empirical correlations relating the friction factor to the Reynolds number, diameter (or hydraulic diameter), and absolute pipe roughness. In general, the values for Reynolds number, diameter, and absolute pipe roughness are known. The classic correlation for the Reynolds number is

$$\mathbf{N_R} = \frac{(D_b - D_p)V}{\nu}, \tag{7-39}$$

where D_b - D_p is the hydraulic diameter for the annulus (ft, m).

Three flow conditions can exist in the annulus. These are laminar, transitional, and turbulent.

The empirical correlation for the friction factor for laminar flow conditions is

$$f = \frac{64}{\mathbf{N_R}}. \tag{7-40}$$

This equation can be solved directly once the Reynolds number is known. Equation (7-40) is valid for values for Reynolds numbers from 0 to 2000.

The empirical correlation for the friction factor for both transitional flow conditions and wholly turbulent flow conditions (for Reynolds numbers greater than 2000) can be determined from the Haaland correlation [6]. This empirical expression is

$$f = \left[\frac{1}{-1.8 \log \left[\left(\frac{\left(\frac{e_{av}}{D_b - D_p}\right)}{3.7} \right)^{1.11} + \frac{6.9}{N_R} \right]} \right]^2, \tag{7-41}$$

where ε_{av} is the approximate average absolute roughness of the annulus surfaces (ft, m). Note that the logarithm in the equation is to the base 10.

Equation (7-42) gives the approximation for ε_{av} for the open hole section of the annulus. This approximation is

$$e_{av} = \frac{e_r\, D_{ob}^2 + e_p\, D_p^2}{D_{ob}^2 + D_p^2}. \tag{7-42}$$

For follow-on calculations for flow in the annulus, the absolute roughness for steel pipe, $e_p = 0.0005$ ft or $e_p = 0.0002$ m, will be used for the outside surfaces of the drill pipe and drill collars, and the inside surface of the casing. The open hole surfaces of boreholes can be approximated with an absolute roughness $\varepsilon_{ob} = 0.01$ ft or $\varepsilon_{ob} = 0.003$ m (i.e., this example value is the same as concrete pipe, which approximates borehole surfaces in limestone and dolomite sedimentary rocks, or in similar competent igneous and metamorphic rocks, see Table 8-1).

Equations (7-38) through (7-42) can be used in sequential integration steps starting at the top of the annulus (with the known exit pressure) and continuing for each subsequent change in the annulus cross-sectional area until the bottom hole pressure is determined. These sequential calculation steps require trial and error solutions. The trial and error process requires selection of the upper limit of the pressure in each integral on the right side of Equation (7-38). This upper limit pressure selection must give a left side integral solution that is equal to the right side integral solution.

7.3 AERATED FLUID DRILLING MODEL

Aerated (or gasified) drilling fluid governing equations are changed very little from the direct circulation general derivation given in Section 7.2. The gases used in aerated fluid drilling are usually either air or membrane generated nitrogen (air stripped of oxygen). The fluids used are usually drilling mud, diesel oil, or formation oil.

The basic mathematical model described in Section 7.2 must be augmented with specialized empirically derived correlation models that take into account changes in flow viscosity and liquid holdup experienced in actual aerated drilling operations [2–4]. In particular, these empirical additions to the mathematical model demonstrate the origins of the increased injection and bottom hole pressures experienced in field operations.

Equations (7-11) through (7-29) describe the flow of aerated drilling fluids in the annulus.

Equations (7-30) through (7-33) describe the flow of aerated drilling fluids through the drill bit orifices or nozzles.

Equations (7-34) through (7-42) describe the flow of aerated drilling fluids through the inside of the drill string.

7.4 STABLE FOAM DRILLING MODEL

As stated earlier, reverse circulation operations are useful in drilling large diameter shallow surface casing boreholes. Stable foam drilling operations are not useful for this type of drilling operation. Stable foam drilling is used in reverse circulation operations in shallow moderate diameter wells only (e.g., depths less that approximately 3000 ft or 900 m), such as deep water wells and other nonpetroleum industry-related wells. The reason for this limitation is because the reverse circulation drill bit with its large open orifice does not create the high shear rate on the drilling fluid needed to generate the stable foam as the foam enters the bottom of the drill string with the rock cuttings. Therefore, the stable foam must be generated at the surface and injected into the top of the annulus; this foam must not break down as it circulates through the entire system. This is not possible in deep large diameter boreholes.

However, it should be noted that stable foam reverse circulation operations are used in petroleum industry deep well work over and related production operations. These operations are feasible because these operations are carried out in moderate diameter wells (e.g., in production casing and production tubing strings).

7.5 AIR AND GAS DRILLING MODEL

Unlike the aerated and stable foam drilling fluid models, the air and gas drilling model requires no special empirical correlations to adjust the results to provide additional results that agree more closely to field data. Chapter 8 will give illustrative examples for this model.

Air (or gas) drilling is a special case of the theory derived in Section 7.2. The governing equations for air (or gas) drilling operations can be obtained by setting $Q_m = 0$ in the equations derived in Section 7.2. The aforementioned assumption restricts the flow in the annulus to two-phase flow (gas and rock cuttings). Setting $Q_m = 0$ in Equation (7-25) yields

$$dP = \left[\frac{\dot{w}_t}{\left(\frac{P_g}{P}\right)\left(\frac{T_{av}}{T_g}\right)Q_g}\right] \left\{1 + \frac{f}{2gD_i}\left[\frac{\left(\frac{P_g}{P}\right)\left(\frac{T_{av}}{T_g}\right)Q_g}{\frac{\pi}{4}D_i^2}\right]^2\right\} dh, \tag{7-43}$$

where

$$\dot{w}_t = \dot{w}_g + \dot{w}_s.$$

The exit pressure in reverse circulation air (or gas) drilling operations is atmospheric pressure, P_{at}, at the top of the inside of the drill string. Separating variables in Equation (7-43) yields

$$\int_{P_{at}}^{P_{ai}} \frac{dP}{B_i(P)} = \int_0^H dh, \tag{7-44}$$

where

$$B_i(P) = \left[\frac{\dot{w}_t}{\left(\frac{P_g}{P}\right)\left(\frac{T_{av}}{T_g}\right)Q_g}\right]\left\{1 + \frac{f}{2g\,D_i}\left[\frac{\left(\frac{P_g}{P}\right)\left(\frac{T_{av}}{T_g}\right)Q_g}{\frac{\pi}{4}D_i^2}\right]^2\right\}.$$

Using Equations (7-6), (7-7), and (7-13), Equation (7-44) can be rearranged to give

$$\int_{P_{at}}^{P_{ai}} \frac{PdP}{\left(P^2 + b_i T_{av}^2\right)} = \frac{a_i}{T_{av}} \int_0^H dh, \tag{7-45}$$

where

$$a_i = \left(\frac{S}{\mathbf{R_e}}\right)\left[1 + \left(\frac{\dot{w}_s}{\dot{w}_g}\right)\right] \tag{7-46}$$

$$b_i = \frac{f}{2g\,D_i}\left(\frac{\mathbf{R_e}}{S}\right)^2 \frac{\dot{w}_g^2}{\left(\frac{\pi}{4}\right)^2 D_i^4}. \tag{7-47}$$

In this form, both sides of Equation (7-45) can be integrated. Using Equations (7-46) and (7-47), the solution to Equation (7-45) is

$$\left|\frac{1}{2}\ln\left(P^2 + b_i T_{av}^2\right)\right|_{P_{at}}^{P_{ai}} = \frac{a_i}{T_{av}}\,|h|_0^H. \tag{7-48}$$

Evaluating Equation (7-48) at the limits and rearranging the results give

$$\ln\left[\frac{P_{ai}^2 + b_i\,T_{av}^2}{P_{at}^2 + b_i\,T_{av}^2}\right] = \frac{2a_i}{T_{av}}H. \tag{7-49}$$

Raising both sides of Equation (7-49) to the natural exponential exponent gives

$$\frac{P_{ai}^2 + b_i T_{av}^2}{P_{at}^2 + b_i T_{av}^2} = e^{\frac{2a_i H}{T_{av}}}. \tag{7-50}$$

Equation (7-50) can be rearranged and a solution obtained for P_{ai}. This will yield

$$P_{ai} = \left[\left(P_{at}^2 + b_i T_{av}^2\right)e^{\frac{2a_i H}{T_{av}}} - b_i T_{av}^2\right]^{0.5}. \tag{7-51}$$

The von Karman empirical correlation for wholly turbulent flow conditions can be used to determine the friction factor in Equation (7-47)[1]. This correlation is

$$f = \left[\frac{1}{2\log\left(\frac{D_i}{e}\right) + 1.14} \right]^2. \tag{7-52}$$

For follow-on calculations for flow in the drill string, the absolute roughness for commercial steel drill pipe, $e_p = 0.0005$ ft or $e_p = 0.0002$ m, will be used for the inside surfaces of the drill pipe and drill collars.

Equations (7-46), (7-47), (7-51), and (7-52) can be used in sequential calculation steps starting at the top of the inside of the drill string and continuing for each subsequent change in cross-sectional area in the inside of the drill string until the pressure above the drill bit inside of the drill string is determined.

There is a single water course in reverse circulation drill bits. Using Equation (7-53), the pressure at the bottom of the inside of the drill string P_{ai} is obtained. The pressure at the bottom of the annulus P_{bb} can be obtained by trial and error using the expression

$$2g\left(\frac{k}{k-1}\right)P_{bb}\,\gamma_{bb}\left[\left(\frac{P_{ai}}{P_{bb}}\right)^{\frac{2}{k}} - \left(\frac{P_{ai}}{P_{bb}}\right)^{\frac{k+1}{k}}\right] = \left(\frac{\dot{w}_g}{A_{bi}}\right)^2. \tag{7-53}$$

These equations will generally yield results that show that the annulus bottom hole pressure P_{bb} differs very little from the pressure above the drill bit inside the drill string P_{ai} for most practical parameters. Therefore, it can usually be assumed that

$$P_{bb} \approx P_{ai}. \tag{7-54}$$

The flow in the annulus is single-phase flow (air or gas). Setting $Q_m = 0$ in Equation (7-37) yields

$$dP = \left[\frac{\dot{w}_g}{\left(\frac{P_g}{P}\right)\left(\frac{T_{av}}{T_g}\right)Q_g} \right] \left\{ 1 - \frac{f}{2g\left(D_h - D_p\right)} \left[\frac{\left(\frac{P_g}{P}\right)\left(\frac{T_{av}}{T_g}\right)Q_g}{\frac{\pi}{4}\left(D_h^2 - D_p^2\right)} \right]^2 \right\} dh. \tag{7-55}$$

Separating variables in Equation (7-59) yields

$$\int_{P_{in}}^{P_{bb}} \frac{dP}{B_a(P)} = \int_0^H dh, \tag{7-56}$$

where

$$B_a(P) = \left[\frac{\dot{w}_g}{\left(\frac{P_g}{P}\right)\left(\frac{T_{av}}{T_g}\right)Q_g}\right]\left\{1 - \frac{f}{2g(D_b - D_p)}\left[\frac{\left(\frac{P_g}{P}\right)\left(\frac{T_{av}}{T_g}\right)Q_g}{\frac{\pi}{4}\left(D_b^2 - D_p^2\right)}\right]^2\right\}.$$

Using Equations (7-6), (7-7), and (7-14), Equation (7-56) can be rearranged to give

$$\int_{P_{in}}^{P_{bb}} \frac{PdP}{\left(P^2 - b_a T_{av}^2\right)} = \frac{a_a}{T_{av}} \int_0^H db, \tag{7-57}$$

where

$$a_a = \frac{S}{\mathbf{R_e}} \tag{7-58}$$

$$b_a = \frac{f}{2g(D_b - D_p)}\left(\frac{\mathbf{R_e}}{S}\right)^2 \frac{\dot{w}_g^2}{\left(\frac{\pi}{4}\right)^2\left(D_b^2 - D_p^2\right)^2}. \tag{7-59}$$

In this form, both sides of Equation (7-57) can be integrated. Using Equations (7-58) and (7-59), the solution to Equation (7-57) is

$$\left|\frac{1}{2}\ \ln\left(P^2 - b_a T_{av}^2\right)\right|_{P_{in}}^{P_{bb}} = \frac{a_a}{T_{av}} |b|_0^H. \tag{7-60}$$

Evaluating Equation (7-60) at the limits and rearranging the results give

$$\ln\left[\frac{P_{bb}^2 - b_a\ T_{av}^2}{P_{in}^2 - b_a\ T_{av}^2}\right] = \frac{2a_a}{T_{av}} H. \tag{7-61}$$

Raising both sides of Equation (7-63) to the natural exponential exponent gives

$$\frac{P_{bb}^2 - b_a T_{av}^2}{P_{in}^2 - b_a T_{av}^2} = e^{\frac{2a_a H}{T_{av}}}. \tag{7-62}$$

Equation (7-62) can be rearranged and a solution obtained for P_{in}. This is

$$P_{in} = \left[\frac{P_{ai}^2 + b_i T_{av}^2\left(e^{\frac{2a_i H}{T_{av}}} - 1\right)}{e^{\frac{2a_i H}{T_{av}}}}\right]^{0.5}. \tag{7-63}$$

The von Karman empirical correlation can be used to determine the friction factor given in Equation (7-59)[1]. This empirical expression is

$$f = \left[\frac{1}{2\ \log\left(\frac{D_b - D_p}{e}\right)}\right]^2, \tag{7-64}$$

where ε_{av} is the approximate average absolute roughness of the annulus surfaces (ft, m). Note that the logarithm in the equation is to the base 10.

Equation (7-65) gives the approximation for e_{av} for the open hole section of the annulus. This approximation is

$$e_{av} = \frac{e_r\, D_{ob}^2 + e_p\, D_p^2}{D_{ob}^2 + D_p^2}. \tag{7-65}$$

For follow-on calculations for flow in the annulus, the absolute roughness for steel pipe, $e_p = 0.0005$ ft or $e_p = 0.0002$ m, will be used for the outside surfaces of the drill pipe and drill collars, and the inside surface of the casing. The open hole surfaces of boreholes can be approximated with an absolute roughness $e_r = 0.01$ ft or $e_r = 0.003$ m (i.e., this example value is the same as concrete pipe, which approximates borehole surfaces in limestone and dolomite sedimentary rocks, or in similar competent igneous and metamorphic rocks, see Table 8-1).

Equations (7-58), (7-59), and (7-63) through (7-65) can be used in sequential calculation steps starting at the bottom of the annulus and continuing for each subsequent change in the cross-sectional area in the annulus until the surface injection pressure is determined.

REFERENCES

1. Daugherty, R. L., Franzini, J. B., and Finnemore, E. J., *Fluid Mechanics with Engineering Applications*, Eighth Edition, McGraw-Hill, 1985.
2. Brown, K. E., and Beggs, H. D., *The Technology of Artificial Lift Methods*, Vol. 1, PennWell Books, 1977.
3. Brown, K. E., *The Technology of Artificial Lift Methods*, Vol. 2a, PennWell Books, 1980.
4. Personal communications with Stefan Miska, Department of Petroleum Engineering, University of Tulsa, Oklahoma, January 1999.
5. Lapedes, D. H., *McGraw-Hill Encyclopedia of the Geological Sciences*, McGraw-Hill, 1978.
6. Kaminski, D. A., and Jensen, M. K., *Introduction to Thermal and Fluid Engineering*, Wiley and Sons, 2005.
7. Gatlin, C., *Petroleum Engineering: Drilling and Well Completions*, Prentice-Hall, 1960.
8. Bourgoyne, A. T., et al., *Applied Drilling Engineering*, First Printing, SPE, 1986.

CHAPTER

8 Air, Gas, and Unstable Foam Drilling

Deep drilling operations with air and gas drilling technology are used in the recovery of oil and natural gas and in the recovery of geothermal steam and hot water. In the late 1970s, it was estimated that air and gas drilling technology was being used on only about 10% of the deep wells drilled and completed [1, 2]. Today, most of the world's oil and natural gas producing fields are in mature sedimentary basins. The application of air and gas drilling technology is usually limited to either (1) performance drilling operations in competent rock formations or (2) underbalanced drilling operations in low pressure reservoirs (i.e., in-fill operations). In the latter operations, the reservoirs are usually sensitive to formation damage caused by traditional mud drilling operations. The advantage of air and gas drilling operations is that the bottom hole annulus pressures can be designed to be below the reservoir bottom hole pressure. The existence of extensive formation damage using traditional drilling technology has led to the development of underbalanced drilling and completion technology. Even though underbalanced drilling and completion operations have applications in low mud weight drilling operations, the vast majority of these operations utilize air and gas drilling technology. Currently, it is estimated that approximately 30% of land-based oil and natural gas recovery drilling and completion operations utilize some form of air and gas drilling technology (either as a performance drilling operation or as an underbalanced drilling operation). As more oil and natural gas producing fields mature throughout the world, this percentage will increase.

This chapter outlines the steps and methods used to plan a successful deep air and gas drilling operation. This chapter also illustrates the application of these steps and methods to the planning of a typical deep drilling operation. The objective of these steps and methods is to allow engineers and scientists to cost-effectively plan their drilling operations and ultimately select their drilling rig, compressor package, and other auxiliary drilling location equipment. The additional benefit of this planning process is that data created by the process can later be used to control the drilling operation as the actual operation progresses. Air and gas drilling operations are different from traditional mud drilling operations and require more intensive attention on the part of the on-site drilling operation supervisor.

Figure 8-1 shows a schematic of drill pipe air injection and mist (unstable foam) injection for the direct circulation configuration. In this configuration, compressible air (or other gas) is injected with some liquids into the top of the drill string (at P_{in}). This gas flows down the inside of the drill string and passes through the drill bit nozzles. When the compressed gas flows into the bottom of the annulus, the rock cuttings (from the advance of the drill bit) are entrained and the resulting mixture flows to the surface in the annulus. The gas and rock cuttings exit the annulus (at P_e) into a horizontal flow line called a "blooey line." The blooey line flows to a burn pit where any hydrocarbons are ignited and burned. In some gas drilling operations, the flow from the end of the blooey line can exit into an open top frac tank. Sealed returns tanks are used to contain contaminated fluids and gases, or hydrocarbons (gas is membrane generated nitrogen).

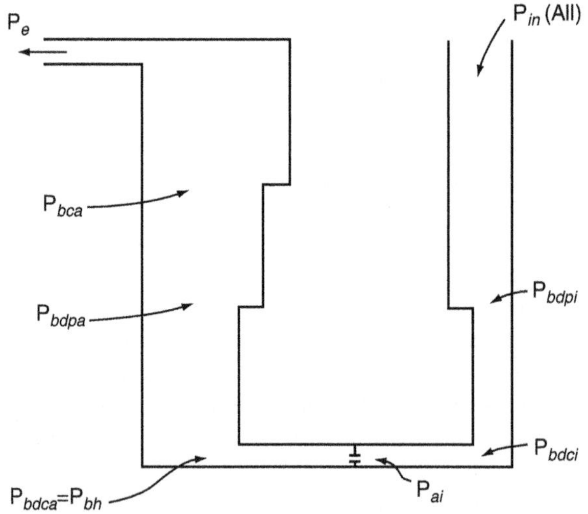

FIGURE 8-1. Schematic of direct circulation. P_{in} is the injection pressure into the top of the drill string, P_{bdpi} is the pressure at the bottom of the drill pipe inside the drill string, P_{bdci} is the pressure at the bottom of drill collars inside the drill string, P_{ai} is the pressure above the drill bit inside the drill string, P_{bdca} is the pressure at the bottom of drill collars in the annulus, P_{bh} is the bottom hole pressure in the annulus, P_{bdpa} is the pressure at the bottom of the drill pipe in the annulus, P_{bca} is the pressure at the bottom of the casing in the annulus, and P_e is the pressure at the top of the annulus.

8.1 DEEP WELL PLANNING

Deep air and gas drilling operations use a variety of compressed gases as the drilling fluid. The majority of the operations use compressed air or membrane generated nitrogen (see Chapter 5). The membrane nitrogen will not support downhole combustion. Since the 1930s, natural gas from pipelines has been used as a drilling gas since it will not combust in the typical downhole environment. Today, natural gas is too expensive to use as a drilling gas. Membrane generated nitrogen is a relatively new technology and has become commercially viable only in the past decade. In this chapter, atmospheric air and membrane generated nitrogen will be used as the example drilling gases.

The basic planning steps for a deep well are as follow:

1. Determine the geometry of the borehole section or sections to be drilled with air or other gases (i.e., open hole diameters, the casing inside diameters, and depths).
2. Determine the geometry of the associated drill strings for the sections to be drilled with air or other gases (i.e., drill bit size and type, the drill collar size, drill pipe size, and maximum depth).
3. Determine the type of rock formations to be drilled and estimate the anticipated drilling rate of penetration. Also, estimate the quantity and depth location of any formation water influxes that might be encountered.
4. Determine the elevation of the drilling site above sea level, the anticipated average operational temperature of the atmospheric air during the drilling operation, and the approximate geothermal temperature gradient.
5. Establish the objective of the air (or other gases) drilling operation:
   - Performance drilling
   - Underbalanced drilling
6. Determine the required approximate minimum volumetric flow rate of air (or other gases) needed to carry the rock cuttings from the well when drilling at the maximum depth section.
7. Select the contractor air compression package that will be used to provide for the drilling operation with the needed volumetric flow rate of air and/or membrane generated nitrogen (use a factor of safety of at least 1.2).
8. Using the compressor(s) air volumetric flow rate to be injected into the well, determine the bottom hole and surface injection pressures as a function of drilling depth (over the interval to be drilled). Also, determine the maximum power required by the compressor(s) and the available maximum derated power from the prime mover(s).
9. Determine the approximate volume of fuel required by the compressor(s) to drill the well.
10. In the event formation water is encountered, determine the approximate volumetric flow rate of "mist" injection water needed to allow formation

water or formation oil to be carried from the well during the drilling operation.

11. Determine the approximate volumetric flow rate of formation water or formation oil that can be carried from the well during the drilling operation (assuming the injected air is saturated for bottom hole conditions).

Chapter 6 derived and summarized the basic direct circulation drilling planning governing equations. In Chapter 7 the basic reverse circulation drilling planning governing equations were derived and summarized. This chapter discusses only direct circulation illustrative examples.

8.2 MINIMUM VOLUMETRIC FLOW RATE AND COMPRESSOR SELECTION

Over the past three decades, various research and commercial organizations have developed mathematical and empirical models used to predict the performance of air and gas drilling operations. Each of the models is based on a variety of engineering assumptions concerning the interaction of the gas flow in the annulus and its rock cuttings and liquid carrying capacities.

8.2.1 Discussion of Theories

In 1957, R. R. Angel developed the first field useful mathematical and empirical model for air and gas drilling operations [1, 2]. This initial work by Angel was supported by industry (i.e., Phillips Petroleum Company) and continues to be useful to drilling supervisors and drilling engineers even today. This modeling effort drew heavily from the large body of engineering knowledge related to industrial pneumatic conveying (the transport of solids by flowing atmospheric air). Thus, from the outset, Angel's model was developed to be an engineering tool. The major air and rock cuttings mixture assumption made in this model was that the rock cutting particles move together from the bottom of the borehole to the surface with the velocity of the local annulus air flow. Through the decades, other researchers have expanded this modeling effort to include other engineering aspects of the drilling operation [3]. This simple hole cleaning theory was demonstrated in Chapter 2 [see Equation (2-1)].

In 1981, interest in air and gas drilling technology found its way into an academic research effort [4, 5]. This research was carried out at the University of Tulsa and later at Pennsylvania State University and was supported by the U.S. Department of Energy. The effort sought to detail the interaction between the gas flow in the annulus and the transport of the rock cuttings. Figures 8-2 and 8-3 illustrate typical slug and nonslug (cluster) motion of particles in industrial pneumatic conveying [6]. This recent experimental work found that in the

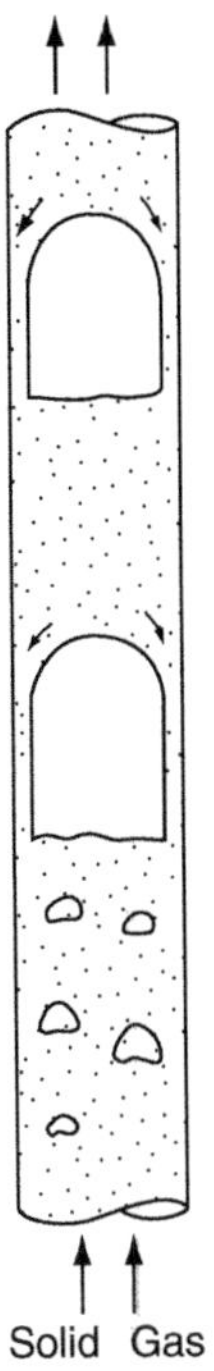

FIGURE 8-2. Schematic shows dense solids phase flow known as "slugging" [10].

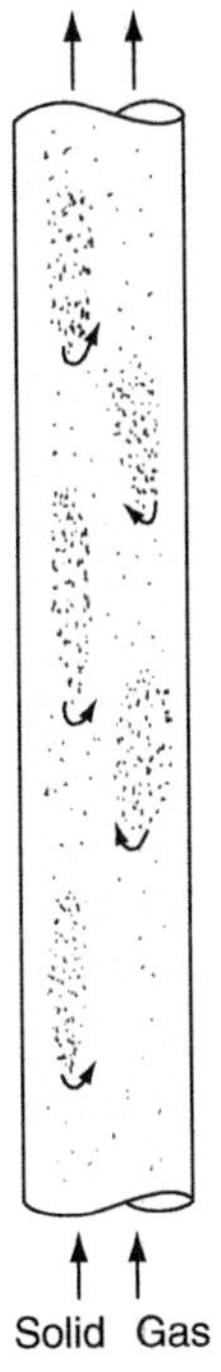

FIGURE 8-3. Schematic showing dense solids phase flow known as "nonslugging" or "clusters" [10].

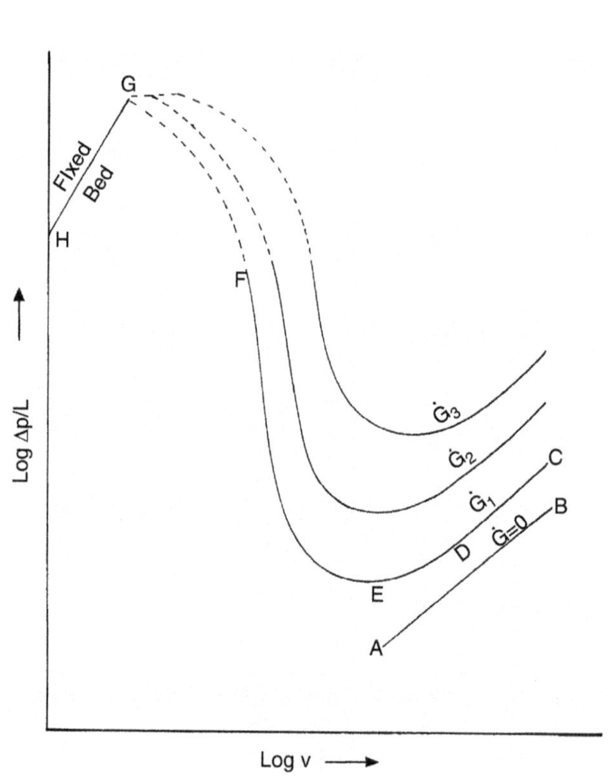

FIGURE 8-4. Solids flow characteristics in vertical pneumatic conveying. Note that the unit of $\dot{G}$ is lb/sec [6].

vertical annulus geometry the upward flow of air will choke with the rock cuttings in much the same manner as industrial vertical pneumatic conveying. This choking effect is shown in Figure 8-4 [6]. This experimental work did verify the original Angel assumption for a limited range of volumetric flow rates with entrained small diameter rock cutting particles. This research effort resulted in future improvements in the predictive models.

In 1983, another predictive model was proposed that took into account a rock cutting particle velocity that was different from that of the air flow [7]. This model gives air and rock mixture values, which can improve the accuracy of the predictions. The problem with this model is that the individual average velocities of the particles are difficult to determine analytically for inclusion in this model.

In 1992, additional theoretical work was carried out to further refine the inclusion of the effect of rock cutting particle velocities [8]. This model also suffered from the difficulty involved in determining average particle velocities.

8.2.2 Engineering Practice

As shown in Figure 8-4, and the theoretical and experimental studies in the pneumatic conveying literature and at the University of Tulsa and Pennsylvania State University, it is clear that the higher the velocity and specific weight of the gas in the vertical flow line, the greater the particle velocities approach the average velocity of the gas flow.

For an air or gas drilling operation, this is also the desirable transport situation. The choke situation occurs when the gas volumetric flow rate decreases and cutting particles begin to slip in to the gas flow transporting them. This causes the rock cutting particles to transition from the dilute phase of solids in the gas flow (where the particles are spread out in the gas) to that of a dense phase of solids (where the particles are clumped together) (see Figures 8-2 and 8-3).

Figure 8-4 can be used to further understand the choking condition and is a schematic representation of empirical data. Line A–B in Figure 8-4 refers to zero solids flow in the pipe (in our case the annulus). The family of curves in Figure 8-4 shows increased solid flow rates as the weight rate of flow increases. At fixed solids weight rate flow $\dot{G}_1$ at a high gas velocity (at point C), the solids volumetric concentration is low (well below 1%) and the particles are generally uniformly dispersed and moving near the average velocity of the gas. This is dilute phase solids flow.

As shown in Chapters 6 and 7, the pressure gradient in fully developed vertical conveying is made up of two components: a wall frictional loss component and a hydrostatic weight component. As the velocity of the gas is decreased for a fixed solids flow rate, the solids volumetric concentration in the pipe increases and the wall frictional loss decreases. Thus, line C–D shows an overall pressure gradient decrease with the decrease in gas velocity. This line indicates that the decrease in frictional loss component is significantly higher than the increase in the hydrostatic component. As the gas velocity is further decreased beyond point D, the hydrostatic component becomes more significant and the resulting curve defines a minimum at point E. A further decrease in the gas flow rate (and velocity) leads to an increase in the pressure gradient. This is shown as the curve E–F and indicates the dominance of the hydrostatic component. The solids concentration along curve E–F is high and is dense phase solids flow.

In general, for air and gas drilling systems, the choking point can be defined as the inflection point along the curve E–F. Fortunately, the choking condition is rarely observed in actual vertical air or gas drilling operations. This is due to two important air and gas drilling operational conditions. (1) As the drill bit advances, the rotating drill string breaks up larger rock particles into smaller, more easily transported particles when the larger particles collide with the drill string surface. This breaking action occurs all along the drill string length, but is likely very pronounced around and just above the drill collars. This mechanism has been observed in vertical drilling operations where downhole pneumatic motors have been tested (drilling with no drill string rotation and with rotation)[9]. (2) As seen in the illustrative examples that follow, the actual volumetric flow rate used in an air drilling operation is determined by the primary compressor output(s) to be used at the drilling location. If the approximate minimum volumetric flow rate to the borehole is determined to be 1000 scfm (*472 standard liters/sec*) and the compressors to be used to supply the compressed air are rated at 700 scfm (*330 standard liters/sec*) each, then two compressors will be used to supply the air. Thus, the actual volumetric flow rate to the well is 1400 scfm (*661 standard liters/sec*). This gives a factor of safety (above the minimum required) of 1.4. It is standard engineering practice to use a volumetric flow rate factor of safety of at least 1.2. This approximate minimum factor of safety (i.e., 1.2) is also used in determining the actual volumetric flow rate to be taken from a natural gas pipeline for a gas drilling operation. The use of primary compressor flow rates above the minimum volumetric flow rate would result in an operational point for the air and gas system somewhere along the curve C–D.

8.2.3 Engineering Planning Graphs

In order to initiate the well planning process given earlier, the geometry of the well must be defined and the anticipated drilling penetration rate estimated. Figures with graphs can be prepared for the approximate minimum volumetric flow rates for a variety of deep well and drill string geometric configurations. These figures are presented in Appendix G. The calculations for these figures are carried out using API Mechanical Equipment Standards standard atmospheric conditions. These are 14.696 psia (*10.134 N/cm^2 abs*) and 60°F (*15.6°C*) (see Chapter 5). Thus, the figures developed will give the minimum volumetric flow rate values for air drilling using air at these API standard atmospheric conditions. Once such figures are developed, the minimum volumetric flow rates can be determined for any other atmospheric conditions (surface locations) from the minimum volumetric flow rates given for API standard conditions. The minimum volumetric flow rate values are calculated using a minimum bottom hole kinetic energy per unit volume value of 3.0 lb-ft/ft^3 (*143.7 N-m/m^3*). Also, it is assumed that the drilling is in sedimentary rock formations with an average specific gravity of 2.7. These deep boreholes figures are developed for a uniform borehole diameter with the top two-thirds of the depth assumed to be cased and the bottom one-third assumed to be open hole. The Appendix G graphs are similar to the original curves of Angel except that they include the open hole roughness that increases the minimum volumetric flow rates by approximately 10 to 13% for the same borehole and drill pipe diameter combinations as those used in the original Angel work. The basic equations used to determine the minimum volumetric flow rate are derived in Chapter 6.

Illustrative Examples 8.1 and 8.2 describe the implementation of the basic planning steps Nos. 1 through 7 given in Section 8.1.

Illustrative Example 8.1 The borehole used in this illustrative example is the same example problem used in the follow-on Chapter 9 (*Aerated Fluids Drilling*) and Chapter 10 (*Stable Foam Drilling*). The $7^7/_8$-in (*200 mm*)-diameter borehole is to be drilled out of the bottom of API $8^5/_8$-in (*219 mm*)-diameter, 32.00-lb/ft (*14.00 kg/m*) nominal, Grade J-55, casing set to 7000 ft (*2134 m*). This is a special clearance casing fabricated by Lonestar Steel that has a drift diameter that will pass a $7^7/_8$-in (*200 mm*)-diameter tricone roller cutter insert drill bit with three open orifices. Figure 8-5 shows the casing and open hole geometric configuration of the well. The inside diameter of this casing is 7.921 in (*201 mm*). The open hole interval below the casing shoe is to be drilled from 7000 ft (*2134 m*) to 10,000 ft (*3048 m*). The drill string is made up of 500 ft

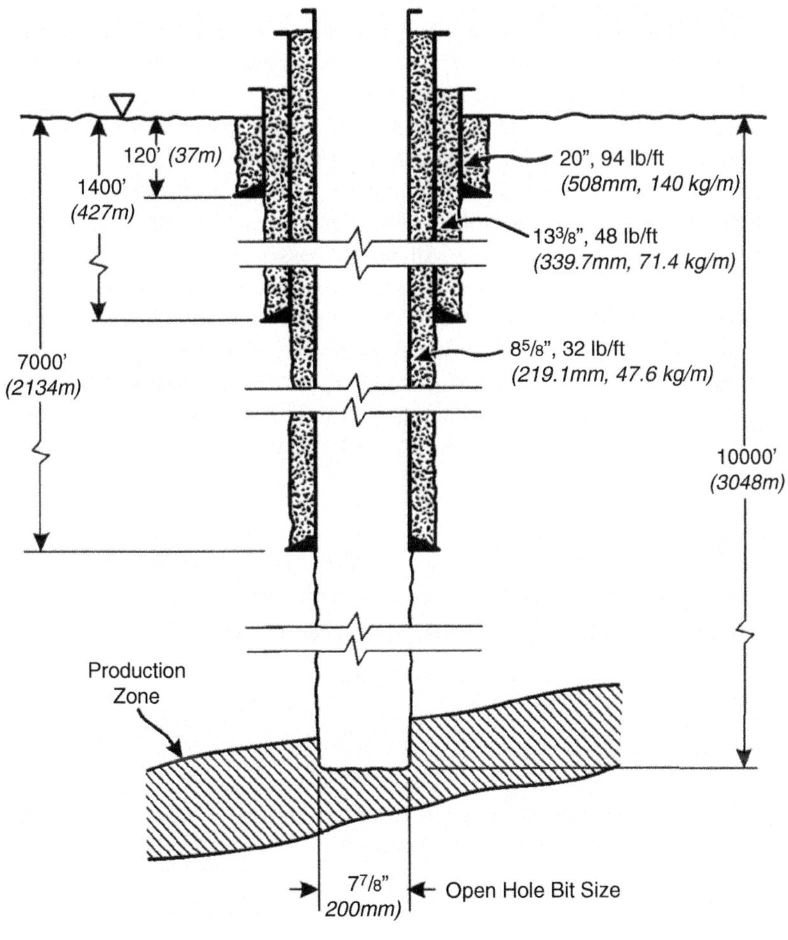

FIGURE 8-5. Illustrative Example 8.1 casing and open hole well geometric configuration.

(*152.4 m*) of $6^1/_4$-in (*158.8 mm*) by $2^{13}/_{16}$-in (*71.4 mm*) drill collars above the drill bit and API $4^1/_2$-in (*190.5 mm*), 16.60-lb/ft (*7.62 kg/m*) nominal, IEU-S135, NC46 to the surface. The tool joints of this drill pipe have an outside diameter of $6^1/_4$ in (*154.8 mm*) and an inner diameter of $2^3/_4$ in (*69.9 mm*). The surface location of the drilling site is 4000 ft (*1219 m*) above sea level in midlatitudes of North America (see Table 5-1 and Appendix B). Also, the regional geothermal gradient is approximately 0.01°F/ft (*0.018°C*). Figure 8.5 shows a schematic of this illustrative example well geometry.

The anticipated drilling rate of penetrations in a sandstone and limestone sequence is approximately 60 ft/hr (*18.3 m/hr*). The blooey line for this drilling operation is a single horizontal section of casing that runs 200 ft (*61 m*) from just above the BOP stack to the burn pit. The casing used for the blooey line is API $8^5/_8$-in (*219 mm*), 24.00-lb/ft (*10.5 kg/m*) nominal, Grade J-55, casing. The inside diameter of this casing is 8.097 in (*205.7 mm*). Two gate valves are in the blooey line at the end of the line attached to the BOP stack. These valves have an inside diameter of $7^9/_{16}$ in (*192 mm*). The objective of this illustrative example is to show how Appendix G can be used to obtain an initial approximation for the minimum volumetric flow rate to clean the well bore while drilling at 10,000 ft (*3048 m*).

The graphs in Appendix G were developed using a simplified model. This model assumes a uniform borehole inside diameter, top to bottom (assumed to be the same as drill bit diameter). All minor losses were ignored. The well is assumed to have casing in the upper two-thirds of the depth and an open hole in the bottom one-third of the well. The annulus is described by the inside diameter of the borehole and the outside diameter of the drill pipe.

Figure G-13 is the appropriate figure in Appendix G that will best approximate the drilling geometry in Illustrative Example 8.1. For a rate of penetration of 60 ft/hr, the approximate minimum volumetric flow rate is 1588.8 scfm (*749.8 standard liters/sec*).

USCS Units

Specific Weight of Air at API Standard Conditions

At API standard conditions, the atmospheric pressure is

$$p_{st} = 14.696 \text{ psia}$$

$$P_{st} = p_{st}\, 144$$

$$P_{st} = 2116.0 \text{ lb/ft}^2 \text{ abs},$$

where p_{st} is atmospheric pressure (psia) and P_{st} is atmospheric pressure (lb/ft^2 abs).

At API standard conditions, the atmospheric temperature is

$$t_{st} = 60^\circ \text{F}$$

$$T_{st} = t_{st} + 459.67$$

$$T_{st} = 519.67^\circ \text{R},$$

where t_{st} is atmospheric temperature (°F) and T_{st} is absolute atmospheric temperature (°R). Using Equation (5-11), the specific weight of the air at standard conditions is

$$\gamma_{st} = \frac{(2,116.0)(1.0)}{(53.36)(519.67)}$$

$$\gamma_{st} = 0.0763\ \mathrm{lb/ft^3},$$

where γ_{st} is the specific weight of the gas (lb/ft^3), $\mathbf{R_e}$ is the gas constant for API standard conditions air (53.36 lb-ft/lb-°R), and S_g is the specific gravity of the particular gas used (for API standard conditions air $S_g = 1.0$).

Specific Weight of Air at 4000 ft Above Sea Level

At 4000 ft above sea level the average atmospheric pressure is

$$p_{at} = 12.685\ \mathrm{psia}$$

$$P_{at} = p_{at}\ 144$$

$$P_{at} = 1826.6\ \mathrm{lb/ft^2\ abs},$$

where p_{at} is atmospheric pressure (psia) and P_{at} is atmospheric pressure (lb/ft^2 abs).

At 4000 ft above sea level the average atmospheric temperature is

$$t_{at} = 44.74^{\circ}\,\mathrm{F}$$

$$T_{at} = t_{at} + 459.67$$

$$T_{at} = 504.41^{\circ}\,\mathrm{R},$$

where t_{at} is atmospheric temperature (°F) and T_{at} is absolute atmospheric temperature (°R).

Using Equation (5-11), the specific weight of the air at standard conditions is

$$\gamma_{at} = \frac{(1826.6)(1.0)}{(53.36)(504.41)}$$

$$\gamma_{at} = 0.0679\ \mathrm{lb/ft^3},$$

where γ_{at} is the specific weight of the gas (lb/ft^3), $\mathbf{R_e}$ is the gas constant for API standard conditions air (53.36 lb-ft/lb-°R), and S_g is the specific gravity of the particular gas used (for API standard conditions air $S_g = 1.0$).

Minimum Volumetric Flow Rate at 4000 ft

From Figure G-13 the minimum volumetric flow rate is

$$q_{\mathrm{minst}} = 1588.8\ scfm$$

$$Q_{\mathrm{minst}} = \frac{q_{\mathrm{minst}}}{60}$$

$$Q_{\mathrm{minst}} = 26.48\ \mathrm{standard}\ ft^3/\mathrm{sec}.$$

The weight rate of flow is

$$\dot{w}_{\text{min}} = \gamma_{st}\, Q_{\text{minat}} = \gamma_{at}\, Q_{\text{minat}}.$$

This becomes

$$\dot{w}_{\text{min}} = (0.0763)(26.48) = (0.0679)\ Q_{\text{minat}}$$

$$Q_{\text{minat}} = \frac{(0.0763)}{(0.0679)}(26.48)$$

$$Q_{\text{minat}} = 29.76 \text{ actual } ft^3/\text{sec}$$

$$q_{\text{minat}} = (60)(29.76)$$

$$q_{\text{minat}} = 1765.6\ acfm.$$

Number of Primary Compressors Units Required

The primary compressor of choice is the four-stage reciprocating piston Dresser Clark CFB-4 compressor described in Chapter 5, Section 5.7.2. This compressor has a rated input of 1200 acfm. The minimum flow rate required to clean the well is 1785.6 acfm. Therefore, to drill this example section of well at the 4000-ft surface location, two Dresser Clark CFB-4 primary compressors will be required.

The factor of safety for the example drilling operations is

$$FS = \frac{(2)(1200)}{(1785.6)}$$

$$FS = 1.34.$$

This exceeds the required minimum factor of safety for volumetric flow rate of 1.2.

SI Units

Specific Weight of Air at API Standard Conditions

At API standard conditions the atmospheric pressure is

$$p_{st} = 10.134 \text{ N/cm}^2\ abs$$

$$P_{st} = p_{st}\ 10000.0$$

$$P_{st} = 101340.0 \text{ N/m}^2 \text{ abs},$$

where p_{st} is atmospheric pressure (N/cm^2 abs) and P_{st} is atmospheric pressure (N/m^2 abs).

At API standard conditions the atmospheric temperature is

$$t_{st} = 15.6^\circ\text{C}$$

$$T_{st} = t_{st} + 273.15$$

$$T_{st} = 288.76 \text{ K},$$

where t_{st} is atmospheric temperature (°C) and T_{st} is absolute atmospheric temperature (K). Using Equation (5-11), the specific weight of the air at standard conditions is

$$\gamma_{st} = \frac{(101340.0)(1.0)}{(29.31)(288.76)}$$

$$\gamma_{st} = 11.97 \text{ N/m}^3,$$

where γ_{st} is the specific weight of the gas (lb/ft^3), $\mathbf{R_e}$ is the gas constant for API standard conditions air (29.31 N-m/m-K), and S_g is the specific gravity of the particular gas used (for API standard conditions air $S_g = 1.0$).

Specific Weight of Air at 1219 m Above Sea Level

At 1219 m above sea level the average atmospheric pressure is

$$p_{at} = 8.749 \text{ N/cm}^2 \; abs$$

$$P_{at} = p_{at} \; 10000.0$$

$$P_{at} = 87490.0 \text{ N/m}^2 \text{ abs},$$

where p_{at} is atmospheric pressure (N/cm^2 abs) and P_{at} is atmospheric pressure (N/m^2 abs).

At API standard conditions, the atmospheric temperature is

$$t_{at} = 7.08^{\circ} \text{ C}$$

$$T_{at} = t_{at} + 273.15$$

$$T_{at} = 280.13 \text{ K},$$

where t_{at} is atmospheric temperature (°C) and T_{at} is absolute atmospheric temperature (K). Using Equation (5-11), the specific weight of the air at standard conditions is

$$\gamma_{at} = \frac{(87490.0)(1.0)}{(29.31)(280.13)}$$

$$\gamma_{at} = 10.66 \text{ N/m}^3,$$

where γ_{at} is the specific weight of the gas (N/m^3), $\mathbf{R_e}$ is the gas constant for API standard conditions air (29.31 N-m/N-K), and S_g is the specific gravity of the particular gas used (for API standard conditions air $S_g = 1.0$).

Minimum Volumetric Flow Rate

From Figure G-13 the minimum volumetric flow rate is

$$q_{\text{minst}} = 749.9 \text{ standard } liters/\text{sec}$$

$$Q_{\text{minst}} = \frac{q_{\text{ minst}}}{1000}$$

$$Q_{\text{minst}} = 0.7499 \text{ standard } m^3/\text{sec}.$$

The weight rate of flow is

$$\dot{w}_{\text{min}} = \gamma_{st} \; Q_{\text{minat}} = \gamma_{at} \; Q_{\text{minat}}.$$

This becomes

$$\dot{w}_{\min} = (12.09)(0.7499) = (10.66)\ Q_{\text{minat}}$$

$$Q_{\text{minat}} = \frac{(12.09)}{(10.66)}(0.7499)$$

$$Q_{\text{minat}} = 0.8505 \text{ actual } m^3/\text{sec}$$

$$q_{\text{minat}} = (1000)(0.8505)$$

$$q_{\text{minat}} = 850.5 \text{ actual } liters/\text{sec.}$$

Number of Primary Compressors Units Required

The primary compressor of choice is the four-stage reciprocating piston Dresser Clark CFB-4 compressor described in Chapter 5, Section 5.7.2. This compressor has a rated input of 566.3 actual liters/sec. The minimum flow rate required to clean the well is 850.5 actual liters/sec. Therefore, to drill this example section of well at the 1219-m surface location, two Dresser Clark CFB-4 primary compressors will be required.

The factor of safety for the example drilling operations is

$$FS = \frac{(2)(566.3)}{(850.5)}$$

$$FS = 1.33.$$

Here again this factor of safety exceeds the required minimum factor of safety for volumetric flow rate (i.e., 1.2).

The differences between the factors of safety of the USCS units solution and the SI units solution are attributable to accumulated round-off errors in conversion constants.

The aforementioned results from Illustrative Example 8.1 can be improved upon. This requires a more extensive treatment of the open hole absolute roughness and inclusion of the major and minor losses at the top of the annulus (e.g., bloy line, Tee and blooey line valves).

Open Hole Absolute Surface Roughness

To carry out this more detailed solution in follow-on solutions, it will be necessary to have more detail of the surface average absolute roughness of the steel tubular surfaces and of the open hole rock surfaces in the open hole section of the well. The steel tubular surfaces can be assumed to have an average surface roughness of $e_s = 0.0005$ ft ($e_s = 0.0002\ m$)[10]. The open hole rock surface average absolute roughness values are approximated in Table 8-1 [3].

In the open hole, the outer surface of the borehole annulus is described by Table 8-1. The inner surface is described by the drill pipe or drill collar absolute average surface roughness of steel tubulars. The average absolute surface

Table 8-1. Open Hole Wall Absolute Surface Roughness for Rock Formation Types

Rock Formation Types	Surface Roughness
Competent, low fracture ■ Igneous (e.g., granite, basalt) ■ Sedimentary (e.g., limestone, sandstone) ■ Metamorphic (e.g., gneiss)	0.01 to 0.02 ft (0.003 to 0.006 m)
Competent, medium fracture ■ Igneous (e.g., granite, basalt) ■ Sedimentary (e.g., limestone, sandstone) ■ Metamorphic (e.g., gneiss)	0.02 to 0.03 ft (0.006 to 0.009 m)
Poor competence, high fracture ■ Igneous (e.g, breccia) ■ Sedimentary (e.g., sandstone, shale) ■ Metamorphic (e.g., schist)	0.02 to 0.04 ft (0.009 to 0.012 m)

roughness of the annulus can be approximated by using a surface area weight average relationship between the open hole surface area and its roughness and the outside surface of the drill string and its roughness. Thus, the value for e_{av} is

$$e_{av} = \frac{e_r\left(\frac{\pi}{4}\right)D_b^2\,H_o + e_s\left(\frac{\pi}{4}\right)D_p^2\,H_o}{\left(\frac{\pi}{4}\right)D_b^2\,H_o + \left(\frac{\pi}{4}\right)D_p^2\,H_o}.$$

The term H_0 cancels and the aforementioned reduces to

$$e_{av} = \frac{e_r\,D_b^2 + e_s\,D_p^2}{D_b^2 + D_p^2}. \tag{8-1}$$

For the sandstone and limestone sequence in the section to be drilled in this well, the open hole absolute surface roughness is assumed to be 0.01 ft (0.003 m).

Blooey Line

The calculation procedure must be initiated with the known atmospheric pressure at the exit to the blooey line. For the illustrative example, it is assumed that the air will exit the well annulus and enter the blooey line with a surface geothermal temperature of 44.74°F (*7.08°C*). It is further assumed that the temperature of the air flow in the blooey line does not change until it exits the blooey line (i.e., the steel blooey line will be at nearly the surface geothermal temperature at steady-state flow conditions). The standard isothermal gas pipeline flow equation can be used to determine the major loss due to pipe wall friction [10]. This equation can be adjusted to also include the minor loss for the Tee turn at the top of the annulus into the blooey line and to include losses due to the two blooey line valves at the entrance end of the line. Therefore, the equation for the pressure in the air (or gas) flow at the entrance end of the blooey line P_b can be approximated as

$$P_b = \left[\left(f_b \frac{L_b}{D_b} + K_t + \Sigma\, K_v\right)\left(\frac{w_g^2\, \mathbf{R}\, T_r}{g\, A_b^2\, S_g}\right) + P_{at}^2\right]^{0.5}, \qquad (8\text{-}2)$$

where f_b is the friction factor for gas flow in the blooey line, L_b is the length of the blooey line (ft, m), D_b is the inside diameter of the blooey line (ft, m), A_b is the cross-sectional area of the inside of the blooey line (ft^2, m^2), K_t is the minor loss factor for the T turn at the top of the annulus, and K_v is the minor loss factor for the valves in the blooey line.

The loss factor for K_t for the single blind Tee at the top of the annulus is approximately 25 (see Figures 8-6 and 8-7, direct circulation only). These approximate minor loss values for Tee's have been obtained from air and gas drilling operations in the San Juan Basin and in the Permian Basin. The gate valves (with the gate in the full open position) in a blooey line have a open inside diameter

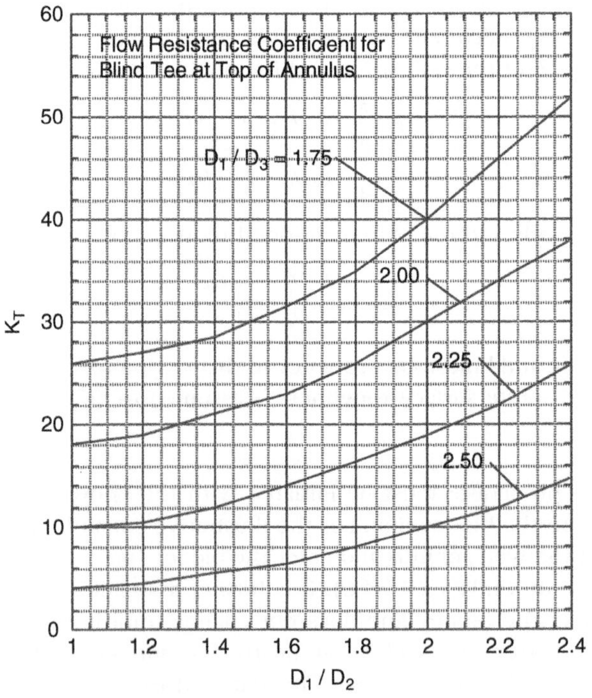

FIGURE 8-6. Flow resistance coefficient for the blind Tee at the top of the annulus.

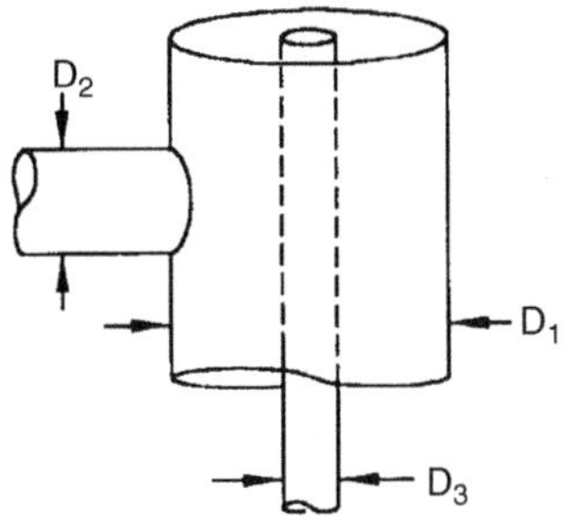

FIGURE 8-7. Dimensions of the blind Tee at the top of the annulus (for Figure 8-6).

that is nearly the same as the inside diameter of the blooey line. The loss factor for K_v for gate valves that have an inside diameter slightly less than the inside diameter of the blooey line is approximately 0.2 [10].

Illustrative Example 8.2 In this illustrative example the actual minimum volumetric flow rate will be found using a Mathcad program that considers all the major and minor friction losses in the flow system described in detail in Illustrative Example 8.1.

This Mathcad solution is given in Appendix C. In this solution, a Mathcad program has been set up to determine the air volumetric flow rate that will give annulus minimum kinetic energy. This is a trial and error solution. The air volumetric flow rate is selected at the start of the program until one of the five annulus kinetic energy per unit volume terms reaches the value of 3.0 lb-ft/ft^3 (*143.7 N-m/m^3*). The air volumetric flow rate that gives a minimum annulus kinetic energy per unit volume is 1994 acfm (*940.96 actual liters/sec*).

USCS Units: Number of Primary Compressors Units Required

As in Illustrative Example 8.1, the primary compressor of choice will be the four-stage reciprocating piston Dresser Clark CFB-4 compressor (see Chapter 5, Section 5.7.2). This compressor has a rated input of 1200 acfm. In this improved Illustrative Example 8.2 solution, the minimum air volumetric flow rate needed to clean the well is 1994 acfm. Therefore, to drill this example section of the well at the 4000-ft surface location, two Dresser Clark CFB-4 primary compressors will be adequate, as they would give a factor of safety slightly greater than 1.2. Therefore, two primary compressors will be used.

The factor of safety for this illustrative example is

$$FS = \frac{(2)(1200)}{(1994)}$$

$$FS = 1.2036.$$

The three Dresser Clark CFB-4 primary compressors will inject 2400 acfm into the circulation system.

SI Units: Number of Primary Compressors Units Required

As in Illustrative Example 8.1, the primary compressor of choice will be the four stage reciprocating piston Dresser Clark CFB-4 (see Chapter 5, Section 5.7.2). This compressor has a rated input of 566.3 actual liters/sec. In this improved Illustrative Example 8.2 solution, the minimum air volumetric flow rate needed to clean the well is 940.96 actual liters/sec. Therefore, to drill this example section of the well at the 1219-m surface location, two Dresser Clark CFB-4 primary compressors will be adequate, as they would give a factor of safety of greater than 1.2. Therefore, two primary compressors must be used.

The factor of safety for this illustrative example is

$$FS = \frac{(2)(566.3)}{(940.96)}$$

$$FS = 1.2036.$$

The two Dresser Clark CFB-4 primary compressors will inject 1132.6 actual liters/sec into the circulation system.

The difference in the factor of safety results between Illustrative Example 8.1 and Illustrative Example 8.2 is due to the inclusion in 8.2 of the major and minor losses of the surface equipment (i.e., blooey line, Tee, blooey line valves) and the inclusion of a more accurate treatment of open hole annulus absolute surface roughness.

Referring back to Chapter 5, Section 5.7, several compressor packages could have been used.

The two stage reciprocating piston Gardner Denver Model WEN primary compressor in Chapter 5, Section 5.7.1 is rated at 700 acfm (*330 actual liters/sec*). Applying the factor of safety of 1.2 would require the use of four of these primary compressors to the example drilling project. This would mean that the volumetric flow rate of air injected into the circulation system of the well would be 2800 acfm (*1321 actual liters/sec*). This would give a factor of safety of 1.4.

The two stage helical lobe (screw) Ingersoll Rand Model XHP 1170 primary compressor in Chapter 5, Section 5.7.3 is rated at 1170 acfm (*552.1 actual liters/sec*). As discussed in Chapter 5, the screw compressor has a seal wear problem that requires that the rated volumetric flow rate be reduced by approximately 7% when these compressors are applied at a field location. This means that the derated primary compressor output is approximately 1088 acfm (*513.4 actual liters/sec*). Applying the factor of safety of 1.2 would require use of three of these primary compressors to the example drilling project. This would mean that the volumetric flow rate of air injected into the circulation system of the well would be 3264 acfm (*1540.3 actual liters/sec*). This would give a factor of safety of 1.46. Here again the operational preference would be to use the four stage reciprocating piston Dresser Clark CFB-4 primary compressor package with a factor of safety of 1.64.

8.3 BOTTOM HOLE AND INJECTION PRESSURES

The aforementioned sections compared the borehole requirements with the compressor capabilities of the compressor packages available in Chapter 5, Section 5.7. The next step in the planning process is to determine the air flow pressure values throughout the circulating system, particularly the bottom hole annulus pressure and the surface injection pressure.

Illustrative Example 8.3 Here again, the basic drilling project well data given in Illustrative Example 8.1 and Illustrative Example 8.2 are used to determine the pressures in the example circulation system. In Illustrative Example 8.2 it was found that two Dresser Clark CFB-4 primary compressors would give the drilling project an appropriate factor of safety with regards to air volumetric flow rate. Therefore, the pressures will be determined using an air volumetric flow rate of 2400 acfm (*1132.7 actual liters/sec*).

Appendix C gives the Mathcad solution for Illustrative Example 8.3 that calculates the pressures in the circulation system using three Dresser Clark CFB-4 primary compressors. Figure 8-8 shows the compressed air flow pressures throughout the circulation system in the well while drilling at 10,000 ft (*3048 m*).

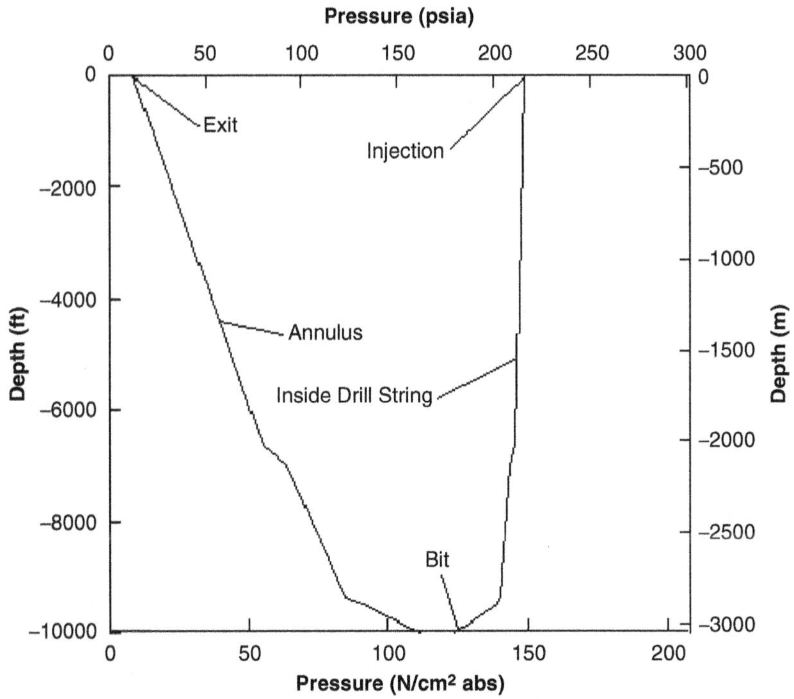

FIGURE 8-8. Illustrative Example 8.3 pressures versus depth.

In Figure 8-8, the bottom hole annulus pressure is 161.7 psia (*111.5 N/cm² abs*) and the injection pressure is 215.4.6 psia (*148.6 N/cm² abs*).

The somewhat erratic nature of the plot in Figure 8-8 reflects use of the lumped geometry method used in the Mathcad calculations. A few sophisticated programs in Fortran or C++ treat each individual tool joint at its location and give a smoother plot. Most commercial programs, however, use an average outside "drill pipe" diameter, which is the weight average (by surface area) of the outside diameter of the tool joint with the outside diameter of the pipe body. This style of program will also give a smoother plot of these circulating system pressures. All of these programs will give injection and bottom hole pressures that are with 1 to 2% of the lumped geometry Mathcad calculations shown in Appendix C.

However, certain important aspects of air drilling can be missed using the weighted average method of treating the drill pipe tool joints. There are very practical deep drilling infill operations in older gas production fields that require sidetrack drilling out of the smaller $5^1/_2$-in (*140 mm*) casing. In such wells, the sidetrack operation can easily utilize $2^3/_8$-in (*60.3 mm*) drill pipes that have inappropriate tool joints that are too large. These tool joints can create choke pressures with the $4^3/_4$-in (*120.7 mm*) open hole and the cased hole sections at the tool joints to actually cause formation damage in the reservoir with static bottom hole pressures as high as 900 psia (*620.7 N/cm² abs*). In essence, it is probably important to retain the tool joint dimensions in some form as either lumped geometry or as the individual tool joints in place in the programs.

Figure 8-9 shows the injection pressure and the bottom hole pressure as drilling takes place from the casing shoe at 7000 ft (*2134 m*) to a final depth of 10,000 ft (*3048 m*). Assuming that this example is more typical of a performance drilling operation, then it is clear that the same air volumetric flow rate (compressor package) can be used to drill the entire 3000 ft (*914 m*) of open hole. The injection pressure increases with increasing drilling depth, but is only slightly higher when drilling at 10,000 ft (*3048 m*) relative to drilling at the top of the open hole section (i.e., approximately 28 psi higher, or 19 N/cm² higher). This lack of change is somewhat a consequence of the choking effect of the drill bit open orifices. This means that the compressed air inside the drill string is hydrostatic pressure dominated and not friction pressure dominated [see Equation (6-1)].

It is important to discuss the time for cuttings to reach the surface. In an air drilled vertical or near vertical borehole, cuttings generated at the bottom of the annulus as the drill bit is advanced are broken up into small fragments as the cuttings collide with the rotating drill string and the borehole wall. This means that within 1000 ft (*305 m*) or so from the bottom of the annulus in most wells, the cuttings are reduced to small fragments. This has been demonstrated by drilling with downhole air motors (with no rotation) and is discussed in Chapter 11. Such small fragments are carried very rapidly up the annulus to the surface and will generally reach the surface in a 10,000 ft (3048 m) hole in approximately 10 min. This delay time is basically a negligible interval when correlating cuttings geology to formations being drilled.

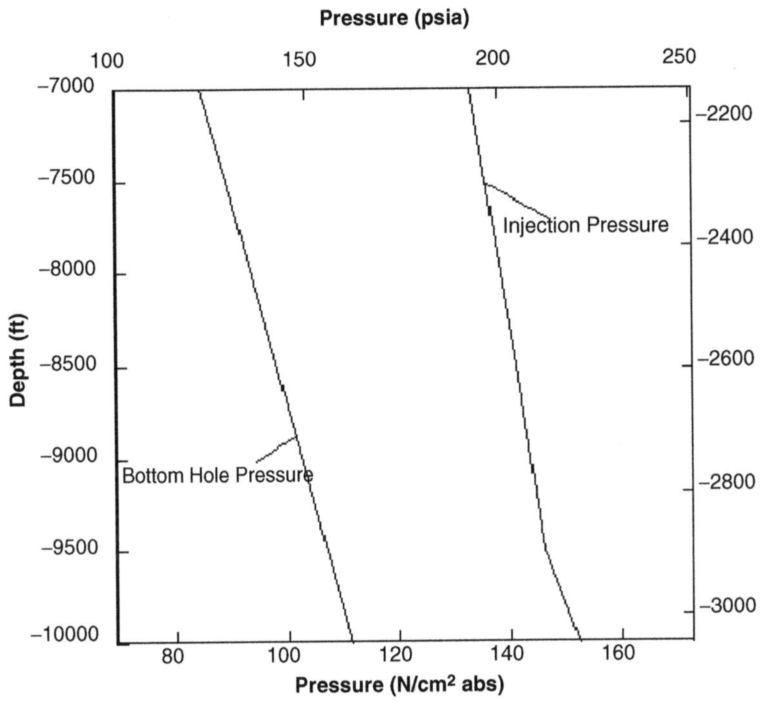

FIGURE 8-9. Drilling ahead, bottom hole and injection pressures versus depth.

8.4 WATER INJECTION AND FORMATION WATER INFLUX

Water is injected into the volumetric flow rate of air (or other gases) flowing from the compressors to the top of the inside of the drill string for the following three important reasons.

1. Saturate the air or other drilling gas with water vapor at bottom hole annulus pressure and temperature conditions.
2. Eliminate the stickiness of the small rock cuttings "flour" generated by the advance of the drill bit.
3. Assist in the suppression of combustion of the mixture of produced hydrocarbons and oxygen-rich air.

Figure 8-10 shows a typical "mist" pump and a typical stand-alone, skid-mounted positive displacement duplex liquid pump driven with a Caterpillar Model 3304 diesel-fueled, naturally aspirated, prime mover. This pump forces liquid into the pressurized air flow line from the compressors to the standpipe on the drill rig (see Figure 3-1).

FIGURE 8-10. Liquid "mist" pump for air (or membrane generated nitrogen) drilling operations.

The liquid pump draws its water from a liquid tank (located at the left back end to the skid-mounted unit shown in Figure 8-10). Often other additives are mixed in the suction tank. These are typically corrosion inhibitors, polymers, and a foamer [11]. Table 8-2 gives the approximate volume mix of these additives.

"Mist" injection is the old term, which was the injection of water with basically no additives. Modern air drilling defines the injection of water with additives as unstable foam drilling operations.

8.4.1 Saturation of Air at Bottom Hole Conditions

Water is injected into the air or other circulation gases at the surface in order to saturate the gas with water vapor at bottom hole conditions. The reason this is done is to assure that the circulation gas, as it flows out of the drill bit orifices into the annulus, will be able to carry formation water coming into the annulus as whole droplets.

Table 8-2. Typical Approximate Additives Volumes per 20 bbls of Water for Unstable Foam Drilling (Actual Commercial Product Volumes May Vary)

Additives	Volume per 20 bbls of Water
Foamer	4.2 to 8.4 gals (16 to 32 liters)
Polymer	1 to 2 quarts (1 to 2 liters)
Corrosion inhibitor	0.5 gal (2 liters)

If water is not injected into the gas at the surface and the gas is dry and hot when it comes out of the drill bit orifices, a portion of the formation water will be absorbed by the gas as water vapor. This would, of course, saturate the dry hot gas with the formation water. This saturation process would, however, decrease the internal energy (i.e., enthalpy) in the gas as the gas enters the annulus. This reduction in internal energy at the bottom of the annulus dramatically reduces the kinetic energy per unit volume of the gas as it flows from the bottom of the annulus to the surface (this reduction in kinetic energy per unit volume is due to a reduction in velocity). The reduction in kinetic energy per unit volume in the annulus reduces the carrying capacity of the gas throughout the annulus. This reduction in carrying capacity of the circulation gas occurs when the circulation system needs all of its carrying capacity to carry to the surface the additional load (beyond the rock cuttings load) of the formation water flowing into the annulus.

By injecting water at the surface into the gas, the gas becomes saturated with water vapor inside the drill string as it flows down the string and is further compressed as it flows to the bottom of the drill string. At the bottom of the inside of the drill string, the gas is saturated, hot, and compressed. When the gas flows into the bottom of the annulus through the drill bit orifices, the gas cannot vaporize the formation water and must lift the formation water as drops. This is the most efficient method of carrying an influx of formation water from the annulus of a well in a gas drilling operation. In general, it is not efficient to try to use dry gas to absorb formation water as a water vapor and, in essence, "dry out" a well.

The empirical formula for determining the saturation of various gases including air can be found in a variety of chemistry handbooks and other literature. The empirical formula for the saturation pressure of air can be written as [12]

$$p_{sat} = 10^{\left[6.39416 - \left(\frac{1,750.286}{217.23 + 0.555 t_{bb}}\right)\right]}, \tag{8-4}$$

where p_{sat} is saturation pressure of the air at annulus bottom hole conditions (psia) and t_{bb} is the bottom hole temperature in the annulus (°F). The approximate volumetric flow rate of injected water to an air drilling operation is determined by the relationship between the above saturation pressure and the bottom hole pressure in the annulus, and the weight rate of flow of air being injected into the top of the drill string. Thus, the flow rate of injected water, q_{iw}, is determined from [13]

$$q_{win} = \left(\frac{p_{sat}}{p_{bb} - p_{sat}}\right)\left(\frac{18.02}{28.96}\right)\left(\frac{3600}{8.33}\right)\dot{w}_g, \tag{8-5}$$

where q_{win} is the volumetric flow rate of injected water (gal/hr).

In Illustrative Example 8.3, the fresh water to be injected at the surface to saturate the air flow at bottom hole conditions is 15 gal/hr (*55 liters/hr*).

8.4.2 Eliminate Stickiness

Surface-injected fresh water may saturate the compressed air flow at bottom hole conditions, but it will create a new problem unless the process is managed correctly. If only a small volumetric flow rate of fresh water is injected or, for that matter, if only a small influx of formation water occurs, the water will combine with rock cuttings fines and flour to create a sticky mixture that will develop into mud rings. The schematic in Figure 8-11 depicts where mud rings are formed in the typical vertical well. The sticky rock flour will form mud rings along the open hole wall at major annulus cross-section transitions where the upward annulus flow of air creates eddy currents.

These mud rings are analogous to a small amount of water being placed in a mixing bowl with wheat flour. The mixture will get very sticky and unworkable. If more water is added, the mixture will reach a point where it will lose its stickiness and become more milky and workable. The addition of injected fresh water to a well through the standpipe must be added at a volumetric flow rate that will ultimately eliminate the bottom hole stickiness, as well as saturate the air at bottom hole annulus conditions. Eliminating bottom hole stickiness will always require more injected water volumetric flow rate than the amount predicted by Equations (8-4) and (8-5). Therefore, injection of fresh water into a well standpipe must not stop at the flow rate predicted by Equations (8-4) and (8-5). The volumetric flow rate must continue until stickiness is curtailed. The

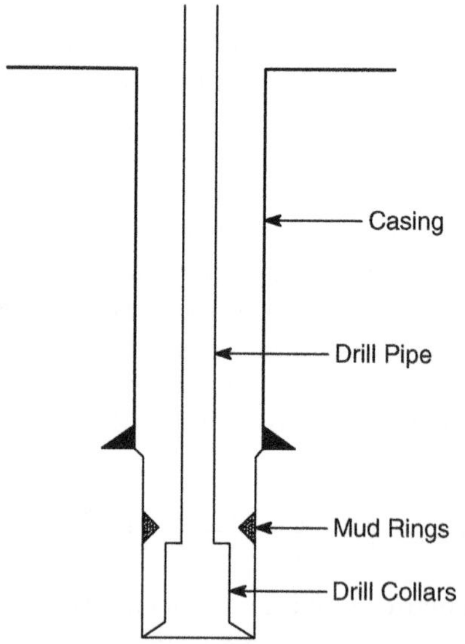

FIGURE 8-11. Mud ring on the open hole section borehole wall.

appropriate volumetric flow rate of fresh water to the well to eliminate stickiness is a field trial and error process.

The actual volumetric flow rate of injected fresh water into the standpipe during a drilling operation will be very dependent on the type of rock being drilled. Certain rock formations when drilled will create more rock flour, which will in turn create more well stickiness problems (i.e., mud rings). The worst formations are dry shale, limestone, and dolomites.

The typical field procedure used to attain the proper water injection flow rate is to watch the surface air standpipe pressure gauge as water injection progresses. Stickiness will usually manifest itself as mud rings and will be seen on the standpipe pressure gauge as sharp increases in pressure. The pressure increases can be as little as 5 or 10 psi (*3 to 7* N/cm^2) and as high as 20 or 30 psi (*15 to 20* N/cm^2). As the water flow rate is being injected into the well (and the drill string is periodically raised and rotated at a higher speed to break up the mud rings), the injected flow rate is increased until the standpipe pressure drops back to near normal. The pressure gauge will always read a slightly higher pressure than before water injection, as the column of air and water in the annulus will be heavier than the previous air column. For the example well described in Illustrative Example 8.3, the typical fresh water injection to the standpipe might be approximately1 to 2 bbl/hr.

Once the appropriate fresh water injection flow rate has been field determined, the drilling operation can go forward and the well can be advanced into rock formations that have formation water influxes. This brings up the problem of how much formation water a typical air drilled well can carry. Figure 8-12 shows results from the Mathcad solution for Illustrative Example 8.3 and shows the critical annulus minimum kinetic energy value as a function of depth for two possible formation water influx volumetric flow rates. For a 10-bbl/hr influx, Figure 8-12 shows that the annulus minimum kinetic energy drops as the open hole section is drilled from 7000 ft (*2134 m*) to 10,000 ft (*3048 m*). However, the situation is worst for a 20-bbl/hr influx (the annulus minimum kinetic energy is lower all along the open hole section). In fact, for the influx of 20 bbl/hr, the minimum kinetic energy drops below 3.0 $lb/ft/ft^3$ (*143.7* $N\text{-}m/m^3$) at a drilling depth of approximately 9200 ft (*2804 m*). This means that cuttings removal as drilling approaches this depth will be impaired, resulting in a variety of bottom hole problems.

8.4.3 Suppression of Hydrocarbon Ignition

The next higher level of water injection volumetric flow rate is the volume needed to suppress the ignition of downhole explosions and fires due to the mixture of circulation air with produced oil, natural gas, or coal dust and fragments as the drill bit is advanced. When drilling into rock formations that are coal seams, or reservoir rock containing oil or natural gas, the steel drill bit action on the rock cutting face can easily cause a spark. If the circulation gas is air, then the three ingredients for downhole ignition are present (i.e., hydrocarbons, a spark, and an oxygen source). Increasing the water injection volumetric flow rate (with additives) to the borehole and creating unstable foam at the bottom of the well can

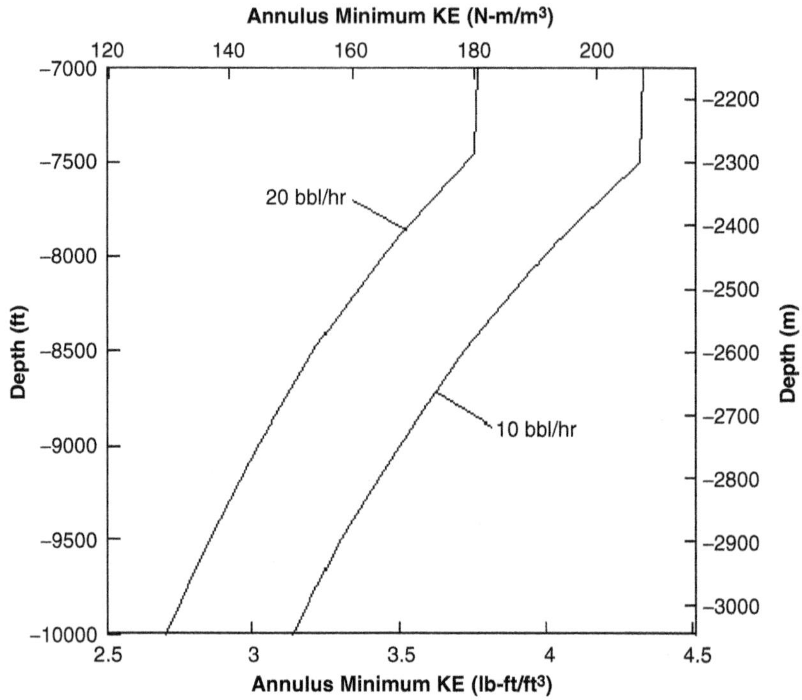

FIGURE 8-12. Drilling ahead, annulus minimum kinetic energy.

suppress most fire and explosion hazards for a vertical well. This has little or no success in horizontal boreholes. This is because vertical wells tend to penetrate the vertical thickness of the hydrocarbon producing reservoir (usually a horizontal sedimentary rock formation). The vertical thickness of the reservoir is of the order of a few hundred feet. Thus, at a drilling rate of 60 ft/hr, the exposure time in the hazardous production zone is only a few hours. However, horizontal boreholes require the drilling of several thousands of feet of open hole in the hydrocarbon producing reservoir. The drilling rates in horizontal boreholes are usually about half the drilling rates in vertical wells. Thus, the exposure time in a horizontal borehole is of the order of 10 to nearly 100 hours. The unsuccessful record of ignition suppression with unstable foam in horizontal boreholes is generally considered the consequence of the long exposure (drilling time) in hydrocarbon producing rock formations. There are, of course, other drilling methods that can be used to suppress or eliminate hydrocarbon ignition. These methods require the use of circulation gases that will not support ignition. These other circulation gases are natural gas for a gas pipeline, industrial liquid nitrogen based gas, and membrane generated nitrogen (created by placing stripper/filter equipment downstream from the compressors to remove oxygen). The use of these other drilling gases can increase drilling operation costs significantly.

Two chemical reactions can occur when oxygen combines with a hydrocarbon such as methane CH_4 and ethane C_2H_6 (and any other hydrocarbons). The lower order combustion is denoted as deflagration and is in an oxygen-lean environment. For methane, the deflagration reaction is

$$CH_4 + 0.5\ O_2 \rightarrow CO + 2\ H_2O \text{ (plus heat).}$$

For ethane, the deflagration process is

$$C_2H_6 + O_2 \rightarrow 2\ CO + 3\ H_2O > \text{ (plus heat).}$$

The higher order combustion is denoted as detonation and is in an oxygen-rich environment. For methane, the detonation reaction is

$$CH_4 + 2\ O_2 \rightarrow CO_2 + 2\ H_2O \text{ (plus heat).}$$

For ethane, the detonation process is

$$C_2H_6 + 3.5\ O_2 \rightarrow 2\ CO_2 + 3\ H_2O \text{ (plus heat).}$$

Figure 8-13 gives the ignition (ignition zone) parameters of pressure versus the percent mixture of natural gas with atmospheric air [14, 15]. In general, natural gas presents the somewhat greater hazard relative to exposure to oil and coal. This is because the mixture of air and natural gas creates an explosive hazard,

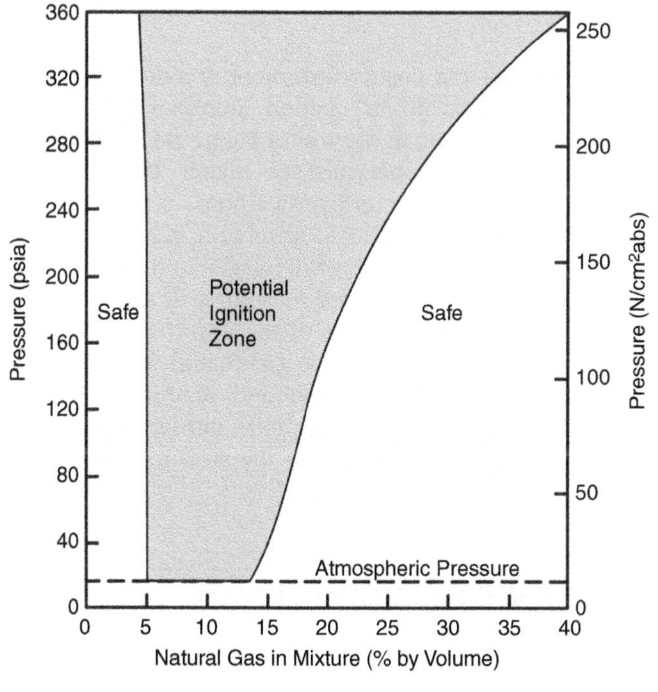

FIGURE 8-13. Ignition mixture by volume of natural gas and atmospheric air [14].

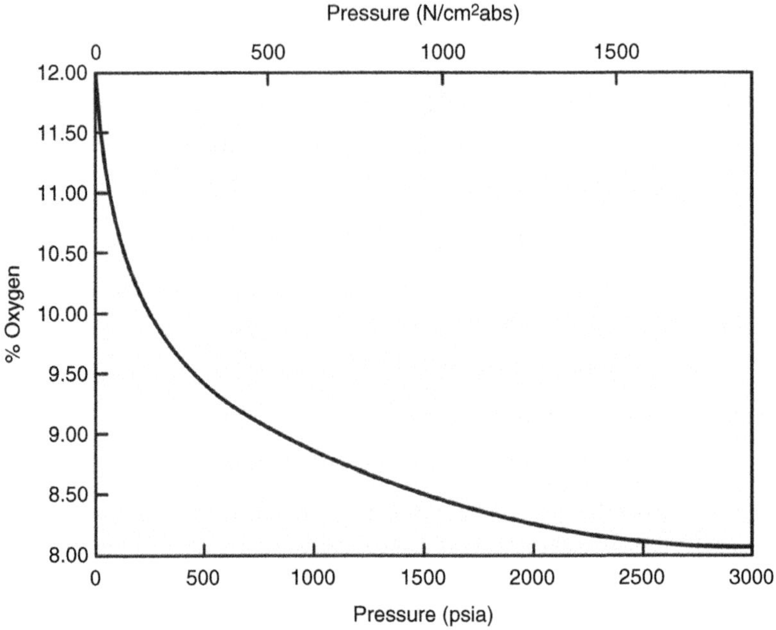

FIGURE 8-14. Content of oxygen to initiate deflagration or detonation [16].

whereas the mixture of oil and coal creates more of a downhole fire hazard. Figure 8-13 shows that the hazard of ignition increases with higher pressures (the wider the region of ignition at the top of Figure 8-13). As the well is drilled deeper, the bottom hole annulus pressure gets higher. This in turn increases the ignition probability in the presence of hydrocarbons.

As discussed in Chapter 5, Section 5.8, membrane generated nitrogen technology can be used to reduce the percentage of oxygen in the filter output stream to approximately 5%. This oxygen deprived inert air will not support deflagration or detonation. Figure 8-14 shows a plot of the approximate minimum percent of oxygen in a mixture with hydrocarbon (methane) at which deflagration or detonation will initiate [16]. Deflagration and detonation will not proceed below these values. This is why membrane filter nitrogen generators are usually calibrated to keep the percent of oxygen in the stream of filter atmospheric air at approximately 5%.

8.5 PRIME MOVER FUEL CONSUMPTION

This section discussed the fuel consumption of the prime mover for the compressor system. Illustrative examples of the fuel consumption were discussed in detail in Chapter 5. Appendix C gives the Mathcad solution for Illustrative Example 8.3. In this solution, the fuel consumption calculations found in Chapter 5 were applied to the Illustrative Example 8.3 solution. The fuel consumption rate required for the Caterpillar Model D398 diesel fueled prime mover of the Dresser Clark CFB-4 compressor operating at an air injection pressure required for drilling at 10,000 ft (*3048 m*) of depth is approximately 32 gal/hr (*120 liters/hr*). The total diesel fuel usage rate for all three compressors on the location is approximately 95 gal/hr.

8.6 CONCLUSIONS

The discussions in this chapter concentrated on direct circulation operations. The Mathcad illustrative examples in Appendix C utilized lumped geometry approximations for the drill pipe body and drill pipe tool joints. These approximations appear to adequately model the overall friction resistance along the circulation system and give accurate results for bottom hole and injection pressures.

Unlike the aerated and stable foam drilling fluids, which are discussed in Chapters 9 and 10, the modeling of air drilling fluids is not a rheology issue. The major issues with air drilling fluids are (1) downhole combustion and (2) the ability of the air drilling fluid to carry formation water from the well. The first issue has been solved by the membrane filter technology used to produce an inert atmospheric air that is defined in this text as membrane generated nitrogen. The second issue can only be solved in the field by experienced drilling personnel.

With all things being equal, air drilling fluid will have the lowest bottom hole pressure of the three drilling fluids being considered in this text: air, aerated, and stable foam.

REFERENCES

1. Angel, R. R., "Volumetric Requirements for Air or Gas Drilling," *Petroleum Transactions, AIME*, 1957.
2. Angel, R. R., *Volume Requirements for Air and Gas Drilling*, Gulf Publishing Company, 1958.
3. Lyons, W. C., *Air and Gas Drilling Manual*, Second Edition, McGraw-Hill Company, 2001.
4. Ikoku, C. U., and Williams, C. R., "Drill Cuttings Transport in Vertical Annuli by Air, Mist and Foam in Aerated Drilling Operations," Contracts for Field Projects and Supporting Research on Enhanced Oil Recovery and Improved Drilling Technology: *Progress Review No. 24*, February 1981.
5. Machado, C. J, and Ikoku, C. U., "Experimental Determination of Solids Fraction and Minimum Volumetric Requirements in Air and Gas Drilling," *Journal of Petroleum Technology*, November 1982.

6. Marcus, R. D., et al., *Pneumatic Conveying of Solids*, Chapman and Hall, 1990.
7. Mitchell, R. F., "Simulation of Air and Mist Drilling for Geothermal Wells," *Journal of Petroleum Technology*, November 1983.
8. Tian, S., and Adewumi, M. A., "Development of Hydrodynamic-Model-Based Air-Drilling Design Procedures, "*SPE Drilling Engineering*, December 1992.
9. Lyons, W. C., Miska, S., and Johnson, P. W., "Downhole Pneumatic Turbine Motor: Testing and Simulation Results," *SPE Drilling Engineering*, September 1990.
10. Daugherty, R. L., Franzini, J. B., and Finnemore, E. J., *Fluid Mechanics with Engineering Applications*, Eighth Edition, McGraw-Hill, 1985.
11. *API Recommended Practice for Testing Foam Agents for Mist Drilling*, First Edition, API RP-46, November 1966.
12. *Handbook of Chemistry*, McGraw-Hill, 1956.
13. Miska, S., "Should We Consider Air Humidity in Air Drilling Operations," *Drill Bit*, July 1984.
14. U.S. Bureau of Mines Report Investigations No. 3798.
15. Coward, H. F., and Jones, G. W., "Limits of Flammability of Gases and Vapors," *Bureau of Mines Bullitin 503*, Washington, DC 1952.
16. Allan, P. D., "Nitrogen Drilling System for Gas Drilling Applications," *SPE 28320*, Presented at the SPE 69th Annual Technical Conference and Exhibition, New Orleans, Louisiana, September 25-28, 1994.

CHAPTER

9 Aerated Fluids Drilling

The term aerated fluids describes the broad category of drilling fluids that are basically incompressible fluids injected with compressed air or other gases. Aerated drilling fluids have been used to drill both shallow and deep boreholes since the advent of air and gas drilling technology in the mid-1930s. The first engineering discussion of an aerated drilling mud project was given in 1953 [1]. Aerated drilling fluids were initially used to drill through rock formations that had fracture and/or pore systems that could drain the incompressible drilling fluids (e.g., fresh water, water- and oil-based drilling muds, formation water, and formation crude oil) from the annulus. These borehole drilling fluid theft rock formations are called lost circulation sections. The injection of air into drilling muds has been considered an important technological tool in countering the detrimental effects of lost circulation sections. The injection of air into drilling mud creates bubbles in the mud and, because of the surface tension properties of the bubbles relative to the properties of rock and drilling mud, the bubbles tend to fill in the fracture or pore openings in the borehole wall as the aerated mud attempts to flow to the thief fractures and pores [2]. This bubble blockage restricts the flow of the drilling mud into these lost circulation sections and thereby allows the drilling operations to progress safely. Aerated fluids have been used to avoid lost circulation in shallow water well drilling, geotechnical drilling, mining drilling, and in deep oil and natural gas recovery drilling operations. Aerated fluids drilling operations are nearly always direct circulation operations.

Since the late 1980s another important application for aerated fluids drilling operations has emerged. This is underbalanced drilling applied to oil and natural gas recovery operations. Over the past two decades practical field research has demonstrated that most oil and natural gas bearing rock formations can be produced more efficiently if they are drilled with drilling fluids that have flowing bottom hole pressures that are slightly less than the pore pressures of the potential producing rock formations being drilled. Underbalanced drilling operations allow the oil or natural gas to be produced into the annulus as the drilling operation progresses. The underbalanced drilling operation allows the natural fracture and pore systems to be kept clear of rock cutting fines and drilling mud filter cake, thereby avoiding formation damage. Formation damage has been a problem in oil and natural gas recovery operations nearly since the discovery of oil and natural gas mineral deposits. Underbalanced drilling operations are often carried out using a variety of incompressible fluids (e.g., crude oil, formation water, or clear water) and a variety of compressible gases (e.g., air, inert atmosphere, or natural gas). An inert atmosphere is created by a filter system (placed downstream of the primary compressor) that strips most of the oxygen from the intake air [3]. This drilling gas is known as membrane generated nitrogen.

This chapter outlines the steps and methods used to plan a successful aerated fluids drilling operation. This chapter also discusses the application of these steps and methods to typical deep drilling operations. The objective of these steps and methods is to allow engineers and scientists to plan their drilling operations cost effectively. The additional benefit of this planning process is that data created by the process can be used later to control the drilling operations as the actual operations progress.

9.1 DEEP WELL DRILLING PLANNING

Aerated drilling operations use a variety of incompressible fluids and compressed gases to develop a gasified drilling fluid. The majority of the operations use standard fresh water-based drilling mud with injected compressed air or membrane generated nitrogen. In this chapter, a performance drilling operation will be assumed and, therefore, standard drilling mud and atmospheric air will be used as the example aerated drilling fluids.

The basic planning steps for a deep well are as follow:

1. Determine the geometry of the borehole sections to be drilled with aerated drilling fluids (i.e., open hole diameters, the casing inside diameters, and section depths).
2. Determine the geometry of the associated drill string for each section to be drilled with aerated drilling fluids (i.e., drill bit size and type, the drill collar size, drill pipe size and description, and maximum depth).
3. Determine the type of rock formations to be drilled in each section and estimate the anticipated drilling rate of penetration (use offset wells).
4. Determine the elevation of the drilling site above sea level, the temperature of the air during the drilling operation, and the approximate geothermal temperature gradient.
5. Establish the objective of the aerated drilling fluids operation:
   - Performance drilling
   - Underbalanced drilling
6. For either of the aforementioned objectives, determine the required approximate volumetric flow rate of the mixture of incompressible fluid and the compressed air (or other gas) to be used in the aerated fluid drilling

operation. This is usually the minimum incompressible volumetric flow rate that would clean the rock cuttings from the bottom of the well and transport the cuttings to the surface. In most aerated drilling operations, the incompressible fluid volumetric flow rate is held constant as drilling progresses through the open hole interval.

7. Using the incompressible fluid and gas volumetric flow rates to be injected into the well, determine the bottom hole annulus pressures and the surface injection pressures as a function of open hole interval drilling depth.
8. Select the contractor compressor(s) that will provide the drilling operation with the appropriate air or gas volumetric flow rate needed to aerate the drilling fluid properly.

Chapter 6 derived and summarized the basic direct circulation drilling planning governing equations. The equations in that chapter are utilized in the discussions and illustrative examples that follow.

9.2 AERATED FLUIDS DRILLING OPERATIONS

Several drill string and well configurations are used for aerated fluid drilling operations. These are divided into two general technique classes of air (or gas) injection operations: drill pipe injection and annulus injection [4].

9.2.1 Drill Pipe Injection

Figure 9-1 shows a schematic of the drill pipe injection-aerated drilling configuration. In this configuration, incompressible fluids and compressible air (or other gas) are injected together into the top of the drill string (at P_{in}). These two fluid streams mix as they flow down the inside of the drill string and pass through the drill bit nozzles. When the mixture of these fluids flows into the bottom of the annulus, the rock cuttings (from the advance of the drill bit) are entrained and the resulting mixture flows to the surface in the annulus. The mixture exits the annulus (at P_e) into a horizontal flow line. This horizontal flow line flows to either conventional open mud tanks or sealed return tanks. Conventional open mud tanks are used when the returning air is not mixed with contaminated fluids or gases or mixed with produced hydrocarbons. Separators are used when hydrocarbon fluids are expected in the return flow from the well.

Figure 9-2 shows a schematic of the jet sub drill string injection drilling configuration. An alternative to flowing the mixture down the entire length of the inside of the drill string to the drill bit is to place a jet sub above the drill collars to allow most of the compressed air (or other gas) to pass from the inside of the drill string into the annulus before the two-phase mixture flows through the inside of the drill collars. The jet sub drill string injection technique is usually used for deep aerated drilling operations where the bottom section of the well is usually drilled with

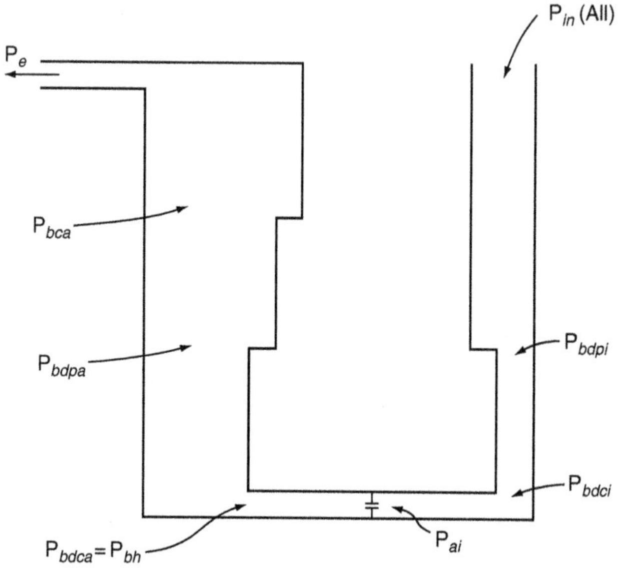

FIGURE 9-1. Schematic of direct circulation. P_{in} is the injection pressure into the top of the drill string, P_{bdpi} is the pressure at the bottom of the drill pipe inside the drill string, P_{bdci} is the pressure at the bottom of drill collars inside the drill string, P_{ai} is the pressure above the drill bit inside the drill string, P_{bdca} is the pressure at the bottom of drill collars in the annulus, P_{bh} is the bottom hole pressure in the annulus, P_{bdpa} is the pressure at the bottom of the drill pipe in the annulus, P_{bca} is the pressure at the bottom of the casing in the annulus, and P_e is the pressure at the top of the annulus.

a small diameter drill bit and corresponding small diameter drill collars. In much the same way as bubbles provide a resistance mechanism to counter the loss of circulation zones, the surface tension of the bubbles in the aerated fluid creates somewhat high pipe friction resistance when flowing through small inside diameter drill collar opens. This increased resistance to flow is not modeled by conventional friction factors derived from homogeneous fluids experiments [5–9]. Thus, in order to reduce circulation pumping pressures, the jet sub is placed in the drill string in the drill pipe section above the drill collars. Usually the jet sub is placed several drill pipe joints above the drill pipe to drill collar transition. There are usually two to three jet nozzles in the jet sub. The jet sub orifices can be sized to allow the compressed gas to be vented to the annulus before the gas-aerated fluid can flow to the drill collars.

When the aerated drill string injection technique is used to drill through loss of circulation zones, a constant volumetric flow rate of incompressible drilling fluid is circulated. The actual volumetric flow rate of the incompressible

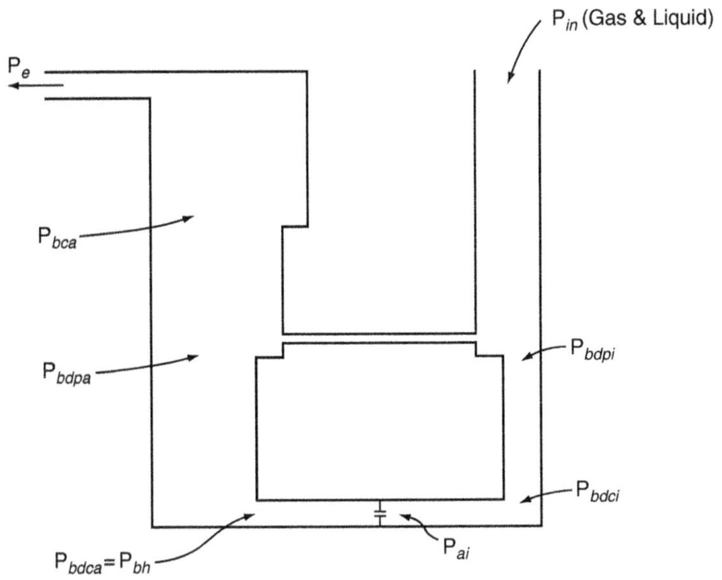

FIGURE 9-2. Schematic of jet sub drill pipe injection-aerated flow.

drilling fluid is usually the minimum flow rate that will adequately clean and carry the rock cuttings from the bottom of the annulus. The volumetric flow rate of injected compressible gas is the amount that will facilitate the performance drilling objectives (e.g., improved rate of penetration, reduce loss of circulation).

In underbalanced drilling situations, these operations usually require that a near constant bottom hole pressure be maintained as the drill bit is advanced through the reservoir. To accomplish this objective, a constant volumetric flow rate of incompressible drilling fluid is injected into the inside of the drill string. Here again, it is desirable to design the volumetric flow rate of the incompressible drilling fluid that will adequately clean and carry the rock cuttings from the bottom of the annulus. In order to maintain a nearly constant bottom hole annulus pressure while the drill bit is advanced, the volumetric flow rates of either the incompressible drilling fluid or the compressed gas (or both) can be adjusted as the well is deepened.

9.2.2 Annulus Injection

Figure 9-3 shows the annulus injection of gas for an aerated drilling operation (can be accomplished with a parasite tubing string drilling behind the casing, a parasite concentric tubing and drill pipe string, or through a completion string) [4]. In this configuration the incompressible drilling fluid is injected into the drill

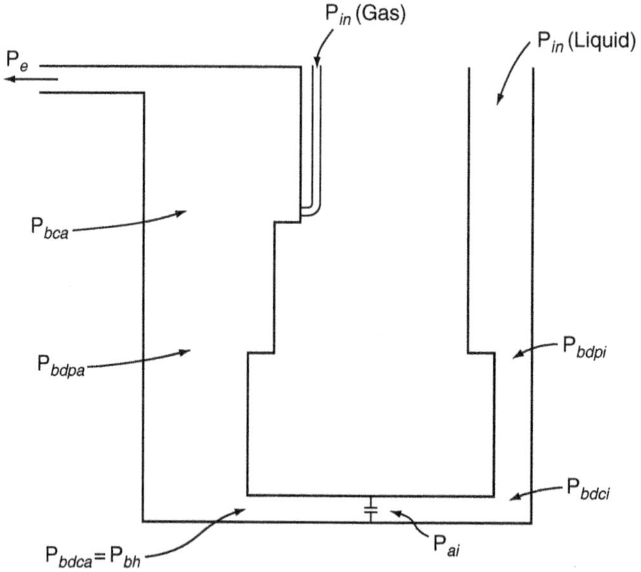

FIGURE 9-3. Schematic of annulus injection-aerated flow (parasite tubing string example).

string and the compressed gas is injected into the annulus. In the early years of aerated drilling, the technique was used to eliminate the threat of formation fracture (and the eventual loss of drilling fluid). More recently, this technique has been used for limited underbalanced drilling operations (usually in the geometry of parasite casing drilling or through completion drilling).

9.2.3 Advantages and Disadvantages

The Drill Pipe Injection Technique

Advantages to the drill pipe injection technique are as follow:

- Because the technique does not require additional downhole equipment, drill pipe injection is less costly than annulus injection.
- Nearly the entire annulus is filled with aerated fluid, thus lower bottom hole pressures can be achieved.
- Because the gas is injected into the annulus at or near the bottom of the annulus, less gas volumetric flow rate is needed to achieve a given bottom hole pressure than via annulus injection.

Disadvantages to the drill pipe injection technique are as follow:

- Aeration of the incompressible drilling fluid cannot be continued when circulation is discontinued for connections and tripping. Therefore, it is difficult to maintain constant bottom hole annulus pressure throughout the entire drilling operations.

- Because the injected gas is trapped under pressure inside the drill string by the various string floats, time must be allowed for the pressure bleed down when making connections and trips. Here again, the bleed down makes it difficult to maintain a constant bottom hole pressure.
- The flow down the inside of the drill string is two-phase flow; therefore, higher pipe wall friction losses are present. The high friction losses result in high pump and compressor pressures during injection. This is somewhat relieved when a jet sub is used.
- The gas phase in the aerated flow attenuates the pulses of conventional measure-while-drilling (MWD) systems. Therefore, conventional mud pulse telemetry MWD cannot be used.

The Annulus Injection Technique

Advantages to the annulus injection technique are as follow:

- Aeration of the incompressible drilling fluid in the annulus above the gas annulus entrance position can continue during connections and trips.
- Flow down the inside of the drill string is single phase; therefore, conventional mud pulse MWD can be used in aerated directional drilling operations.
- The compressor pressure used to maintain gas injection will usually be low when compared to the pressure required for direct circulation (to the drill bit) aerated drilling operations.

Disadvantages to the annulus injection technique are as follow:

- Because the parasite tubing string or the temporary casing is placed at a particular fixed location in the well, the aeration technique is more inflexible than the drill pipe injection techniques.
- Initiating (kickoff) gas flow to the annulus requires very high compressor pressures.
- Because the gas is injected into the annulus at fixed depths that are well above the bottom of the annulus, higher gas volumetric flow rates are required to maintain constant bottom hole pressures than in the drill pipe injection technique.

9.3 MINIMUM VOLUMETRIC FLOW RATES

Most aerated drilling operations are planned with a constant volumetric flow rate of incompressible drilling fluid and only the volumetric flow rate of the compressed gas is allowed to vary. The volumetric flow rate of gas is usually increased as the depth is increased in order to maintain the same aerated fluid properties in the annulus column. The drill pipe injection technique requires that both the incompressible drilling fluid injection and the compressible gas injection be suspended when connections and trips are made. Similarly, the annulus injection technique requires that the incompressible drilling fluid injection be suspended when connections and trips are made. Further, the cleaning, lifting, and suspension capabilities of the incompressible drilling mud is generally independent of the depth of drilling. Conversely, the cleaning and lifting capabilities of compressed gas are dependent of the depth of drilling. Also, it must be noted that compressed gas drilling fluids have little or no suspension capabilities. Therefore, when designing an aerated drilling fluid, the injected compressed gas should not be assumed to contribute to bottom hole cleaning, lifting, and suspension of rock cuttings in the annulus. The additional cleaning and lifting properties of the compressed gas to the aerated drilling fluid should be considered bonuses. This argument requires that the incompressible drilling fluid properties and circulation characteristics be designed to provide the aerated drilling operations with stand-alone cleaning, lifting, and suspension capabilities of the rock cuttings in the annulus.

9.3.1 Discussion of Theories

A variety of minimum volumetric flow rate theories can be used to design incompressible drilling fluid properties and circulation characteristics for direct circulation drilling operations [10–12]. In order to formulate a workable simple procedure for determining the minimum volumetric flow rates of incompressible fluid and ultimately the gas, this chapter will confine its attention to performance drilling operations.

The average annulus velocity of the fluid, V_f, in the largest annulus cross section will be the sum of the critical concentration velocity, V_c, and the terminal velocity, V_t, of the average size rock cutting particle in the incompressible drilling fluid.

Thus, the average fluid velocity in the annulus is

$$V_f = V_c + V_t, \tag{9-1}$$

where V_f is the incompressible drilling fluid (ft/sec, m/sec), V_c is the critical concentration velocity (ft/sec, m/sec), and V_t is the terminal velocity of the rock cuttings particle (ft/sec, m/sec).

The critical concentration velocity is the additional velocity needed to distribute the rock cuttings through the incompressible drilling fluid at a predetermined concentration factor. The usual concentration factor is 0.04 or lower [13]. Therefore, the critical concentration velocity V_c is

$$V_c = \frac{\kappa}{3,600\,C}, \tag{9-2}$$

where κ is the instantaneous drilling rate (ft/hr, m/hr) and C is the concentration factor (usually assumed to be 0.04).

Equations (9-1) and (9-2) can be used with any consistent set of units.

Terminal Velocities

For direct circulation operations the terminal velocity of the rock cutting particle is assessed in the annulus section of the borehole where the cross-sectional area is the largest. It is assumed that the rock cuttings average particle will fall in the drilling fluid with a terminal velocity (and, therefore, a particle velocity Reynolds number) that indicates turbulent flow conditions around the particle.

Aerated drilling operations can utilize either Newtonian or non-Newtonian incompressible fluid components. In this section, only Newtonian fluids will be used in examples. The Stokes' law describes the terminal velocity of a particle where the flow around the particle is laminar [11]. This equation in consistent units is

$$V_{tl} = \left(\frac{D_c^2}{18\,\mu}\right)\left(\gamma_s - \gamma_f\right), \tag{9-3}$$

where V_{tl} is the terminal velocity for laminar flow conditions (ft/sec, m/sec), D_c is the approximate average diameter of the rock cutting particle (ft, m), γ_s is the specific weight of the solid rock cutting (lb/ft^3, N/m^3), γ_f is the specific weight of the incompressible drilling fluid (lb/ft^3, N/m^3), and μ is the absolute viscosity of the fluid (lb-sec/ft^2, N-sec/m^2).

The Rittinger empirical correlation has been modified to take into account friction loss in the flow around the particle [11]. This modified correlation is given in Equation (9-4). This equation can be used for both transitional and turbulent flows depending on the nondimensional particle flow Reynolds number value and the associated flow friction factor. This correlation in consistent units is

$$V_{tt} = \left[\frac{4}{3}\left(\frac{\gamma_s - \gamma_f}{\gamma_f}\right)\frac{g\,D_c}{f_p}\right]^{0.5}, \tag{9-4}$$

where V_{tt} is the terminal velocity for transitional/turbulent conditions (ft/sec, m/sec), g is the acceleration of gravity (32.2 ft/sec^2, 9.81 m/sec^2), and f_p is the particle friction factor.

The nondimensional Reynolds number for flow around the particle in consistent units is

$$\mathbf{N}_{\mathbf{R}_c} = \frac{D_c V_t}{\nu}, \tag{9-5}$$

where ν is the kinematic viscosity of the flowing fluid (ft^2/sec, m^2/sec).

Equation (9-3) is valid for particle Reynolds number ≤ 3 (laminar flow conditions).

Equation (9-4) is valid for transitional flow conditions for particle Reynolds number >3 and ≤ 300 and valid for turbulent flow conditions for particle Reynolds number >300.

Figure 9-4 shows the relationship between particle Reynolds number and particle friction factor, f_p, for particles having various coefficients of sphericity, ϕ. The sphericity coefficient is used to describe the geometry of the outer surface

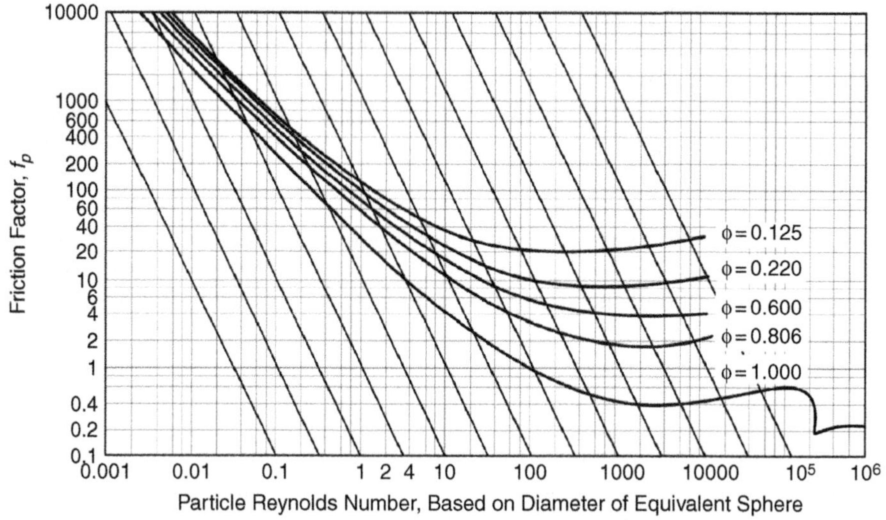

FIGURE 9-4. Cuttings particle Reynolds number for non-Newtonian terminal velocity determinations.

of the particle (the perfect sphere $\phi = 1.0$). The lower the sphericity value, the higher the particle friction factor. This also means that the lower the sphericity value, the lower the terminal velocity. Using a sphericity of 1.0 will give the highest, and therefore the most, engineering conservative drilling fluid lifting properties.

9.3.2 Engineering Practice

The engineering practice for aerated drilling operations is to design the incompressible drilling fluid to have minimum lifting capability for the planned open hole interval. Modern aerated drilling operations utilize a variety of incompressible drilling fluids. These can be fresh water and salt water drilling muds, oil-based drilling muds, production fluids, fresh waters, and formation waters.

The minimum volumetric flow rate of the incompressible drilling fluid will be determined using Equations (9-1) through (9-5). When assessing the lifting capability of Newtonian incompressible drilling fluids, analyses should include whether the flow conditions around the particle are laminar, transitional, or turbulent.

Once the incompressible drilling fluid minimum volumetric flow rate has been determined, a volumetric flow rate of injected compressed gas can be determined for a variety of performance or underbalanced drilling objectives. The basic equations derived in Chapter 6 for direct circulations will be used to solve the illustrative examples in this section.

Illustrative Example 9.1 The borehole to be used in this illustrative example is the basic example used in Chapter 8. The $7^7/_8$-in (*200 mm*)-diameter borehole is drilled out of the bottom of API $8^5/_8$-in (*219 mm*)-diameter 32.00-lb/ft (*14.00 kg/m*) nominal casing set to 7000 ft (*2134 m*)(see Figure 9-4 for well casing and open hole geometric configuration). This is special clearance casing fabricated by Lonestar Steel, which has a drift diameter that will pass a $7^7/_8$-in (*200 mm*)-diameter drill bit. The drill bit used to drill the interval is a $7^7/_8$-in (*200 mm*)-diameter tricone roller cutter insert type with three $^{11}/_{32}$-in (*8.7 mm*) nozzles. The anticipated drilling rate in a sandstone and limestone sequence (sedimentary rock) is approximately 60 ft/hr (*18.3 m/hr*). The open hole interval below the casing shoe to be drilled is from 7000 ft (*2134 m*) to 10,000 ft (*3047 m*).

The drill string for this illustrative example is made up of 500 ft (*152.4 m*) of $6^1/_4$-in (*158.8 mm*) by $2^{13}/_{16}$-in (*71.4 mm*) drill collars above the drill bit and

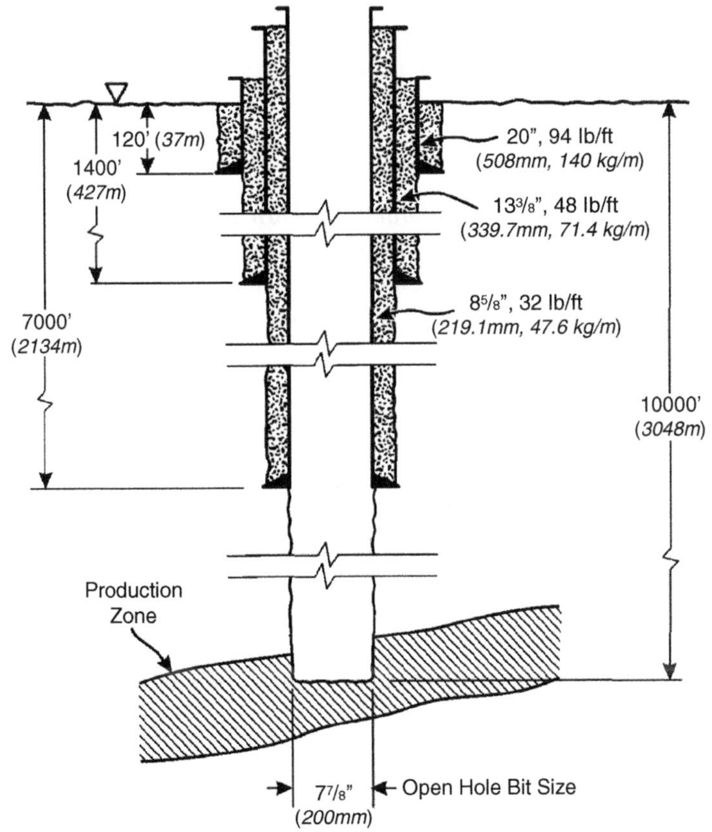

FIGURE 9-5. Illustrative Example 9.1 casing and open hole well geometric configuration.

API $4^1/_2$-in (*114 mm*), 16.60-lb/ft (*7.52 kg/m*) nominal, IEU-S135, NC46 drill pipe to the surface. The tool joints of this drill pipe have an outside diameter of $6^1/_4$ in (*154.8 mm*) and an inner diameter of $2^3/_4$ in (*69.9 mm*). The drilling is carried out at a surface location of 4000 ft (*2129 m*) above sea level (see Table 5-1 and Appendix B). The regional geothermal gradient is approximately 0.01°F/ft (*0.018°C/m*).

The incompressible drilling fluid has a specific weight of 10 lb/gal (*1200.0 kg/m*^{3}) and a viscosity of 30 cp. The drill pipe injection technique is used, and the drilling operation is carried out to maintain a bottom hole pressure of 3360 psig (*2414 N/cm*^{2} *gauge*) at the bottom of the open hole interval at 10,000 ft (*3048 m*). A horizontal section of API line pipe that runs 100 ft (*30.5 m*) from just above the BOP stack to the mud tank is used as the return flow line. The surface return line is an API $6^5/_8$-in (*168 mm*) line pipe, 32.71 lb/ft (*49.0 kg/m*) nominal, Grade B. The inside diameter of this line pipe is 5.625 in (*143 mm*). Two gate valves are installed in the return flow line at the end of the line that is attached to the BOP stack. These valves have an inside diameter of $5^9/_{16}$ in (*141 mm*). An average drilling cuttings particle diameter of 0.36 in (*9.14 mm*) and a sphericity coefficient of 1.0 are used to determine the approximate minimum volumetric flow rate of the incompressible drilling fluid that is required for this aerated drilling operation.

Minimum Incompressible Fluid Volumetric Flow Rate (USCS Units)

The incompressible fluid used in this illustrative example is a water-based drilling mud with plastic properties. Therefore, the minimum volumetric flow rate of the incompressible drilling fluid is determined using Equations (9-1) through (9-5) above.

The anticipated drilling rate is 60 ft/hr. The critical concentration velocity is

$$V_c = \frac{60}{3,600\ (0.04)}$$

$$V_c = 0.417\ \text{ft/sec}.$$

The specific weight of the sedimentary rock to be drilled is approximately

$$\gamma_s = (2.7)\ (62.4)$$

$$\gamma_s = 168.5\ \text{lb/ft}^3,$$

where the specific gravity of sedimentary rock is assumed to be approximately 2.7 (see Chapter 6).

The specific weight of the 10-lb/gal incompressible drilling fluid in consistent units is

$$\gamma_f = \frac{10\ (12)^3}{231}$$

$$\gamma_f = 75.0\ \text{lb/ft}^3.$$

The viscosity of the incompressible drilling fluid in consistent units is

$$\mu_f = 30\,(0.001)(0.02089)$$

$$\mu_f = 0.0006267\ \frac{\text{lb} - \text{sec}}{\text{ft}^2}.$$

Assuming laminar flow conditions around the particle, Stokes' law will be used to determine the initial value of the particle terminal velocity. Equation (9-3) is

$$V_{tl} = \left[\frac{\left(\frac{0.36}{12}\right)^2}{18(0.0006267)}\right](168.5 - 75.0)$$

$$V_{tt} = 7.46\ ft/\text{sec}.$$

The incompressible drilling fluid density ρ_f is

$$\rho_f = \frac{75.0}{32.2}$$

$$\rho_f = 2.329\ \frac{\text{lb} - \text{sec}^2}{\text{ft}^4}.$$

The general equation for kinematic viscosity of the fluid is

$$\nu = \frac{\mu}{\rho}, \tag{9-6}$$

where ν is kinematic viscosity (ft^2/sec). The kinematic viscosity is

$$\nu = \frac{0.0006267}{2.329}$$

$$\nu = 0.0002691\ \text{ft}^2/\text{sec}.$$

The Reynolds number of the fluid flow around the average cuttings particle [Equation (9-5)] is

$$\mathbf{N}_{\mathbf{R_c}} = \frac{7.46\left(\frac{0.36}{12}\right)}{0.0002691}$$

$$\mathbf{N}_{\mathbf{R_c}} = 832.$$

This Reynolds number is greater than 3 and therefore the flow around the particle is not laminar.

Assuming that the flow conditions around the particle are transitional, Equation (9-4) becomes

$$V_{tt} = \left[\frac{4}{3}\left(\frac{168.5 - 75.0}{75.0}\right)\frac{(32.2)\left(\frac{0.36}{12}\right)}{f_p}\right]^{0.5}.$$

The value of f_p in the aforementioned equation is determined by trial and error using the result of the aforementioned equation, Equation (9-5) (for $\mathbf{N_{R_c}}$), and Figure 9-4. The value of f_p must be on the $\psi = 1.0$ curve. These analyses yield a value of $f_p = 0.80$ and a $\mathbf{N_{R_c}} = 158$. This clearly establishes that the fluid flow around the particle is transitional. The transitional terminal velocity of the average cuttings particle from the aforementioned equation for $f_p = 0.80$ is

$$V_{tt} = 1.417\ ft/\text{sec.}$$

Equation (9-1) for the average total fluid velocity in annulus becomes

$$V_f = 1.417 + 0.417$$

$$V_f = 1.834\ ft/\text{sec.}$$

The total velocity of the fluid must be the minimum average velocity of the incompressible fluid in the borehole annulus section where the cross-sectional area is the largest. In this illustrative example, the largest cross-sectional area of the annulus is in the cased section of the well where the inside diameter of the casing is 7.921 in and the outside of the diameter of the drill pipe is 4.50 in. Thus, this annulus cross-sectional area, A_a, is

$$A_a = \left(\frac{\pi}{4}\right)\left[\left(\frac{7.921}{12}\right)^2 - \left(\frac{4.50}{12}\right)^2\right]$$

$$A_a = 0.232\ \text{ft}^2.$$

The volumetric flow rate in the aforementioned annulus section is

$$Q_a = (0.232)(1.834)$$

$$Q_a = 0.426\ \text{ft}^3/\text{sec.}$$

In field units, this is

$$q_a = \frac{(0.426)(60)(12)^3}{231}$$

$$q_a = 191.0\ \text{gpm.}$$

Minimum Incompressible Fluid Volumetric Flow Rate (SI Units)

The incompressible fluid used in this illustrative example is a water-based drilling mud with plastic properties. Therefore, the minimum volumetric flow rate of the incompressible drilling fluid is determined using Equations (9-1) through (9-5).

The anticipated drilling rate is 18.3 m/hr. The critical concentration velocity is

$$V_c = \frac{18.3}{3,600\ (0.04)}$$

$$V_c = 0.127\ \text{m/sec.}$$

The specific weight of the sedimentary rock to be drilled is approximately

$$\gamma_s = (2.7)(1000.0)(9.81)$$

$$\gamma_s = 26487.0 \text{ N/m}^3,$$

where the specific gravity of sedimentary rock is assumed to be approximately 2.7 (see Chapter 6).

The specific weight of the 1200.0-kg/m^3 incompressible drilling fluid in consistent units is

$$\gamma_f = 1200.0\ (9.81)$$

$$\gamma_f = 11772.0 \text{ N/m}^3.$$

The viscosity of the incompressible drilling fluid in consistent units is

$$\mu_f = 30\ (0.001)$$

$$\mu_f = 0.030\ \frac{\text{N} - \text{sec}}{\text{m}^2}.$$

Assuming laminar flow conditions around the particle, Stokes' law will be used to determine the initial value of the particle terminal velocity. Equation (9-3) is

$$V_{tt} = \left[\frac{(0.00914)^2}{18\ (0.030)}\right](28487.0 - 11772.0)$$

$$V_{tt} = 2.28\ m/\text{sec}.$$

The incompressible drilling fluid density ρ_f is

$$\rho_f = 1200.0\ N - \text{sec}^2/m^4.$$

The kinematic viscosity is

$$v = \frac{0.030}{1200.0}$$

$$v = 0.0000250 \text{ m}^2/\text{sec}.$$

The Reynolds number of the fluid flow around the average cuttings particle [Equation (9-6)] is

$$\mathbf{N}_{\mathbf{R}_c} = \frac{2.28(0.00914)}{0.0000250}$$

$$\mathbf{N}_{\mathbf{R}_c} = 832.$$

This Reynolds number verifies that the flow around the particle is not laminar.

Assuming that the flow conditions around the particle is transitional, Equation (9-4) becomes

$$V_{tt} = \left[\frac{4}{3}\left(\frac{28487.7 - 11772.0}{11772.0}\right)\frac{(9.81)(0.00914)}{f_p}\right]^{0.5}.$$

The value of f_p in the aforementioned equation is determined by trial and error using the result of the aforementioned equation, Equation (9-5) (for $\mathbf{N}_{\mathbf{R_c}}$), and Figure 9-4. The value of f_p must be on the $\psi = 1.0$ curve. These analyses yield a value of $f_p = 0.80$ and a $\mathbf{N}_{\mathbf{R_c}} = 158$. This clearly establishes that the fluid flow around the particle is transitional. The transitional terminal velocity of the average cuttings particle from the above equation for $f_p = 0.80$ is

$$V_{tt} = 0.432m/\text{sec}.$$

Equation (9-1) for the average total fluid velocity in annulus becomes

$$V_f = 0.432 + 0.127$$

$$V_f = 0.559\ m/\text{sec}.$$

The total velocity of the fluid must be the minimum average velocity of the incompressible fluid in the borehole annulus section where the cross-sectional area is the largest. In this illustrative example, the largest cross-sectional area of the annulus is in the cased section of the well where the inside diameter of the casing is 201.2 mm and the outside of the diameter of the drill pipe is 114.3 mm. Thus, this annulus cross-sectional area, A_a, is

$$A_a = \left(\frac{\pi}{4}\right)\left[(0.2012)^2 - (0.1143)^2\right]$$

$$A_a = 0.02153\ \text{m}^2.$$

The volumetric flow rate in the aforementioned annulus section is

$$Q_a = (0.0215)(0.559)$$

$$Q_a = 0.0121\ \text{m}^3/\text{sec}.$$

In field units, the aforementioned is

$$q_a = 12.04\ liters/\text{sec}$$

$$q_a = 722.7\ \text{lpm}.$$

9.4 NONFRICTION AND FRICTION ILLUSTRATIVE EXAMPLES

Over the past two decades, the analyses of aerated fluid vertical drilling problems have been carried out by two distinct analytic methodologies.

The first methodology ignores the major and minor friction losses due to fluid flow inside the drill string and in the annulus. This methodology includes only the fluid column weight [14, 15]. This methodology was originally derived and adapted for aerated drilling from the large body of literature pertaining to multiphase flow of oil and natural gas in production tubing [16, 17].

The second methodology can include all the complexity of fluid flow friction losses. The initial application of this methodology also came from adapting multiphase oil and gas flow tubing production theory to aerated drilling annulus problems. This production theory application includes only major friction losses and was not applicable to complicated borehole geometry. New additions to this methodology, which do not come from production literature, have included major and minor losses and can be applied to complicated borehole geometry. Also included in this second methodology is the inclusion of liquid holdup correlations, which were first examined in the production literature [16, 17].

Chapter 6 derived the basic aerated fluid drilling governing equations and presented their auxiliary friction factor and nozzle flow equations. These equations form the foundation for both methodologies as discussed in this treatise.

9.4.1 Nonfriction Approximation

The simple nonfriction methodology allows straightforward deterministic approximate solutions of aerated drilling problems. However, the practical applicability of these nonfriction solutions is limited to shallow (generally less than 3000 ft of depth) wells with simple geometric profiles. The nonfriction solution will be applied to a deep well example only as a demonstration and ultimate comparison of the results to those obtained from the friction solution.

In what follows, the basic equations in Chapter 6 for aerated drilling are used to derive the nonfriction governing equation. Letting $f \rightarrow 0$ in Equation (6-48) yields

$$\int_{P_e}^{P_{bb}} \frac{dP}{\left[\dfrac{\dot{w}_t}{\left(\dfrac{P_g}{P}\right)\left(\dfrac{T_{av}}{T_g}\right)Q_g + Q_f}\right]} = \int_0^H dh,$$

where P_{bb} is the pressure at the bottom of the annulus (lb/ft^2, abs) and P_e is the pressure at the exit from the annulus (lb/ft^2, abs). This equation can be rearranged and integrated to yield

$$\left| P_g\left(\frac{T_{av}}{T_g}\right)\left(\frac{Q_g}{\dot{w}_t}\right)\ln P + \left(\frac{Q_f}{\dot{w}_t}\right)P \right|_{P_e}^{P_{bb}} = \left| h \right|_0^H .$$

Evaluating the aforementioned equation at the limits, rearranging the result, and solving for gas volumetric flow rate, Q_g, yields

$$P_g\left(\frac{T_{av}}{T_s}\right)Q_g \ln\left(\frac{P_{bb}}{P_s}\right) + Q_f(P_{bb} - P_s) = \dot{w}_t\, H. \qquad (9\text{-}7)$$

The nonfriction solution is usually not applied to a deep well like our example. This is only done as a demonstration. It is instructive to compare the nonfriction solution results to the full friction solution results obtained later in this chapter. The approximate values for the volumetric flow rates of the incompressible fluid and the compressible gas (air) can be used as initial values for the complicated trial and error solution required for the full friction solution.

Illustrative Example 9.2 Using the nonfriction method and the given data and results obtained in Illustrative Example 9.1, determine the approximate volumetric flow rate of air required for the incompressible drilling fluid (drilling mud) volumetric flow rate of approximately 191 gpm (*723 lpm*) while drilling at 10,000 ft (*3048 m*).

USCS Units

Table 5-1 gives an average atmospheric pressure of 12.685 psia for a surface location of 4000 ft above sea level (also see Appendix B). The actual atmospheric pressure for the air at the drilling location (that will be used by the compressor) is

$$p_{at} = 12.685 \text{ psia}$$

$$P_{at} = p_{at}\ 144$$

$$P_{at} = 1826.6 \text{ lb/ft}^2 \text{ abs.}$$

The actual atmospheric temperature of the air at the drilling location (used by the compressor), T_{at}, is

$$t_{at} = 44.74^\circ\ F$$

$$T_{at} = t_{at} + 459.67$$

$$T_{at} = 504.41^\circ\ R.$$

Thus, P_g and T_g become

$$P_g = P_{at} = 1826.6 \text{ lb/ft}^2 \text{ abs}$$

$$T_g = T_{at} = 504.41^\circ\ R.$$

Using Equation (4-11), the specific weight of the gas entering the compressor is

$$\gamma_g = \frac{(1826.6)(1.0)}{(53.36)(504.41)}$$

$$\gamma_g = 0.0679 \text{ lb/ft}^3.$$

The bottom hole pressure in absolute pressure is

$$p_{bb} = 3360 + 12.685$$

$$p_{bb} = 3372.685 \text{ psia}$$

$$P_{bb} = p_{bb}\ 144$$

$$P_{bb} = 485666.6 \text{ lb/ft}^2 \text{ abs.}$$

The temperature of the rock formations near the surface (geothermal surface temperature) is estimated to be the approximate average year-round temperature at that location on the earth's surface. Table 5-1 gives 44.74°F for the average year-round temperature for a surface elevation location of 4000 ft above sea level (see

Appendix B). Therefore, the absolute temperature of the rock formations at the surface is

$$t_r = 44.74^\circ\ F$$

$$T_r = t_r + 459.67$$

$$T_r = 504.41^\circ\ R.$$

The depth of the well is

$$H = 10000 \text{ ft.}$$

The bottom hole temperature, T_{bh}, is

$$T_{bh} = T_r + 0.01\ H$$

$$T_{bh} = 604.41^\circ\ R.$$

The borehole average temperature is

$$T_{av} = \frac{504.41 + 604.41}{2}$$

$$T_{av} = 554.41^\circ\ R.$$

The open hole diameter is

$$d_h = 7.875 \text{ inches}$$

$$D_h = \frac{d_h}{12}$$

$$D_h = 0.656 \text{ ft.}$$

The estimated drilling rate of penetration is 60 ft/hr. The weight rate of flow of solids from the advance of the drill bit is

$$\dot{w}_s = \left(\frac{\pi}{4}\right) D_h^2\,(62.4)\ S_s \left[\frac{\kappa}{(60)(60)}\right] \tag{9-8}$$

$$\dot{w}_s = \left(\frac{\pi}{4}\right) (0.656)^2\ (62.4)\ (2.7)\left[\frac{60}{(60)(60)}\right]$$

$$\dot{w}_s = 0.949 \text{ lb/sec.}$$

The volumetric flow rate of drilling mud in consistent units is

$$Q_f = \frac{(191)(231)}{(12)^3(60)}$$

$$Q_f = 0.426 \text{ ft}^3/\text{sec.}$$

The weight rate of flow of the drilling mud is

$$\dot{w}_f = (75.0)(0.426)$$

$$\dot{w}_f = 31.92 \text{ lb/sec.}$$

Equation (9-7) can be solved by trial and error. The value of q_g is selected and $\dot{w}_g$ determined from

$$\dot{w}_g = \gamma_g Q_g.$$

Substituting the values of H, $\dot{w}_f$, $\dot{w}_g$, $\dot{w}_s$, Q_f, P_s, P_{bh}, T_g, and T_{av} into Equation (9-7), a value of Q_g can be determined. The trial and error process is carried out until the right side and the left side equal each other. This gives an air volumetric flow rate of

$$Q_g = 6.617 \text{ actual ft}^3/\text{sec}.$$

The air volumetric flow rate in field units is

$$q_g = (6.617)(60)$$

$$q_g = 397 \text{ acfm}$$

or

$$q_g = 353 \text{ scfm}.$$

SI Units

Table 5-1 gives an average atmospheric pressure of 8.749 N/cm^2 abs for a surface location of 1219.1 m above sea level (also see Appendix B). The actual atmospheric pressure for the air at the drilling location (used by the compressor) is

$$P_{at} = 8.749 \; N/cm^2 \; abs$$

$$P_{at} = P_{at} \; 10000$$

$$P_{at} = 87490.0 \text{ N/m}^2 \text{ abs}.$$

The actual atmospheric temperature of the air at the drilling location (used by the compressor) is

$$t_{at} = 7.08^\circ \; C$$

$$T_{at} = t_{at} + 273.15$$

$$T_{at} = 280.23 \; K.$$

Thus, P_g and T_g become

$$P_g = P_{at} = 8.749 \; N/cm^2 \text{ abs}$$

$$T_g = T_{at} = 280.23 \; K.$$

Using Equation (4-11), the specific weight of the gas entering the compressor is

$$\gamma_g = \frac{(87490.0)(1.0)}{(29.31)(280.23)}$$

$$\gamma_g = 10.652 \text{ N/m}^2.$$

The bottom hole pressure in absolute pressure is

$$p_{bb} = 2317.39 + 8.749$$

$$p_{bb} = 2329.14 \text{ N/cm}^2 \text{ abs}$$

$$P_{bb} = P_{bb}\ 10000$$

$$P_{bb} = 23261410.0 \text{ N/m}^2 \text{ abs.}$$

The temperature of the rock formations near the surface (geothermal surface temperature) is estimated to be the approximate average year-round temperature at that location on the earth's surface. Table 5-1 gives 7.08°C for average year-round temperature for a surface elevation location of 1219.1 m above sea level (see Appendix B). Therefore, the absolute temperature of the rock formations at the surface is

$$t_r = 7.08^\circ\ C$$

$$T_r = t_r + 273.15$$

$$T_r = 280.23\ K.$$

The depth of the well is

$$H = 3048 \text{ m.}$$

The bottom hole temperature is

$$T_{bb} = T_r + 0.018\ H$$

$$T_{bb} = 334.96\ K.$$

The borehole average temperature is

$$T_{av} = \frac{280.23 + 334.96}{2}$$

$$T_{av} = 307.60\ K.$$

The open hole diameter is

$$d_b = 200.0\ mm$$

$$D_b = \frac{d_b}{1000}$$

$$D_b = 0.200 \text{ m.}$$

The estimated drilling rate of penetration is 60 ft/hr. The weight rate of flow of solids from the advance of the drill bit is

$$\dot{w}_s = \left(\frac{\pi}{4}\right) D_b^2\ (9810.0)\ S_s \left[\frac{\kappa}{(60)(60)}\right] \tag{9-8}$$

$$\dot{w}_s = \left(\frac{\pi}{4}\right) (0.200)^2\ (9810.0)\ (2.7) \left[\frac{18.3}{(60)(60)}\right]$$

$$\dot{w}_s = 4.222 \text{ N/sec.}$$

The volumetric flow rate of drilling fluid in consistent units is

$$Q_f = \frac{(0.723)}{(60)}$$

$$Q_f = 0.01215 \text{ m}^3/\text{sec.}$$

The weight rate of flow of the drilling mud is

$$\dot{w}_f = (1.2)(9810.0)(0.01215)$$

$$\dot{w}_f = 143.0 \text{ N/sec.}$$

Equation (9-7) can be solved by trial and error. The value of q_g is selected and $\dot{w}_g$ determined from

$$\dot{w}_g = \gamma_g Q_g.$$

Substituting the values of H, $\dot{w}_f$, $\dot{w}_g$, $\dot{w}_s$, Q_f, P_s, P_{bh}, T_g, and T_{av} into Equation (9-7), a value of Q_g can be determined. The trial and error process is carried out until the right side and the left side equal each other. This gives an air volumetric flow rate of

$$Q_g = 0.187 \text{ actual m}^3/\text{sec.}$$

The air volumetric flow rate in field units is

$$q_g = 187.0 \text{ actual liters/sec.}$$

9.4.2 Major and Minor Losses for Homogeneous Multiphase Flow

The governing equations for the second methodology are also presented in Chapter 6 [see Equation (6-26)]. These equations require much complicated trial and error solutions to obtain solutions that can be used in operational applications. The basic assumption for these equations is that the flow of the incompressible fluid and the gas proceed in the annulus and inside drill string as a homogeneous multiphase flow. This means that the theory does not account for any relative motion between the two fluids. This homogeneous flow assumption extends to the exclusion of the cuttings stream relative movement with respect to fluids in the annulus.

Illustrative Example 9.3 carries out the solution of the examples described in Illustrative Examples 9.1 and 9.2. This solution assumes a homogeneous flow of fluids and cuttings in the annulus and a homogeneous flow of fluids inside the drill string.

Illustrative Example 9.3 In this illustrative example, data given in Illustrative Examples 9.1 and 9.2 are used for solutions using the friction solution method given in Chapter 6. The objective of this example is to determine the approximate volumetric flow rate of the air required to be injected with the incompressible drilling fluid volumetric flow rate of 191 gpm (*723 lpm*) in the top of the inside of the drill string while drilling at a depth of 10,000 ft (*3048 m*). This is to be done while maintaining a bottom hole annulus pressure of 3360 psig (*2317 N/cm² gauge*).

The detailed Mathcad solution for this example is given in Appendix D. Figure 9-6 shows multiphase flow pressures throughout the circulation system in the project well while drilling at 10,000 ft (*3048 m*). In Figure 9-6, the bottom hole annulus pressure is 3373 psia (*2326 N/cm² abs*) in the annulus and the injection pressure at the standpipe is 637 psia (*439 N/cm² abs*).

The solution requires a trial and error selection of the volumetric flow rate of gas injected into the top of the drill string. This gas injection rate is approximately 1225 scfm (*596 standard liters/sec*). This injection rate, together with the incompressible volumetric flow rate of 191 gpm (*723 lpm*), is required to maintain the bottom hole annulus pressure of 3360 psig (*3214 N/cm² gauge*) at a drilling depth of 10,000 ft (*3048 m*). The relatively high injection pressure (for both the liquid and the gas) reflects the homogeneous hydrostatic weight of the fluid column inside the drill string above the bit nozzle chokes. This homogeneity of the multiphase flows in the annulus and inside the drill string are also reflected by the nearly equal slopes of the annulus and inside the drill string pressure plots shown in Figure 9-6.

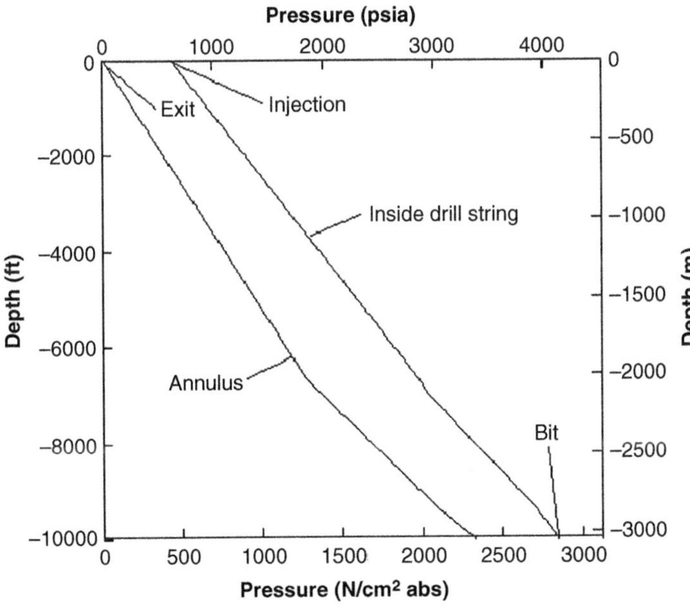

FIGURE 9-6. Illustrative Example 9.3 pressure versus depth.

9.4.3 Major and Minor Losses and the Effect of Fluid Holdup

The mathematical models described by Equation (6-26) for the fluid flow in the annulus was formulated on the assumption of homogeneous multiphase flow where all the fluids are assumed to flow at the same velocity. In reality, in vertical upward multiphase flow streams, the liquid and solid phases always flow slower (due to gravity) than the gas phase, resulting in liquid holdup. In vertical downward multiphase flow streams, the liquid phase (due to gravity) flows faster than the gas phase, which results in gas holdup.

In upward flow (in the annulus), differences in phase velocities result in annulus volume fractions of liquids and solids that are different from the volume fractions at the injection point (surface). To be specific, the amount of pipe occupied by a phase is often different from its proportion of the total volumetric flow rate at the injection point. This allows the annulus to essentially "store" more liquid in it relative to the homogeneous model discussed in the previous subsection. The most important empirical model for this type of flow is the generalized Hagedorn-Brown model [5].

Empirical models similar to those just described for downward multiphase flow (inside the drill pipe) and for multiphase flow through the drill bit nozzles do not exist for petroleum applications.

Liquid Holdup

The term liquid holdup is defined mathematically as

$$y_L = \frac{V_L}{V}, \tag{9-9}$$

where y_L is the incompressible (liquid) fluid fraction, V_L is the liquid phase volume (ft^3, m^3), and V is the total volume in the annulus (ft^3, m^3).

Liquid holdup depends on flow regime, fluid properties, and pipe size and configuration. Its value can only be determined quantitatively through experimental measurements. A direct application of the liquid holdup is to use it for estimating mixture specific weight in two-phase flow. This mixed specific weight was used in Equation (6-26). This new mixed specific weight is

$$\gamma_{mix} = y_L \gamma_L + (1 - y_L)\gamma_G, \tag{9-10}$$

where γ_L is the liquid-specific weight (lb/ft^3, N/m^3) and γ_G is the gas-specific weight (lb/ft^3, N/m^3).

The generalized Hagedorn-Brown correlation is widely used in petroleum production engineering for multiphase flow performance calculations [5]. Most applications pertain to multiphase production fluids flowing up the inside of the production tubing. This discussion applies the generalized Hagedorn-Brown correlation to multiphase drilling fluids up the annulus.

The generalized Hagedorn-Brown correlation determines liquid holdup values from three empirical charts [5]. All of these charts are accessed by using the following additional dimensionless numbers:

Liquid velocity number N_{vL}:

$$N_{vL} = F_{vL}\, u_{SL} \sqrt[4]{\frac{\gamma_L}{\sigma}}. \tag{9-11}$$

Gas velocity number N_{vG}:

$$N_{vG} = F_{vG}\, u_{SG} \sqrt[4]{\frac{\gamma_L}{\sigma}}. \tag{9-12}$$

Pipe diameter number N_D:

$$N_D = F_D\, D \sqrt{\frac{\gamma_L}{\sigma}}. \tag{9-13}$$

Liquid viscosity number N_L:

$$N_L = F_L\, \mu_L \sqrt[4]{\frac{1}{\gamma_L \sigma^3}}. \tag{9-14}$$

where u_{SL} is the superficial velocity of the liquid phase (ft/sec, m/sec), u_{SG} is the superficial velocity of the gas phase (ft/sec, m/sec), σ is the liquid-gas interfacial tension (dyne/cm, dyne/cm), D is the hydraulic diameter of the flow cross section (ft, m), and μ_L is the liquid viscosity (cp, cp).

USCS unit conversion constants are

$$F_{vL} = 1.938$$

$$F_{vG} = 1.938$$

$$F_D = 120.872$$

$$F_L = 0.15726$$

SI unit conversion constants are

$$F_{vL} = 1.7964$$

$$F_{vG} = 1.7964$$

$$F_D = 31.664$$

$$F_L = 0.55646$$

Figure 9-7 is the first chart and is an empirical graphic representation of the relationship between parameter CN_L versus N_L [5]. Guo and colleagues [19] found that Figure 9-7 can be replaced by the following correlation equation:

$$CN_L = 10^Y, \tag{9-15}$$

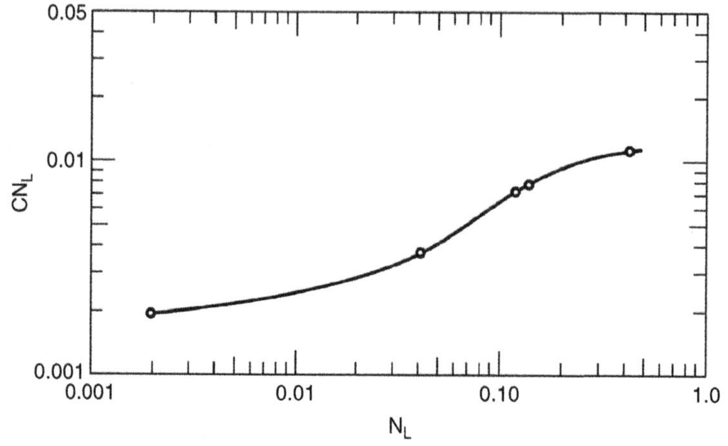

FIGURE 9-7. Nondimensional CNL versus nondimensional NL [5].

where

$$Y = -2.69851 + 0.15841X_1 - 0.55100X_1^2 + 0.54785X_1^3 - 0.12195X_1^4$$

and

$$X_1 = \log(N_L) + 3.$$

Once the value of parameter CN_L has been determined, it is used to calculate the value of the nondimensional group (at depth and pressure position along the annulus pressure gradient)

$$\frac{N_{vL}(CN_L)}{N_{vG}^{0.575}N_D}\left(\frac{P}{P_a}\right)^{0.1},$$

where P_a is the atmospheric pressure (lb/ft^2 abs, N/m^2 abs) and P is the pressure at the calculation location at depth (lb/ft^2 abs, N/m^2 abs).

The second empirical correlation chart is given in Figure 9-8 [5]. Guo and colleagues [19] found that Figure 9-8 can be replaced by the following correlation equation:

$$\left(\frac{y_L}{\psi}\right) = -0.10307 + 0.61777\left[\log(X_2) + 6\right] - 0.63295\left[\log(X_2) + 6\right]^2$$
$$+ 0.29598\left[\log(X_2) + 6\right]^3 - 0.0401\left[\log(X_2) + 6\right]^4, \quad (9\text{-}16)$$

where

$$X_2 = \frac{N_{vL}(CN_L)}{N_{vG}^{0.575}N_D}\left(\frac{P}{P_a}\right)^{0.1}. \quad (9\text{-}17)$$

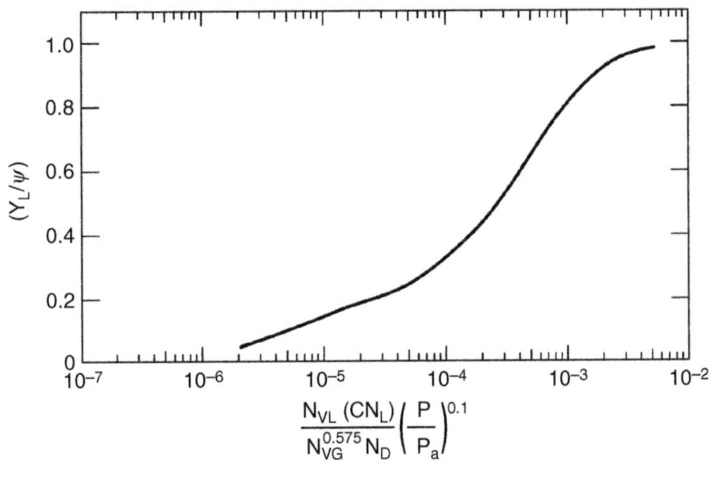

FIGURE 9-8. (y_L/ψ) versus the nondimensional group $\frac{N_{VL}(CN_L)}{N_{VG}^{0.575} N_D}\left(\frac{P}{P_a}\right)^{0.1}$ [5].

The third empirical correlation chart is given in Figure 9-9 [5]. Guo and coworkers [19] found that Figure 9-9 can be represented by the following correlation equation:

$$\psi = 0.91163 - 4.82176X_3 + 1232.25X_3^2 - 22253.6X_3^3 + 116174.3X_3^4, \tag{9-18}$$

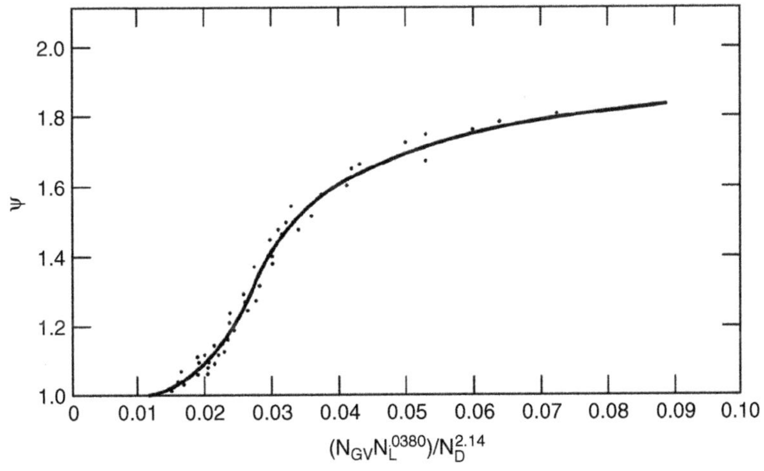

FIGURE 9-9. ψ versus nondimensional group $\frac{N_{GV} N_L^{0.380}}{N_D^{2.14}}$ [5].

where

$$X_3 = \frac{N_{vG} N_L^{0.38}}{N_D^{2.14}}. \tag{9-19}$$

Equations (9-18) and (9-19) are only valid for

$$\frac{N_{vG} N_L^{0.38}}{N_D^{2.14}} > 0.01.$$

Note that for $\frac{N_{GV} N_L^{0.380}}{N_D^{2.14}} \le 0.01$, $\psi = 1.0$ [19].

The liquid holdup y_L can be calculated using

$$y_L = \psi\left(\frac{y_L}{\psi}\right). \tag{9-20}$$

The interfacial tension is a function of pressure and temperature. Interfacial tension is determined using the following empirical correlation:

$$\sigma = \sigma_{74} - \frac{(\sigma_{74} - \sigma_{280})(t - 74)}{206}, \tag{9-21}$$

where the temperature t is in °F.

For temperatures in °C, the correlation is the following:

$$\sigma = \sigma_{74} - \frac{(\sigma_{74} - \sigma_{280})(1.8t_C - 42)}{206}, \tag{9-22}$$

where temperature t_C is in °C, and

$$\sigma_{74} = 75 - 1.108\, p^{0.349} \tag{9-23}$$

and

$$\sigma_{280} = 53 - 0.1048\, p^{0.637}, \tag{9-24}$$

where pressure p is in psia, or

$$\sigma_{74} = 75 - 6.323\, P_{MPa}^{0.349} \tag{9-25}$$

and

$$\sigma_{280} = 53 - 2.517\, P_{MPa}^{0.637} \tag{9-26}$$

where pressure P_{MPa} is in MPa abs.

Illustrative Example 9.4 In this illustrative example, data in Illustrative Example 9.3 will be resolved, but in this example, liquid holdup empirical correlations will be considered for the flow up the annulus.

The detailed Mathcad solution for this example is given in Appendix D. Figure 9-10 shows multiphase flow pressures throughout the circulation system in the project well while drilling at 10,000 ft (*3048 m*). In Figure 9-10, the bottom hole annulus pressure is 3373 psia (*2326 N/cm² abs*) in the annulus and the surface injection pressure at the standpipe is 62.4 psia (*43.0 N/cm² abs*).

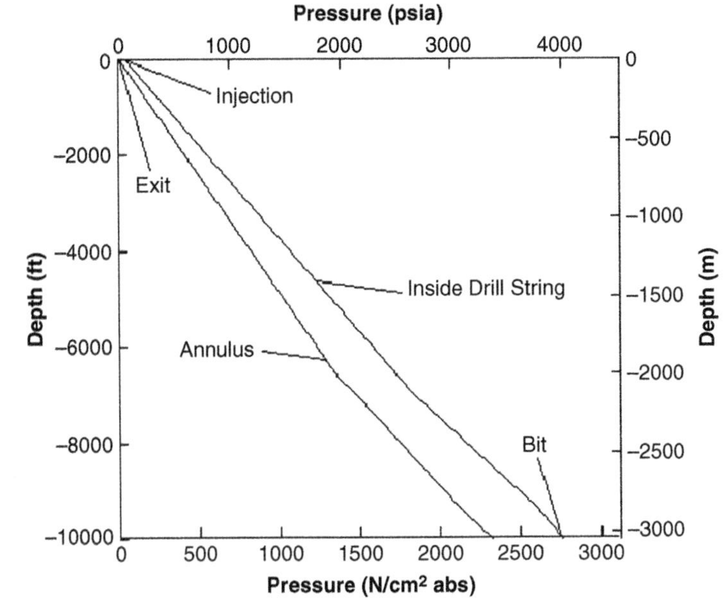

FIGURE 9-10. Illustrative Example 9.4 pressure versus depth.

The solution requires a trial and error selection of the volumetric flow rate of gas injected into the top of the drill string. This gas injection rate is approximately 301 scfm (*142 standard liters/sec*). This injection rate, together with the incompressible volumetric flow rate of 191 gpm (*723 lpm*), is required to maintain the bottom hole annulus pressure of 3360 psig (*3214 N/cm^2 gauge*) at a drilling depth of 10,000 ft (*3048 m*).

Comparing the gas injection rate in this example with the gas injection rate obtained in Illustrative Example 9.3 (i.e., the multiphase homogeneous flow solution) shows that the effect of liquid holdup is to allow storage of a significant quantity of liquid in the annulus at steady state flow conditions. This is reflected in the low incompressible fluid and gas injection pressure at the standpipe. This low gas and liquid injection pressure at the standpipe is the result of the higher liquid content inside the drill string (since the gas injection rate is so low). This means that the hydrostatic column of liquid places a higher pressure on the bottom of the inside of the drill string.

Figure 9-11 shows the injection pressure and the bottom hole pressure as drilling takes place from the casing shoe at 7000 ft (*2134 m*) to the final depth of 10,000 ft (*3048 m*). Figure 9-11 shows the injection pressure and the bottom hole pressure as a function of drilling depth and assumes that the same gas injection of 301 scfcm (*142 liters/sec*) will be used as the open hole section is drilled. This would be consistent with a performance drilling operation. Figure 9-11

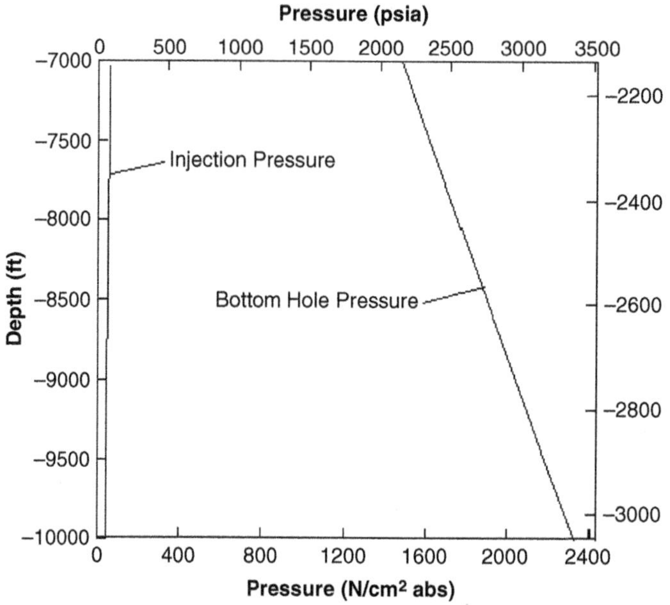

FIGURE 9-11. Drilling ahead, bottom hole annulus, and injection pressures versus depth.

shows that the injection changes very little as the open hole is drilled. The injection pressure is 87.4 psia (*60.4 N/cm^2 abs*) at the casing shoe and 62.4 psia (*43.0 N/cm^2 abs*) at the bottom of the section at 10,000 ft (*3048 m*). The bottom hole annulus pressure variation is much larger, changing from 2151.2 psia (*1483.7 N/Cm^2 abs*) at the casing shoe to 3372.2 psia (*2326.2 N/cm^2 abs*) at the bottom of the open hole section.

9.5 CONCLUSIONS

The discussions in this chapter have concentrated on direct circulation performance drilling operations. The Mathcad illustrative examples in Appendix D utilized lumped geometry approximations for the drill pipe body and drill pipe tool joints. These approximations appear to quite adequately model the overall friction resistance along the circulation system and give reasonable results, particularly for the bottom hole annulus pressures. The large variations in pressure versus depth in the annulus and inside the drill string are masked in the aerated fluid model (due to large pressure magnitudes) as compared to the similar air drilling model plots in Chapter 8 (small pressure magnitudes). Field results compare favorably with liquid holdup model bottom hole annulus pressure calculations (given similar incompressible and compressible fluid injection rates).

The injection pressures are not as close to field operations data. Since there are few reliable gas holdup models for the downward multiphase flow inside the drill string and for multiphase flow through the drill bit nozzles, this is expected.

REFERENCES

1. Bobo, R. A., and Barrett, H. M., "Aeration of Drilling Fluids," *World Oil*, Vol 145, No. 4, 1953.
2. Graves, S. L., Niederhofer, J. D., and Beavers, W. M., "A Combination Air and Fluid Drilling Technique for Zones of Lost Circulation in the Black Warrior Basin," *SPE Drilling Engineering*, February 1986.
3. Allan, P. D., "Nitrogen Drilling System for Gas Drilling Applications," *SPE 28320,* Presented at the SPE 69th Annual Technical Conference and Exhibition, New Orleans, Louisiana, September 25–28, 1994.
4. *Underbalanced Drilling Manual*, Gas Research Institute Publication, GRI Reference No. GRI-97/0236, 1997.
5. Hegedorn, A. R., and Brown, K. E., "Experimental Study of Pressure Gradients Occurring During Continuous Two-Phase Flow in Small Diameter Vertical Conduits," *Journal of Petroleum Technology*, April 1965.
6. Langlinais, J. P., Bourgoyne, A. T., and Holden, W. R., "Frictional Pressure Losses for the Flow of Drilling Mud and Mud/Gas Mixtures," *SPE 11993,* Presented at the 58th Annual Technical Conference and Exhibition, San Francisco, California, October 5–8, 1983.
7. Mitchell, B. J., "Test Data Fill Theory Gap on Using Foam as a Drilling Fluid," *Oil and Gas Journal*, September 1971.
8. Krug, J. A., and Mitchell, B. J., "Charts Help Find Volume, Pressure Needed for Foam Drilling," *Oil and Gas Journal*, February 7, 1972.
9. Blauer, R. E., Mitchell, B. J., and Kohlhaas, C. A., "Determination of Laminar, Turbulent and Transitional Foam Flow Friction Losses in Pipes," *SPE 4885,* Presented at the 1974 Annual SPE California Regional Meeting, San Francisco, California, November 2–4, 1974.
10. Gatlin, C., *Petroleum Engineering: Drilling and Well Completions*, Prentice-Hall, 1960.
11. Bourgoyne, A. T., et al, *Applied Drilling Engineering*, *SPE*, First Printing, 1986.
12. Moore, P. L., *Drilling Practices Manual*, The Petroleum Publishing Company, 1974.
13. Guo, B., Hareland, G., and Rajtar, J., "Computer Simulation Predicts Unfavorable Mud Rate and Optimum Air Injection Rate for Aerated Mud Drilling," *SPE 26892,* Presented at the 1993 Eastern Regional Conference and Exhibition held in Pittsburgh, Pennsylvania, November 2–4, 1993.
14. Poettman, F. H., and Carpenter, P. G., *Drilling and Production Practice*, API, 1952.
15. Poettman, F. H., and Bergman, W. E., "Density of Drilling Muds Reduced by Air Injection," *World Oil*, August 1955.

16. Brown, K. E., and Beggs, H. D., *The Technology of Artificial Lift Methods*, Vol. 1, PennWell Books, 1977.

17. Brown, K. E., *The Technology of Artificial Lift Methods*, Vol. 2a, PennWell Books, 1980.

18. Daugherty R. L., Franzini, J. B., and Finnemore, E. J., *Fluid Mechanics with Engineering Applications*, Eighth Edition, McGraw-Hill, 1985.

19. Guo, B., Lyons, W. C., and Ghalambor, A., *Petroleum Production Engineering*, Elsevier, 2007.

CHAPTER

Stable Foam Drilling

10

The term stable foam from the mathematical modeling point of view describes a special class of aerated drilling fluids and work over completion fluids. This class of drilling fluid is made up of a specific mixture of incompressible fluids injected with compressed air or other gases. To create a stable foam drilling fluid, the incompressible component is usually a mixture of fresh water with a surfactant foaming agent. However, stable foam drilling fluids can use formation water with dissolved salts. The surfactant foaming agent usually comprises about 1 to 5% by volume of the treated water being injected (depending on the surfactant product).

The term "stiff foam" refers to the use of viscosified (aqueous polymer) water instead of fresh nonviscosified water as the incompressible fluid component (typical viscosity additives are polyanionic cellulose, xanthan gum polymers, and carboxymethyl cellulose). In essence, stiff foam is drilling mud with a surfactant additive. The subject of stiff foam is beyond the scope of this chapter.

Figure 10-1 shows a schematic representation of a direct circulation foam drilling operation. In a typical deep drilling operation, the foam mixture is injected into the top of the inside of the drill string (i.e., direct circulation, see Chapter 6). The mixture of the incompressible fluid (with surfactant) and compressed air (or other gas) flows as an aerated fluid mixture down the inside of the drill string to the bottom of the string just above the drill bit nozzles. The nozzles in the drill bit are usually required in order to allow the foam to be generated by the high shearing action in the fluid as the mixture passes through the nozzles. In this manner, the foam is formed at the bottom of the annulus and then flows as stable foam up the annulus.

In deep wells, it is often necessary to place a back pressure valve and pressure gauge in the return line from the annulus. This valve and gauge are used to modulate the flow from the well in such a manner as to assure that the flow remains as a stable foam throughout the annulus and return line up to the valve. The pressure gauge allows rig personnel to obtain periodic pressure readings, which in turn can be converted into foam quality data.

There are some shallow direct circulation drilling situations where the foam can be preformed at the surface and injected into the inside of the drill string.

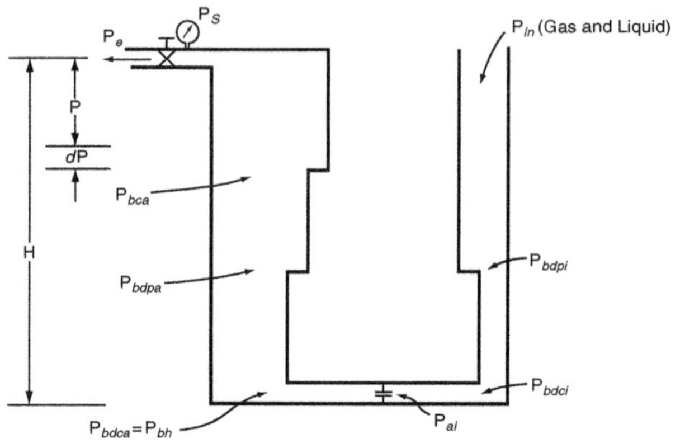

FIGURE 10-1. Schematic of direct circulation. P_{in} is the injection pressure into the top of the drill string, P_{bdpi} is the pressure at the bottom of the drill pipe inside the drill string, P_{bdci} is the pressure at the bottom of drill collars inside the drill string, P_{ai} is the pressure above the drill bit inside the drill string, P_{bdca} is the pressure at the bottom of drill collars in the annulus, P_{bh} is the bottom hole pressure in the annulus, P_{bdpa} is the pressure at the bottom of the drill pipe in the annulus, P_{bca} is the pressure at the bottom of the casing in the annulus, P_s is the surface pressure at the top of the annulus (usually back pressure created by an adjustable valve), and P_e is the exit pressure at the end of the surface return flow line.

In reverse circulation drilling or work over operations the performed foam can be injected into the top of the annulus. Flowing foam will create large friction forces and these forces are a function of the pressure conduit diameter. The smaller the diameter, the larger the resisting friction forces of flow. This foam property precludes performing foam for injection into the inside of the drill string in deep direct circulation operations. For discussions in this chapter, stable foam drilling fluids will be assumed to be an aerated drilling fluid as it flows down the inside of the drill string and a stable foam drilling fluid as it flows up the annulus.

A new term foam quality Γ (or gas fraction) must be used to discuss the rheology of the stable foam drilling fluid. Rheology describes the bubble and liquid structures that can form in stable foam and the mechanisms that create friction resistance as the foam flows in the conduits. Stable foam rheology will be combined with the basic derived equations and the empirical friction factor expressions in Chapter 6 (or Chapter 7) to give a usable theory for stable foam drilling predictive calculation models. Foam quality is defined as

$$\Gamma = \frac{Q_g}{Q_g + Q_f}, \tag{10-1}$$

where Q_g is the volumetric flow rate of the compressible gas (ft^3/sec, m^3/sec) and Q_f is the volumetric flow rate of the incompressible fluid (ft^3/sec, m^3/sec).

10.1 STABLE FOAM RHEOLOGY

Stable foam work over and drilling fluids have been used by the oil and gas industry since the 1950s [1–4]. Since the introduction of foam drilling fluids there have been efforts to mathematically describe the rheology of stable foam fluids. In the 1970s, the first serious experimental and practical mathematical modeling studies were carried out to describe the static and dynamic physical characteristics of stable foam [5–8]. In the 1980s, successful foam rheology and flow models were developed. One practical power law mathematical model was developed by Sanghani for foam flowing in pipe and in annulus geometry pressure conduits [9]. In addition, the most extensive foam experiments were conducted by Ikoku, Okpobiri, and Machado, which described the frictional resistance of the foam structure flow in pipes and the motion of solids cutting in these structures [10–13]. During the 1990s, field operational data became available that could be compared with the existing rheologic and flow models [14, 15]. In the early 2000s, Guo, Sun, and Ghalambor correlated the Sanghani power law model with the extensive experimental data of Ikoku and colleagues [16].

Laboratory and field experiments show that stable foam will exist within certain limits of the foam quality value. These limits are approximately the foam qualities of 0.60 and 0.98 [9–11, 13]. If the foam quality value falls below approximately 0.60, the foam will separate into its two phases. If the foam quality value is above 0.98, the stable continuous foam becomes unstable in that the foam flows as slugs of foam (which is denoted as "mist" as described in Chapter 8). For stable foam deep drilling operations the lower foam quality values are usually at the bottom of the annulus and the higher foam quality values at the top of the annulus.

Stable foam is a complex flowing structure, which has viscosities that change as a function of pressure, temperature, and pressure conduit geometry. Understanding the viscosity change mechanism of the foam drilling fluid is critical to the successful development of practical predictive flow models. In general, the parameters of pressure, temperature, and geometry can be represented more conveniently by foam quality, foam average velocity, and the effective diameter of the pipe or annulus. Experiments have shown that viscosity magnitude can increase 10 to 15 times as the foam quality is increased from 0.60 to 0.98. A similar change can accompany changes in velocity and effective diameter [17].

The development of a power law rheologic model for flowing stable foam has allowed major improvements in flow modeling efforts. The power law equation for the effective viscosity for the flow of foam inside a pipe pressure conduit is [9]

$$\mu_e = \frac{K\,C_{vis}}{g}\left(\frac{3n+1}{4n}\right)^n\left(\frac{8\ V}{D}\right)^{n-1}, \tag{10-2}$$

where K is the consistency index, n is the power law exponent, V is the average velocity of the foam inside the pipe (ft/sec, m/sec), D is the inside diameter of the pipe (ft, m), g is the acceleration of gravity (32.2 ft/sec^2, 9.81 m/sec^2), and C_{vis} is a conversion constant (equals 1.0 for USCS, equals 14.5727 for SI).

The power law equation for the effective viscosity for the flow of foam in an annulus pressure conduit is [9]

$$\mu_e = \frac{K\ C_{vis}}{g}\left(\frac{2n+1}{3n}\right)^n\left(\frac{12\ \ V}{D_2 - D_1}\right)^{n-1}, \tag{10-3}$$

where V_{an} is the average velocity of the foam in the annulus (ft/sec, m/sec), D_1 is the inner surface diameter of the annulus (ft, m), and D_2 is the outer surface diameter of the annulus (ft, m).

The correlation expression for the consistency index as a function of the foam quality Γ is [16]

$$\begin{aligned} K = -0.15626 + 56.147\ \ \Gamma - 312.77\ \ \Gamma^2 + 576.65\ \ \Gamma^3 \\ + 63.960\ \ \Gamma^4 - 960.46\ \ \Gamma^5 - 154.68\ \ \Gamma^6 + 1670.2\ \ \Gamma^7 \\ -937.88\ \ \Gamma^8. \end{aligned} \tag{10-4}$$

Also, the correlation expression for the power law exponent as a function of foam quality Γ is [16]

$$\begin{aligned} n = 0.096932 + 2.3654\ \ \Gamma - 10.467\ \ \Gamma^2 + 12.955\ \ \Gamma^3 \\ +14.467\ \ \Gamma^4 - 39.673\ \ \Gamma^5 + 20.625\ \ \Gamma^6 \end{aligned} \tag{10-5}$$

These correlation expressions were developed for the fresh water and surfactant combination used in the experiments by Ikoku and colleagues. The surfactant foaming agent used in the experiments was a 1.0% solution (with fresh water) of Adofoam BF-1 (Nalco Chemical Co.)[9–13].

The following is a limited list of the commercial foaming agents available through various drilling service and chemical companies. These are

DuraFoam	Weatherford
SMA-100	Weatherford
DrillFoam PDS	Air Drilling Associates
DrillFoam MSDS	Air Drilling Associates
Adofoam BF-1	Nalco Chemical Company
AirFoam AP-50	Aqua-Clear
AirFoam B	Aqua-Clear
F-52 surfactant	Dowell
F-78 surfactant	Dowell

10.1.1 Laboratory Stable Foam Screening

The flow performance of a stable foam drilling fluid is very dependent on the quality of the fresh water (or formation water), the surfactant, and what other additives would be used to create the foam. In order to assess the performance of a proposed mixture of water, surfactant, and additives in a future drilling project, it is necessary to (1) establish a nominal baseline laboratory assessment of a stable foam mixture that has undergone extensive laboratory experimental assessments and has a record of good performance in the field; (2) devise and use a simple laboratory test that could be used to measure other potential stable foam mixtures against the nominal baseline mixture; and (3) create an empirical relationship that can be used to predict the future performance and, more importantly, the future nonperformance of nonbaseline drilling project mixtures. In this stable foam performance screening, the quality of the water used in a drilling project cannot be overemphasized. Therefore, it is very important that the actual water used in a forthcoming drilling operation be used in these laboratory screening tests.

A screening test used by a few of the companies is a highly modified version of an ASTM foam properties test [17]. The test requires the use of a commercial blender and a tall 1000-ml beaker. The actual mixture of drilling project water, surfactants, and additives is to be used in the test. One hundred milliliters of the drilling project water with its additives (excluding the surfactant) is placed in the blender. The appropriate quantity of surfactant is placed in the blender with the water plus additives mixture. The blender is operated with an open top for 10 min. Stable foam will be created as the blender is operated. At the end of the blending time, a spatula is used to pour all the stable foam in the blender into the beaker. The total "height" in milliliters of the mixture is recorded (e.g., 500 ml). A stop watch is used to obtain the time needed for 50 ml of liquid to accumulate at the bottom of the beaker. This time is known as the "half-lifetime" of the foam mixture and is recorded in seconds (e.g., 200 sec). For illustration, these values will be considered "nominal baseline" for a stable foam mixture that has shown good performance characteristics in extensive laboratory dynamic flow tests and in field drilling project results. Using Equation (10-1), the foam quality of the nominal baseline stable foam mixture would be $\Gamma = 0.80$.

10.1.2 Empirical Algorithm for Screening

Most screening empirical algorithms are constructed on the basis of the height and half-life ratios of the proposed drilling project mixture values to the nominal baseline mixture values. The expression given here is an example of such an empirical algorithm. In this expression, Λ is the value used to multiple the effective viscosity of the nominal baseline stable foam mixture.

$$\Lambda = C_{test}\left(\frac{h_t}{h_n}\right)^a \left(\frac{\tau_t}{\tau_n}\right)^b, \tag{10-4}$$

where h_t is the height of the test mixture (ml), h_n is the height of the nominal baseline mixture (ml), τ_t is the half-life time of the test mixture (sec), τ_n is the half-life of the nominal baseline mixture (sec), and C_{test}, a, and b are constants determined by dynamic laboratory experiments or field data.

This nominal baseline effective viscosity would have been determined previously for a variety of foam quality values using dynamic laboratory test and evaluations using Equation (10-2) or (10-3).

10.2 DEEP WELL DRILLING PLANNING

Stable foam drilling operations can use a variety of incompressible fluids and compressed gases to develop stable foam. The majority of the operations use fresh water and commercial surfactants with injected compressed air or membrane generated nitrogen. In this chapter, fresh water, surfactant, and atmospheric air will be used as an example of a stable foam drilling fluid.

The basic planning steps for a deep well are as follow:

1. Determine the geometry of the borehole sections to be drilled with the stable foam drilling fluids (i.e., open hole diameters, the casing inside diameters, and section depths).
2. Determine the geometry of the associated drill strings for each section to be drilled with stable foam drilling fluids (i.e., drill bit size and type, drill collar size, drill pipe size and description, and maximum depths).
3. Determine the type of rock formations to be drilled in each section and estimate the anticipated drilling rate of penetration (use offset wells).
4. Determine the elevation of the drilling location above sea level, the estimated temperature of the atmospheric air during the drilling operation, and the approximate geothermal temperature gradient.
5. Establish the objective of the stable foam drilling fluids operation:
   - Performance drilling
   - Underbalanced drilling
6. For either of the aforementioned objectives, determine the required approximate volumetric flow rate of the mixture of incompressible fluid (with surfactant) and the compressed air (or other gas) to be used to create the stable foam drilling fluid. This required mixture volumetric flow rate is governed by (a) the foam quality at the top of annulus (i.e., return flow line back pressure) and (b) the rock cutting carrying capacity of the flowing mixture in the critical annulus cross-sectional area (usually the largest cross-sectional area of the annulus).
7. Using the incompressible fluid and gas volumetric flow rates to be injected into the well, determine the bottom hole pressure (and foam quality) and required back pressure (and foam quality) at the top of the annulus as a function of drilling depth. Also determine the surface injection pressure as a function of drilling depth (over the open hole interval to be drilled).
8. Select the contractor compressor(s) that will provide the drilling operation with the appropriate air or gas volumetric flow rate needed to create the stable foam drilling fluid.

Chapter 6 derived and summarized the basic direct circulation drilling planning governing equations. The equations in this chapter will be utilized in the discussions and illustrative examples that follow.

10.3 STABLE FOAM DRILLING OPERATIONS

Stable foam drilling operations are carried out using only drill string injection configurations. Figure 10-1 shows a schematic of the drill pipe injection stable foam drilling configuration. Both incompressible fluid (with surfactant) and compressible air (or other gas) are injected together into the top of the drill string (at P_{in}). These fluid streams mix as they go down the inside of the drill string and pass through the drill bit nozzles. Stable foam is created when the fluids pass through the drill bit nozzles. As stable foam is generated in the bottom of the annulus, the rock cuttings (from the advance of the drill bit) are entrained in the foam and the resulting mixture flows to the surface in the annulus.

The mixture exits the annulus into a horizontal surface return flow line just upstream of the back pressure valve and gauge (upstream of P_s). The pressure readings just upstream of the back pressure valve can be converted to foam quality values. In order to assure that there is a continuous column of stable foam from the bottom of the annulus to the return line back pressure valve, the foam quality upstream of the valve should be held at 0.98 or less. As conditions change in the well (e.g., an influx of formation water, wall caving), the back pressure valve can be adjusted by rig personnel to give pressure readings required to give the needed foam quality at the surface. Some operators design their return line systems so that the system itself provides sufficient back pressure resistance to the foam flowing from the top of the annulus. In these return line systems, the stable foam transitions to an unstable foam as the flow reaches the top of the annulus (or in the horizontal return line itself) and the flow from the return line at its exit (at P_e) is in slugs of unstable foam.

This horizontal return flow line flows to either a burn pit or sealed returns tanks. The burn pit is used when the returning air is mixed with hydrocarbons that can be burned off as they enter the pit. Sealed returns tanks are also used to contain contaminated fluids and gases, or hydrocarbons. The use of sealed tanks can only be carried out where the stable foam drilling operation is utilizing membrane-generated inert nitrogen gas as the foaming gas.

Advantages and disadvantages of the stable foam direct circulation drill are as follow:

Advantages

- The technique does not require any additional downhole equipment.
- Nearly the entire annulus is filled with the stable foam drilling fluid, thus a variety of low bottom hole pressures can be achieved.
- Since the bubble structures of stable foam drilling fluids have a high fluid yield point, these structures can support the small rock cuttings in

suspension when drilling operations are discontinued to make connections. Stable foams have seven to eight times the rock cutting carrying capacity of water.

- Rock cuttings retrieved from the foam at the surface are easy to analyze for rock properties information.

Disadvantages

- Stable foam fluids injection cannot be continued when circulation is discontinued during connections and tripping. Therefore, it can be difficult to maintain underbalanced conditions during connections and trips.
- Because the injected gas is trapped under pressure inside the drill string by the various string floats, time must be allowed for the pressure bleed down when making connections and trips. Here again the bleed down makes it difficult to maintain an underbalanced condition.
- The flow down the inside of the drill string is two-phase flow and, therefore, high pipe friction losses are present. The high friction losses result in high pump and compressor pressures during injection.
- The gas phase in the stable foam attenuates the pulses of conventional MWD systems. Therefore, conventional mud pulse telemetry MWD cannot be used. Often EM MWD systems can be used.

10.4 MINIMUM VOLUMETRIC FLOW RATES

Most stable foam vertical performance drilling operations are drilled over a constant cross-section interval with constant incompressible fluid volumetric flow rates, constant compressible gas volumetric flow rates, and a constant back pressure setting at the surface. Therefore, the minimum volumetric flow rates of incompressible fluid and compressible gas to carry drill bit cuttings from the well can be determined at the maximum true measured and/or vertical depth of the constant cross-section well bore.

Because most potential producing formations are usually of the order of a few hundred feet or tens of meters thick, vertical drilling of these formations also requires little changes in back pressure or flow rates as the section is drilled. Here again, the maximum depth can be used to determine minimum volumetric flow rates to carry the drill bit cuttings for the well. This will not be the case for drilling horizontal boreholes underbalanced. Horizontal drilling will be discussed in Chapter 12.

There are few minimum volumetric flow rate theories available for use in stable foam drilling modeling. Stable foam drilling fluids have high effective viscosities and yield points. Early experiments have found that spherical rock particles will fall at terminal velocities of the order of 10 to 20 ft/min [18]. These terminal velocities are low when compared to terminal velocities in water. Early experiments also indicated that terminal velocities tend to increase with increasing foam quality. These experimental data were used to develop plots that could be used to approximate the minimum volumetric flow rate of stable foam [8]. However, these plots were not easy to use in stable foam flow models. Moore developed empirical relationships that were more model friendly [19]. These relationships are the basis of the example calculations used in this chapter.

For stable foam drilling applications the minimum volumetric flow rate will be determined using a rather simple, straightforward procedure, which requires that the circulating incompressible fluid be capable, on its own, of maintaining a minimum concentration of rock cuttings in the largest annulus section of the well [6]. This requires that the average velocity of the fluid, V_f, in the largest annulus section be equal to or greater than the sum of the critical concentration velocity, V_c, and the terminal velocity, V_t, of the average size rock cutting particle in the drilling fluid.

Thus, the average fluid velocity in the annulus is

$$V_f = V_c + V_t, \tag{10-5}$$

where V_f is the incompressible drilling fluid (ft/sec), V_c is the critical concentration velocity (ft/sec), and V_t is the terminal velocity of the rock cuttings particle (ft/sec).

It is tacitly assumed that if the incompressible drilling fluid can carry the rock cuttings on its own, then the injection of gas into the fluid will enhance the overall carrying capacity of the aerated fluid.

The critical concentration velocity is the additional velocity needed to distribute the rock cuttings through the incompressible drilling fluid at a predetermined concentration factor. The usual maximum concentration factor is 0.04. Therefore, the critical concentration velocity V_c is

$$V_c = \frac{\kappa}{3,600 \quad C_{cf}}, \tag{10-6}$$

where κ is the instantaneous drilling rate (ft/hr, m/hr) and C_{cf} is the concentration factor.

Equations (10-5) and (10-6) can be used with any consistent set of units.

Terminal Velocities (USC Units)

For direct circulation operations the terminal velocity of the rock cutting particle is assessed in the annulus section of the borehole where the cross-sectional area is the largest.

Empirical data indicate that laminar flow conditions exist when the nondimensional Reynolds number for the flow around the particle is between 0 and 1. The empirical relationship for the terminal velocity of a rock cutting particle in an annulus with laminar flow V_{t1} is

$$V_{t1} = 0.0333 \quad D_c^2 \left(\frac{\gamma_s - \gamma_f}{\mu_e} \right), \tag{10-7a}$$

where V_{t1} is the terminal velocity of the particle in laminar flow (ft/sec), D_c is the approximate diameter of the rock cutting particle (ft), γ_s is the specific weight of the solid rock cutting (lb/ft^3), γ_f is the specific weight of the incompressible drilling fluid (lb/ft^3), and μ_e is the effective absolute viscosity (lb-sec/ft^2).

Empirical data indicate that transition flow conditions exist when the nondimensional Reynolds number for the flow around the particle is between 1 and 2000. The empirical relationship for the terminal velocity of a rock cutting particle in an annulus with transition flow V_{t2} is

$$V_{t2} = 0.492 \quad D_c \left[\frac{(\gamma_s - \gamma_f)^{\frac{2}{3}}}{(\gamma_f \quad \mu_e)^{\frac{1}{3}}} \right], \tag{10-8a}$$

where V_{t2} is the terminal velocity of the particle in transition flow (ft/sec).

Empirical data indicate that turbulent flow conditions exist when the nondimensional Reynolds number for the flow around the particle is greater than 2000. The empirical relationship for the terminal velocity of a rock cutting particle in an annulus with turbulent flow V_{t3} is

$$V_{t3} = 5.35 \left[D_c \left(\frac{\gamma_s - \gamma_f}{\gamma_f} \right) \right]^{\frac{1}{2}}, \tag{10-9a}$$

where V_{t3} is the terminal velocity of the particle in turbulent flow (ft/sec).

Terminal Velocities (SI Units)

Empirical data indicate that laminar flow conditions exist when the nondimensional Reynolds number for the flow around the particle is between 0 and 1. The empirical relationship for the terminal velocity of a rock cutting particle in an annulus with laminar flow, V_{t1}, is

$$V_{t1} = 0.0333 \quad D_c^2 \left(\frac{\gamma_s - \gamma_f}{\mu_e} \right), \tag{10-7b}$$

where V_{t1} is the terminal velocity of the particle in laminar flow (m/sec), D_c is the approximate diameter of the rock cutting particle (m), γ_s is the specific weight of the solid rock cutting (N/m^3), γ_f is the specific weight of the incompressible drilling fluid (N/m^3), and μ_e is the effective absolute viscosity (N-sec/m^2).

Empirical data indicate that transition flow conditions exist when the nondimensional Reynolds number for the flow around the particle is between 1 and 2000. The empirical relationship for the terminal velocity of a rock cutting particle in an annulus with transition flow V_{t2} is

$$V_{t2} = 0.331 \quad D_c \left[\frac{(\gamma_s - \gamma_f)^{\frac{2}{3}}}{(\gamma_f \quad \mu_e)^{\frac{1}{3}}} \right], \tag{10-8b}$$

where V_{t2} is the terminal velocity of the particle in transition flow (m/sec).

Empirical data indicate that turbulent flow conditions exist when the nondimensional Reynolds number for the flow around the particle is greater than 2000. The empirical relationship for the terminal velocity of a rock cutting particle in an annulus with turbulent flow V_{t3} is

$$V_{t3} = 2.95\left[D_c\left(\frac{\gamma_s - \gamma_f}{\gamma_f}\right)\right]^{\frac{1}{2}}, \tag{10-9b}$$

where V_{t3} is the terminal velocity of the particle in turbulent flow (m/sec).

Note that Equations (10-7a) to (10-9a) and (10-7b) and (10-9b) were originally developed in field units [19]. To be consistent with most of the other equations in this text, these equations have been restated in consistent USC units [Equations (10-7a) to (10-9a)] and consistent SI units [Equations (10-7b) to (10-9b)].

The nondimensional Reynolds number $\mathbf{N_{Rc}}$ for the flow around the particle is

$$\mathbf{N_{Rc}} = \frac{D_c\, V_t}{\nu}, \tag{10-10}$$

where D_c is the diameter of the particle (ft, m), V_t is the terminal velocity of the particle (ft/sec, m/sec), and ν is the kinematic viscosity of the flowing fluid (ft^2/sec, m^2/sec).

The aforementioned nondimensional Reynolds number equation can be used with any consistent set of units.

10.5 NONFRICTION AND FRICTION ILLUSTRATIVE EXAMPLES

Early analyses of stable foam drilling fluids utilized nonfriction solutions to obtain an initial approximate solution for these complex problems. Later efforts to simulate stable foam drilling situations introduced flow friction (major losses only) additions to the simple nonfriction theory [20, 21]. The nonfriction and friction theories are outlined mathematically in Chapter 6 (direct circulation) and Chapter 7 (reverse circulation).

The nonfriction ignores all major and minor friction losses due to fluid flow inside the drill string and in the annulus. This methodology includes only pressure due to fluid column weight.

The friction theories initially included the complexity of the major fluid flow friction losses [15, 21, 22]. The initial drilling application theories came from adaptations of multiphase oil and gas flow in production tubing. These production-based theories included only major friction losses and were not directly applicable to complicated drilling borehole geometry. In the past two decades, new additions to friction foam flow theories have included extensive experimental and field data-based empirical correlations. These new foam flow drilling fluids theories include both major and minor losses.

10.5.1 Nonfriction Approximation

The simple nonfriction methodology allows straightforward deterministic approximate solutions of stable foam drilling problems. However, the practical applicability of these nonfriction solutions is limited to simple geometric borehole profiles.

The simple nonfriction solution is obtained from Equation (10-8) by letting $f \rightarrow 0$. The nonfriction approximation is

$$\left| P_g \left(\frac{T_{av}}{T_g}\right)\left(\frac{Q_g}{\dot{w}_t}\right) \ln P + \left(\frac{Q_f}{\dot{w}_t}\right) P \right|_{P_s}^{P_{bb}} = | \, b \, |_0^H . \tag{10-11}$$

The aforementioned equation can be integrated into the following equation:

$$P_g \left(\frac{T_{av}}{Tg}\right) Q_g \ln\left(\frac{P_{bb}}{P_s}\right) + Q_f (P_{bb} - P_s) = \dot{w}_t \; H. \tag{10-12}$$

Illustrative Example 10.1 The borehole used in this illustrative example is the basic example used in Chapters 8 and 9. The $7^7/_8$-in (*200 mm*)-diameter borehole is drilled out of the bottom of API $8^5/_8$-in (*219 mm*)-diameter 32.00-lb/ft (*14.00 kg/m*) nominal casing set to 7000 ft (*2134 m*) (see Figure 10-2 for well casing and open hole geometric configuration). This is special clearance casing fabricated by Lonestar Steel that has a drift diameter that will pass a $7^7/_8$-in (*200 mm*)-diameter drill bit. The drill bit used to drill the interval is a $7^7/_8$-in (*200 mm*)-diameter tricone roller cutter insert type with three $^{12}/_{32}$-in (*9.5 mm*) nozzles. The anticipated drilling rate in a sandstone and limestone sequence (sedimentary rock) is approximately 60 ft/hr (*18.3 m/hr*). The open hole interval below the casing shoe to be drilled is from 7000 ft (*2134 m*) to 10,000 ft (*3047 m*). A back pressure of 80 psig (*55.2 N/cm² gauge*) is applied to the end of the return line to control the foam quality at the top of the annulus. This back pressure is selected so that the foam quality just upstream from the back pressure valve will be approximately 0.98.

The drill string for this illustrative example is made up of 500 ft (*152.4 m*) of $6^1/_4$-in (*158.8 mm*) by $2^{13}/_{16}$-in (*71.4 mm*) drill collars above the drill bit and API $4^1/_2$-in (*114 mm*), 16.60-lb/ft (*7.52 kg/m*) nominal, IEU-S135, NC46 to the surface. The tool joints of this drill pipe have an outside diameter of $6^1/_4$ in (*154.8 mm*) and an inner diameter of $2^3/_4$ in (*69.9 mm*). The drilling is carried out at a surface location of 4000 ft (*2129 m*) above sea level (see Table 5-1 and Appendix B). The regional geothermal gradient is approximately 0.01°F/ft (*0.018°C/m*). The borehole is drilled with a stable foam drilling fluid composed of fresh water (with a surfactant) and compressed air.

Appendix E gives the Mathcad detailed solution of Illustrative Example 10.1. The nonfriction solution is sometimes used to obtain the lower boundary of the bottom hole pressure. This value is useful for the efficient initiation of the friction solution. In this nonfriction solution, the return line and the small geometric

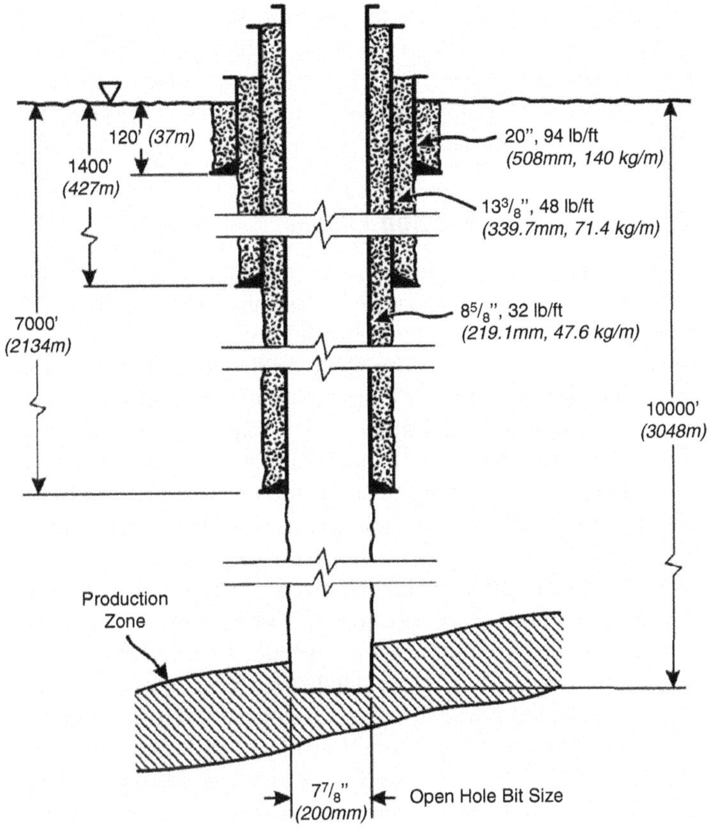

FIGURE 10-2. Illustrative Example 10.1 casing and open hole well geometric configuration.

changes in the inside diameter of the annulus casing and open hole are ignored. Also, the effective foam viscosity is not relevant for this solution. In this solution, a fresh water injection rate of 50 gpm (*189 liters/min*) and an air injection rate of 2120 scfm (*1000 standard liters/sec*) are used to obtain the nonfriction solution. It should be understood that volumetric flow rates of air or gas to be injected into the standpipe for a foam drilling operation will be specified in either standard cubic feet per minute (scfm) or standard liters per minute (for a specified set of standard conditions, e.g., ISO, API, ASME).

As discussed in Chapter 9, unlike air drilling operations, the volumetric flow rate of air for a stable foam operation is selected to create, together with the fresh water volumetric flow rate, stable foam throughout the annulus volume. In this event, a surface bypass valve and line would be used to vent excess air flow from the feed line from the compressors to the injection standpipe operation.

In this special illustrative example, no friction due to flow is considered. Therefore, the bottom hole pressure reflects only an effective static weight of the foam column in the annulus. The nonfriction solution gives a bottom hole pressure of 386 psia (*264 N/cm² abs*). The foam quality at the top of the annulus is 0.98 (at the back pressure valve) and 0.93 at the bottom of the annulus.

10.5.2 Major and Minor Friction Losses

The governing equations for the direct circulation foam drilling friction solution were presented in Chapter 6. These require complex trial and error solutions that can be applied to deep wells with complicated borehole geometry. The adding of major and minor friction losses to the foam drilling fluid theory allows analytic solutions that simulate the actual drilling situation more closely.

In general, the adding of flow friction losses requires the addition of volumetric flow rates of incompressible fluid and compressed gas to compensate for the added energy losses in the system (relative to the nonfriction solution above). This is reflected in a higher bottom hole pressure in the annulus and elsewhere in the system.

Illustrative Example 10.2 In this illustrative example, data given in Illustrative Example 10.1 are used for a solution that considers both major and minor friction losses for direct circulation foam drilling. In this solution the pressure and foam quality for annulus foam flow will be determined as a function of depth. Also, the pressure inside the drill string as a function of depth will be determined for this aerated flow region. The bottom hole cleaning characteristics of the foam flow are also determined. For this example, it is assumed that a 100-ft (*30.5 m*)-long, 5.625-in (*143.9 mm*) inside diameter return flow line allows foam from the top of the annulus to flow through the back pressure valve to the separator. The back pressure at the end of the return line is 80 psig (*55.2 N/cm² gauge*). Also, it is assumed that the foam screening test run on the foam to be used in the forthcoming drilling operation has been given a beaker height of 610 ml and a half-life of 280 sec. It will be assumed that the constants C, a, and b for Equation (10-4) are all 1.0.

Appendix E gives the Mathcad detailed solution of Illustrative Example 10.2. Figure 10-3 shows the flowing foam pressures in the annulus and the pressures in the aerated flow of the mixture of incompressible fluid and incompressible gas inside of the drill string. The injected fresh water flow rate and compressed air flow rate to the standpipe are 50 gpm (*189 liters/min*) and 2120 scfm (*standard 1000 liters/sec*).

It is clear from Figure 10-3 that the foam drilling fluid in this example can provide a significant side wall and bottom hole pressure in the open hole section. The value of 1315 psia (*907 N/cm² abs*) bottom hole pressure at 10,000 ft (*3047 m*) is higher than the bottom hole pressures calculated in the air drilling

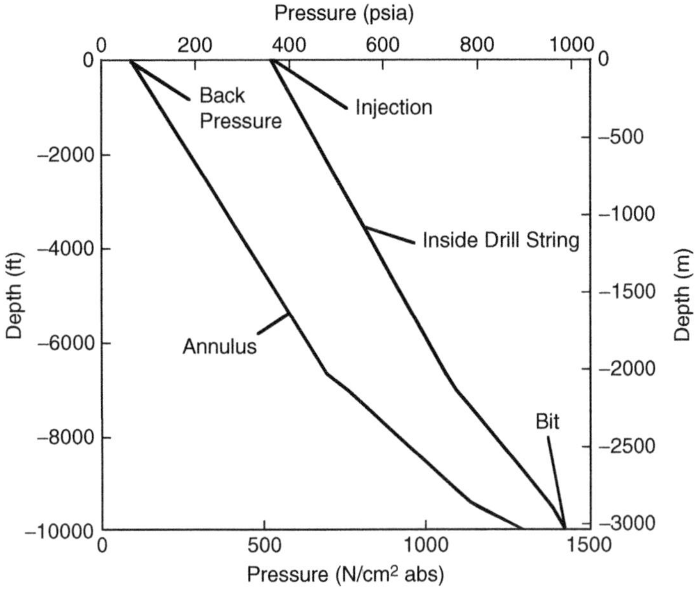

FIGURE 10-3. Illustrative Example 10.2 pressures versus depth.

illustrative example in Chapter 8 and lower than the aerated drilling illustrative example in Chapter 9.

In foam drilling, use of a surfactant with the water and air allows the development of a foam structure in the annulus, which in turn provides resistance to the flow of the foam from the bottom of the annulus. This resistance is demonstrated by the high bottom hole pressure of 1315 psia *(907 N/cm² abs)*. In this friction solution the foam quality upstream of the back pressure value is 0.98 (the same as in the nonfriction solution), but the bottom hole foam quality in this friction solution is 0.81 (the nonfriction annulus bottom hole quality is 0.93). This lower bottom hole annulus foam quality reflects the friction resistance to foam flow in the annulus. It should be noted that the nonfriction solution approximates the bottom hole pressure and foam quality under nonflowing or static conditions. This means that the difference between the friction bottom hole pressure, 1315 psia *(907 N/cm² abs)*, and the nonfiction bottom hole pressure, 386 psia *(266 N/cm² abs)*, is 929 psi *(641 N/cm²)* of friction pressure created by the flow of the foam structure in the annulus.

Figure 10-4 shows the injection pressure and bottom hole pressure as drilling takes place from the casing shoe at 7000 ft *(2134 m)* to the final depth of 10,000 ft *(3048 m)*. Assuming that this example is more typical of a performance

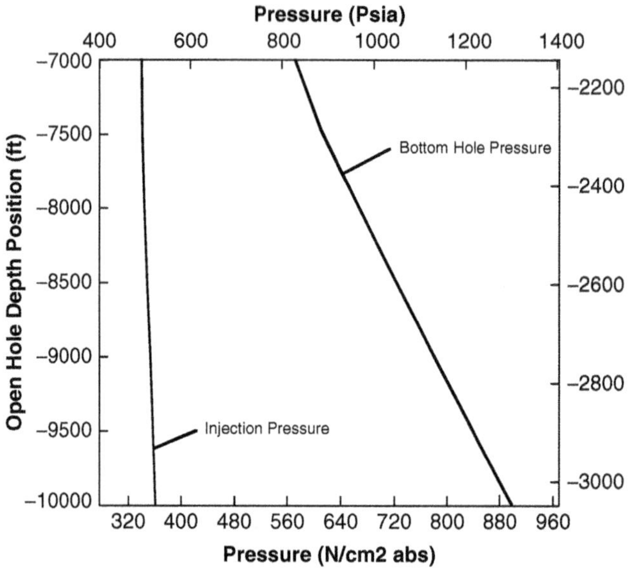

FIGURE 10-4. Drilling ahead, bottom hole, and injection pressures versus depth.

drilling operation, then it is clear that the same water (and additives), air injection rates, and back pressure setting could be used to drill the entire 3000 ft *(914 m)* of open hole. Injection pressure changes very little with increasing drilling depth, staying at around 500 psia *(345 N/cm^2 abs)*. However, the bottom hole pressure nearly doubles as drilling progresses from 7000 ft *(2134 m)* to the final depth of 10,000 ft *(3048 m)*. At the 10,000-ft (3048 m) depth the bottom hole pressure is 1315 psia *(907 N/cm^2 abs)*. This is a significant bottom hole pressure when drilling into possible water producing formations. This magnitude of pressure could curtail some formation water and natural gas influxes.

Figure 10-5 shows the foam quality as drilling takes place from the casing shoe at 7000 ft *(2134 m)* to the bottom of the open hole section at 10000 ft *(3047 m)*. Also shown is the transit time for the rock cuttings as a function of the drilling depth below the casing shoe.

The foam quality decreases from approximately 0.86 to 0.81 as the drilling depth increases. This is a small change over the open hole interval. The decrease in foam quality reflects the increase in bottom hole pressure with depth. Also shown in Figure 10-5 is the time to get rock cuttings to the surface as a function of drilling depth. The time to surface or "bottoms up" time increases as the drilling depth increases.

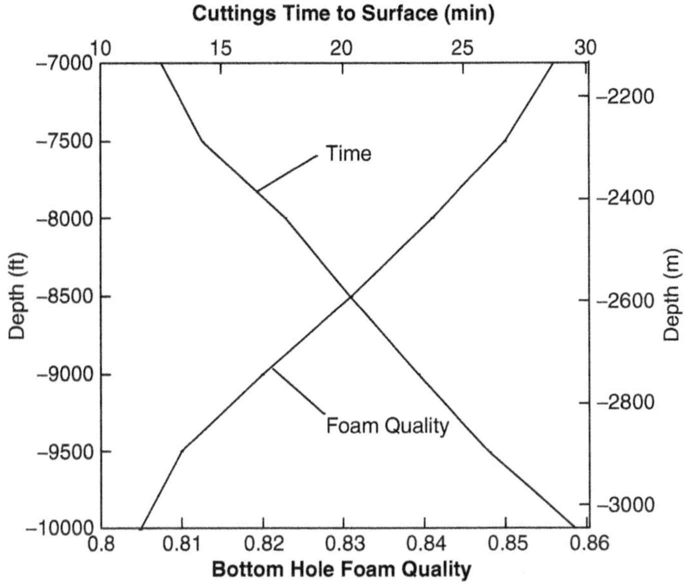

FIGURE 10-5. Drilling ahead, bottom hole foam quality, and cuttings time to surface versus depth.

10.6 CONCLUSIONS

The discussions in this chapter concentrated on direct circulation operations. Stable foam drilling operations are generally restricted to direct circulation operations (note: reverse circulation techniques are used extensively in work over and production operations).

The Mathcad illustrative example calculations in Appendix E have utilized lumped geometry approximations for the drill pipe body and drill pipe tool joints. Such approximations appear to adequately model the overall friction resistance along the circulation system and give accurate results for bottom hole and injection pressures. The Mathcad solution program illustrated in Appendix E and results discussed in this chapter have been correlated successfully with the other programs used in industry and with field data from foam drilled wells.

The importance of obtaining accurate foam rheologic data for forthcoming foam drilling operations cannot be overemphasized. In order for predictive calculations to be relevant and useful for planning and operations, these programs must use foam rheology data that reflect the actual components to be used in the operation. Further, these predictive programs will be only as good as the rheology models within them. These rheology models must be correlated with real experimental data. The correlation process can be made efficient by obtaining quality experimental data on a combination of waters, surfactants, and additives that have been successful in laboratory tests and in field drilling operations (i.e., the nominal baseline foam). Once this baseline nominal foam has been selected, a screening process must be set up to measure all other combinations against this baseline mixture.

New foams are being developed for use with oil-based fluids used in synthetic oil-based mud drilling operations. These foams are being developed for operations in high temperature wells and possibly in future drilling and work over offshore operations.

REFERENCES

1. Fried, A. N., "The Foam Drive for Increasing the Recovery of Oil," *Report of Investigation No. 5866*; U.S. Bureau of Mines, Washington, DC, 1961.
2. Raza, S. H., and Marsden, S. S., "The Streaming Potential and the Rheology of Foam," *SPE Journal*, Vol. 7, No. 4, 1967.
3. Hutchison, S. O., "Foam Workovers Cut Costs 50%," *World Oil*, January 1969.
4. David, A., and Marsden, S. S., "The Rheology of Foam," *SPE Paper 2544*, SPE Annual Meeting, Denver, September 28–20, 1969.
5. Mitchell, B. J., *Viscosity of Foam*, Ph.D. Dissertation, University of Oklahoma, 1969.
6. Blauer, R. E., Mitchell, B. J., and Kohlhaas, C. A., "Determination of Laminar, Turbulent, and Transitional Foam Flow Losses in Pipes," *SPE Paper 4885*, SPE Annual California Regional Meeting, San Francisco, April 4–5, 1974.
7. Beyer, A. H., Millhone, R. S., and Foote, R. W., "Flow Behavior of Foam as a Well Circulating Fluid," *SPE Paper 3986*, SPE Annual Conference and Exhibition, San Antonio, October 8–11, 1972.
8. Krug, J. A., and Mitchell, B. J., "Charts Help Find Volume, Pressure Needed for Foam Drilling," *Oil and Gas Journal*, February 7, 1972.
9. Sanghani, V., *Rheology of Foam and its Implications in Drilling and Cleanout Operations*, Master of Science Thesis, University of Tulsa, Oklahoma, 1982.
10. Machado, C. J., and Ioku, C. U., "Experimental Determination of Solids Friction Factor and Minimum Volumetric Requirements in Air and Gas Drilling," *SPE Paper 9938*, SPE California Regional Meeting, Bakersfield, March 25–26, 1981.
11. Okpobiri, G. A., and Ioku, C. U., "Experimental Determination of Friction Factors for Mist and Foam Drilling and Well Cleanout Operations," *ASME Paper AO-204*, ASME Energy Sources Technology Conference and Exhibition, Houston, January 30–February 3, 1983.
12. Okpobiri, G. A., and Ioku, C. U., "Volumetric Requirements for Foam and Mist Drilling Operations," *SPE Paper 11723*, SPE California Regional Meeting, Ventura, March 23–25, 1983.
13. Okpobiri, G. A., and Ioku, C. U., "Volumetric Requirements for Foam and Mist Drilling Operations," *SPE Drilling Engineering*, February 1986.

14. Giffin, D. R., and Lyons, W. C., "Case Studies of Design and Implementation of Underbalanced Wells," *SPE Paper 55060*, SPE Rocky Mountain Regional Meeting, Gillette, Wyoming, May 15–18, 1999.

15. Ozbayoglu, M. E., Kuru, E., Miska, S., and Takach, N., "A Comparative Study of Hydraulic Models for Foam Driling," *SPE Paper 65489*, SPE/PS CIM International Conference on horizontal Well Technology, Calgary, Alberta, Canada, November 6–8, 2000.

16. Guo, B., Sun, K., and Ghalambor, A., "A Closed Form Hydraulics Equation for Predicting Bottom-Hole Pressure in UBD with Foam," Presented at IADC/SPE Underbalanced Technology Conference and Exhibition, Houston, Texas, March 25–26, 2003.

17. ASTM D 1173-53, "Standard Test Method for Foaming Properties of Surface-Active Agents," reapproved 2001.

18. Abbott, W. K., *An Analysis of Slip Velocities of Spherical Particles in Foam Drilling Fluids*, Master of Science Thesis, Colorado School of Mines, 1974.

19. Moore, P. L., *Drilling Practices Manual*, Petroleum Publishing Company, Tulsa, 1974.

20. Lord, D. L., "Mathematical Analysis of Dynamic and Static Foam Behavior." *SPE 7927*, SPE Symposium on Low-Permeability Reservoirs, Denver, Colorado, May 20–22, 1979.

21. Okpobiri, G. A., and Ikoku, C. U., "Volumetric Requirements for Foam and Mist Drilling Operations," *SPEDE*, February 1986.

22. Chen, Z. et al., "Rheology and Hydraulicsof Polymer (HEC)-Based Drilling Foams at Ambient Temperature Conditions," *SPE Journal*, March 2007.

CHAPTER 11

Specialized Downhole Drilling Equipment

There are two downhole drilling devices used with air and gas drilling technology. These are the down-the-hole hammer (DTH) and the progressive cavity positive displacement motor (PDM) drilling motor. The DTH is unique to air and gas drilling. It can only be used with an air or other gas and with unstable (mist) drilling fluids. The PDM was originally developed for use with incompressible drilling fluids, but will operate on unstable foam, aerated, and stable foam drilling fluids. PDMs that are operated on these drilling fluids have sealed bearing packages and bypasses that are unique to light drilling fluid operations.

11.1 DTH

There are two basic designs for the DTH. One design utilizes a flow path of the compressed air through a control rod (or feed tube) down the center hammer piston (or through passages in the piston) and then through the hammer bit. The other design utilizes a flow path in the wall of the hammer housing. This type of hammer design allows for the flow of gas in this wall small annulus passage (around the piston) and then through the hammer bit. Some hammers combine these two design concepts. DTHs require rotation of the drill string in order to allow advance of the drill bit in the rock formation. This rotation allows the insert teeth to cover the rock face with rock impacting drill bit inserts and remove the rock cuttings with the exhaust gas from the piston chamber.

Figure 11-1 shows a schematic of a typical control rod flow design. The hammer action of the piston on the top of the drill bit shank provides an impact force that is transmitted down the shank to the bit interface with the rock formation. This transmitted impact force crushes the rock at the rock face. The gas flow through DTH moves in an intermittent flow. In shallow boreholes there is little annulus back pressure and a low pressure on the top of the hammer inside the drill string. In these situations, the piston will impact the top of the bit shank at a rate of from about 800 to 2000 strikes per minute (depending on the

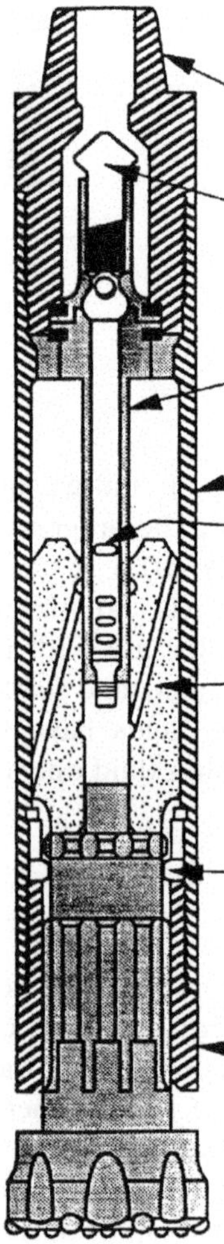

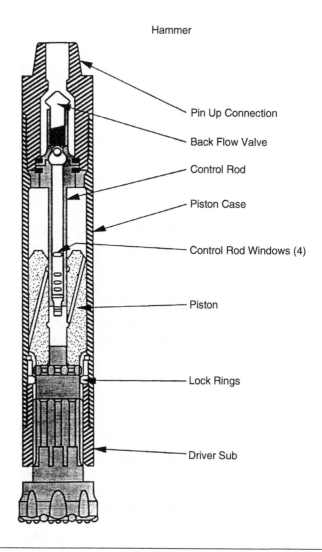

FIGURE 11-1. Schematic cutaway of a typical air hammer (courtesy of Diamond Air Drilling Services, Incorporated).

volumetric flow rate of gas and the detailed internal design of the hammer). However, in deep boreholes where the annulus back pressure is usually high, impacts can be as low as 100 to 400 strikes per minute [1].

Figure 11-1 shows the air hammer suspended from a drill string with the shoulder of the bit not in contact with the shoulder of the driver sub (housing) of the DTH. In this position, compressed air flows through the pin connection at the top of the hammer to the bit without actuating the piston action (i.e., to blow the borehole clean). When the hammer is placed on the bottom of the borehole and weight is placed on the hammer, the bit shank will be pushed up inside the hammer housing until the bit shoulder is in contact with the shoulder of the driver sub. This action aligns one of the piston ports (of one of the flow passage through the piston) with one of the control rod windows. This allows the compressed air to flow to the space below the piston, which in turn forces the piston upward in the hammer housing. During this upward stroke of the piston, no air passes through the bit shank to the rock face. In essence, rock cutting transport is suspended during this upward stroke of the piston.

When the piston reaches the top of its stroke, another one of the piston ports aligns with one of the control rod windows and supplies compressed air to the open space above the piston. This air flow forces the piston downward until it impacts the top of the bit shank. At the same instant the air flows to the space above the piston, the foot valve at the bottom of the control rod opens, and air inside the drill string is exhausted through the control rod, bit shank, and the bit orifices to the rock face. This compressed air exhaust entrains the rock cuttings created by the drill bit for transport up the annulus to the surface. This impact force on the bit allows the rotary action of the drill bit to be very effective in destroying rock at the rock face. This in turn allows the air hammer to drill with low WOB (or the order of a few 1000 lbs or a few 100 kg). Typically, a 6.9-in (*175 mm*) outside diameter DTH is used to drive a $7^7/_8$-in (*200 mm*). This generally requires a weight on bit of 1500 lb (*682 kg*). DTHs must have some type of lubricant injected into the injected gas during the drilling operation. This lubricant is needed to lubricate the piston surfaces as it moves in the hammer housing. DTHs are used exclusively for vertical drilling operations. There are attempts underway to develop a gas-driven hammer that can be used in a directional borehole.

DTHs are available in housing outside diameters from 3 in (*76.2 mm*) to 16 in (*406.4 mm*). The 3-in (*76.2 mm*) housing outside diameter hammer can drill a borehole as small as $3^5/_8$ in (*92.1 m*). The 16-in (*406.4 mm*) housing outside diameter hammer can drill a borehole from $17^1/_2$ in (*444.5 mm*) to 33 in (*838.2 mm*). For shallow drilling operations, conventional air hammer bits are adequate. For deep drilling operations (usually oil and gas recovery wells), higher quality oil field drill bits are required (see Chapter 4).

There are a variety of manufacturers of DTHs. These manufacturers use several different designs for their respective products. The air hammer utilizes very little power in moving the piston inside the hammer housing. For example, a typical 6.9-in (*175 mm*) outside diameter air hammer will use less than 2 hp driving the piston at around 600 strikes per minute. This is a very small amount of power relative to the total needed for the actual rotary drilling operation. Thus, it is clear that the vast majority of the power to the drill string is provided by the rotary table. Therefore, any pressure loss (i.e., energy loss) due to the piston lifting effort can usually be ignored. The major pressure loss in the flow through an air hammer is due to the flow energy losses from the constrictions in the flow path when the air is allowed to exit the hammer (on the down stroke of the piston).

All air hammer designs have internal flow constrictions. These flow constrictions can be used to model the flow losses through the hammer.

Figure 11-2 shows the pressure drop through two typical DTH models as a function of the volumetric flow rate of air through the hammers [2]. These data can be obtained from the hammer manufacturer, although it often takes some research. In most designs the constrictions can be approximately represented by a set of internal orifice diameters in the hammer housing, piston, and flow passages to the drill bit exhaust holes. In order to obtain the near linear plots shown in Figure 11-2, each position on the plot represents a slightly different internal setup within the hammer to allow that hammer to operate with a different volumetric flow rate.

Combining the complexity of these flow passages and the unsteady state flow conditions within the hammer, this type of analysis is probably not worth attempting in detail. From data shown in Figure 11-2, it is clear that flow conditions a few tens of feet (*meters*) above the hammer inside the drill string are steady state.

The best way to analyze gas flow pressures in a drill string with a DTH is to analyze the unsteady state dynamics of the hammer by itself and then combine

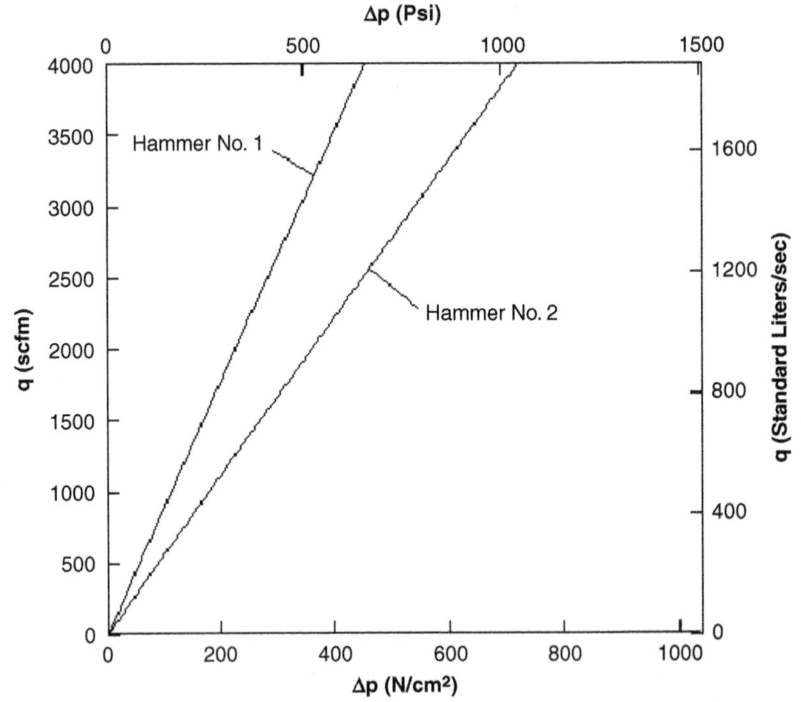

FIGURE 11-2. Two DTH models and their pressure losses through their flow constrictions.

the essential properties of the study to the steady-state flow analysis already developed in Chapter 8 for air and gas drilling fluids.

The unsteady state dynamics of a DTH is governed by the volume of the piston displacement. As the piston moves in its chamber, the gas flow reverses the piston before it strikes the top of the chamber. The chamber volume is described by the open length of the chamber and the inside diameter of the chamber. Thus, the volume is

$$Vol = A_{pc}\, L_c$$

or

$$Vol = \left(\frac{\pi}{4}\right) D_{pc}^2\, L_c, \tag{11-1}$$

where *Vol* is the total volume of the piston chamber (ft^3, m^3), A_p is the cross-sectional area of the inside of the piston chamber (ft^2, m^2), D_{pc} is the inside diameter of the piston chamber (ft, m), and L_c is the length of the chamber above the piston (ft, m).

The length of the chamber just given is the distance from the top of the piston when the piston is in its full down position in contact with the top of the shank of the drill bit. Thus, the above volume describes the maximum possible displacement of the hammer. As stated earlier, in operation the piston does not move all the way up to the top of the chamber. Therefore, an effective fraction of chamber usage friction is introduced to define the actual DTH displacement volume. The DTH volume is

$$Vol_{DTH} = \left(\frac{\pi}{4}\right) D_{pc}^2\, L_c\, e_s, \tag{11-2}$$

where Vol_{DTH} is the effective hammer displacement volume (ft^3, m^3) and e_s is the effective fraction of the length of the piston motion (taken here as 0.80).

Knowing the volumetric flow rate of the gas entering the top of the DTH, then the number of strokes per minute of the piston becomes

$$N_{ps} = \frac{Q_{in}\, 60}{2\, Vol_{DTH}}, \tag{11-3}$$

where N_{ps} is the piston strokes (strokes per minute) and Q_{in} is the volumetric flow rate of the gas through the DTH (ft^3/sec, m^3/sec).

The time for the piston to complete a stroke is

$$\Delta\tau_{ts} = \frac{60}{N_{ps}}, \tag{11-4}$$

where $\Delta\tau_{ts}$ is the total time for a piston stroke (sec).

The velocity of the piston as it impacts the top of the drill bit shank is

$$V_p = \left[2 \left(\frac{P_{in}\, A_p + W_p}{\left(\frac{W_p}{g}\right)} \right) e_s\, L_s \right]^{0.5}, \tag{11-5}$$

where V_p is the velocity of the piston (ft/sec, m/sec), P_{in} is the pressure in the gas as it enters the top of the DTH (lb/ft^2, N/m^2), W_p is the weight of the piston (lb, N), and g is the acceleration of gravity (32.2 ft/sec^2, 9.81 m/sec^2).

The total kinetic energy generated when the piston strikes the top of the shank of the drill bit is

$$KE_{ps} = \frac{1}{2}\left(\frac{W_p}{g}\right)V_p^2, \tag{11-6}$$

where KE_{ps} is kinetic energy (lb-ft, N-m).

Illustrative Example 11.1 For the two DTH examples given in Figure 11-2, determine the number of strokes per minute and the total kinetic energy transferred to the top of the shank of the drill bit assembly. Also determine the total stroke time and the downward impact travel time. Use the example well that was used in the illustrative example discussed in Chapters 8, 9, and 10. Figure 11-3 shows this example well geometry. In this illustrative example,

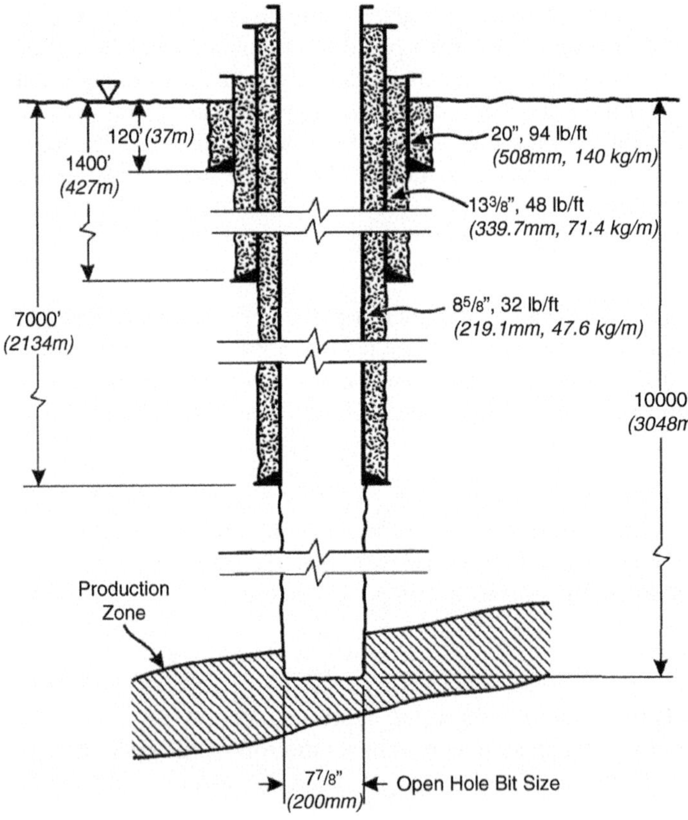

FIGURE 11-3. Illustrative Example 11.1 casing and open hole well geometric configuration.

assume that the hammer is being used to drill at a depth of 10,000 ft (*3048 m*). Assume that 3600 acfm (*1698 liters/sec*) is the circulation volumetric flow rate of air. Table 11-1 gives the required geometric and physical properties data for hammers No. 1 and No. 2.

From Illustrative Example 8.4 (in Chapter 8 and Appendix C), the bottom hole annulus pressure while drilling at 10,000 ft (*3048 m*) is 251.8 psia (*173.6 N/cm² abs*). The details of Illustrative Example 8.4 are found in Chapter 8 and Appendix C. The approximate pressure drops across the hammer models are given in Figure 11-2.

Appendix F gives the Mathcad detailed solutions for hammer No. 1 and No. 2. Table 11-2 gives a summary of the results.

It is instructive to show a portion of the aforementioned results in graphic form. Figure 11-4 shows where hammer No. 1 and No. 2 are located on a KE_{ps} versus N_{ps} plot. Other candidate DTHs could be potted on the same map to facilitate some rational decisions regarding which DTH might be better for a particular deep drilling application.

The information just given can be used to calculate circulation pressures for a DTH drilling operation similar to that given in the Mathcad solution for Illustrative Example 8.4 (see Appendix C). The annulus flow pressures will remain the same as in Illustrative Example 8.4. To obtain the pressures on the inside of the drill string, it is only necessary to replace the "flow through the bit nozzles" subprogram in Illustrative Example 8.4 to a pressure drop value appropriate to the applicable DTH used in the calculation (from data similar to Figure 11-2).

Because the Figure 11-2 vertical axis is given in scfm (standard liters/sec), it is necessary to enter the plot with a volumetric flow rate of 2136 scfm (*1008 standard liters/sec*). This volumetric flow rate is equivalent to the total compressor output of 2400 acfm (*1133 actual liters/sec*). For hammer No. 1 at 2136 scfm (*1008 standard liters/sec*), the approximate pressure drop through this hammer

Table 11-1. Hammer Dimensional and Physical Property Data

Hammer	Length (in, mm)	Diameter (in, mm)	Weight (lb, kg)
No. 1	7.11 (181)	5.90 (150)	97 (44)
No. 2	4.69 (119)	6.00 (152)	78 (36)

Table 11-2. Illustrative Example 11.1 Summary of Results

Hammer	N_{ps} (spm)	KE_{ps} (lb-ft, N-m)	$\Delta\tau_t$ (sec)	$\Delta\tau_{dn}$ (sec)
No. 1	405	6538 (8864)	0.148	0.014
No. 2	418	6313 (8559)	0.141	0.009

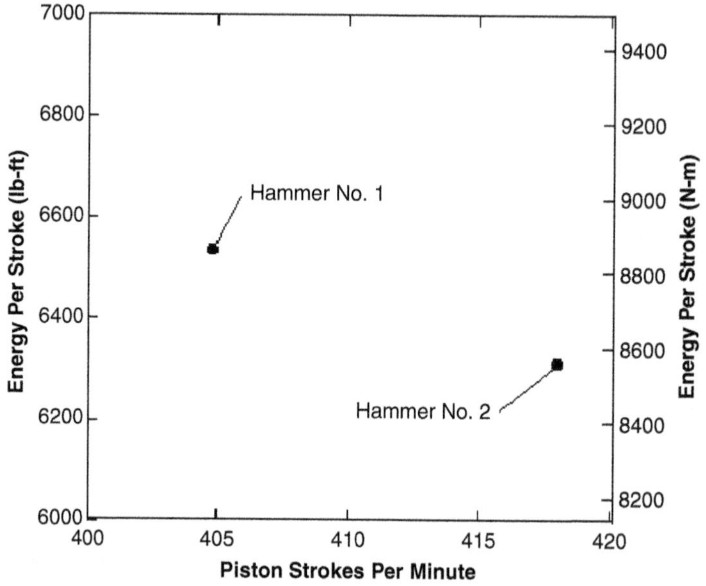

FIGURE 11-4. Graphic representation of Illustrative Example 11.1 results.

is 340 psi. Figure 11-5 shows the drilling ahead plot for bottom hole annulus pressures and injection pressures as a function of open hole depths from 7000 ft (*2134 m*) to 10,000 ft (*3047 m*). Comparing this figure with Figure 8-9, the bottom hole annulus pressures are unchanged, but the injection pressures are significantly higher. These higher injection pressures reflect the additional pressure losses through the DTH at the respective depths.

11.2 PDM

The most commercially successful positive displacement fluid downhole motor has been the progressive displacement "cavity" motor. This motor design is based on the work of French engineer Rene Moineau, who patented numerous variants of these devices for hydraulic (incompressible fluid) pumps (and hydraulic motors) between 1930 and 1948 [3]. The driving shaft of these motors is a rigid shaft composed of helical lobe repeating sections (like a screw thread) and is denoted as the rotor. This rigid helical lobe shaft is inserted into a flexible helical lobe cavity sheath (tight fitting) denoted as the stator. The outside surface of the sheath is affixed to the inside surface of a rigid cylindrical housing. As fluid is pumped under pressure into one end of this device motor in the space between

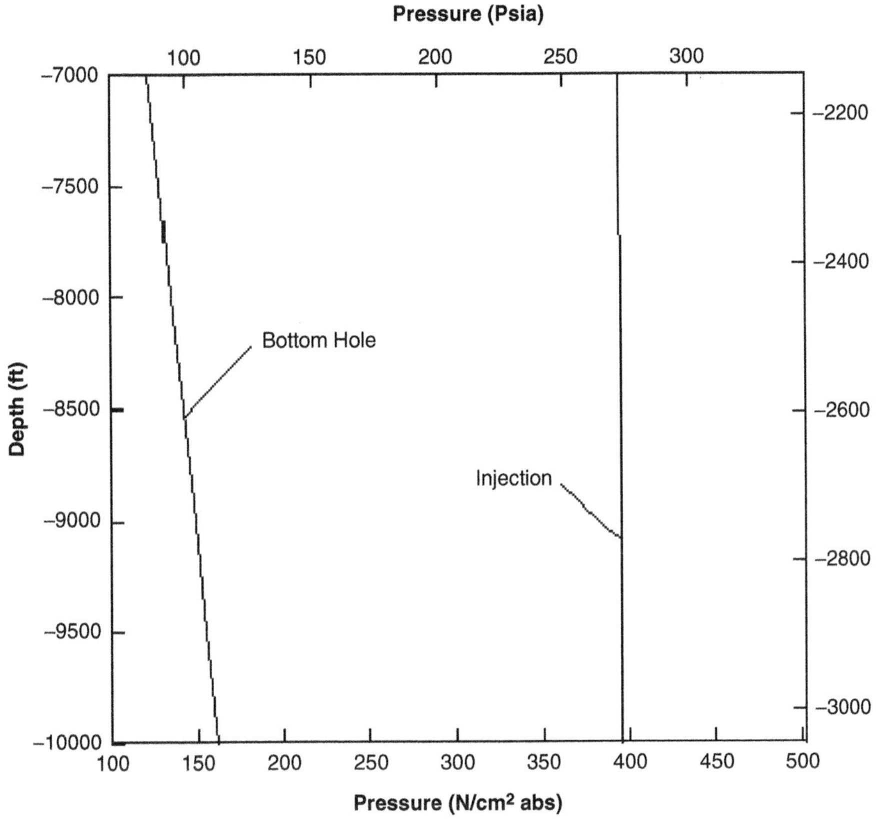

FIGURE 11-5. Illustrative Example 11.1 and hammer No. 1 injected volumetric flow rates versus piston strokes per minute.

the rigid drive shaft and the flexible helical lobe cavity, the rigid shaft is forced to rotate as the fluid passes through the progression of cavities in the motor.

Figure 11-6 shows a cutaway view of the helical lobe section of a typical downhole positive displacement mud motor based on this Moineau design [4]. The flow of the drilling mud through the motor would force the rotor to rotate. The bottom end of the rotor is connected to a flexible universal joint coupling and shaft with a bearing assembly and drive sub. Through this connection the drill bit is turned at the same rotational speed as the rotor. The drilling mud exits the motor at the bottom end and flows through the drill bit open orifices (or nozzles). At the bottom of the annulus, the drilling mud entrains the rock cuttings generated by the drill bit and carries these cuttings to the surface in the annulus.

The example positive displacement motor shown in Figure 11-6 has a five lobe rigid helical shaft and a six lobe flexible helical cavity sheath configuration.

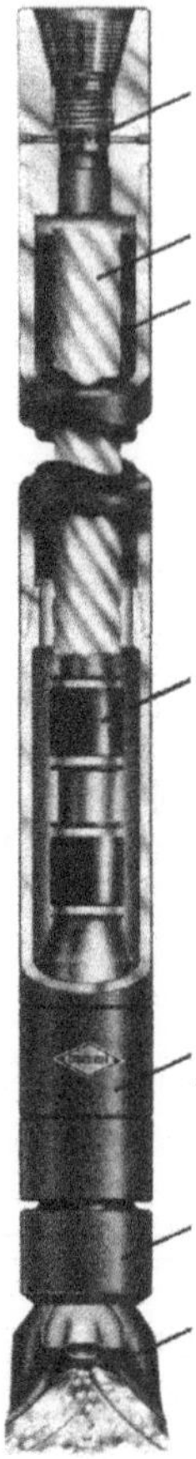

FIGURE 11-6. Cutaway view of a typical 5:6 lobe configuration positive displacement mud motor (courtesy of Baker Oil Tools).

A cross-section view of this configuration is shown in Figure 11-7, together with several other typical configurations [3].

Assuming the same downhole motor size and fluid volumetric flow rate through the motor (same hydraulic power), the basic differences among the five typical lobe configurations given in Figure 11-7 are as follow:

- The higher the lobe configuration, the higher the rotor shaft output torque.
- The higher the lobe configuration, the lower the rotor shaft output speed.

Depending on the drilling applications, one motor size and lobe configuration may be more effective than another.

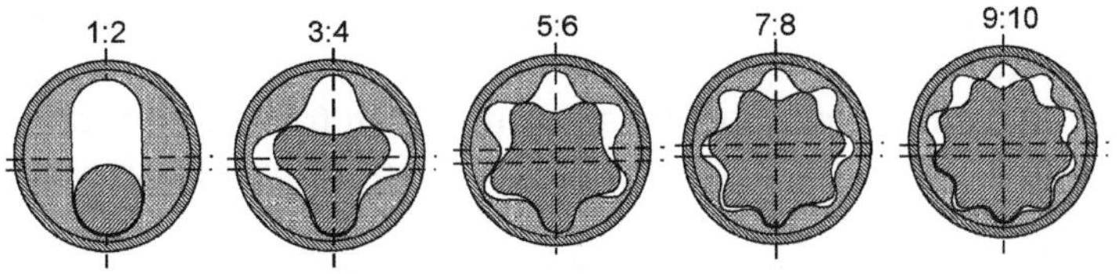

FIGURE 11-7. Different lobe configurations for progressive helical lobe positive displacement motors.

USCS Units

PDMs are defined by their displacement. For USCS units, the displacement per revolution is

$$s = \frac{q_{PDM}\,231}{N_r}, \tag{11-7a}$$

where s is the displacement (in^3), q is the volumetric flow rate (gpm), and N_r is the rotation speed of the motor and drill bit (rpm).

SI Units

For SI units the displacement per revolution is

$$s = \frac{q_{PDM}}{1000\,N_r}, \tag{11-7b}$$

where s is the displacement (m^3), q is the volumetric flow rate (lpm), and N_r is the rotation speed of the motor/drill bit (rpm).

Downhole positive displacement motors have been very useful in directional drilling operations. For directional drilling operations, the downhole positive displacement motor is made up to a bent sub directly above the motor with an MWD system made up above the bent sub. The conventional drill string is made up to the top of the MWD.

These drilling mud-actuated positive displacement motors have been adapted for use in air and gas drilling operations. Positive displacement motors converted for air and gas drilling operations usually involve replacing the conventional mud motor bearing assembly (which uses drilling mud as the lubricant) with a sealed grease-lubricated bearing assembly. Also, the dimensional tolerances between the rigid helical lobe shaft and the helical lobe flexible sheath are relaxed to provide a looser fit between these elements. To operate these downhole motors in an air drilling operation, a liquid lubricant must be injected into the compressed air flow being injected into the top of the drill string. These downhole motors can be operated with aerated drilling fluids (usually drilling mud with air or gas aeration) or with stable foam with little or no special motor preparations.

The primary operational concern when drilling with a downhole positive displacement motor using compressed air (or gas) is the tendency of the rotor shaft to rotate at too high a speed and, thus, destroy (with friction heat) the elastomer flexible helical cavity sheath of the motor. One operational situation where this can occur is when there is excessive expansion of the air as it passes through the progressive cavities of the motor. Such excessive expansion can allow runaway rotor speeds, which can only be controlled by inserting appropriately sized nozzles into the drill bit (or a single nozzle inside motor flow passage above the drill bit connection)[5]. The small diameter nozzles choke the air flow from the motor and provide a back pressure at the bottom of the positive displacement motor (above the drill bit). This back pressure controls the air expansion and, thereby, controls the output rotational speed of the rotor.

Another runaway speed situation can occur when the drilling load is taken off the motor by lifting the drill string. Under these conditions, the rotor can again go to high speeds and destroy the elastomer flexible sheath. This must be controlled by installing a bypass valve above the motor section. This bypass valve is actuated when weight is taken off the drill bit. When the weight is removed, the valve is opened and most of the flow down the inside of the drill string is diverted directly to the annulus. Little fluid flow goes through the motor cavity and, therefore, the speed of the rotor is kept under control.

The downhole positive displacement motor is actuated by the volumetric flow rate of the fluid passing through it. The output speed of the motor rotor shaft is directly proportional to the volumetric flow rate of the fluid. The torque that the rotor shaft can produce is nearly directly proportional to the pressure drop through the motor. Thus, the power the motor can generate is also directly proportional to the pressure drop through the motor. Figures 11-8a and 11-8b show the torque and power outputs versus pressure drop plots for a typical $6^3/_4$-in (*171.5 mm*) outside diameter, 5:6 lobe, downhole positive displacement motor with a volumetric flow rate of 400 gpm (*1514 lpm*) of drilling mud or other incompressible drilling fluid.

The plots shown in Figures 11-8a and 11-8b are for the drilling fluid volumetric flow rate of 400 gpm (*1514 lpm*). This volumetric flow rate gives a nearly constant motor rotor rotation speed of 200 rpm (the drill bit speed). There are similar families of plots that can be prepared for this motor for other drilling fluid volumetric flow rates. Higher volumetric flow rates yield higher drill bit speeds, and lower volumetric flow rates yield lower drill bit speeds. The torque and power plots are similarly altered for increases and decreases in volumetric flow rates.

Illustrative Example 11.2 Determine the injection pressure and resulting compressor fuel consumption when a Baker Oil Tools Model Mach 1, $6^3/_4$-in (*171.5 mm*) outside diameter, 5:6 lobe, downhole positive displacement motor is utilized to drill the open hole interval from 7000 ft (*2133 m*) to 10,000 ft (*3048 m*)

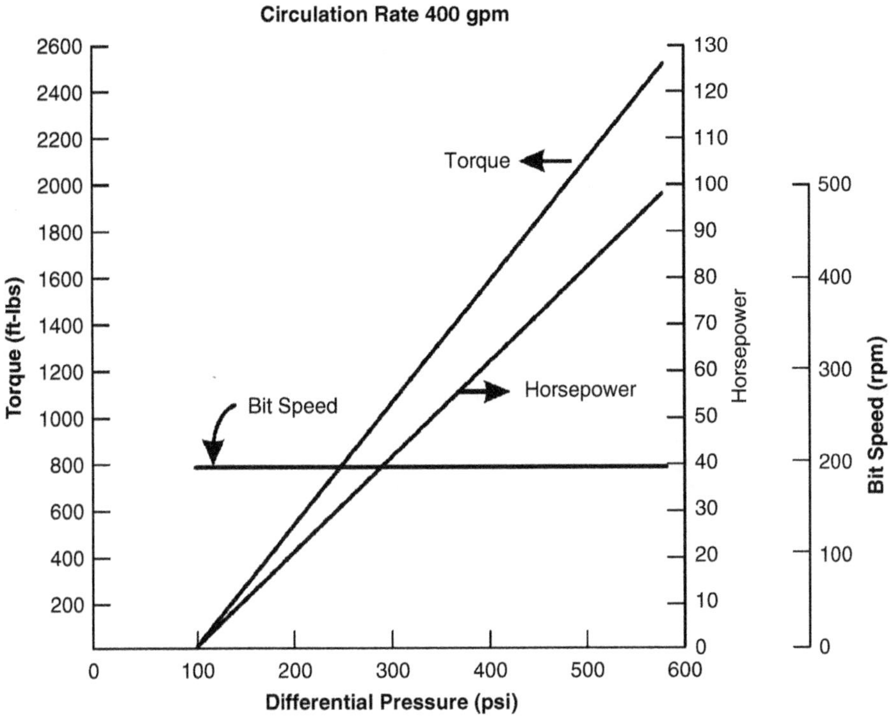

FIGURE 11-8a. Horsepower and torque rotor shaft output versus pressure drop across a Baker Oil Tools Model Mach 1, 6 3/4-in (171.4 mm) outside diameter positive displacement motor (courtesy of Baker Oil Tools).

of the Illustrative Example 8.3 series in Chapter 8. The positive displacement motor uses a $7^7/_8$-in tricone roller cutter drill bit to drill at 10,000 ft (*3048 m*). It is required that the downhole motor develops 25 hp when drilling under load. In the Illustrative Example 8.3 series, two semitrailer-mounted Dresser Clark Model CFB-4, four-stage reciprocating piston compressors were selected. Each of these compressors was driven by a Caterpillar Model D398 diesel-fueled prime mover. These two compressor units provide a total of 2400 acfm (*1133 actual liters/sec*) to the borehole. The drilling location is at 4000 ft (*1219 m*) above sea level. The annulus solution given in Illustrative Example 8.3 will be used as the starting point for this example. The horsepower/torque versus pressure plots are shown for the aforementioned PDM in Figures 11-8a and 11-8b.

Appendix F gives the detailed Mathcad solution for Illustrative Example 11.2. Figure 11-9 gives the pressures versus depth for the circulation system given in

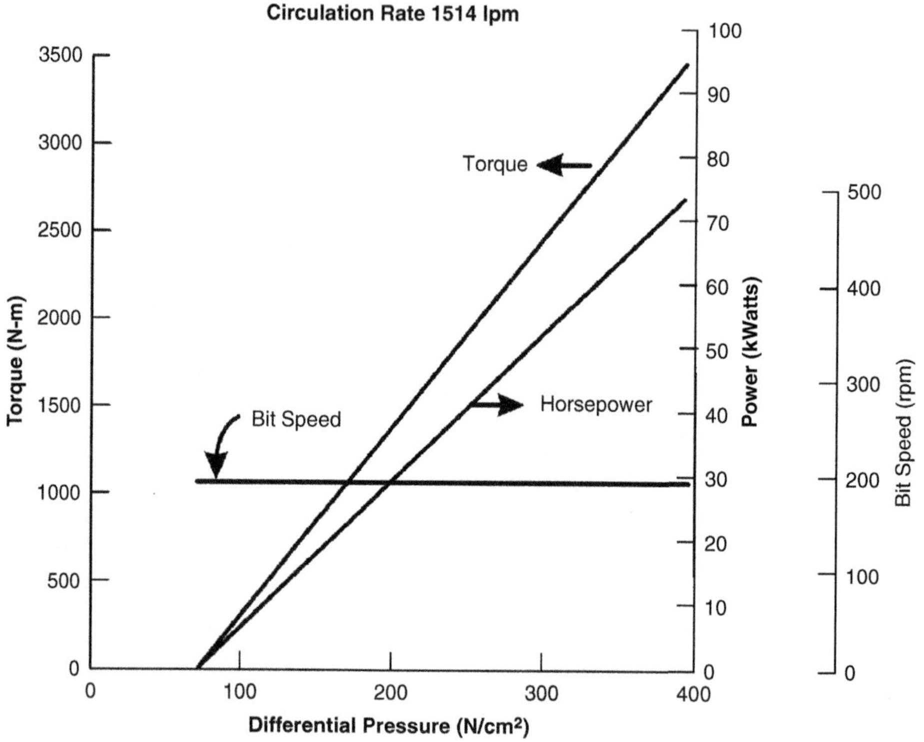

FIGURE 11-8b. Watts and torque rotor shaft output versus pressure drop across a Baker Oil Tools Model Mach 1, 171.5-mm outside diameter positive displacement motor (courtesy of Baker Oil Tools).

this PDM example and shows that the bottom hole annulus pressure is 161.7 psia (*122.1 N/cm² abs*) while drilling at 10,000 ft (*3048 m*). The injection pressure for this PDM example is 781.5 psia (*539.0 N/cm² abs*).

Figure 11-10 gives bottom hole annulus pressures and injection pressures for drilling ahead from the casing shoe at 7000 ft (*2133 m*) to the target depth of the open hole at 1000 ft (*3048 m*). Unlike most other examples in this text, this example shows that the injection pressure actually decreases as the open hole depth is increased. This clearly shows the dominance of the hydrostatic head portion of the pressure [see Equation (6-1)]. This is caused by the high pressure above the PDM required to operate the PDM at approximately 25 hp (*33.5 kW*). Note also that it was necessary to reduce the diameter of the drill bit nozzles in order to control the rotation speed of the PDM drive shaft and, thereby, the power output of the PDM (see Appendix F)[5].

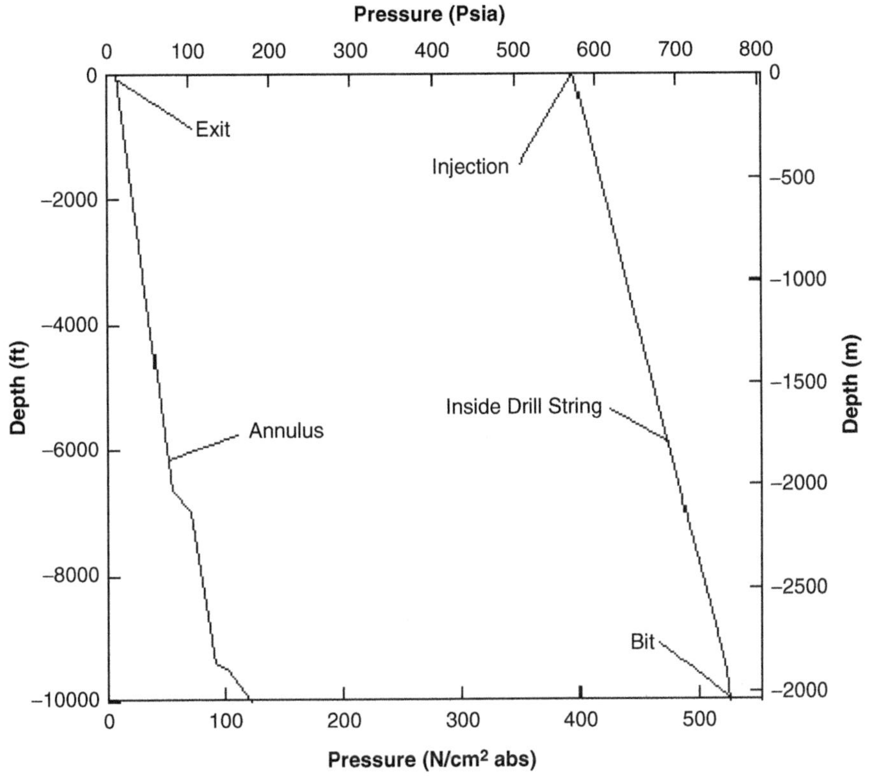

FIGURE 11-9. Illustrative Example 11.2 pressure versus depth.

11.3 CONCLUSIONS

The specialized surface downhole equipment discussion has general applications to vertical and directional drilling operations. The specialized downhole air hammer is only used for straight hole vertical drilling. This is because the drill string must be rotated in order for the drill bit to impact all of the drilling face. Due to the low weight on bit for DTH drilling, the wells are very straight (relative to the rather cork screw shape of a conventional rotary drilled borehole). However, there are several efforts underway in industry to develop a DTH directional capability.

PDMs are nearly always used to drill directional boreholes. Normal operations require nonrotation of the drill string when making course corrections with a bent sub in the BHA. This necessitates sliding of the BHA as the PDM driven drill bit makes the course correction. When holding an angle the drill string is rotated to nullify the bent sub action. Just as in conventional drilling fluid operations,

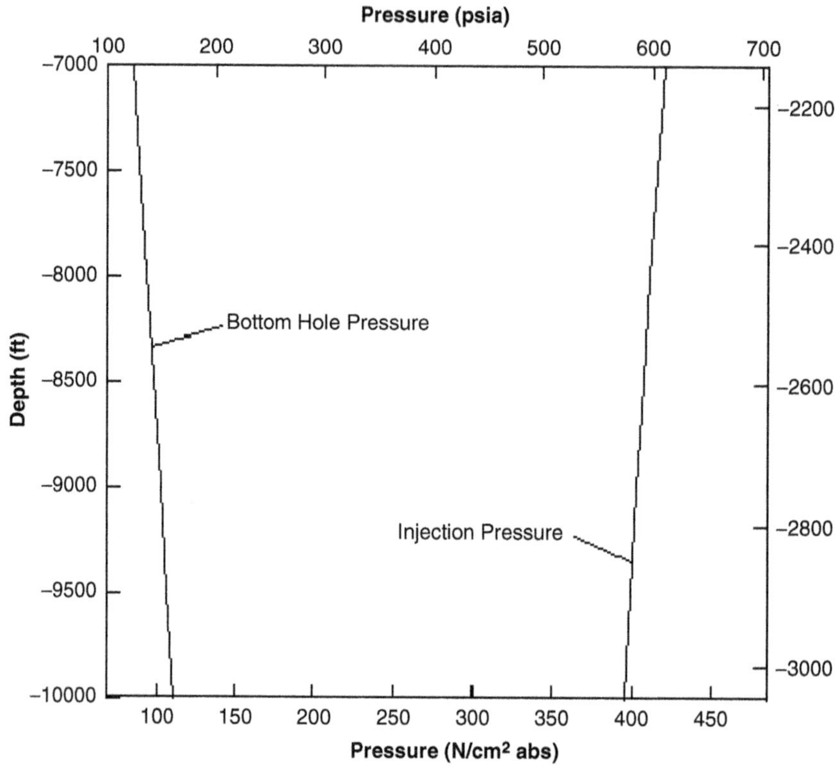

FIGURE 11-10. Drilling ahead, bottom hole annulus, and injection pressure versus depth.

drilling with a PDM in the BHA with gas or foam will require high compressor injection pressures.

An important issue for specialized downhole equipment is that of lubrication. There are some environmental drilling applications that require no lubricants be injected into the compressed air flow to the drill string. Also, there are underbalanced oil and natural gas recovery drilling applications that do not allow water to be injected into the compressed air or gas flow to the drill string. Downhole pneumatic turbine motors are the only specialized downhole drilling equipment that can be used with dry compressed air (or other drilling gases).

REFERENCES

1. *Atlas Copco Manual*, Atlas Copco Company, Fourth Edition, 1983.

2. *A to Z of DTH Drilling*, Halco USA, 2000.

3. Tiraspolsky, W., *Hydraulic Downhole Drilling Motors*, Gulf Publishing, 1981.

4. Bourgoyne, A. T., Millheim, K. K., Chenevert, M. E., and Young, F. S., *Applied Drilling Engineering*, SPE, First Printing, 1986.

5. Lyons, W. C., U. S. Patent No. 4,553,611; issued November 19, 1985; "Pressure Regulator for Downhole Turbine."

CHAPTER

Underbalanced Drilling

12

12.1 INTRODUCTION

Underbalanced drilling is an oil field term. These drilling operations require that the bottom hole annulus pressure during circulation of the drilling fluid be maintained at a magnitude that is less that the static bottom hole pore pressure of the reservoir producing pressure. These underbalanced operational requirements extend to the follow-on completions. Oil and gas drilling operations are generally the only drilling operations that require the drill bit to advance into rock formations that contain high pore pressure reservoir fluids. When underbalanced conditions are maintained during drilling or completions, reservoir fluids will enter the well bore and be carried out as the drilling fluids are circulated through the well. Note that the only other operations that drill into high pore pressure rock formations are geothermal drilling operations and artesian water well drilling. Geothermal drilling and completion operations are usually carried out with the oil industry.

Conventional drilling operations are either balanced or overbalanced. These operations require that the circulating fluid exert bottom hole pressures that are equal to or greater than the fluid pore pressure of the reservoir. In these operations the drilling fluid is the primary control barrier that keeps the high pressure reservoir fluids in the reservoir. Thus, the circulating drilling fluid is not cut with reservoir fluids. The secondary control barrier in these operations is the mechanical BOP. In underbalanced drilling operations the primary control barriers that restrain the reservoir fluid from flowing to the surface in an uncontrolled manner are the RCH, the blooey line, and so on. Basically, these primary control barriers restrict the flow of the reservoir fluids that are entrained in the annulus flow of the circulating drilling fluid. The secondary control barrier in these operations is also the mechanical BOP.

Figures 12-1 and 12-2 show graphic schematics of the physical fluid flow from the producing reservoir formation during conventional drilling and underbalanced drilling operations.

Low head drilling is the most utilized conventional drilling technique. Care must be taken when using this technique when transitioning from circulating

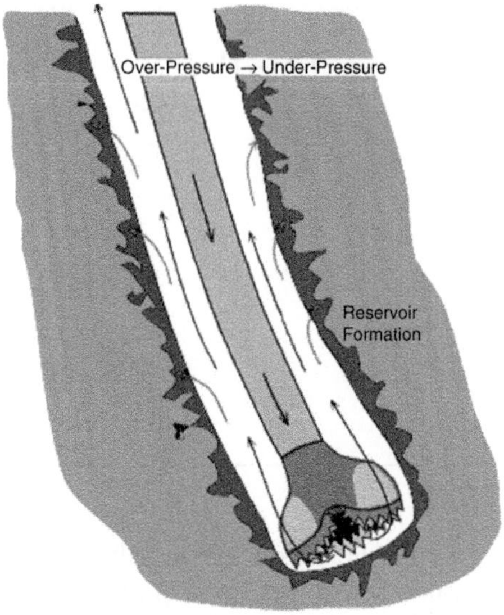

FIGURE 12-1. Conventional drilling with drilling fluid and fines flowing into the reservoir.

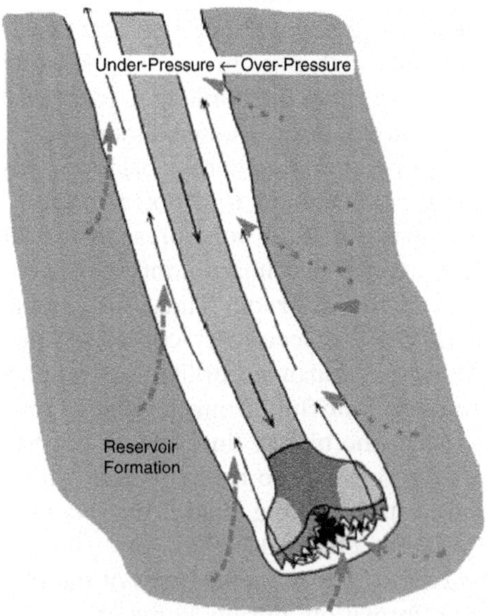

FIGURE 12-2. Underbalanced drilling with formation fluids flowing from the reservoir.

to not circulating. If the drilling fluid is designed to keep the bottom hole pressure just above the pore pressure while circulating, then when there is no circulation the well could take a kick. Obviously, low head drilling is used to reduce excessive formation damage (that results when drilling is overbalanced). Underbalanced drilling allows the bottom hole pressure to always remain below the pore pressure of the producing reservoir. This allows the produced fluids to continually flow and flush the near borehole pores and permeable passages, thus reducing the possibility of formation damage.

A number of drilling fluids types that can be used in underbalanced drilling operations have been discussed in Chapter 2. These are

- Conventional drilling muds
- Gasified drilling fluids
- Stable foam drilling fluids
- Unstable foam ("mist") fluids
- Air and gas drilling fluids

All of the drilling fluids just mentioned allow the design of bottom hole circulation pressures that can be below nearly any reservoir pressure. This is accomplished by adjusting the density (or weight) of the drilling fluid for the actual drilling operation application. Oil and gas deposits are found in sedimentary basins. These basins are formed over geologic time of hundreds of millions of years. These basins go through many geologic processes (e.g., submergence, sedimentation, uplifting, tectonics). When the basin-forming geologic time is long, these basin sediments become well cemented (usually with $CaCO_3$)[1]. These basins form competent rock formations at depth and are referred to as *mature* sedimentary basins. These basins usually have dispersed hydrostatic columns of retained ancient sea waters or fresh lake waters and often have no retained water in their formations at depth. In young or *immature* sedimentary basins, the cementing processes have not yet fully cured. Such immature rock structures cannot fully support themselves in the buoyant environment of retained ancient sea and fresh water. In general, underbalanced drilling and completions operations are basically restricted to mature sedimentary basins. Immature basins can only be drilled underbalanced with conventional drilling muds.

It has long been known that the longer an open hole section is left open during both drilling or completion operations, the more hole-related problems occur. Because of the increased penetration rates while drilling underbalanced, hole problems can be minimized. The most important impact of utilizing underbalanced drilling and completion techniques is the prospect of improved long-term production from completed wells in producing formations [2]. Underbalanced drilling and completion of a well reduce or eliminate formation damage to the well bore. This often eliminates or reduces the requirements for well stimulation. Overall, underbalanced drilling and completion operations hold the prospect of enhancing the production of oil and gas from reservoirs.

Drilling engineers must deal with the existing subsurface conditions. As the drill bit is advanced, new and possibly unknown conditions can exist. Careful planning for underbalanced (and conventional) operations must utilize offset data in order to minimize subsurface problems.

12.2 VERTICAL WELLS

The typical underbalanced drilling operation is the drilling vertical or near vertical well in an existing producing oil and/or natural gas field. These wells can be air or gas, aerated, or foam drilling fluid operations. The air or gas drilled wells can be rotary with roller cone, PDC, or air hammer drill bits. In aerated and foam drilled, wells are drilled with roller cone or PDC drill bits.

These wells are drilled at a closer spacing than the original field (i.e., infill drilling). Infill drilling is carried out in older producing fields to allow for the production of portions of the field that could not be produced efficiently by the original wells.

Since many of these fields have been produced for 30 or more years, their bottom hole shut in reservoir pressures in the original wells (and in any new wells) can be as low as 400 to 900 psia (*276 to 621 N/cm² abs*). Assuming that the typical infill drilling operation will be at depths greater than 3000 ft (*914 m*), then air and gas drilling technology must be used in order to maintain a bottom hole pressure that will be below the reservoir static bottom hole shut in pressure. In some older producing fields that have higher bottom hole shut-in reservoir pressures, aerated and foam drilling fluids are the operational choices. It should be noted that some of the new vertical wells in these fields are sidetrack boreholes drilled out of the bottom of the original wells in the field.

12.2.1 Underbalanced Example

In the illustrative examples that follow, the example well that was used in Chapters 8, 9, and 10 is altered to demonstrate underbalanced drilling mathematical modeling. Figure 12-3 shows a schematic of a well that has a 300-ft (*91.4 m*) reservoir thickness. Above the reservoir the well is drilled to a depth of approximately 9700 ft (*2956 m*) where the API $8^5/_8$-in (*219 mm*), 32.00-lb/ft (*14.00 kg/m*) nominal, Grade J-55, casing is set. The trap or seal formation above the reservoir is shale and the reservoir formation is assumed to be a fractured limestone. The casing set is placed in the harder more competent limestone formation. Therefore, the casing borehole is drilled around 30 ft (*9.1 m*) into the reservoir and casing run and cemented to 9700 ft (*2956 m*). The desire is to drill out of the bottom of the casing shoe and to drill through the reservoir underbalanced.

The drill string is initially made up of 9200 ft (*2804 m*) of API 4 1/2-in (*114.3 mm*) drill pipe and 500 ft (*152 m*) of $6^1/_4$-in by $2^{13}/_{16}$-in (*159 mm by 71.4 mm*) drill collars. The drill bit will be a $7^7/_8$-in (*200 mm*) roller cone. As required by IADC standards, the inside of the drill string will be fitted with several

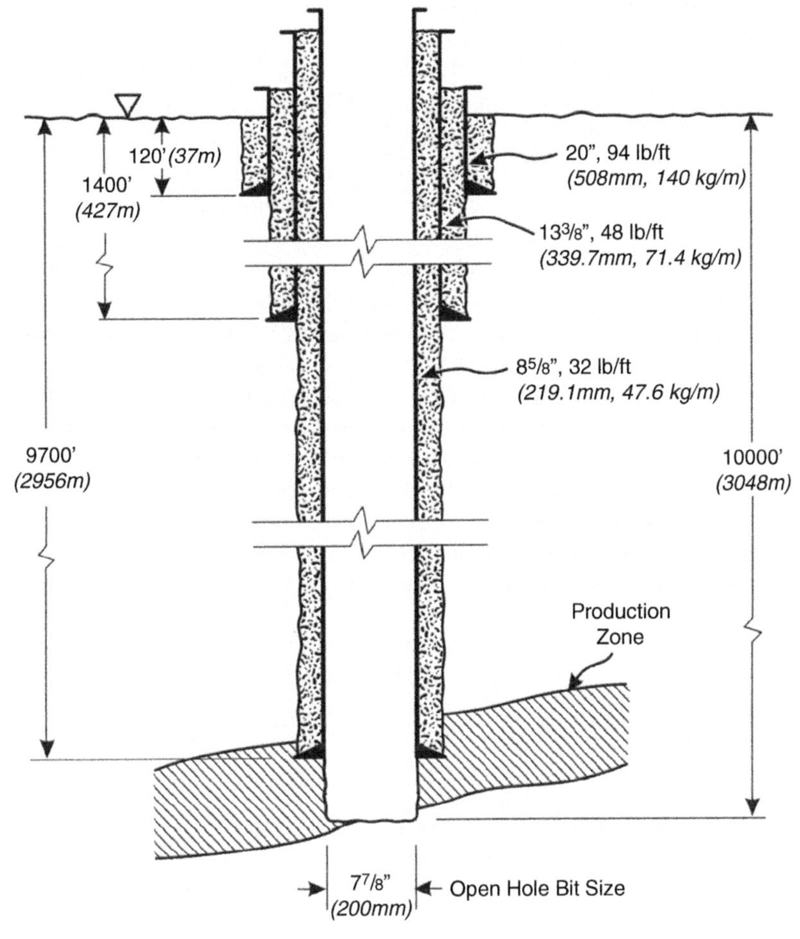

FIGURE 12-3. Underbalanced drilling of example well.

float valves. The well head at the surface will have a rotating control head (i.e., RCH) capable of controlling surface annulus pressures up to 1500 psia (*1035 N/cm² abs*). This RCH is essential to plans for drilling through the reservoir underbalanced.

As drilling progresses through the reservoir formation, connections will need to be made. To keep the well underbalanced while connections are made, the drill string will be stripped from the well through the RCH. The well will be allowed to flow during connection operations. When the drill bit has been advanced to the bottom of the reservoir or, if necessary, slightly into the rock formation below the reservoir, the drill string can be removed from the well via stripping and snubbing operations.

It is necessary during an underbalanced drilling operation to not allow the well to become overbalanced during the entire operation. It is also necessary to complete the well without allowing the well to become overbalanced. The completion operation can be very complex if elaborate production tubing and bottom hole production assemblies are required. In general, most underbalanced drilled wells are open hole completions and simple production strings with bottom hole preslotted liners. The production tubing in bottom hole assembly must be snubbed and stripped into the well.

Reservoir formation vertical thicknesses are usually small relative to the overall depth of the well. Therefore, when applying the mathematical modeling techniques demonstrated in Chapter 8 (air and gas drilling), Chapter 9 (aerated drilling), and Chapter 10 (stable foam drilling), only one depth in the reservoir is used for the calculations. It is evident when examining the illustrative examples in Chapters 8, 9, and 10 that the bottom hole pressures change very little over a few hundred feet or meters of depth. Figure 12-4 shows a schematic of the "operative envelope" concept used to summarize the results of underbalanced planning calculations [3].

The vertical stripped region in Figure 12-4 is the underbalanced drilling target design objective. The vertical axis is the bottom hole pressure at the reservoir, and the horizontal axis is the gas injected into the well via the top of the inside

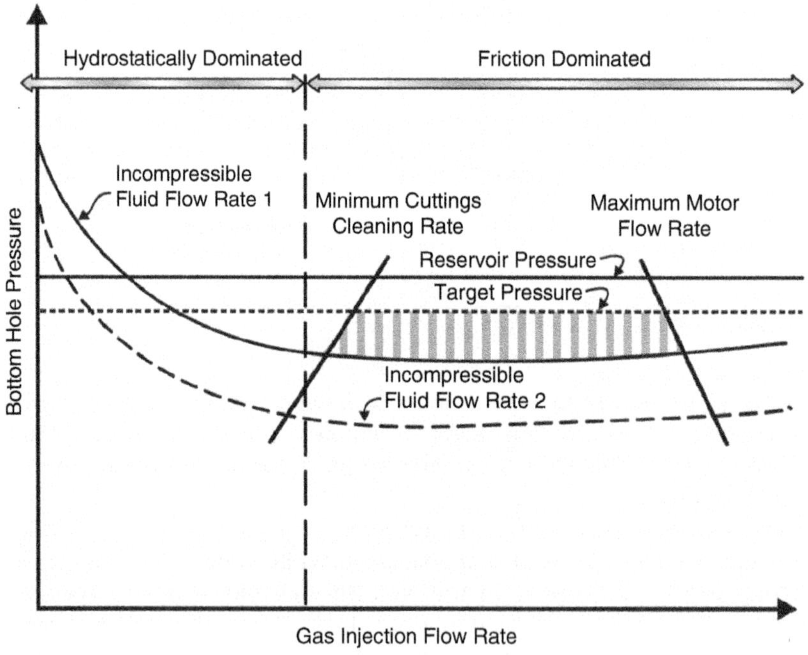

FIGURE 12-4. Underbalanced drilling operative envelope concept.

of the drill string. The anticipated reservoir shut-in pressure is shown as a horizontal straight line. The concaved curved lines show two different incompressible fluid injection rates into the top of the inside of the drill string, together with various gas injection rates (i.e., incompressible fluid flow rate 1 and incompressible fluid flow rate 2). The incompressible fluid flow rate 2 is lower than incompressible fluid flow rate 1 (i.e., at the same gas injection rate, rate 2 has a lower bottom hole pressure at the reservoir). The envelope in Figure 12-4 is limited on the left side by the flow rates needed for bottom hole cleaning. On the right side of the envelope the limit is the maximum flow rate needed to operate a downhole motor (e.g., PDM or turbine) or the DTH.

The actual drilling engineers, objective will be to select the surface and downhole equipment that will give a bottom hole pressure that is on the horizontal target pressure line. This line is usually 100 to 200 psi (*67 to 138* N/cm^2) below the anticipated reservoir shut-in pressure. This guarantees that the well will flow while the reservoir formation is penetrated with the advancing drill bit. If the actual productivity index correlation equation (know as the PI curve) is known for the target reservoir, this equation can be coupled with the underbalanced drilling model and a more precise design selection of surface and subsurface equipment can be made since this will also give the actual volumetric flow rates of reservoir fluids expected in the annulus return flow.

12.2.2 Air, Gas, and Unstable Foam

Illustrative Example 12.1 This example applies the direct circulation air and gas drilling model reviewed in Chapter 6 and demonstrated in Chapter 8 (and Appendix C) to the underbalanced example shown in Figure 12-3 and discussed earlier. In this illustrative example, the objective of the operation is to drill the last 300 ft (*91.4 m*) of the limestone reservoir while purposely allowing the reservoir fluids to flow to the well annulus where they are removed by the circulating gas. The results of these calculations are presented in the operative envelope form of Figure 12-4.

Figure 12-5 shows the operative envelope for drilling of the example limestone reservoir with membrane nitrogen and mist drilling fluids. Note that the specific gravity of membrane nitrogen is taken in these example calculations as $SG_{mn} = 0.972$.

In Figure 12-5, the dry nitrogen curve is the lower limit of the operative envelope (fresh water incompressible flow rate of 0 gpm). The unstable foam (mist) upper limit is given by the fresh water incompressible flow rate of 50 gpm. The fresh water incompressible flow rate includes a surfactant and any other liquid additives. The left limit of the envelope is the minimum cuttings lifting flow rates.

It is clear from Figure 12-5 that the membrane nitrogen and unstable foam (mist) drilling for this example well geometry could be applied to drill underbalanced through reservoir candidates with bottom hole shut-in pressures from approximately 60 to 400 psia (*41 to 275* N/cm^2 *abs*).

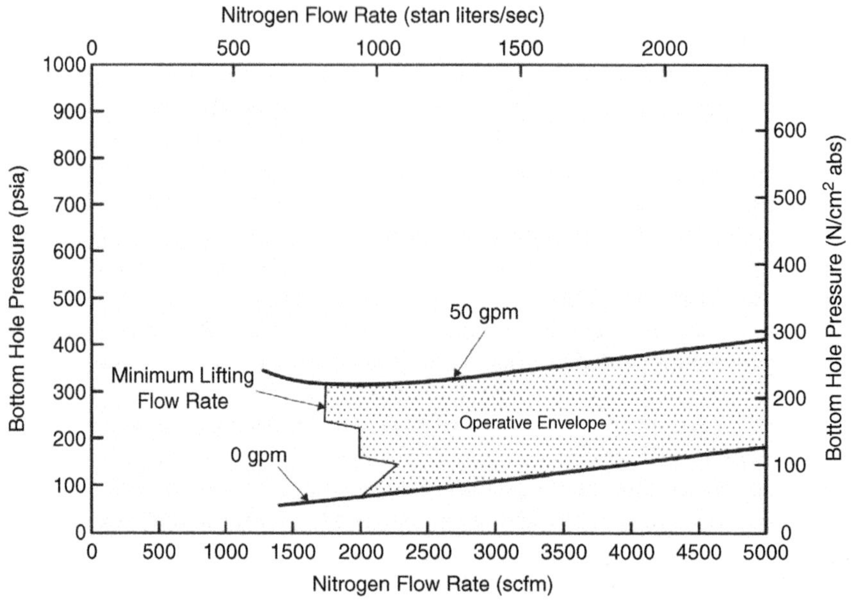

FIGURE 12-5. Membrane nitrogen and unstable foam (mist) operative envelope.

12.2.3 Aerated Fluid

Illustrative Example 12.2 This example applies the direct circulation aerated (gasified) drilling model reviewed in Chapter 6 and demonstrated in Chapter 9 (and Appendix D) to the underbalanced example shown in Figure 12-3. As in the example given earlier, in this illustrative example the objective of the operation is to drill the last 300 ft (*91.4 m*) of the limestone reservoir while purposely allowing the reservoir fluids to flow to the well annulus where they are removed by the circulating gasified drilling fluid.

Figure 12-6 shows the operative envelope for drilling of the example limestone reservoir with gasified drilling fluids (combination of membrane nitrogen and an incompressible fresh water drilling fluid). In practice, gasified drilling fluids can utilize a variety of liquids as the incompressible fluid component. This can include fresh water and formation water to more compressible liquids such as formation oil and diesel. The major concern in underbalanced drilling operations is the compatibility of these liquids to the chemistry of the reservoir rock and its produced fluids.

In Figure 12-6 the 50-gpm incompressible drilling fluid flow rate is the lower limit of the operative envelope. The 200-gpm incompressible drilling fluid flow rate is the upper limit. The left limit of the envelope is the minimum cuttings lifting flow rate and a right limit is given for the maximum flow rate to operate a PDM.

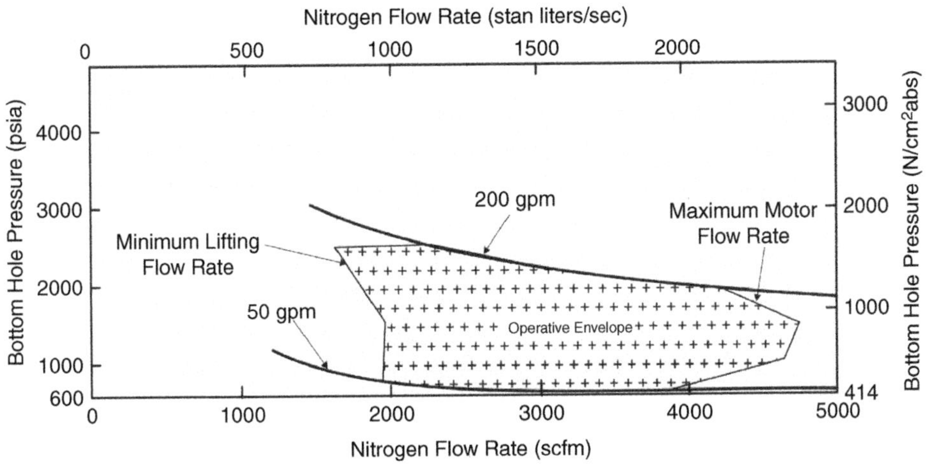

FIGURE 12-6. Gasified drilling fluids operative envelope.

It is clear from Figure 12-6 that gasified drilling fluids for this example well geometry could be applied to drill underbalanced through reservoir candidates with bottom hole shut-in pressures from approximately 600 to 2700 psia (*414 to 1862 N/cm^2 abs*).

12.2.4 Stable Foam

Illustrative Example 12.3 This example applies the direct circulation stable foam drilling model reviewed in Chapter 6 and demonstrated in Chapter 10 (and Appendix E) to the underbalanced example shown in Figure 12-3. As in the examples given earlier, in this illustrative example the objective of the operation is to drill the last 300 ft (*91.4 m*) of the limestone reservoir while purposely allowing the reservoir fluids to flow to the well annulus where they are removed by the circulating stable foam drilling fluid.

Figure 12-7 shows the operative envelope for drilling of the example limestone reservoir with stable foam drilling fluids (combination of membrane nitrogen and incompressible fresh water).

In Figure 12-7 the 7-gpm incompressible drilling fluid flow rate is the lower limit of the operative envelope. The 50-gpm incompressible drilling fluid flow rate with 0 psig (*0 N/cm^2 gauge*) is the upper limit of the shaded area. Stable foam is often developed in the annulus of the well by utilizing a back pressure at the surface on the horizontal return flow line. Using a back pressure on the annulus extends the upper limit. Figure 12-7 shows an alternate upper limit with a back pressure of 80 psig (*55 N/cm^2 gauge*). Figure 12-7 shows that the left limit of the envelope is the minimum cuttings lifting flow rate.

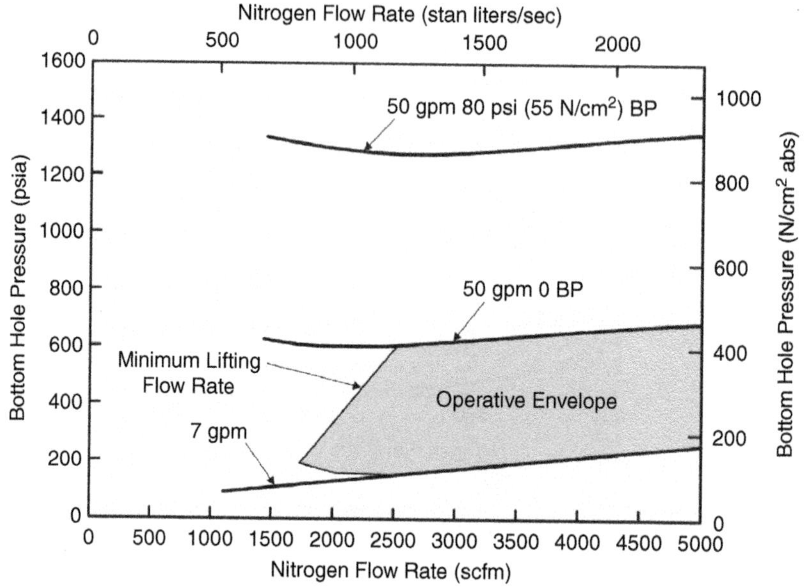

FIGURE 12-7. Stable foam drilling fluids operative envelope.

It is clear from Figure 12-7 that stable foam drilling fluids for this example well geometry could be applied to drill underbalanced through reservoir candidates with bottom hole shut-in pressures from approximately 100 to 1400 psia (*70 to 966 N/cm² abs*).

12.2.5 Summary

Figure 12-8 shows a combination of the three illustrative example operative envelopes for the example geometry shown in Figure 12-3. This summary figure indicates that air and gas drilling technology can be utilized to drill underbalanced through a variety of reservoir candidates that might have shut-in reservoir pressures from as low as atmospheric to as high as 4000 psia (*2760 N/cm² abs*). For reference, the figure shows a horizontal static hydrostatic head for a well filled with crude oil, which is basically the lightest possible liquid drilling fluid used in a drilling operation.

12.3 DIRECTIONAL WELLS

Directional drilling is generally categorized by the build angle rate of the directional portion of the borehole. These categories are usually stated in borehole angle build rate in angle degrees per vertical depth distance (ft, m). As an

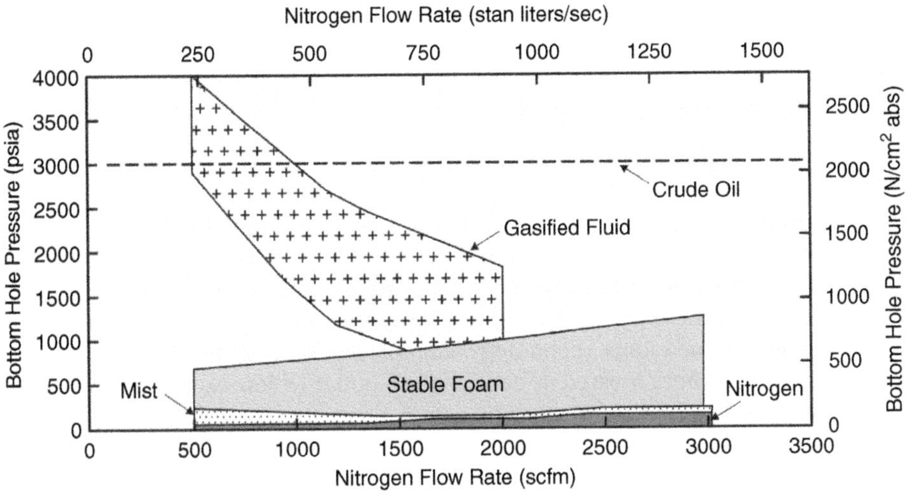

FIGURE 12-8. Summary of operative envelopes for air and gas drilling technology.

alternative, specifications for long radius, medium radius, and short radius drilling tools and auxiliary equipment are presented in tabular form in Table 12-1. All three of these directional drilling build rate categories have unique downhole and surface equipment.

Directional drilling technology as applied to underbalanced drilling operations is used in two distinct situations.

Table 12-1. Directional Drilling Technology Specification (Baker Tool Company)

	Tool Sizes (in, mm)	Minimum Bit Diameter (in, mm)	Minimum Radius (ft, m)
Long radius	4 3/4 (120.7)	6 (152.4)	1000 (300)
	6 3/4 (171.5)	8 1/2 (215.9)	1000 (300)
Med radius	3 3/4 (95.3)	4 1/2 (114.3)	286 (87)
	4 3/4 (120.7)	6 (152.4)	286 (87)
	6 3/4 (171.5)	8 1/2 (215.9)	400 (122)
	8 (203.2)	12 1/4 (311.2)	400 (122)
Short radius	3 3/4 (95.3)	4 1/2 (114.3)	19 (6)
	4 3/4 (120,7)	5 7/8 (149.2)	32 (9)
	4 3/4 (120.7)	6 1/4 (158.7)	38 (9)

1. The drilling of sidetracks through the vertical thickness of the reservoir formation.
2. The drilling of horizontal or near horizontal boreholes parallel to the reservoir formation bedding plains.

Long Radius Drilling

In the early years of rotary drilling (i.e., the late 1920s), long radius directional operations were the only directional methods available for the control of deviated (from the vertical) boreholes. Much of the early constraint on deviation of boreholes was due to the availability of high-quality steel tubulars for downhole operations. Present long radius technology can place the drill bit in a 3-ft (*1 m*)-diameter target sphere located at depths of 15,000 ft (*4500 m*) and at horizontal displacements up to 30,000 ft (*9000 m*). It is generally accepted that long radius boreholes are drilled with a build rate of approximately 2° to 6° per 100 ft (*30 m*). This build rate translates to radii of curvature of boreholes of approximately 1000 ft (*300 m*) to 3000 ft (*900 m*). Long radius boreholes are limited to kicked-off depths of 1000 ft (*300 m*) or greater. Long radius boreholes are applied most frequently in underbalanced operations to the drilling sidetrack boreholes in infill operations. This type of directional drilling is often denoted as "extended reach" drilling.

Medium Radius Drilling

Medium radius directional drilling is characterized by a build rate of approximately 6° to 40° per 100 ft (*90 m*). This build rate translates to radii of curvature of boreholes of approximately 300 ft (*90 m*) to 700 ft (*210 m*). The development of medium radius drilling technology in the mid 1980s was the direct result of improved high-quality tubulars and sophisticated downhole gyroscopic surveying equipment. Medium radius directional drilling technology is mostly applied to land directional drill operations. Medium radius directional boreholes can be initiated at nearly any depth in a vertical well. Present medium radius technology can place the drill bit in a 3-ft (*1 m*)-diameter target sphere located at a depth of 15,000 ft (*4500 m*) and with horizontal displacements from the vertical of 3000 ft (*900 m*). Medium radius boreholes are applied most frequently in underbalanced operations to the drilling of horizontal or near horizontal boreholes in infill operations.

Short Radius Drilling

Short radius directional drilling is characterized by a build rate of approximately 40° to 70° per 100 ft (*30 m*). This build rate translates to radii of curvature of boreholes of approximately 82 ft (*25 m*) to 140 ft (*43 m*) for intermediate short radius technology and 40 ft (*12 m*) to 82 ft (*25 m*) for ultrashort radius technology. Present short radius technology can place the drill bit in a 10-ft (*3 m*)-diameter target sphere located at a depth of 5000 ft (*1500 m*) and with horizontal displacements from the vertical of 1000 ft (*300 m*). Short radius boreholes are applied most frequently in underbalanced operations to the drilling of horizontal or near horizontal boreholes in infill operations.

12.3.1 Directional Control and Surveying

Many technologies have been developed through the past seven decades directed at improving directional drilling using conventional incompressible drilling fluids (e.g., water-based, salt water-based, oil-based, and synthetic oil-based drilling muds). Two decades ago air and gas drilling technology was a small niche area of the drilling industry. Up until the late 1980s little attention was given to the development of directional drilling technologies for air and gas drilling operations.

Present-day air and gas drilling directional operations have relied heavily on existing conventional directional technologies for their operations. Field operations over the past 20 years have shown that there are few directional control technologies that are successful with air and gas drilling technology.

Single and Multishot Borehole Survey Instruments

Magnetic single shot and multishot instruments cannot be used while drilling is progressing. These downhole instruments require nonmagnetic drill collars to be effective. These instruments are run in the well when all drilling operations have ceased. Thus, when used simply as survey tools, these downhole instruments are not subjected to the high drill string vibrations that characterize compressed air (or other gas) drilling operations. Magnetic downhole instruments are usually used to obtain a three-dimensional plot of the borehole. Once the magnetic instruments are retrieved, the photos developed (usually at the rig site), and the position readings entered into a computer, calculations are made and the borehole trajectory is plotted. This borehole trajectory plot can be used by a directional driller to make the appropriate mechanical corrections to the drill string (i.e., orientation of the tool face of the bent sub) to improve the accuracy of a directional drill operation. However, when using magnetic instruments, such corrections cannot be made rapidly (as with an MWD system). Using magnetic survey instruments in a directional drilling operation is not a real-time process. However, in some directional drilling operations, particularly small diameter boreholes, the use of magnetic survey instruments to control borehole trajectory is quite cost-effective. In these drilling situations, even the single shot instrument can be used to give position data that can be plotted with a computer to give a good quality borehole trajectory. This trajectory can then be used to make tool face corrections to improve subsurface target intercept accuracy. This simple technology is often used in sidetrack operations in infill drilling projects.

The basic downhole gyroscopic survey instrument can be used in much the same manner as the magnetic survey instruments discussed earlier. These instruments can be run into the well when drilling has ceased and borehole deviation versus depth readings taken as the instrument is withdrawn from the well. These simple gyroscopic survey instruments are hardwired to the surface with an active wire line or are operated downhole with batteries. In the hardwire case, data coming from the well can be used in real time to update a borehole trajectory plot. In the battery-operated case, the downhole survey tool must be retrieved and the information downloaded to a computer to prepare the borehole trajectory plot. Once the trajectory is known, appropriate mechanical corrections can be made to the drill string by the directional driller to improve the accuracy of the directional drilling operation. The main advantage of the gyroscopic survey instruments is that they can be used in any part of the well (cased and open hole) and do not require nonmagnetic drill collars in the drill string.

Conventional MWD Equipment

MWD downhole and surface equipment utilize survey data in a real-time analysis process to give an updated trajectory of a directional borehole and, with appropriate software, predict the future borehole path. This knowledge allows the directional driller to make appropriate mechanical corrections in drill string orientation (called "steering" the drill string) that will allow the advancing drill bit to hit an intended subsurface target area. An MWD requires that the downhole survey instrument operate in the drill string as drilling progresses. Both magnetic and gyroscopic survey instruments can be used as basic survey tools for MWD systems. Once survey data have been obtained (either by magnetic or gyroscopic survey), the MWD system must provide a way to get this information to the surface where the directional driller can act on the information and make appropriate drill string orientation adjustments.

Conventional drilling mud operations have MWD systems that make use of an acoustic mud pulse signal communication system that provides nearly real-time survey information to the surface operators (via a pulse-generated binary code). However, this mud pulse system will only work in drilling fluids with up to approximately 25% by volume of compressible gas content. This restricts its usage in air and gas drilling technology to aerated drilling fluids. In drilling fluids with greater than 25% compressible gas content, the acoustic signals are dampened and scattered before they reach the surface receivers.

Steering Tools

The steering tool is an early type of MWD system. Steering tools have a hardwire connection that runs from the downhole survey instrument package in the BHA to a surface computer and output printer and plotter. Figure 12-9 shows a schematic of this type of system. The active hardwire is run down the outside of the drill pipe near the surface (secured to the drill pipe) to a side-entry sub. The side-entry sub is in the drill string near its top. The hardwire plugs into a plug connection in the side-entry sub. In the inside of the side-entry sub is a similar plug connection that connects to another active wire line that runs inside the drill string to the survey instruments in the BHA. The steering tool shown in Figure 12-9 uses magnetic survey instruments to give compass/pendulum or

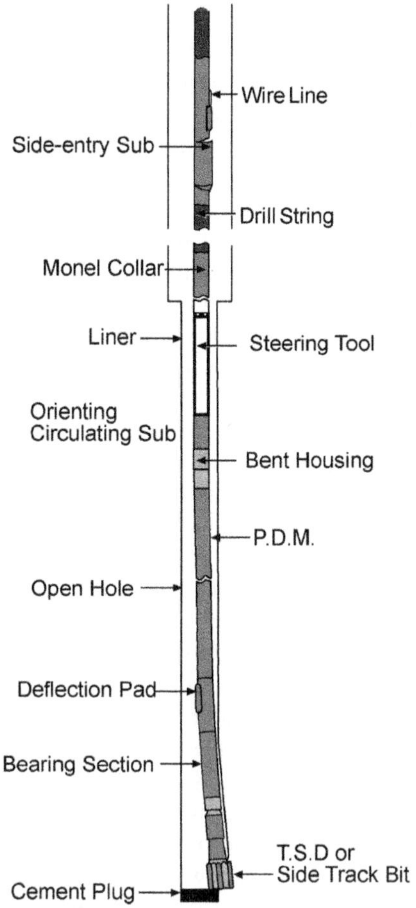

FIGURE 12-9. Schematic of a steering tool system.

accelerometer readings to the surface in real time. The survey package in the BHA is located just above the bent housing (sub) and is referenced to the tool face of the bent housing. Because this is a magnetic survey instrument, the survey package must be in a nonmagnetic drill collar or heavyweight drill pipe. With the survey package referenced to the bent housing tool face, the surface operator always knows how the compass/pendulum readings relate to the position of the bent sub. The steering tool example in Figure 12-9 makes use of a PDM to rotate the drill bit. The downhole motor, together with the bent housing, allows the housing tool face to be oriented. When the tool face is oriented properly, the entire drill string can be slid along the low side of the borehole as the drill bit is advanced in a predetermined direction. When it is necessary to drill straight after the direction corrections have been made and the directional portion of the well completed, the drill string can be raised to the side-entry sub and the hardwire section on the outside of the drill string removed. The drill string can then be lowered into the well and slowly rotated as the combined PDM and drill string rotation advance the drill bit. In this manner, the effect of the bent housing can be averaged out and the borehole advanced along a more or less straight directional course. Obviously, any downhole motor that can operate on compressed air (or other gas), aerated drilling mud, or stable foam can be used with a steering tool to provide directional drilling capabilities for air and gas drilling operations.

Steering tools have had moderate success in providing directional drilling capabilities for air and gas drilling operations. These successes have generally been confined to larger diameter borehole directional operations. Steering tools are available in tool outside diameters from $9^1/_2$ in *(241.3 mm)* to $3^3/_8$ in *(85.8 mm)* in diameter [i.e., for borehole diameters from $17^1/_2$ in *(444.5 mm)* to $4^3/_4$ in *(120.7 mm)*]. Presently, steering tools are used extensively to drill large diameter relatively shallow wells to extract methane from near surface and deep coal mines.

Electromagnetic MWD

The electromagnetic MWD downhole tool transmits its downhole survey measurements by emitting electromagnetic waves, which are received at a surface antenna, processed by a computer, and outputted as printouts and trajectory plots that can be used by the directional driller to make drill string corrections. The electromagnetic waves carrying survey data transmit through the rock formations between the downhole tool and the surface (see Figure 12-10). Because the electromagnetic transmission does depend on the rock types being traversed by the waves, the operational capability of this MWD system can be depth limited. This electromagnetic telemetry system does not depend on a particular type of drilling fluid being in the well, thus the system can be used with any rotary drilling fluid system. The electromagnetic telemetry system operates on a long-life battery subsystem; therefore, no drilling fluid driven on-board generator is needed.

These electromagnetic MWD tools are available with either magnetometer/accelerometer-based survey subsystems or gyroscope-based survey subsystems. The three-axis magnetometer and accelerometer survey subsystem must utilize nonmagnetic housings and drill collars in much the same manner as the downhole magnetic survey instruments discussed earlier. The alternative gyroscope-based survey subsystem utilizes two, two-axis gyroscopes, and three-axis accelerometers. This gyroscope-based subsystem does not require nonmagnetic housings or drill collars. Here again, any downhole motor that can operate on compressed air (or other gas), aerated drilling mud, or stable foam can be used with a steering tool to provide directional drilling capabilities for air and gas drilling operations.

The most serious problems are the depth transmission limitation of the electromagnetic system and its reduced transmission capability through halite rock formations. This can be relieved somewhat by using relay subs placed in the drill string at strategic intervals to boost the transmission signal from the MWD to the

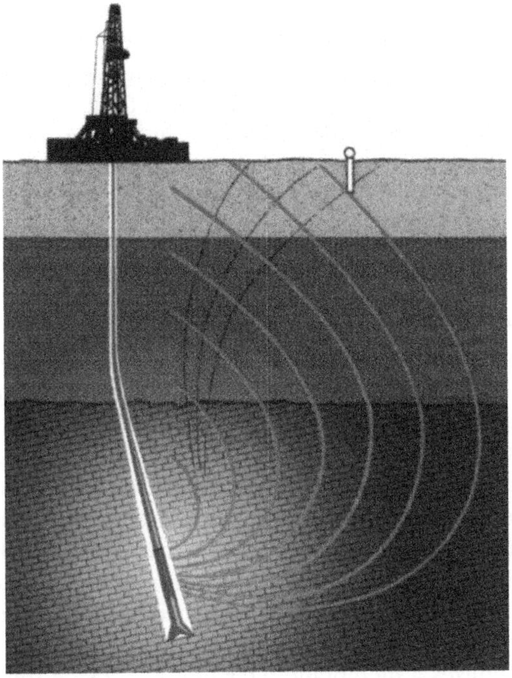

FIGURE 12-10. Schematic of an electromagnetic MWD transmission of data to a surface antenna (courtesy of Geoservices Inc.).

surface antenna. It is not clear if this signal boosting technique will be feasible in the small diameter boreholes that are used most often in compressed air (or other gas) drilling operations.

12.3.2 Horizontal Drilling with Membrane Nitrogen

The very recent development of membrane filter units to reduce the oxygen percentage in atmospheric air has been driven by the need to eliminate the risk of downhole fires and explosions when drilling boreholes in rock formations containing hydrocarbons (also see Section 5.8 in Chapter 5). This problem was recognized in the early years of the development of air and gas drilling technology. In those early years, the solution was to use natural gas as the drilling fluid. However, using natural gas as a drilling fluid increases the risk of surface fire or explosions in and around the drill rig. Also, although in early years natural gas was inexpensive, today natural gas has a sizable share of the energy market and the cost of using natural gas for drilling operations has become prohibitive.

The risk of downhole fires and explosions exists for both vertical and horizontal drilling operations. However, this risk is far more acute for horizontal drilling operations. This is due to the fact that during a typical horizontal underbalanced drilling operation, the horizontal interval drilled in the reservoir rock formations is many times longer than in typical vertical or sidetrack underbalanced drilling operations. Further, the drilling rate of penetration for a horizontal drilling operation will be about half (or less) that of a vertical drilling operation (assuming a similar rock type). This means that the risk for downhole fires and explosions is many times higher than for a vertical underbalanced drilled well. Because of the high productivity potential for new horizontal drilling technology, a concerted effort was made starting around the late 1980s to develop a safer drilling gas.

Allowable Oxygen Concentrations

For the past three decades membrane filter technologies have been used to separate oxygen (and some other molecules) from gas mixtures, particularly atmospheric air. The early developments in membrane filter technology were directed at medical applications. In the mid-1990s, membrane filter technology was adapted to utilize high volumetric flow rates of atmospheric air and to strip a high fraction of the oxygen from the air flow to produce an inert atmospheric air, which is known more commonly as membrane generated nitrogen [4]. For drilling operations, membrane filter technology has been incorporated in portable skid-mounted units that can be placed in series with a primary compressor(s) and a drill rig (see Section 5.8 in Chapter 5). Figure 12-11 shows a schematic of a basic drilling location plan that utilizes a membrane filter unit to provide the drilling operation with membrane nitrogen.

In drilling operations, membrane units are available in input flow rate capacities that are rated as 750, 1500, and 3000 scfm. This is approximately the same rating system used for most primary compressors. As can be seen in Figure 12-11, the membrane units are fed compressed air from the primary compressors. The filtered output of the membrane filter units is in turn fed to the booster compressor, if required.

Minimum Volumetric Flow Rates for Horizontal Boreholes

There is general confusion regarding determination of the minimum volumetric flow rates for drilling operations that include a final long horizontal borehole increment. There are two different operational situations that affect how the minimum volumetric flow rate for a particular horizontal well situation is determined.

Drilling a horizontal interval at the bottom of a well is usually accomplished by using a downhole motor that is used in conjunction with some type of downhole drilling control equipment. This downhole drilling control equipment can be either a sophisticated MWD system referenced to a bent sub above the downhole motor or a simple single shot survey referenced to the bent sub tool face. Regardless of the drilling control method used, when direction corrections are not being made, the entire drill string is rotated slowly to average out the effect of the bent sub to allow "straight" drilling. When drilling a horizontal borehole, the drill bit cuttings generated by the bit advance will have a tendency to fall and accumulate

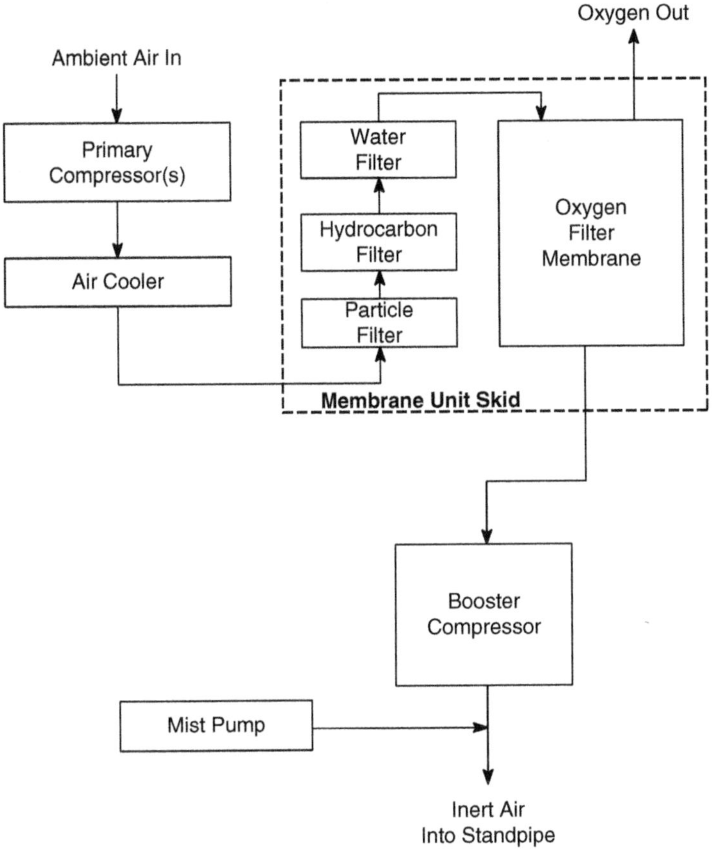

FIGURE 12-11. Schematic of location plan utilizing a membrane filter unit [4].

on the low side of the hole. The rotation of the drill string mechanically agitates these accumulated rock cuttings. This agitation will force the rock cuttings into the drilling fluid return flow stream in the annulus. Thus, drill string rotation promotes more efficient horizontal borehole cleaning.

When directional corrections are being made, rotation is stopped, the drill string is oriented so that the bent sub can be effective in correcting the directions of drilling, and the drill string is slid along the low side of the borehole as the downhole motor rotates the drill bit (allowing the drill bit to advance). Under this operational situation, the rock cuttings that accumulate on the low side of the hole will not be mechanically agitated and forced into the return flow stream above the drill string. This operational situation is probably the worst case from the viewpoint of horizontal borehole cleaning. To analyze this particular operational situation, it will be necessary to draw on the pneumatic conveying literature [5]. For horizontal pneumatic conveying, the phenomenon of accumulation or the low side of a horizontal flow line is known as "saltation." To further understand saltation, Figure 12-12 shows the solids flow characteristics for horizontal pneumatic conveying.

The near vertical dashed line in the center of Figure 12-12 shows the saltation velocity of the air flow for various solid flow rates ($\dot{G}$). The right side of the saltation line is the dilute phase flow, and the left side of the saltation line is dense phase flow. Schematics a and b in Figure 12-12 show different air/solids flow situations for the dilute phase flow. Schematic a shows all solids entrained in the air flow. Schematic b shows a small volume of the solids lying on the low side of the horizontal flow duct, but the air flow with entrained solids flowing at steady-state conditions above slower moving low side steady-state solids/air flow. Schematics c, d, e, and f show various stages of unsteady dense solids/air flow.

Clearly the minimum volumetric flow rate for the sliding drill string situation in horizontal drilling should attempt to avoid excessive saltation. Sliding of the drill string in the horizontal borehole is complicated by the actual borehole cross-section geometry. When sliding the drill string, the string will lay on the low side of the borehole. Thus, the return flow in the borehole is not flow in

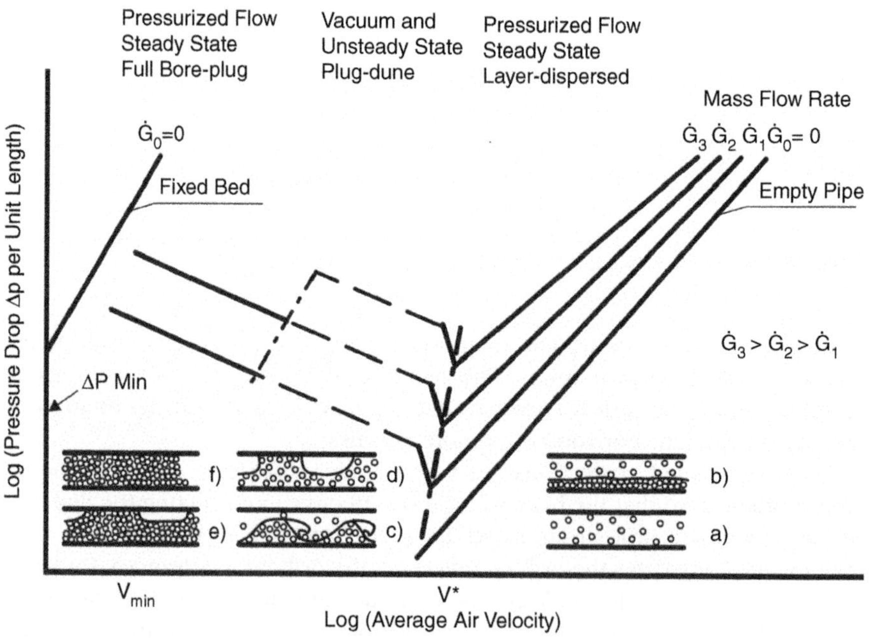

FIGURE 12-12. Solids flow characteristics in horizontal pneumatic conveying. Note that the unit of $\dot{G}$ is in the units (lb/sec, N/sec)[4].

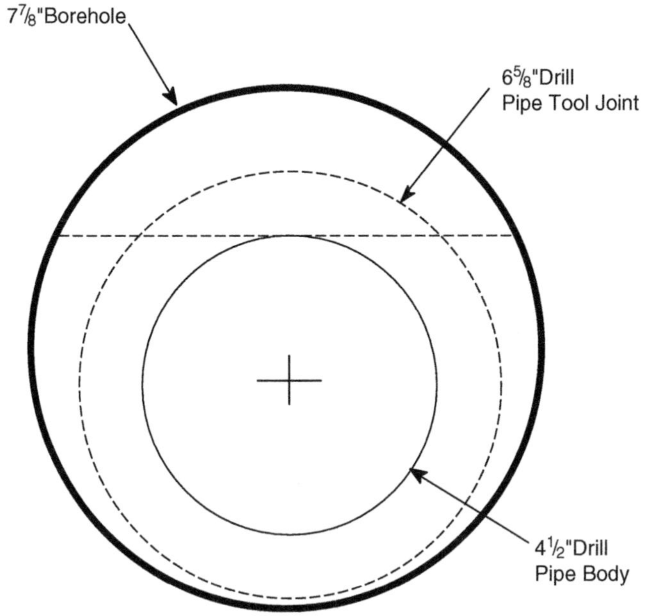

FIGURE 12-13. Cross-section geometry of a $7^7/_8$-in (*200 mm*) borehole with a sliding $4^1/_2$-in (*114.3 mm*) drill pipe on the low side of the hole.

the full annulus cross section. To give some perspective, Figure 12-13 shows an example of the borehole cross-section geometry for a $7^7/_8$-in (*200 mm*) borehole with API $4^1/_2$-in (*114.3 mm*), 16.60-lb/ft (*7.62 kg/m*) nominal, EU-S135, NC50 drill pipe. This drill pipe has tool joints with a $6^5/_8$-in (*168.3 mm*) outside diameter. As can be seen in Figure 12-13, the tool joint larger diameter holds the drill pipe body off the bottom of the borehole (the tool joint is the dashed circle).

The actual return flow of the gas with entrained rock cuttings will take the path of least resistance around the drill string. This means that most of the flow will be above the horizontal dashed line in Figure 12-13.

Illustrative Example 12.1 Determine the cross-section area and the hydraulic diameter of the assumed flow path opening described earlier for the drill pipe borehole geometry shown in Figure 12-13. Ignore the protrusion of the drill pipe tool joints into this flow area.

The hydraulic radius is defined for this flow path opening (area above the horizontal dashed line) as [6]

$$R_b = \frac{\text{flow cross} - \text{sectional area}}{\text{wetted perimeter}}. \tag{12-1}$$

Note that the hydraulic radius term is not a true geometric radius. The flow path opening cross-section area is (in USCS)

$$a_{fp} = 12.0\ in^2$$

or (in SI units)

$$a_{fp} = 77.4\ cm^2.$$

The wetted perimeter is (in USCS units)

$$s_{wp} = 16.2 \text{ inches}$$

or (in SI units)

$$s_{wp} = 41.1 \text{ cm}.$$

Equation (12-1) becomes (in USCS units)

$$R_h = \frac{12.0}{16.2}$$

$$R_h = 0.740 \text{ inch}$$

or (in SI units)

$$R_h = \frac{77.4}{41.1}$$

$$R_h = 1.88 \text{ cm}.$$

The effective hydraulic diameter of the flow path opening is (in USCS units)

$$d_h = 4\,R$$

$$d_h = 4(0.740)$$

$$d_h = 2.96 \text{ inches}$$

or (in SI units)

$$d_h = 4(1.88)$$

$$d_h = 7.52 \text{ cm}.$$

The return flow of circulation gas, gasified, or stable foam drilling fluids will be channeled to the annulus space above the dashed line above the drill string body outer surface. For the same volumetric flow rate, the flow velocity in this upper cross section will be higher than if the flow were channeled through an ideal annulus space with equal openings on all sides of the drill string (assuming constant volumetric flow rate). Thus, the carrying capacity of this higher velocity channel above the drill string should be high. However, there are protrusions of tool joints into this "path of least resistance" that will cause cuttings to be dropped into the low side of the borehole in spaces that do not have high velocity flows. Therefore, saltation of cuttings into the low side of the borehole is unavoidable. Thus, the only way to avoid unmanageable saltation accumulations in the low side of the borehole will be to rotate the drill string as much as possible. This mechanical agitation is essential. Then drilling directionally with a PDM (by sliding the drill string tool face-oriented BHA), the drill string must be pulled back and forth while rotating the drill string to mechanically agitate the cuttings on the low side of the open hole. This agitation must be accompanied by circulation to get the cuttings entrained and removed. This careful cleaning process while sliding the PDM BHA drill string appears to unavoidable. The adaptation of rotary steerable technology for air and gas drilling technology operations will be welcomed.

12.3.3 Equations for Radius and Slant (or Horizontal) Drilling

Direct circulation equations have been developed for air or gas drilling operations that can be used to model the radius segment and the horizontal (or slant) segment of a directionally drilled borehole [7]. These equations have been developed using the basic equations given in Chapter 6.

The equations that model the radius and straight sections of a directional borehole neglect the column weight of the air or gas in these borehole segments.

Air and Gas Drilling Radius and Horizontal Equations

a. Annulus: The equation for the pressure at the bottom of the radius segment P_{br} in the annulus is

$$P_{br} = \left[P_{tr}^2 + 2\, a_a\, b_a\, T_{av}\, L\, e^{\frac{2\, a_a\, R}{T_{av}}}\right]^{0.5}, \tag{12-2}$$

where P_{tr} is the pressure at the top of the radius segment (lb/ft^2 abs, N/cm^2 abs) and R is the radius of the curved segment (ft, m).

The values of a_a and b_a are the same as those given in Chapter 6 for air and gas drilling.

The arc length of the curved segment L is

$$L = \theta_m R \tag{12-3}$$

where θ_m is the maximum angle the borehole axis makes with vertical (radians).

The equation for the pressure at the bottom (toe) of the horizontal segment P_{ex} in the annulus is

$$P_{ex} = \left[P_{en}^2 + 2\, a_a\, b_a\, T_{av}\, L\right]^{0.5}, \tag{12-4}$$

where P_{en} is the pressure at the entrance to the horizontal segment (lb/ft^2 abs, N/m^2 abs) and L is the length of the horizontal segment (ft, m).

b. Inside the Drill String: The equation for the pressure at the top of the radius segment P_{tr} inside the drill string is

$$P_{tr} = \left[P_{br}^2 + 2\, a_i\, b_i\, T_{av}\, L\, e^{\frac{2\, a_i\, R}{T_{av}}}\right]^{0.5}, \tag{12-5}$$

where P_{br} is the pressure at the bottom of the radius segment (lb/ft^2 abs, N/cm^2 abs). The values of a_i and b_i are the same as those given in Chapter 6 for air and gas drilling.

The equation for the pressure at the bottom of the horizontal segment P_{en} inside the drill string is

$$P_{en} = \left[P_{ex}^2 + 2\, a_i\, b_i\, T_{av}\, L\right]^{0.5}, \tag{12-6}$$

where P_{ex} is the pressure at the exit to the horizontal segment (lb/ft^2 abs, N/cm^2 abs).

Minimum Volumetric Flow Rate

As was discussed earlier, the minimum volumetric flow rate of gas to clean the bottom of the well in a horizontal drilling operation is very dependent on its changing angle (sliding) or its holding angle (and rotating). The mechanical agitation of the cuttings at the low side of the open borehole is critical to the removal of the cuttings. Figure 12-14 shows the example that has been used throughout Chapters 8 to 11 adapted for a horizontal drilling operation. In this example, the horizontal portion of the well is kicked off at a point just below the casing

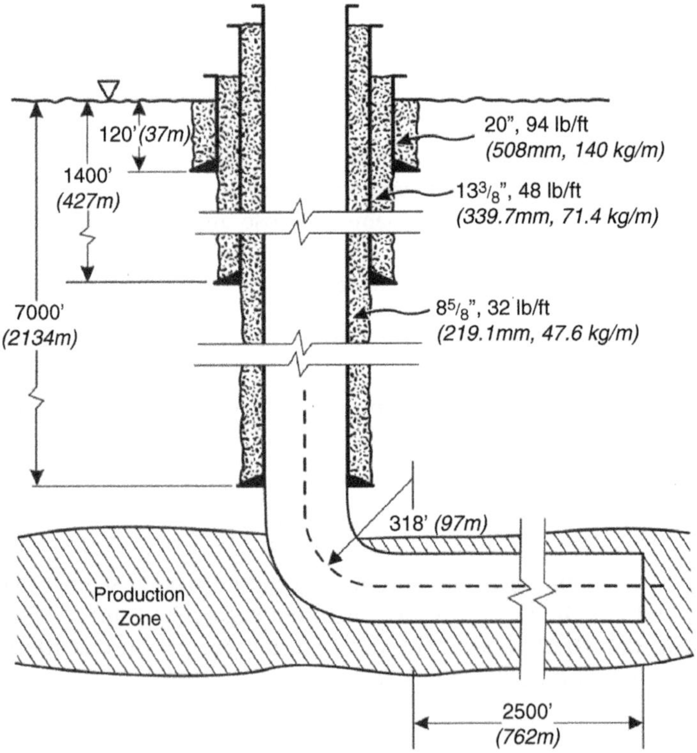

FIGURE 12-14. Medium radius horizontal well profile.

shoe (this is called the KOP). As already discussed in Chapter 8, the required volumetric flow rate dictated three Dresser Clark CFB-4 compressors with a combined volumetric flow rate of 3600 acfm (*1699 actual liters/sec*). This flow is probably pushing the high limit of the PDM discussed in Chapter 11. Even if a more volumetric flow rate of gas was required to drill the open hole in Figure 12-14, the motor maximum would be the controlling limit.

Therefore, cleaning of the horizontal section must depend on judicious use of rotating and pulling the string back and forth while rotating when sliding the PDM.

Illustrative Example 11.2 (i.e., the PDM example) can be altered easily with Equations (12-2) to (12-5) to configure that illustrative example to the horizontal profile in Figure 12-14.

The aerated drilling fluid and the stable foam basic equations in Chapter 6 can be altered to obtain equations similar to Equations (12-2) to (12-6) for these drilling fluids.

Field Operations

The following case histories of directional drilling operations demonstrate the accuracy of the planning calculation procedures discussed in earlier chapters and in this chapter that utilize complete major and minor friction loss terms.

Case History No. 1: This well was drilled in the San Juan Basin in northwestern New Mexico. This was a directional borehole drilled to an MD of 7240 ft (*4993 m*) and a TVD of 5568 ft (*3840 m*). The well was cased to an MD of 5173 ft (*3568 m*) with an API 7-in (*177.8 mm*) casing. The KOP was at 5183 ft (*3575 m*). A $4^3/_4$-in (*120.7 mm*)-diameter PDM with a $6^1/_4$-in (*158.8 mm*) drill bit was used to drill a medium radius curved segment of the borehole with an arc length of 310 ft (*94.5 m*). A PDM was also used to drill the horizontal segment of the borehole to a length of 1428 ft (*435.2 m*) in the Dakota sand formation. The drill string was made up of $3^1/_2$-in (*88.9 mm*) heavyweight drill pipe in the vertical section of the well and an API 3 1/2-in (*88.9 mm*) drill pipe in the curved and horizontal section of the well. The drilling rate in the horizontal segment of the borehole was 23 ft/hr (*7.0 m/sec*). Membrane nitrogen was used as the drilling fluid. The volumetric flow rate injected into the well was 2400 acfm (*1132.6 actual liters/sec*). The surface location elevation was approximately 6100 ft (*1858 m*) above mean sea level. The predicted surface injection pressure was 492 psig (*339.3 N/cm^2 gauge*) and the actual injection pressure was 500 psig (*344.9 N/cm^2 gauge*).

Case History No. 2: This well was also drilled in the San Juan Basin in northwestern New Mexico. This was a directional borehole drilled to an MD of 8065 ft (*2458 m*) and a TVD of 5800 ft (*1768 m*). The well was cased to an MD of 5461 ft (*1664 m*) with API 7-in (177.8 mm) casing. The cased hole was kicked off at an MD of 5557 ft (*1694 m*) so that the bottom of the casing had a build angle of 20° to vertical. A $4^3/_4$-in (*120.7 mm*)-diameter PDM with a $6^1/_4$-in (*158.8 mm*) drill bit was used to drill a medium radius curved segment of the borehole with an arc length of 666 ft (*203 m*) to a final angle to vertical of 88°. Also in this well, a PDM was used to drill the horizontal segment of the borehole to a length of 1937 ft (*590 m*) in the Dakota sand formation. The drill string was made up of 3 1/2-in (*88.9 mm*) heavyweight drill pipe in the vertical section of the well and an API $3^1/_2$-in (*88.9 mm*) drill pipe in the curved and horizontal section of the well. The drilling rate in the horizontal segment of the borehole was 25 ft/hr (*7.6 m/sec*). Membrane nitrogen was used as the drilling fluid. The volumetric flow rate injected into the well was 2800 acfm (*1321 actual liters/sec*). The surface location elevation was approximately 6200 ft (*1890 m*). The predicted surface injection pressure was 539 psig (*372 N/cm² gauge*) and the actual injection pressure was 550 psig (*379 N/cm²* gauge).

12.4 CONCLUSIONS

The analytic models developed in Chapters 8, 9, and 10 for performance and underbalanced drilling applications are accurate predictive tools. These tools are useful for the well-planning process as well as for assessment activities as the well is being drilled. Underbalanced drilling operations represent a technology that is finding great acceptance for the infill drilling of onshore fields. Air and gas drilling technology is a vital component of the parent technology.

The Mathcad detailed solutions given in the appendices show how the various solutions are organized. These should allow the reader to either oversee the development of an in-house program or allow an individual to develop such a program on their own. At the very least, an understanding of these programs should allow an engineer to assess whether a contracted program is giving reasonable results.

Three improvements need to be made to allow more efficient drilling of underbalanced wells.

1. Improvement in MWD for high gas content drilling fluids.
2. Methods and equipment that will allow more efficient and more reliable bottom hole cleaning of high angle directionally drilled boreholes.
3. More output reliable and cost-efficient primary compressors.

REFERENCES

1. Leet, L. D., and Judson, S., *Physical Geology*, Prentice-Hall, 1954.
2. *Underbalanced Drilling Manual*, GRI Reference No. GRI-97/0236, Gas Research Institute, 1997.
3. Lunan, B., and Boote, K. S., "Underbalanced Drilling Techniques Using a Closed System to Control Live Wells—Western Canadian Basin Case Histories," *SPE/CIM*, Annual One Day Conference, Calgary, Alberta, Canada, November 1994.

4. Allan, P. D., "Nitrogen Drilling System for Gas Drilling Applications," *SPE 28320*, Presented at the SPE 69th Annual Technical Conference and Exhibition, New Orleans, Louisiana, September 25–28, 1994.

5. Marcus, R. D., Leung, L. S. , Klinzing, G. E., and Rizk, F., *Pneumatic Conveying of Solids*, Chapman and Hall, 1990.

6. Daugherty, R. L., Franzini, J. B., and Finnemore, E. J., *Fluid Mechanics with Engineering Applications*, Eighth Edition, McGraw-Hill, 1985.

7. Guo, B., Miska, S. Z., and Lee, R., "Volume Requirements for Directional Air Drilling," IADC/SPE 27510, Presented at the 1994 IADC/SPE Drilling Conference, Dallas, Texas, February 15–18, 1994.

APPENDIX

Dimensions and Units, Conversion Factors

A

The systems of dimensions and units used in mechanics are based on Newton's second law of motion, which is force equals mass multiplied by acceleration, or

$$F = ma \tag{A-1}$$

for consistent systems of units, where F is force, m is mass, and a is acceleration.

In the English unit system, engineers define a pound of force as the force required to accelerate 1 slug of mass at the rate of 1 foot per second per second. One slug of mass has a weight of approximately 32.2 lb when acted upon by the acceleration of gravity present at the surface of the Earth. Thus, Equation (A-1) in English units is

$$1 \text{ lb} = (1 \text{ slug})(1 \text{ ft/sec}^2).$$

In the International System of Units (or SI metric), engineers define a newton of force as the force required to accelerate 1 kilogram of mass at the rate of 1 meter per second per second. Thus, Equation (A-1) in the SI metric is

$$1 \text{ N} = (1 \text{ kg})(1 \text{ m/sec}^2).$$

Physicists, however, utilize a version of the SI metric that defines a dyne of force as the force required to accelerate 1 gram of mass at the rate of 1 centimeter per second per second.

Unfortunately, these different systems tend to create confusion. In many parts of the world engineers use the kilogram for both force and mass units. With universal adoption of metric SI, however, this confusion should gradually disappear.

Any system based on length (L), mass (M), and time (T) is absolute because it is independent of the gravitational acceleration g. A system based on length (L), weight, i.e., force (F), and time (T) is referred to as a gravitational system, as weight depends on the value of g, which in turn varies with location (i.e., altitude and latitude). Hence the weight (W) of a certain mass varies with its location. This variation is not generally considered in this text as the variation in the value of g is small as long as we are analyzing a problem on or quite near the Earth's surface. Fluid mechanics problems for other locations, such as the Moon, where g is very different than on the Earth, can be handled by the methods presented in this text if proper consideration is given to the value of g.

The metric SI is known as a mass system, as mass is not dependent on the gravitational system. Indeed, the gravitational acceleration g at the location either on the Earth's surface or on the Moon's surface needs to be known in order to handle specific problems.

Table A-1. Definitions of USC system quantities

Force	1 lb =1 slug-ft/sec^2
Area	1 acre = 43,560 ft^2
Energy	1 Btu = 778 ft-lb
Flow rate	1 cfs = 448.83 gpm
Length	1 ft = 12 in 1 yd = 3 ft 1 statute mile = 5,280 ft 1 nautical mile = 6,000 ft
Mass	1 slug = 1 lb-sec^2/ft
Power	1 hp = 550 ft-lb/sec = 0.708 Btu/sec
Velocity	1 mph = 1.467 ft/sec 1 knot = 1.689 ft/sec = 1.152 mph
Volume	1 ft^3 = 7.48 U.S. gal 1 U.S. gal = 231 in^3 = 0.1337 ft^3 (8.34 lb of water) 1 British imperial gal = 1.2 U.S. gal (10 lb of water)
Weight	1 U.S. (short) ton = 2,000 lb 1 British (long) ton = 2,240 lb

Table A-2. Definitions of SI quantities

Force	1 N = 1 kg-m/sec^2
Area	1 hectare (ha) = 10^4 m^2 = 100 m square
Energy (or work)	1 joule (J) = 1 N-m
Flow rate	m^3/sec = 60,000 lpm
Length	1 centimeter = 10 millimeter 1 m = 100 centimeter 1 kilometer = 1,000 m

Table A-2. Definitions of SI quantities—cont'd	
Mass	1 kg = 1,000 gram 1 metric ton = 1,000 kg
Power	1 Watt = 1 N-m/sec
Velocity	1 kph = 1,000 m/hr = 0.277 m/sec
Volume	1 m^3 = 1,000 liter 1 liter = 10^3 cm
Weight	See Force

Table A-3. Basic quantities USC system to SI conversions

	English unit	**SI unit**
Acceleration of gravity	32.2 ft/sec^2	9.81 m/sec^2
Density of water (at 39.4°F or 4°C)	1.94 slug/ft^3	1000 kg/m^3
Specific weight of water	62.4 lb/ft^3	9810 N/m^3
Standard sea level atmospheric pressure API (at 60°F) ASME (at 68°F) U.K. (at 60°F) Continental Europe (at 15°C)	 14.966 psia 14.7 psia 30.0 inches of Hg 750 mm of Hg	

Table A-4. Other important USC system and SI quantities and conversions	
Engineering gas constant $\mathbf{R_e}$ English Metric	 $\mathbf{R_e}$ = 53.36 ft-lb/lb-°R $\mathbf{R_e}$ = 29.31 N-m/N-K
Energy English Metric SI	 1 ft-lb = 1 lb-ft 1 N-m = 1 joule (J)
Heat English Metric	 1 Btu = 252 cal (heat required to raise 1.0 lb of water 1.0° R) 1 cal = 4.187 J (heat required to raise 1.0 g of water 1.0 K)

(continued)

Table A-4. Other important USC system and SI quantities and conversions—cont'd	
Temperature	
English	$°R = 459.67° + °F$
Metric SI	$K = 273.15° + °C$
Pressure	
English	1 lb/ft^2 = 144 psi
Metric SI	1 pascal (Pa) = N/m^2
Power	
English	1 horsepower = 550 ft-lb/sec
Metric SI	1 watt (W) = 1 J/sec (or 1 N-m/sec)
Absolute viscosity	
English	1 lb-sec/ft^2 = 47.88 N-sec/m^2
Metric SI	1 poise (P) = 10^{-1} N-sec/m^2
Kinematic viscosity	
English	1 ft^2/sec = 0.0929 m^2/sec
Metric SI	1 stoke (St) = 10^{-4} m^2/sec

Table A-5. Commonly used prefixes for SI units

Factor by which unit is multiplied	Prefix	Symbol
10^9	giga	G
10^6	mega	M
10^3	kilo	k
10^{-2}	centi	c
10^{-3}	milli	m
10^{-6}	micro	μ
10^{-9}	nano	n

Table A-6. Conversion factors (USC system to SI)

	To convert English units	Multiply by	To obtain metric (SI) units
Acceleration	ft/sec^2	0.3048	m/sec^2
Area	ft^2 Acre (43,560 ft^2) Hectare (100 m^2)	0.0929	m^2 0.4047
Density	$slug/ft^3$	515.4	kg/m^3
Energy (work or quantity of heat)	ft-lb ft-lb Btu (778 ft-lb)	1.356 3.77×10^{-7} 1,055	N-m (joule) kWh N-m (joule)
Flow rate	ft^3/sec	0.02832	m^3/sec (10^3 liters/sec)
Force	lb	4.4484	N
Length	in ft mile (5,280 ft)	25.4 0.3048 1.609	mm m km
Mass	slug	14.59	kg
Power	ft-lb/sec hp (550 ft-lb/sec)	1.356	W (N-m/sec) 745.7 W (N-m/sec)
Pressure	psi psi lb/ft^2	6,895 0.6895 47.88	N/m^2 (pascal) N/cm^2 N/m^2 (pascal)
Specific heat	ft-lb/slug-°R	0.1672	N-m/kg-K
Specific weight	lb/ft^3	157.1	N/m^3
Velocity	ft/sec	0.3048	m/sec
Viscosity Absolute Kinematic	 $lb\text{-}sec/ft^2$ ft^2/sec	 47.88 0.0929	 $N\text{-}sec/m^2$ m^2/sec (10^4 St)
Volume	ft^3 gallon (U.S.)	0.02832 3.785	m^3 liters (10^{-3} m^3)
Volumetric flow rate	ft^3/sec ft^3/min	0.0283 0.4719	m^3/sec liters/sec

Table A-7. Conversion factors (SI to USC system)

	To convert metric SI units	Multiply by	To obtain English units
Acceleration	m/sec^2	3.281	ft/sec^2
Area	m^2 Hectare (100 m^2)	10.76 2.471	ft^2 Acre (43,560 ft^2)
Density	kg/m^3	0.001940	slug/ft^3
Energy (work or quantity of heat)	N-m (joule) kWh N-m (joule)	0.7376 2.650×10^6 0.0009480	ft-lb ft-lb Btu (778 ft-lb)
Flow rate	m^3/sec (10^3 liters/sec) ft^3/sec		35.31
Force	N	0.2248	lb
Length	mm m km	0.03937 3.281 0.6215	in ft mile (5280 ft)
Mass	kg	0.06854	slug
Power	W (N-m/sec) W (N-m/sec)	0.7375 0.001341	ft-lb/sec hp (550 ft-lb/sec)
Pressure	N/m^2 (pascal) N/cm^2 N/m^2 (pascal)	0.0001450 1.4499 0.02089	psi psi lb/ft^2
Specific heat	N-m/kg-K	5.981	ft-lb/slug-°R
Specific weight	N/m^3	0.006365	lb/ft^3
Velocity	m/sec	3.281	ft/sec
Viscosity Absolute Kinematic	 N-sec/m^2 m^2/sec (10^4 St)	 0.02089 10.76	 lb-sec/ft^2 ft^2/sec
Volume	m^3 liters (10^{-3} m^3)	35.31 0.2642	ft^3 gallons (U.S.)
Volumetric flow rate	m^3/sec liters/sec	35.34 2.119	ft^3/sec ft^3/min

APPENDIX

Average Annual Atmospheric Conditions

B

This appendix gives the graphic representation of the average atmospheric conditions for midlatitudes (30° N to 60° N) of the North American continent.

Figure B-1 gives the average annual atmospheric pressure of air for midlatitudes of the North American continent as a function of surface elevation location above mean sea level. These average annual atmospheric pressures are of critical importance in predicting the actual weight rate of flow of air (or other gases) at an actual drilling location (see Chapters 5 to 12). Figure B-1a is given in USCS units and Figure B-1b is given in SI units.

Figure B-2 gives the average annual atmospheric temperature of air for midlatitudes of the North American continent as a function of surface elevation location above mean sea level. These average annual atmospheric temperatures are of critical importance in predicting the approximate geothermal temperature at an actual drilling location (see Chapters 5 to 12). Figure B-2a is given in USCS units and Figure B-2b is given in SI units.

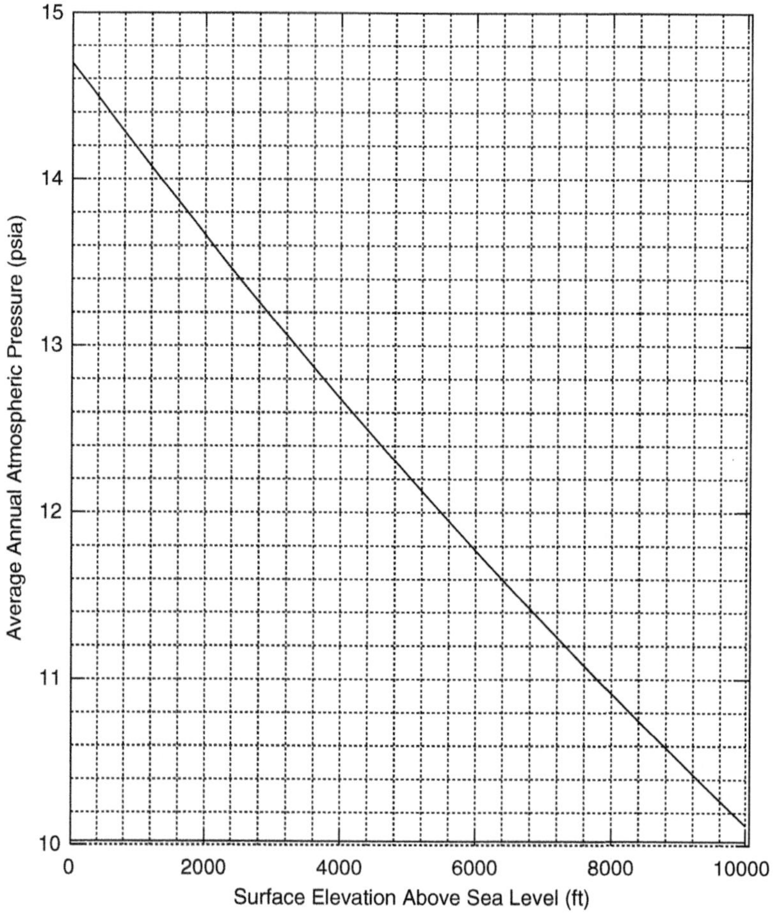

FIGURE B-1a. Average annual atmospheric pressure versus surface elevation above mean sea level (USCS units).

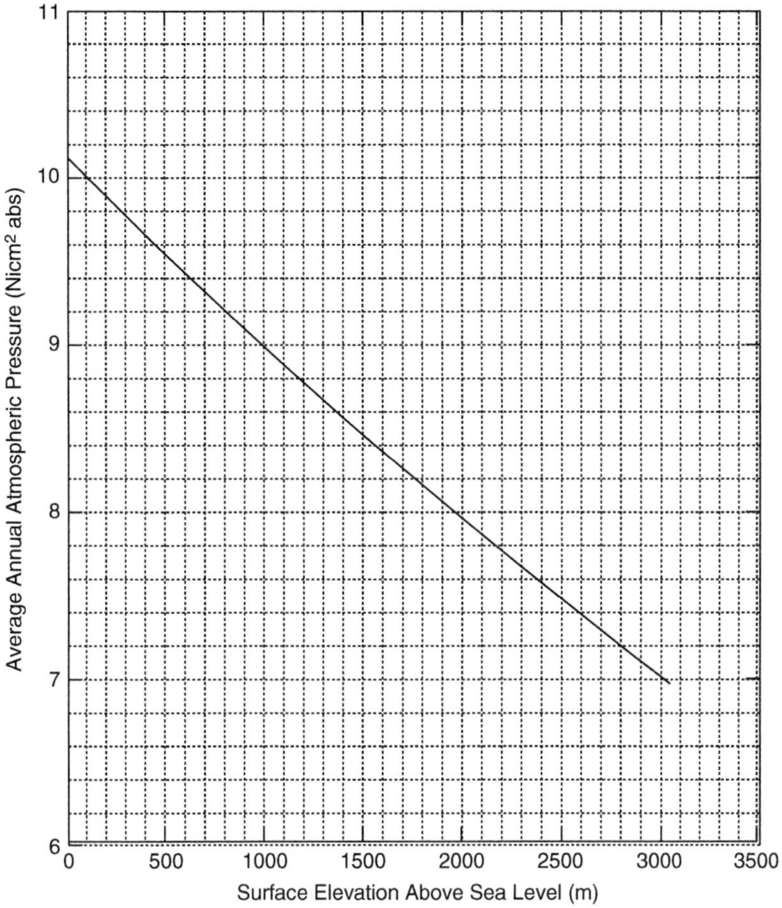

FIGURE B-1b. Average annual atmospheric pressure versus surface elevation above mean sea level (SI units).

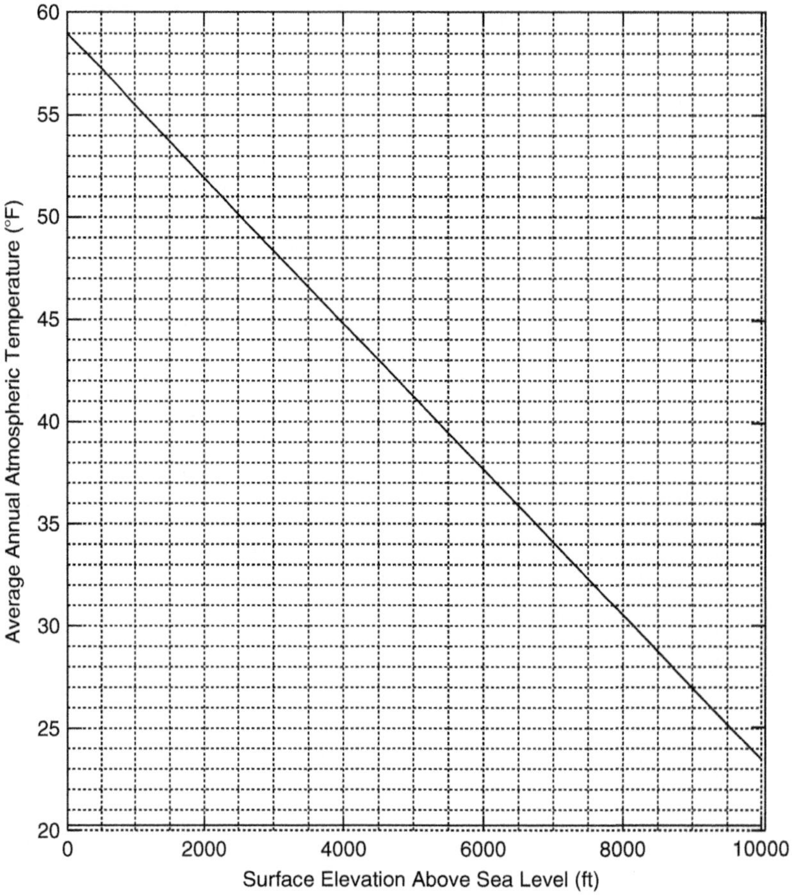

FIGURE B-2a. Average annual atmospheric temperature versus surface elevation above sea level (USCS units).

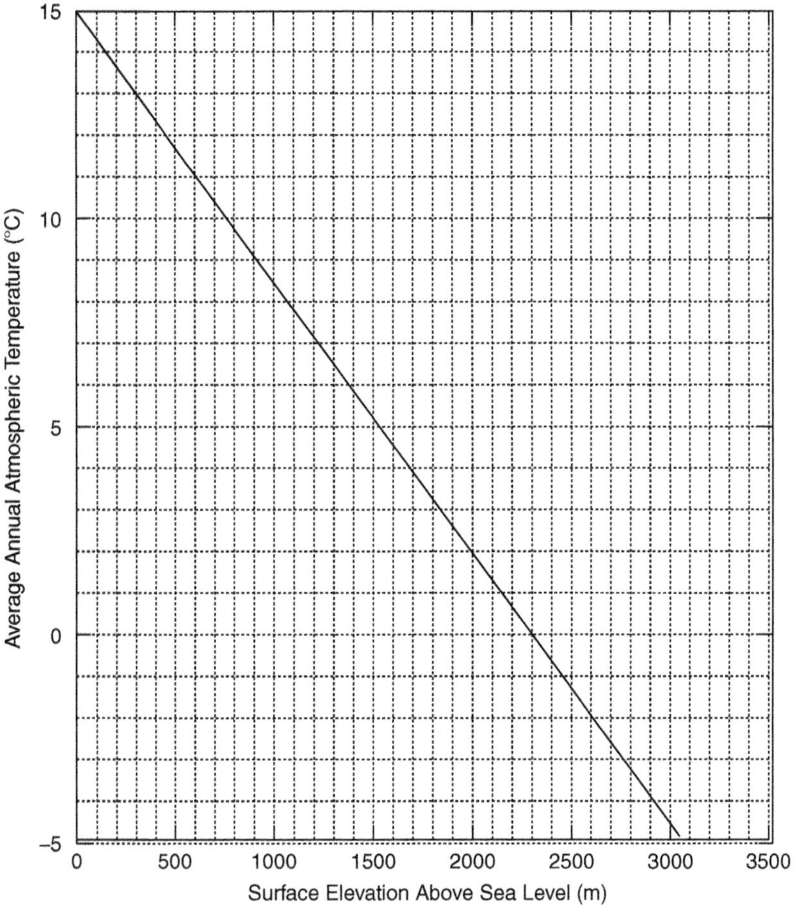

FIGURE B-2b. Average annual atmospheric temperature versus surface elevation above sea level (SI units).

APPENDIX

Chapter 8 Illustrative Example MathCad™ Solutions

C

This appendix presents the detailed MathCad solutions for the illustrative examples in Chapter 8.

Illustrative Example 8.2: USCS version, Minimum Volumetric Flow Rate - Drilling Fluid is Air. Determining the bottom hole and injection pressures while drilling at 10,000 ft of depth for the drilling operations described in Illustrative Examples 8.1.

Minimum volumetric flow rate (trial and error to attain minimum KE)

$q_a := 1994$	flow rate produced from one compressor (acfm)

Basic standard and elevation atomospheric conditions

$p_{st} := 14.696$	API standard atmospheric pressure (psia)
$t_{st} := 60.0$	API standard atmospheric temperature (°F)
$alt := 4000$	surface elevation of drilling site (ft)
$p_{at} := 12.685$	pressure from table for surface elevation (psia)
$t_{at} := 44.74$	temperature from table for surface elevation (°F)
$P_{st} := p_{st} \cdot 144$	convert API standard pressure to lb/ft^2
$P_{st} = 2116.224$	(lb/ft^2 abs)
$T_{st} := t_{st} + 459.67$	convert API standard temp to °R
$T_{st} = 519.67$	(°R)
$P_{at} := p_{at} \cdot 144$	convert actual atm pressure to lb/ft^2
$P_{at} = 1826.64$	(lb/ft^2 abs)
$T_{at} := t_{at} + 459.67$	convert actual atm temp to °R

Borehole geometry

$H_c := 7000$	total depth of cased hole (ft)
$H_{oh} := 3000$	total depth of open hole (ft)

$ROP := 60$	estimated rate of penetration (ft/hr)
$d_{bh} := 7.875$	borehole (drill bit) diameter (in)
$od_{dc} := 6.25$	outer diameter in drill collar (in)
$id_{dc} := 2.8125$	inner diameter in drill collar (in)
$od_{dp} := 4.5$	outer diameter in drill pipe (in)
$id_{dp} := 3.826$	inner diameter in drill pipe (in)
$od_{tj} := 6.25$	outer diameter in drill pipe tool joint (in)
$id_{tj} := 2.75$	inside diameter in drill pipe tool joint (in)
$od_{c} := 8.625$	outer diameter in drill casing (in)
$id_{c} := 7.921$	inside diameter in drill casing (in)
$od_{bl} := 8.625$	outer diameter in blooey line (in)
$id_{bl} := 8.097$	inside diameter in blooey line (in)
$L_{bl} := 200$	length of blooey line (ft)
$e_{cs} := 0.0005$	roughness of commercial steel (ft)
$e_{r} := 0.01$	roughness of rock table (ft)
$L_{dc} := 500$	total length of drill collars (ft)
$L_{tj} := 1.5$	average length of one tool joint (ft)
$L_{dp} := 30$	average length of one drill pipe section [ft]
$d_{n} := 0.7$	nozzle diameter (in)
$N_{n} := 3$	number of nozzles in the bit

$\beta := 0.01$ temperature gradient (°F/ft)

Section height calculations

$H_1 := H_c - L_{tj} \cdot \left(\frac{H_c}{L_{dp}} \right)$ calculate length of drill pipe in cased section

$H_1 = 6650$ (ft)

$H_2 := L_{tj} \cdot \left(\frac{H_c}{L_{dp}} \right)$ calculate length of tool joint in cased section

$H_2 = 350$ (ft)

$H_3 := H_{oh} - L_{dc} - L_{tj} \cdot \left(\frac{H_{oh} - L_{dc}}{L_{dp}} \right)$ calculate length of drill pipe in open hole section

$H_3 = 2375$ (ft)

$H_4 := L_{tj} \cdot \left(\frac{H_{oh} - L_{dc}}{L_{dp}} \right)$ calculate length of tool joints in open hole section

$H_4 = 125$ (ft)

$H_5 := L_{dc}$ length of drill collar

$H_5 = 500$ (ft)

Formation natural gas and water influx

$q_w := 0.0$ injected "mist" water (bbl/hr)

$q_{ng} := 0$ natural gas influx (scfm)

$q_{fw} := 0$ formation water influx (bbl/hr)

Constants

$SG_a := 1.0$ specific gravity of air

$SG_w := 1.0$ specific gravity of fresh water

$SG_{ng} := 0.70$ specific gravity of natural gas

$SG_{fw} := 1.07$ specific gravity of formation water

$SG_r := 2.7$ specific gravity of rock

$SG_{df} := 0.8156$ specific gravity of diesel fuel

$k := 1.4$ gas ratio of specific heats

$R := 53.36$ engineering gas constant (ft-lb/lb-°R)

$\gamma_w := 62.4$ specific weight of fresh water (lb/ft^3)

$\gamma_{wg} := 8.33$ specific weight of water (lb/gal)

$g := 32.2$ gravitational constant [ft/sec^2]

Preliminary calculations

$D_{bh} := \frac{d_{bh}}{12}$ convert borehole diameter to ft

$D_{bh} = 0.6563$ (ft)

$rop := \frac{ROP}{60 \cdot 60}$ convert ROP to ft/sec

$rop = 0.0167$ (ft/sec)

$w_s := \left[\left(\frac{\pi}{4}\right) \cdot D_{bh}^{2} \cdot 62.4 \cdot SG_r \cdot rop\right]$ weight rate of solid flow out of blooey line (lb/sec)

$w_s = 0.9498$ (lb/sec)

$Q_a := \frac{q_a}{60}$ convert air flow rate to ft^3/ sec

$Q_a = 33.2333$ (ft^3/sec)

$Q_w := \frac{(q_w \cdot 42 \cdot 231)}{60 \cdot 60 \cdot 12^3}$ convert "mist" water flow to ft^3/sec

$Q_w = 0$ (ft^3/sec)

$Q_{ng} := \frac{q_{ng}}{60}$ convert natural gas flow rate to ft^3/sec

$Q_{ng} = 0$ (ft^3/sec)

$Q_{gt} := Q_a + Q_{ng}$ total flow of the gas (ft^3/sec)

$Q_{gt} = 33.2333$ (ft^3/sec)

$Q_{fw} := \frac{(q_{fw} \cdot 42 \cdot 231)}{60 \cdot 60 \cdot 12^3}$ convert formation water flow to ft^3/sec

$Q_{fw} = 0$ (ft^3/sec)

$\gamma_a := \frac{(P_{at} \cdot SG_a)}{R \cdot T_{at}}$ specific weight of air (lb/ft^3)

$\gamma_a = 0.0679$ (lb/ft^3)

$\gamma_{ng} := \frac{(P_{at} \cdot SG_{ng})}{R \cdot T_{at}}$ specific weight of natural gas (lb/ft^3)

$\gamma_{ng} = 0.0475$ (lb/ft^3)

$w_a := \gamma_a \cdot Q_a$ weight rate of flow of air (lb/sec)

$w_a = 2.2554$ (lb/sec)

$w_w := SG_w \cdot \gamma_w \cdot Q_w$ weight rate of flow of "mist" water (lb/sec)

$w_w = 0$ (lb/sec)

$w_{ng} := \gamma_{ng} \cdot Q_{ng}$ weight rate of flow of natural gas (lb/sec)

$w_{ng} = 0$ (lb/sec)

$w_{fw} := SG_{fw} \cdot \gamma_w \cdot Q_{fw}$ weight rate of flow of formation water (lb/sec)

$w_{fw} = 0$ (lb/sec)

$w_{gt} := w_a + w_{ng}$ total weight rate of flow of gas (lb/sec)

$w_{gt} = 2.2554$ (lb/sec)

Blooey line

$ID := \frac{id_{bl}}{12}$ convert blooey line inner diameter to ft

$ID = 0.6747$ (ft)

$A := \frac{\pi}{4} \cdot ID^2$ cross sectional area (ft^2)

$A = 0.3576$ (ft^2)

$$f := \left(\frac{1}{2 \cdot \log\left(\frac{ID}{e_{cs}}\right) + 1.14} \right)^2$$ friction factor using von Karman correlation

$f = 0.0183$

$\frac{id_c}{od_{dp}} = 1.7602$ compute D1/D3 ratio for Fig. 8-5

$\frac{id_c}{id_{bl}} = 0.9783$ compute D1/D2 ratio for Fig. 8-5

Figure 8.5 gives the following flow resistance coefficient for the blind Tee:

$K_t := 25$

$K_v := 0.2$ loss factor for gate valves in blooey line

$$P_b := \left[\left[\left(f\cdot\frac{L_{bl}}{ID}\right) + K_t + 2\cdot K_v\right]\cdot\left[\frac{\left(w_{gt}^{\ 2}\cdot R\cdot T_{at}\right)}{g\cdot A^2}\right] + P_{at}^{\ 2}\right]^{0.5}$$ calculate pressure of the gas at the entrance to the blooey line

$P_b = 2088.3615$ (lb/ft^2)

$p_b := \frac{P_b}{144}$ convert to lb/in^2

$p_b = 14.5025$ (psia)

Annulus cased hole drill pipe body section

$H := H_1$ (ft)

$ID := \frac{id_c}{12}$ convert inner diameter of casing to ft

$ID = 0.6601$ (ft)

$OD := \frac{od_{dp}}{12}$ convert outer diameter of drill pipe to ft

$OD = 0.375$ (ft)

$A := \frac{\pi}{4}\cdot\left(ID^2 - OD^2\right)$ annulus area between casing and drill pipe (ft^2)

$A = 0.2318$ (ft^2)

$$a_a := \frac{SG_a}{R}\cdot\left[1 + \frac{(w_s + w_w + w_{fw})}{w_{gt}}\right]$$ calculate constant

$a_a = 0.0266$

$$f := \left[\frac{1}{2\cdot\log\left[\frac{(ID - OD)}{e_{cs}}\right] + 1.14}\right]^2$$ friction factor using von Karman correlation

$f = 0.0226$

$$b_a := \left[\left[\frac{f}{2\cdot g\cdot(ID - OD)}\right]\cdot\left(\frac{R}{SG_a}\right)^2\cdot\frac{w_{gt}^2}{\left(\frac{\pi}{4}\right)^2\cdot\left(ID^2 - OD^2\right)^2}\right]$$ calculate constant

$b_a = 331.9324$

$T := T_{at} + \beta\cdot H_1$ temperature at bottom of section (°R)

$T = 570.91$ (°R)

$$T_{ave1} := \frac{(T_{at} + T)}{2}$$ average temperature of section (°R)

$T_{ave1} = 537.66$ (°R)

$T_{ave} := T_{ave1}$

$$P_1 := \left[\left[\left(P_b^2 + b_a\cdot T_{ave}^2\right)\cdot e^{\left[\frac{(2\cdot a_a\cdot H)}{T_{ave}}\right]}\right] - b_a\cdot T_{ave}^2\right]^{0.5}$$ calculate pressure

$P_1 = 9894.6307$ (lb/ft²)

$p_1 := \frac{P_1}{144}$ convert to psia

$p_1 = 68.7127$ (psia)

$\gamma := \frac{P_1 \cdot SG_a}{R \cdot T}$ specific weight of gas (lb/ft³)

$\gamma = 0.3248$ (lb/ft³)

$V := \frac{w_{gt}}{\gamma \cdot A}$ velocity (ft/sec)

$V = 29.9622$ (ft/sec)

$KE_1 := \frac{1}{2} \cdot \frac{\gamma}{g} \cdot V^2$ kinetic energy of gas flow (lb-ft/ft³)

$KE_1 = 4.5277$ (lb-ft/ft³)

Annulus cased hole tool joint section

$H := H_2$

$ID := \frac{id_c}{12}$ convert inner diameter of casing to ft

$ID = 0.6601$ (ft)

$OD := \frac{od_{tj}}{12}$ convert outer diameter of tool joint to ft

$OD = 0.5208$ (ft)

$OD_t := \frac{od_{dp}}{12}$ use drill pipe OD for kinetic energy to ft

$OD_t = 0.375$ (ft)

$A := \frac{\pi}{4}\left(ID^2 - OD_t^{\,2}\right)$ annulus area for tool joint KE (ft^2)

$A = 0.2318$ (ft^2)

$a_a := \frac{SG_a}{R} \cdot \left[1 + \frac{\left(w_s + w_w + w_{fw}\right)}{w_{gt}}\right]$ calculate constant

$a_a = 0.0266$

$f := \left[\frac{1}{2 \cdot \log\left[\frac{(ID - OD)}{e_{cs}}\right] + 1.14}\right]^2$ friction factor using von Karman correlation

$f = 0.0275$

$b_a := \left[\left[\frac{f}{2 \cdot g \cdot (ID - OD)}\right] \cdot \left(\frac{R}{SG_a}\right)^2 \cdot \frac{w_{gt}^{\,2}}{\left(\frac{\pi}{4}\right)^2 \cdot \left(ID^2 - OD^2\right)^2}\right]$ calculate constant

$b_a = 2663.2525$

$T_{temp} := T$

$T := T_{temp} + \beta \cdot H_2$ temperature at bottom of section (°R)

$T = 574.41$ (°R)

$T_{ave2} := \frac{\left(T_{temp} + T\right)}{2}$ average temperature of section (°R)

$T_{ave2} = 572.66$ (°R)

$T_{ave} := T_{ave2}$

$$P_2 := \left[\left[\left(P_1{}^2 + b_a \cdot T_{ave}{}^2\right) \cdot e^{\left[\frac{(2 \cdot a_a \cdot H)}{T_{ave}}\right]}\right] - b_a \cdot T_{ave}{}^2\right]^{0.5}$$ calculate pressure

$P_2 = 11403.6873$ (lb/ft^2)

$p_2 := \frac{P_2}{144}$ convert to psia

$p_2 = 79.1923$ (psia)

$\gamma := \frac{P_2 \cdot SG_a}{R \cdot T}$ specific weight of gas (lb/ft^3)

$\gamma = 0.3721$ (lb/ft^3)

$V := \frac{w_{gt}}{\gamma \cdot A}$ velocity of gas (ft/sec)

$V = 26.1567$ (ft/sec)

$KE_2 := \frac{1}{2} \cdot \frac{\gamma}{g} \cdot V^2$ kinetic energy of gas flow (lb-ft/ft^3)

$KE_2 = 3.9526$ (lb-ft/ft^3)

Annulus of open hole drill pipe section

$H := H_3$

$ID := D_{bh}$ open hole diameter of borehole (ft)

$ID = 0.6563$ (ft)

$OD := \frac{od_{dp}}{12}$ convert diameter of drill pipe to ft

$OD = 0.375$ (ft)

$$A := \frac{\pi}{4}\left(ID^2 - OD^2\right)$$ area between borehole and drill pipe (ft^2)

$A = 0.2278$ (ft^2)

$$a_a := \frac{SG_a}{R}\cdot\left[1 + \frac{\left(w_s + w_w + w_{fw}\right)}{w_{gt}}\right]$$ calculate constant

$a_a = 0.0266$

$$e_{ave} := \frac{\left(e_r ID^2 + e_{cs} OD^2\right)}{ID^2 + OD^2}$$ weighted average roughness (ft)

$e_{ave} = 0.0077$ (ft)

$$f := \left[\frac{1}{2\cdot\log\left[\frac{(ID - OD)}{e_{ave}}\right] + 1.14}\right]^2$$ friction factor using von Karman correlation

$f = 0.0549$

$$b_a := \left[\left[\frac{f}{2\cdot g\cdot(ID - OD)}\right]\cdot\left(\frac{R}{SG_a}\right)^2\cdot\frac{w_{gt}^2}{\left(\frac{\pi}{4}\right)^2\cdot\left(ID^2 - OD^2\right)^2}\right]$$ calculate constant

$b_a = 845.3769$

$T_{temp} := T$

$T := T_{temp} + \beta\cdot H_3$ temperature at bottom of section (°R)

$T = 598.16$

$$T_{ave3} := \frac{\left(T_{temp} + T\right)}{2}$$ average temperature of section (°R)

$T_{ave3} = 586.285$ (°R)

$T_{ave} := T_{ave3}$

$$P_3 := \left[\left[\left(P_2^{\,2} + b_a \cdot T_{ave}^{\,2}\right) \cdot e^{\left[\frac{(2 \cdot a_a \cdot H)}{T_{ave}}\right]}\right] - b_a \cdot T_{ave}^{\,2}\right]^{0.5}$$ calculate pressure

$P_3 = 15209.8496$ (lb/ft²)

$p_3 := \frac{P_3}{144}$ convert to psia

$p_3 = 105.624$ (psia)

$\gamma := \frac{P_3 \cdot SG_a}{R \cdot T}$ specific weight of gas (lb/ft³)

$\gamma = 0.4765$ (lb/ft³)

$V := \frac{w_{gt}}{\gamma \cdot A}$ velocity of gas (ft/sec)

$V = 20.7773$ (ft/sec)

$KE_3 := \frac{1}{2} \cdot \frac{\gamma}{g} \cdot V^2$ kinetic energy of gas flow (lb-ft/ft³)

$KE_3 = 3.1944$ (lb-ft/ft³)

Annulus open hole tool joint section

$H := H_4$

$ID := D_{bh}$ open hole diameter of borehole (ft)

$ID = 0.6563$ (ft)

$OD := \frac{od_{tj}}{12}$ convert tool joint diameter to ft

$OD = 0.5208$ (ft)

$OD_t := \frac{od_{dp}}{12}$ drill pipe diameter (ft)

$OD_t = 0.375$ (ft)

$A := \frac{\pi}{4}\left(ID^2 - OD_t^{\,2}\right)$ annulus area for tool joint KE (ft^2)

$A = 0.2278$ (ft^2)

$a_a := \frac{SG_a}{R}\cdot\left[1 + \frac{(w_s + w_w + w_{fw})}{w_{gt}}\right]$ calculate constant

$a_a = 0.0266$

$e_{ave} := \frac{\left(e_r ID^2 + e_{cs} OD^2\right)}{ID^2 + OD^2}$ weighted average roughness (ft)

$e_{ave} = 0.0063$ (ft)

$f := \left[\frac{1}{2\cdot\log\left[\frac{(ID - OD)}{e_{ave}}\right] + 1.14}\right]^2$ friction factor using von Karman correlation

$f = 0.0692$

$b_a := \left[\left[\frac{f}{2\cdot g\cdot(ID - OD)}\right]\cdot\left(\frac{R}{SG_a}\right)^2\cdot\frac{w_{gt}^{\,2}}{\left(\frac{\pi}{4}\right)^2\cdot\left(ID^2 - OD^2\right)^2}\right]$ calculate constant

$b_a = 7335.9699$

$T_{temp} := T$

$T := T_{temp} + \beta \cdot H_4$ temperature at bottom of section (°R)

$T = 599.41$

$T_{ave4} := \frac{(T_{temp} + T)}{2}$ average temperature of section (°R)

$T_{ave4} = 598.785$ (°R)

$T_{ave} := T_{ave4}$

$$P_4 := \left[\left[\left(P_3^2 + b_a \cdot T_{ave}^2\right) \cdot e^{\left[\frac{(2 \cdot a_a \cdot H)}{T_{ave}}\right]}\right] - b_a \cdot T_{ave}^2\right]^{0.5}$$ calculate pressure

$P_4 = 16227.6442$ (lb/ft²)

$p_4 := \frac{P_4}{144}$ convert to psia

$p_4 = 112.692$ (psia)

$\gamma := \frac{P_4 \cdot SG_a}{R \cdot T}$ specific weight of gas (lb/ft³)

$\gamma = 0.5074$ (lb/ft³)

$V := \frac{w_{gt}}{\gamma \cdot A}$ velocity of gas (ft/sec)

$V = 19.5149$ (ft/sec)

$KE_4 := \frac{1}{2} \cdot \frac{\gamma}{g} \cdot V^2$ kinetic energy of gas flow (lb-ft/ft³)

$KE_4 = 3.0003$ (ft-lb/ft^3)

Annulus open hole drill collar section

$H := H_5$

$ID := D_{bh}$ open hole diameter of borehole (ft)

$ID = 0.6563$ (ft)

$OD := \frac{od_{dc}}{12}$ convert drill collar diameter to ft

$OD = 0.5208$ (ft)

$A := \frac{\pi}{4}\left(ID^2 - OD^2\right)$ drill pipe diameter (ft)

$A = 0.1252$ (ft)

$a_a := \frac{SG_a}{R}\cdot\left[1 + \frac{\left(w_s + w_w + w_{fw}\right)}{w_{gt}}\right]$ calculate constant

$a_a = 0.0266$

$e_{ave} := \frac{\left(e_r ID^2 + e_{cs} OD^2\right)}{ID^2 + OD^2}$ weighted average roughness (ft)

$e_{ave} = 0.0063$ (ft)

$f := \left[\frac{1}{2\cdot\log\left[\frac{(ID - OD)}{e_{ave}}\right] + 1.14}\right]^2$ friction factor using von Karman correlation

$f = 0.0692$

$$b_a := \left[\left[\frac{f}{2\cdot g\cdot(ID - OD)}\right]\cdot\left(\frac{R}{SG_a}\right)^2\cdot\frac{w_{gt}^2}{\left(\frac{\pi}{4}\right)^2\cdot\left(ID^2 - OD^2\right)^2}\right]$$ calculate constant

$b_a = 7335.9699$

$T_{temp} := T$

$T := T_{temp} + \beta\cdot H_5$ temperature at bottom of section (°R)

$T = 604.41$

$$T_{ave5} := \frac{(T_{temp} + T)}{2}$$ average temperature of section (°R)

$T_{ave5} = 601.91$ (°R)

$T_{ave} := T_{ave5}$

$$P_5 := \left[\left[\left(P_4^2 + b_a\cdot T_{ave}^2\right)\cdot e^{\left[\frac{(2\cdot a_a\cdot H)}{T_{ave}}\right]}\right] - b_a\cdot T_{ave}^2\right]^{0.5}$$

$P_5 = 19886.9003$ (lb/ft²)

$$p_5 := \frac{P_5}{144}$$

$p_5 = 138.1035$ (psia)

$p_{bh} := p_5$ annulus bottom hole pressure (psia)

$$\gamma := \frac{P_5 \cdot SG_a}{R \cdot T}$$ specific weight of gas (lb/ft^3)

$\gamma = 0.6166$ (lb/ft^3)

$$V := \frac{w_{gt}}{\gamma \cdot A}$$ velocity of gas (ft/sec)

$V = 29.2172$ (ft/sec)

$$KE_5 := \frac{1}{2} \cdot \frac{\gamma}{g} \cdot V^2$$ kinetic energy of gas flow (lb-ft/ft^3)

$KE_5 = 8.1736$ (ft-lb/ft^3)

Bottom hole through the bit nozzles

$$D_n := \frac{d_n}{12}$$ convert nozzle diameter to ft

$D_n = 0.0583$ (ft)

$$A_n := N_n \cdot \left(\frac{\pi}{4}\right) \cdot D_n^{\;2}$$ total flow area of nozzles in bit (ft^2)

$A_n = 0.008$ (ft^2)

$$\gamma := \frac{\left(P_5 \cdot SG_a\right)}{R \cdot T}$$ specific weight of air (lb/ft^3)

$\gamma = 0.6166$ (lb/ft^3)

$$P := P_5 \cdot \left[1 + \frac{\left(\frac{w_{gt}}{A_n}\right)^2}{2 \cdot g \cdot \left(\frac{k}{k-1}\right) \cdot P_5 \cdot \gamma}\right]^{\frac{k}{k-1}}$$ calculate pressure

$P = 21952.024$ (lb/ft^2)

$$p := \frac{P}{144}$$ convert to psia

$$p = 152.4446$$ (psia)

Open hole inside drill string drill collar section

$$D := \frac{id_{dc}}{12}$$ convert diameter to ft

$$D = 0.2344$$ (ft)

$$a_i := \left(\frac{SG_a}{R}\right)\cdot\left(1 + \frac{w_w}{w_{gt}}\right)$$ calculate constant

$$a_i = 0.0187$$

$$f := \left(\frac{1}{2\cdot\log\left(\frac{D}{e_{cs}}\right) + 1.14}\right)^2$$ friction factor using von Karman correlation

$$f = 0.0238$$

$$b_i := \left(\frac{f}{2\cdot g\cdot D}\right)\cdot\left(\frac{R}{SG_a}\right)^2\cdot\frac{w_{gt}^{\,2}}{\left(\frac{\pi}{4}\right)^2\cdot D^4}$$ calculate constant

$$b_i = 12270.4923$$

$$P := \left[\frac{\left[P^2 + b_i\cdot T_{ave5}^{\,2}\cdot\left[e^{\frac{(2\cdot a_i\cdot H_5)}{T_{ave5}}} - 1\right]\right]}{e^{\frac{(2\cdot a_i\cdot H_5)}{T_{ave5}}}}\right]^{(0.5)}$$ calculate pressure

$$P = 24564.1894$$ (lb/ft^2)

$$p := \frac{P}{144}$$ convert to psia

$p = 170.5846$ (psia)

Open hole inside drill string tool joint section

$D := \frac{id_{tj}}{12}$ convert diameter to ft

$D = 0.2292$ (ft)

$a_i := \left(\frac{SG_a}{R}\right)\cdot\left(1 + \frac{w_w}{w_{gt}}\right)$ calculate constant

$a_i = 0.0187$

$f := \left(\frac{1}{2\cdot\log\left(\frac{D}{e_{cs}}\right) + 1.14}\right)^2$ friction factor using von Karman correlation

$f = 0.0239$

$b_i := \left(\frac{f}{2\cdot g\cdot D}\right)\cdot\left(\frac{R}{SG_a}\right)^2\cdot\frac{w_{gt}^{\,2}}{\left(\frac{\pi}{4}\right)^2\cdot D^4}$ calculate constant

$b_i = 13812.7708$

$$P := \left[\frac{\left[p^2 + b_i\cdot T_{ave4}^{\,2}\cdot\left[e^{\frac{(2\cdot a_i\cdot H_4)}{T_{ave4}}} - 1\right]\right]}{e^{\frac{(2\cdot a_i\cdot H_4)}{T_{ave4}}}}\right]^{(0.5)}$$ calculate pressure

$P = 25244.7168$ (lb/ft^2)

$p := \frac{P}{144}$ convert to psia

$p = 175.3105$ (psia)

Open hole inside drill string drill pipe body section

$$D := \frac{id_{dp}}{12}$$ convert diameter to ft

$D = 0.3188$ (ft)

$$a_i := \left(\frac{SG_a}{R}\right)\cdot\left(1 + \frac{w_w}{w_{gt}}\right)$$ calculate constant

$a_i = 0.0187$

$$f := \left(\frac{1}{2\cdot\log\left(\frac{D}{e_{cs}}\right) + 1.14}\right)^2$$ friction factor using von Karman correlation

$f = 0.022$

$$b_i := \left(\frac{f}{2\cdot g\cdot D}\right)\cdot\left(\frac{R}{SG_a}\right)^2\cdot\frac{w_{gt}^2}{\left(\frac{\pi}{4}\right)^2\cdot D^4}$$ calculate constant

$b_i = 2429.4007$

$$P := \left[\frac{\left[P^2 + b_i\cdot T_{ave3}^2\cdot\left[e^{\frac{(2\cdot a_i\cdot H_3)}{T_{ave3}}} - 1\right]\right]}{e^{\frac{(2\cdot a_i\cdot H_3)}{T_{ave3}}}}\right]^{(0.5)}$$ calculate pressure

$P = 25790.5863$ (lb/ft^2)

$$p := \frac{P}{144}$$ convert to psia

$p = 179.1013$ (psia)

Cased hole inside drill string tool joint section

$D := \frac{id_{tj}}{12}$ convert diameter to ft

$D = 0.2292$ (ft)

$a_i := \left(\frac{SG_a}{R}\right)\cdot\left(1 + \frac{w_w}{w_{gt}}\right)$ calculate constant

$a_i = 0.0187$

$f := \left(\frac{1}{2\cdot\log\left(\frac{D}{e_{cs}}\right) + 1.14}\right)^2$ friction factor using von Karman correlation

$f = 0.0239$

$b_i := \left(\frac{f}{2\cdot g\cdot D}\right)\cdot\left(\frac{R}{SG_a}\right)^2\cdot\frac{w_{gt}^{\ 2}}{\left(\frac{\pi}{4}\right)^2\cdot D^4}$ calculate constant

$b_i = 13812.7708$

$$P := \left[\frac{\left[P^2 + b_i\cdot T_{ave2}^{\ 2}\cdot\left[e^{\frac{(2\cdot a_i\cdot H_2)}{T_{ave2}}} - 1\right]\right]}{e^{\frac{(2\cdot a_i\cdot H_2)}{T_{ave2}}}}\right]^{(0.5)}$$

calculate pressure

$P = 27434.9755$ (lbs/ft^2)

$p := \frac{P}{144}$ convert to psia

$p = 190.5207$ (psia)

Cased hole inside drill string pipe body section

$$D := \frac{id_{dp}}{12}$$ convert diameter to ft

$D = 0.3188$ (ft)

$$a_i := \left(\frac{SG_a}{R}\right)\cdot\left(1 + \frac{w_w}{w_{gt}}\right)$$ calculate constant

$a_i = 0.0187$

$$f := \left(\frac{1}{2\cdot\log\left(\frac{D}{e_{cs}}\right) + 1.14}\right)^2$$ friction factor using von Karman correlation

$f = 0.022$

$$b_i := \left(\frac{f}{2\cdot g\cdot D}\right)\cdot\left(\frac{R}{SG_a}\right)^2\cdot\frac{w_{gt}^{\,2}}{\left(\frac{\pi}{4}\right)^2\cdot D^4}$$ calculate constant

$b_i = 2429.4007$

$$P := \left[\frac{\left[P^2 + b_i\cdot T_{ave1}^{\,2}\cdot\left[e^{\frac{(2\cdot a_i\cdot H_1)}{T_{ave1}}} - 1\right]\right]}{e^{\frac{(2\cdot a_i\cdot H_1)}{T_{ave1}}}}\right]^{0.5}$$ calculate pressure

$P = 27092.1415$ (lb/ft^2)

$$p := \frac{P}{144}$$ convert to psia

$p = 188.1399$ (psia)

$p_{in} := p$

p_{in} = 188.1399 drilling string injection pressure (psia)

Kinetic energy summary of each section (KE greater than 3.0 lb-ft/ft^3)

KE_1 = 4.5277 (lb-ft/ft^3)

KE_2 = 3.9526 (lb-ft/ft^3)

KE_3 = 3.1944 (lb-ft/ft^3)

KE_4 = 3.0003 (lb-ft/ft^3)

KE_5 = 8.1736 (lb-ft/ft^3)

Illustrative Example 8.3: USCS version, Two Dresser Clark CFB-4 Compressors - Drilling Fluid is Air. Determining the bottom hole and injection pressures while drilling at 10,000 ft of depth for the drilling operations described in Illustrative Examples 8.1 and 8.2.

Compressor data - Dresser Clark Model CFB-4 with Caterpillar Model D398 prime mover

$q_c := 1200$	flow rate produced from one compressor (acfm)
$num_c := 2$	number of compressors (required for min KE)
$\varepsilon_m := 0.8$	mechanical efficiency
$n_s := 4$	number of stages
$HP_{max} := 760$	max sea level HP rating of prime mover (HP)
$c := 0.02$	clearance volume ratio for compressor

Basic standard and elevation atomospheric conditions

$p_{st} := 14.696$	API standard atmospheric pressure (psia)
$t_{st} := 60.0$	API standard atmospheric temperature (°F)
$alt := 4000$	surface elevation of drilling site (ft)
$p_{at} := 12.685$	pressure from table for surface elevation (psia)
$t_{at} := 44.74$	temperature from table for surface elevation (°F)
$P_{st} := p_{st} \cdot 144$	convert API standard pressure to lb/ft^2
$P_{st} = 2116.224$	(lb/ft^2 abs)
$T_{st} := t_{st} + 459.67$	convert API standard temp to °R
$T_{st} = 519.67$	(°R)

$P_{at} := p_{at} \cdot 144$ convert actual atm pressure to lb/ft^2

$P_{at} = 1826.64$ (lb/ft^2 abs)

$T_{at} := t_{at} + 459.67$ convert actual atm temp to °R

Borehole geometry

$H_c := 7000.0$ total depth of cased hole (ft)

$H_{oh} := 3000.0$ total depth of open hole (ft)

$ROP := 60$ estimated rate of penetration (ft/hr)

$d_{bh} := 7.875$ borehole (drill bit) diameter (in)

$od_{dc} := 6.25$ outer diameter in drill collar (in)

$id_{dc} := 2.8125$ inner diameter in drill collar (in)

$od_{dp} := 4.5$ outer diameter in drill pipe (in)

$id_{dp} := 3.826$ inner diameter in drill pipe (in)

$od_{tj} := 6.25$ outer diameter in drill pipe tool joint (in)

$id_{tj} := 2.75$ inside diameter in drill pipe tool joint (in)

$od_c := 8.625$ outer diameter in drill casing (in)

$id_c := 7.921$ inside diameter in drill casing (in)

$od_{bl} := 8.625$ outer diameter in blooey line (in)

$id_{bl} := 8.097$ inside diameter in blooey line (in)

$L_{bl} := 200$ length of blooey line (ft)

$e_{cs} := 0.0005$ roughness of commercial steel (ft)

$e_r := 0.01$ roughness of rock table (ft)

$L_{dc} := 500$ total length of drill collars (ft)

$L_{tj} := 1.5$ average length of one tool joint (ft)

$L_{dp} := 30$ average length of one drill pipe section [ft]

$d_n := 0.7$ nozzle diameter (in)

$N_n := 3$ number of nozzles in the bit

$\beta := 0.01$ temperature gradient (°F/ft)

Section height calculations

$H_1 := H_c - L_{tj} \cdot \left(\frac{H_c}{L_{dp}} \right)$ calculate length of drill pipe in cased section

$H_1 = 6650$ (ft)

$H_2 := L_{tj} \cdot \left(\frac{H_c}{L_{dp}} \right)$ calculate length of tool joint in cased section

$H_2 = 350$ (ft)

$H_3 := H_{oh} - L_{dc} - L_{tj} \cdot \left(\frac{H_{oh} - L_{dc}}{L_{dp}} \right)$ calculate length of drill pipe in open hole section

$H_3 = 2375$ (ft)

$H_4 := L_{tj} \cdot \left(\frac{H_{oh} - L_{dc}}{L_{dp}} \right)$ calculate length of tool joints in open hole section

$H_4 = 125$ (ft)

$H_5 := L_{dc}$ length of drill collar

$H_5 = 500$ (ft)

Formation natural gas and water influx

$q_w := 0.5$	injected "mist" water (bbl/hr)
$q_{ng} := 0.0$	natural gas influx (scfm)
$q_{fw} := 0.00$	formation water influx (bbl/hr)

Constants

$SG_a := 1.0$	specific gravity of air
$SG_w := 1.0$	specific gravity of fresh water
$SG_{ng} := 0.70$	specific gravity of natural gas
$SG_{fw} := 1.07$	specific gravity of formation water
$SG_r := 2.7$	specific gravity of rock
$SG_{df} := 0.8156$	specific gravity of diesel fuel
$k := 1.4$	gas ratio of specific heats
$R := 53.36$	engineering gas constant (ft-lb/lb-°R)
$\gamma_w := 62.4$	specific weight of fresh water (lb/ft^3)
$\gamma_{wg} := 8.33$	specific weight of water (lb/gal)
$g := 32.2$	gravitational constant [ft/sec^2]

Preliminary calculations

$D_{bh} := \dfrac{d_{bh}}{12}$	convert borehole diameter to ft

$D_{bh} = 0.6563$ (ft)

$rop := \frac{ROP}{60 \cdot 60}$ convert ROP to ft/sec

$rop = 0.0167$ (ft/sec)

$w_s := \left[\left(\frac{\pi}{4}\right) \cdot D_{bh}^{2} \cdot 62.4 \cdot SG_r \cdot rop\right]$ weight rate of solid flow out of blooey line (lb/sec)

$w_s = 0.9498$ (lb/sec)

$q_a := q_c \cdot num_c$ air flow rate from all compressors (acfm)

$Q_a := \frac{q_a}{60}$ convert air flow rate to ft^3/ sec

$Q_a = 40$ (ft^3/sec)

$Q_w := \frac{(q_w \cdot 42 \cdot 231)}{60 \cdot 60 \cdot 12^3}$ convert "mist" water flow to ft^3/sec

$Q_w = 0.0008$ (ft^3/sec)

$Q_{ng} := \frac{q_{ng}}{60}$ convert natural gas flow rate to ft^3/sec

$Q_{ng} = 0$ (ft^3/sec)

$Q_{gt} := Q_a + Q_{ng}$ total flow of the gas (ft^3/sec)

$Q_{gt} = 40$ (ft^3/sec)

$Q_{fw} := \frac{(q_{fw} \cdot 42 \cdot 231)}{60 \cdot 60 \cdot 12^3}$ convert formation water flow to ft^3/sec

$Q_{fw} = 0$ (ft^3/sec)

$\gamma_a := \frac{(P_{at} \cdot SG_a)}{R \cdot T_{at}}$ specific weight of air (lb/ft^3)

$\gamma_a = 0.0679$ (lb/ft³)

$\gamma_{ng} := \frac{(P_{at} \cdot SG_{ng})}{R \cdot T_{at}}$ specific weight of natural gas (lb/ft³)

$\gamma_{ng} = 0.0475$ (lb/ft³)

$w_a := \gamma_a \cdot Q_a$ weight rate of flow of air (lb/sec)

$w_a = 2.7146$ (lb/sec)

$w_w := SG_w \cdot \gamma_w \cdot Q_w$ weight rate of flow of "mist" water (lb/sec)

$w_w = 0.0487$ (lb/sec)

$w_{ng} := \gamma_{ng} \cdot Q_{ng}$ weight rate of flow of natural gas (lb/sec)

$w_{ng} = 0$ (lb/sec)

$w_{fw} := SG_{fw} \cdot \gamma_w \cdot Q_{fw}$ weight rate of flow of formation water (lb/sec)

$w_{fw} = 0$ (lb/sec)

$w_{gt} := w_a + w_{ng}$ total weight rate of flow of gas (lb/sec)

$w_{gt} = 2.7146$ (lb/sec)

Blooey line

$ID := \frac{id_{bl}}{12}$ convert blooey line inner diameter to ft

$ID = 0.6747$ (ft)

$A := \frac{\pi}{4} \cdot ID^2$ cross sectional area (ft²)

$A = 0.3576$ (ft²)

$$f := \left(\frac{1}{2\cdot\log\left(\frac{ID}{e_{cs}}\right) + 1.14}\right)^2$$

friction factor using von Karman correlation

$f = 0.0183$

$$\frac{id_c}{od_{dp}} = 1.7602$$

compute D1/D3 ratio for Fig.8-5

$$\frac{id_c}{id_{bl}} = 0.9783$$

compute D1/D2 ratio for Fig. 8-5

Figure 8.5 gives the following flow resistance coefficient for the blind Tee:

$K_t := 25$

$K_v := 0.2$ loss factor for gate valves in blooey line

$$P_b := \left[\left[\left(f\cdot\frac{L_{bl}}{ID}\right) + K_t + 2\cdot K_v\right]\cdot\left[\frac{\left(w_{gt}{}^2\cdot R\cdot T_{at}\right)}{g\cdot A^2}\right] + P_{at}{}^2\right]^{0.5}$$

calculate pressure of the gas at the entrance to the blooey line

$P_b = 2195.6749$ (lb/ft²)

$$p_b := \frac{P_b}{144}$$

convert to lb/in²

$p_b = 15.2477$ (psia)

Annulus cased hole drill pipe body section

$H := H_1$ (ft)

$$ID := \frac{id_c}{12}$$ convert inner diameter of casing to ft

$ID = 0.6601$ (ft)

$$OD := \frac{od_{dp}}{12}$$ convert outer diameter of drill pipe to ft

$OD = 0.375$ (ft)

$$A := \frac{\pi}{4} \cdot \left(ID^2 - OD^2\right)$$ annulus area between casing and drill pipe (ft^2)

$A = 0.2318$ (ft^2)

$$a_a := \frac{SG_a}{R} \cdot \left[1 + \frac{\left(w_s + w_w + w_{fw}\right)}{w_{gt}}\right]$$ calculate constant

$a_a = 0.0256$

$$f := \left[\frac{1}{2 \cdot \log\left[\frac{(ID - OD)}{e_{cs}}\right] + 1.14}\right]^2$$ friction factor using von Karman correlation

$f = 0.0226$

$$b_a := \left[\left[\frac{f}{2 \cdot g \cdot (ID - OD)}\right] \cdot \left(\frac{R}{SG_a}\right)^2 \cdot \frac{w_{gt}^{\ 2}}{\left(\frac{\pi}{4}\right)^2 \cdot \left(ID^2 - OD^2\right)^2}\right]$$ calculate constant

$b_a = 480.8635$

$$T := T_{at} + \beta \cdot H_1$$ temperature at bottom of section (°R)

$T = 570.91$ (°R)

$T_{ave1} := \frac{(T_{at} + T)}{2}$ average temperature of section (°R)

$T_{ave1} = 537.66$ (°R)

$T_{ave} := T_{ave1}$

$$P_1 := \left[\left[\left(P_b^{\;2} + b_a \cdot T_{ave}^{\;2}\right) \cdot e^{\left[\frac{(2 \cdot a_a \cdot H)}{T_{ave}}\right]}\right] - b_a \cdot T_{ave}^{\;2}\right]^{0.5}$$ calculate pressure

$P_1 = 11495.7701$ (lb/ft²)

$p_1 := \frac{P_1}{144}$ convert to psia

$p_1 = 79.8317$ (psia)

$\gamma := \frac{P_1 \cdot SG_a}{R \cdot T}$ specific weight of gas (lb/ft³)

$\gamma = 0.3774$ (lb/ft³)

$V := \frac{w_{gt}}{\gamma \cdot A}$ velocity (ft/sec)

$V = 31.04$ (ft/sec)

$KE_1 := \frac{1}{2} \cdot \frac{\gamma}{g} \cdot V^2$ kinetic energy of gas flow (lb-ft/ft³)

$KE_1 = 5.6456$ (lb-ft/ft³)

Annulus cased hole tool joint section

$H := H_2$

$$ID := \frac{id_c}{12}$$ convert inner diameter of casing to ft

$ID = 0.6601$ (ft)

$$OD := \frac{od_{tj}}{12}$$ convert outer diameter of tool joint to ft

$OD = 0.5208$ (ft)

$$OD_t := \frac{od_{dp}}{12}$$ use drill pipe OD for kinetic energy to ft

$OD_t = 0.375$ (ft)

$$A := \frac{\pi}{4}\left(ID^2 - OD_t^{\,2}\right)$$ annulus area for tool joint KE (ft^2)

$A = 0.2318$ (ft^2)

$$a_a := \frac{SG_a}{R}\cdot\left[1 + \frac{\left(w_s + w_w + w_{fw}\right)}{w_{gt}}\right]$$ calculate constant

$a_a = 0.0256$

$$f := \left[\frac{1}{2\cdot\log\left[\frac{(ID - OD)}{e_{cs}}\right] + 1.14}\right]^2$$ friction factor using von Karman correlation

$f = 0.0275$

$$b_a := \left[\left[\frac{f}{2\cdot g\cdot(ID - OD)}\right]\cdot\left(\frac{R}{SG_a}\right)^2\cdot\frac{w_{gt}^{\,2}}{\left(\frac{\pi}{4}\right)^2\cdot\left(ID^2 - OD^2\right)^2}\right]$$ calculate constant

$b_a = 3858.1981$

$T_{temp} := T$

$T := T_{temp} + \beta \cdot H_2$ temperature at bottom of section (°R)

$T = 574.41$ (°R)

$T_{ave2} := \frac{(T_{temp} + T)}{2}$ average temperature of section (°R)

$T_{ave2} = 572.66$ (°R)

$T_{ave} := T_{ave2}$

$$P_2 := \left[\left[\left(P_1^{\,2} + b_a \cdot T_{ave}^{\,2}\right) \cdot e^{\left[\frac{(2 \cdot a_a \cdot H)}{T_{ave}}\right]}\right] - b_a \cdot T_{ave}^{\,2}\right]^{0.5}$$ calculate pressure

$P_2 = 13290.2784$ (lb/ft^2)

$p_2 := \frac{P_2}{144}$ convert to psia

$p_2 = 92.2936$ (psia)

$\gamma := \frac{P_2 \cdot SG_a}{R \cdot T}$ specific weight of gas (lb/ft^3)

$\gamma = 0.4336$ (lb/ft^3)

$V := \frac{W_{gt}}{\gamma \cdot A}$ velocity of gas (ft/sec)

$V = 27.0135$ (ft/sec)

$KE_2 := \frac{1}{2} \cdot \frac{\gamma}{g} \cdot V^2$ kinetic energy of gas flow (lb-ft/ft^3)

$KE_2 = 4.9133$ (lb-ft/ft^3)

Annulus of open hole drill pipe section

$H := H_3$

$ID := D_{bh}$ open hole diameter of borehole (ft)

$ID = 0.6563$ (ft)

$OD := \frac{od_{dp}}{12}$ convert diameter of drill pipe to ft

$OD = 0.375$ (ft)

$A := \frac{\pi}{4}\left(ID^2 - OD^2\right)$ area between borehole and drill pipe (ft[2])

$A = 0.2278$ (ft[2])

$a_a := \frac{SG_a}{R} \cdot \left[1 + \frac{\left(w_s + w_w + w_{fw}\right)}{w_{gt}}\right]$ calculate constant

$a_a = 0.0256$

$e_{ave} := \frac{\left(e_r ID^2 + e_{cs} OD^2\right)}{ID^2 + OD^2}$ weighted average roughness (ft)

$e_{ave} = 0.0077$ (ft)

$f := \left[\frac{1}{2 \cdot \log\left[\frac{(ID - OD)}{e_{ave}}\right] + 1.14}\right]^2$ friction factor using von Karman correlation

$f = 0.0549$

$$b_a := \left[\left[\frac{f}{2\cdot g\cdot(ID - OD)}\right]\cdot\left(\frac{R}{SG_a}\right)^2\cdot\frac{w_{gt}^{\ 2}}{\left(\frac{\pi}{4}\right)^2\cdot\left(ID^2 - OD^2\right)^2}\right]$$ calculate constant

$b_a = 1224.6798$

$T_{temp} := T$

$T := T_{temp} + \beta\cdot H_3$ temperature at bottom of section (°R)

$T = 598.16$

$$T_{ave3} := \frac{\left(T_{temp} + T\right)}{2}$$ average temperature of section (°R)

$T_{ave3} = 586.285$ (°R)

$T_{ave} := T_{ave3}$

$$P_3 := \left[\left[\left(P_2^{\ 2} + b_a\cdot T_{ave}^{\ 2}\right)\cdot e^{\left[\frac{\left(2\cdot a_a\cdot H\right)}{T_{ave}}\right]}\right] - b_a\cdot T_{ave}^{\ 2}\right]^{0.5}$$ calculate pressure

$P_3 = 17736.0029$ (lb/ft²)

$$p_3 := \frac{P_3}{144}$$ convert to psia

$p_3 = 123.1667$ (psia)

$$\gamma := \frac{P_3\cdot SG_a}{R\cdot T}$$ specific weight of gas (lb/ft³)

$\gamma = 0.5557$ (lb/ft³)

$V := \frac{w_{gt}}{\gamma \cdot A}$	velocity of gas (ft/sec)
$V = 21.4459$	(ft/sec)
$KE_3 := \frac{1}{2} \cdot \frac{\gamma}{g} \cdot V^2$	kinetic energy of gas flow (lb-ft/ft^3)
$KE_3 = 3.9685$	(lb-ft/ft^3)

Annulus open hole tool joint section

$H := H_4$	
$ID := D_{bh}$	open hole diameter of borehole (ft)
$ID = 0.6563$	(ft)
$OD := \frac{od_{tj}}{12}$	convert tool joint diameter to ft
$OD = 0.5208$	(ft)
$OD_t := \frac{od_{dp}}{12}$	drill pipe diameter (ft)
$OD_t = 0.375$	(ft)
$A := \frac{\pi}{4}\left(ID^2 - OD_t^2\right)$	annulus area for tool joint KE (ft^2)
$A = 0.2278$	(ft^2)
$a_a := \frac{SG_a}{R} \cdot \left[1 + \frac{\left(w_s + w_w + w_{fw}\right)}{w_{gt}}\right]$	calculate constant
$a_a = 0.0256$	

$$e_{ave} := \frac{\left(e_r ID^2 + e_{cs} OD^2\right)}{ID^2 + OD^2}$$ weighted average roughness (ft)

$e_{ave} = 0.0063$ (ft)

$$f := \left[\frac{1}{2\cdot\log\left[\frac{(ID - OD)}{e_{ave}}\right] + 1.14}\right]^2$$ friction factor using von Karman correlation

$f = 0.0692$

$$b_a := \left[\left[\frac{f}{2\cdot g\cdot(ID - OD)}\right]\cdot\left(\frac{R}{SG_a}\right)^2\cdot\frac{w_{gt}^{\;2}}{\left(\frac{\pi}{4}\right)^2\cdot\left(ID^2 - OD^2\right)^2}\right]$$ calculate constant

$b_a = 10627.4659$

$T_{temp} := T$

$T := T_{temp} + \beta\cdot H_4$ temperature at bottom of section (°R)

$T = 599.41$

$$T_{ave4} := \frac{\left(T_{temp} + T\right)}{2}$$ average temperature of section (°R)

$T_{ave4} = 598.785$ (°R)

$T_{ave} := T_{ave4}$

$$P_4 := \left[\left[\left(P_3^{\;2} + b_a\cdot T_{ave}^{\;2}\right)\cdot e^{\left[\frac{(2\cdot a_a\cdot H)}{T_{ave}}\right]}\right] - b_a\cdot T_{ave}^{\;2}\right]^{0.5}$$ calculate pressure

$P_4 = 18945.9611$ (lb/ft²)

$p_4 := \frac{P_4}{144}$ convert to psia

$p_4 = 131.5692$ (psia)

$\gamma := \frac{P_4 \cdot SG_a}{R \cdot T}$ specific weight of gas (lb/ft³)

$\gamma = 0.5923$ (lb/ft³)

$V := \frac{w_{gt}}{\gamma \cdot A}$ velocity of gas (ft/sec)

$V = 20.1183$ (ft/sec)

$KE_4 := \frac{1}{2} \cdot \frac{\gamma}{g} \cdot V^2$ kinetic energy of gas flow (lb-ft/ft³)

$KE_4 = 3.7228$ (ft-lb/ft³)

Annulus open hole drill collar section

$H := H_5$

$ID := D_{bh}$ open hole diameter of borehole (ft)

$ID = 0.6563$ (ft)

$OD := \frac{od_{dc}}{12}$ convert drill collar diameter to ft

$OD = 0.5208$ (ft)

$A := \frac{\pi}{4}\left(ID^2 - OD^2\right)$ drill pipe diameter (ft)

$A = 0.1252$ (ft)

$$a_a := \frac{SG_a}{R}\cdot\left[1 + \frac{(w_s + w_w + w_{fw})}{w_{gt}}\right]$$ calculate constant

$a_a = 0.0256$

$$e_{ave} := \frac{\left(e_r ID^2 + e_{cs} OD^2\right)}{ID^2 + OD^2}$$ weighted average roughness (ft)

$e_{ave} = 0.0063$ (ft)

$$f := \left[\frac{1}{2\cdot\log\left[\frac{(ID - OD)}{e_{ave}}\right] + 1.14}\right]^2$$ friction factor using von Karman correlation

$f = 0.0692$

$$b_a := \left[\left[\frac{f}{2\cdot g\cdot(ID - OD)}\right]\cdot\left(\frac{R}{SG_a}\right)^2\cdot\frac{w_{gt}^{\;2}}{\left(\frac{\pi}{4}\right)^2\cdot\left(ID^2 - OD^2\right)^2}\right]$$ calculate constant

$b_a = 10627.4659$

$T_{temp} := T$

$T := T_{temp} + \beta\cdot H_5$ temperature at bottom of section (°R)

$T = 604.41$

$$T_{ave5} := \frac{\left(T_{temp} + T\right)}{2}$$ average temperature of section (°R)

$T_{ave5} = 601.91$ (°R)

$T_{ave} := T_{ave5}$

$$P_5 := \left[\left[\left(P_4^{\ 2} + b_a \cdot T_{ave}^{\ 2}\right) \cdot e^{\left[\frac{(2 \cdot a_a \cdot H)}{T_{ave}}\right]}\right] - b_a \cdot T_{ave}^{\ 2}\right]^{0.5}$$

$P_5 = 23282.5904$ (lb/ft²)

$P_{bh} := P_5$

$$p_5 := \frac{P_5}{144}$$

$p_5 = 161.6847$ (psia)

$p_{bh} := p_5$ annulus bottom hole pressure (psia)

$$\gamma := \frac{P_5 \cdot SG_a}{R \cdot T}$$ specific weight of gas (lb/ft³)

$\gamma = 0.7219$ (lb/ft³)

$$V := \frac{w_{gt}}{\gamma \cdot A}$$ velocity of gas (ft/sec)

$V = 30.0373$ (ft/sec)

$$KE_5 := \frac{1}{2} \cdot \frac{\gamma}{g} \cdot V^2$$ kinetic energy of gas flow (lb-ft/ft³)

$KE_5 = 10.1139$ (ft-lb/ft³)

Bottom hole through the bit nozzles

$$D_n := \frac{d_n}{12}$$ convert nozzle diameter to ft

$D_n = 0.0583$ (ft)

$$A_n := N_n \cdot \left(\frac{\pi}{4}\right) \cdot D_n^{\ 2}$$ total flow area of nozzles in bit (ft²)

$A_n = 0.008$ (ft^2)

$\gamma := \frac{(P_5 \cdot SG_a)}{R \cdot T}$ specific weight of air (lb/ft^3)

$\gamma = 0.7219$ (lb/ft^3)

$$P := P_5 \cdot \left[1 + \frac{\left(\frac{w_{gt}}{A_n}\right)^2}{2 \cdot g \cdot \left(\frac{k}{k-1}\right) \cdot P_5 \cdot \gamma}\right]^{\frac{k}{k-1}}$$ calculate pressure

$P = 25843.134$ (lb/ft^2)

$P_{sonic} := \frac{P_{bh}}{0.528}$ verify subsonic flow

$P_{sonic} = 44095.8152$ (lb/ft^2) - flow is sonic since $P < P_{sonic}$

$p := \frac{P}{144}$ convert to psia

$p = 179.4662$ (psia)

Open hole inside drill string drill collar section

$D := \frac{id_{dc}}{12}$ convert diameter to ft

$D = 0.2344$ (ft)

$a_i := \left(\frac{SG_a}{R}\right) \cdot \left(1 + \frac{w_w}{w_{gt}}\right)$ calculate constant

$a_i = 0.0191$

$$f := \left(\frac{1}{2 \cdot \log\left(\frac{D}{e_{cs}}\right) + 1.14}\right)^2$$ friction factor using von Karman correlation

$f = 0.0238$

$$b_i := \left(\frac{f}{2\cdot g\cdot D}\right)\cdot\left(\frac{R}{SG_a}\right)^2\cdot\frac{w_{gt}^{\ 2}}{\left(\frac{\pi}{4}\right)^2\cdot D^4}$$ calculate constant

$b_i = 17776.0049$

$$P := \left[\frac{\left[P^2 + b_i\cdot T_{ave5}^{\ 2}\cdot\left[e^{\frac{(2\cdot a_i\cdot H_5)}{T_{ave5}}} - 1\right]\right]}{e^{\frac{(2\cdot a_i\cdot H_5)}{T_{ave5}}}}\right]^{(0.5)}$$ calculate pressure

$P = 29119.4535$ (lb/ft^2)

$$p := \frac{P}{144}$$ convert to psia

$p = 202.2184$ (psia)

Open hole inside drill string tool joint section

$$D := \frac{id_{tj}}{12}$$ convert diameter to ft

$D = 0.2292$ (ft)

$$a_i := \left(\frac{SG_a}{R}\right)\cdot\left(1 + \frac{w_w}{w_{gt}}\right)$$ calculate constant

$a_i = 0.0191$

$$f := \left(\frac{1}{2\cdot\log\left(\frac{D}{e_{cs}}\right) + 1.14}\right)^2$$ friction factor using von Karman correlation

$f = 0.0239$

$$b_i := \left(\frac{f}{2 \cdot g \cdot D}\right) \cdot \left(\frac{R}{SG_a}\right)^2 \cdot \frac{w_{gt}^2}{\left(\frac{\pi}{4}\right)^2 \cdot D^4}$$ calculate constant

$b_i = 20010.2715$

$$P := \left[\frac{\left[P^2 + b_i \cdot T_{ave4}^2 \cdot \left[e^{\frac{(2 \cdot a_i \cdot H_4)}{T_{ave4}}} - 1\right]\right]}{e^{\frac{(2 \cdot a_i \cdot H_4)}{T_{ave4}}}}\right]^{(0.5)}$$ calculate pressure

$P = 29968.8494$ (lb/ft^2)

$$p := \frac{P}{144}$$ convert to psia

$p = 208.117$ (psia)

Open hole inside drill string drill pipe body section

$$D := \frac{id_{dp}}{12}$$ convert diameter to ft

$D = 0.3188$ (ft)

$$a_i := \left(\frac{SG_a}{R}\right) \cdot \left(1 + \frac{w_w}{w_{gt}}\right)$$ calculate constant

$a_i = 0.0191$

$$f := \left(\frac{1}{2 \cdot \log\left(\frac{D}{e_{cs}}\right) + 1.14}\right)^2$$ friction factor using von Karman correlation

$f = 0.022$

$$b_i := \left(\frac{f}{2\cdot g\cdot D}\right)\cdot\left(\frac{R}{SG_a}\right)^2\cdot\frac{w_{gt}^{\,2}}{\left(\frac{\pi}{4}\right)^2\cdot D^4}$$ calculate constant

$b_i = 3519.4219$

$$P := \left[\frac{\left[P^2 + b_i\cdot T_{ave3}^{\,2}\cdot\left[e^{\frac{(2\cdot a_i\cdot H_3)}{T_{ave3}}} - 1\right]\right]}{e^{\frac{(2\cdot a_i\cdot H_3)}{T_{ave3}}}}\right]^{(0.5)}$$ calculate pressure

$P = 30704.3034$ (lb/ft^2)

$$p := \frac{P}{144}$$ convert to psia

$p = 213.2243$ (psia)

Cased hole inside drill string tool joint section

$$D := \frac{id_{tj}}{12}$$ convert diameter to ft

$D = 0.2292$ (ft)

$$a_i := \left(\frac{SG_a}{R}\right)\cdot\left(1 + \frac{w_w}{w_{gt}}\right)$$ calculate constant

$a_i = 0.0191$

$$f := \left(\frac{1}{2\cdot\log\left(\frac{D}{e_{cs}}\right) + 1.14}\right)^2$$ friction factor using von Karman correlation

$f = 0.0239$

$$b_i := \left(\frac{f}{2\cdot g\cdot D}\right)\cdot\left(\frac{R}{SG_a}\right)^2\cdot\frac{w_{gt}^{\ 2}}{\left(\frac{\pi}{4}\right)^2\cdot D^4}$$ calculate constant

$b_i = 20010.2715$

$$P := \left[\frac{\left[P^2 + b_i\cdot T_{ave2}^{\ 2}\cdot\left[e^{\frac{(2\cdot a_i\cdot H_2)}{T_{ave2}}} - 1\right]\right]}{e^{\frac{(2\cdot a_i\cdot H_2)}{T_{ave2}}}}\right]^{(0.5)}$$ calculate pressure

$P = 32745.6024$ (lbs/ft^2)

$$p := \frac{P}{144}$$ convert to psia

$p = 227.4$ (psia)

Cased hole inside drill string pipe body section

$$D := \frac{id_{dp}}{12}$$ convert diameter to ft

$D = 0.3188$ (ft)

$$a_i := \left(\frac{SG_a}{R}\right)\cdot\left(1 + \frac{w_w}{w_{gt}}\right)$$ calculate constant

$a_i = 0.0191$

$$f := \left(\frac{1}{2\cdot\log\left(\frac{D}{e_{cs}}\right) + 1.14}\right)^2$$ friction factor using von Karman correlation

$f = 0.022$

$$b_i := \left(\frac{f}{2\cdot g\cdot D}\right)\cdot\left(\frac{R}{SG_a}\right)^2\cdot\frac{w_{gt}^{\,2}}{\left(\frac{\pi}{4}\right)^2\cdot D^4}$$ calculate constant

$b_i = 3519.4219$

$$P := \left[\frac{\left[P^2 + b_i\cdot T_{ave1}^{\,2}\cdot\left[e^{\frac{(2\cdot a_i\cdot H_1)}{T_{ave1}}} - 1\right]\right]}{e^{\frac{(2\cdot a_i\cdot H_1)}{T_{ave1}}}}\right]^{0.5}$$ calculate pressure

$P = 32428.8057$ (lb/ft2)

$$p := \frac{P}{144}$$ convert to psia

$p = 225.2$ (psia)

$p_{in} := p$

$p_{in} = 225.2$ drilling string injection pressure (psia)

Determine compressor system fuel composition rate

$p_i := p_{at}$ input pressure (psia)

$p_o := p_{in}$ output pressure (psia)

$p_o = 225.2$ (psia)

$q_i := q_c$ input flow rate for one compressor

$$r_s := \left(\frac{p_o}{p_i}\right)^{\frac{1}{n_s}}$$ compression ratio for each stage

$r_s = 2.0527$

$$\varepsilon_v := 0.96\left[1 - c\cdot\left(r_s^{\frac{1}{k}} - 1\right)\right]$$ volumetric efficiency (recip only)

$\varepsilon_v = 0.9471$

$$W_s := \left[\frac{(n_s\cdot k)}{k-1}\right]\cdot\left[\frac{(p_i\cdot q_i)}{229.17}\right]\cdot\left[\left(\frac{p_o}{p_i}\right)^{\frac{(k-1)}{n_s\cdot k}} - 1\right]$$ theoretical shaft horsepower (HP)

$W_s = 212.1141$ (HP)

$$W_{as} := \frac{W_s}{\varepsilon_m\cdot\varepsilon_v}$$ actual shaft power (HP)

$W_{as} = 279.9495$ (HP)

Derate turbocharged prime mover for surface elevation of 4000 ft above sea

$\%R := 0.10$ % reduction required due to altitude (see Fig 5-15)

$HP_{alt} := HP_{max} - \%R\cdot HP_{max}$ maximum power available prime mover (HP)

$HP_{alt} = 684$ (HP)

Actual shaft power is less than prime mover maximum avialable, so OK

$$PR := \left(\frac{W_{as}}{HP_{alt}}\right)\cdot 100$$ power ratio of prime mover

$PR = 40.9283$

Use above PR with Fig. 5-16 (extrapolate) and obtain approximate rate of fuel consumption rate

$FC := 0.69$ fuel consumption (lb/HP-hr)

$w_f := FC \cdot W_{as}$ weight rate of fuel consumption (lb/hr)

$w_f = 193.1651$ (lb/hr)

$q_f := \frac{w_f}{SG_{df} \cdot \gamma_{wg}}$ fuel volume per hour for one compressor (gal/hr)

$q_f = 28.4319$ (gal/hr)

$q_{ftotal} := q_f \cdot num_c$ total fuel consumed for all compressors (gal/hr)

$q_{ftotal} = 56.8639$ diesel fuel rate (gal/hr)

Kinetic energy summary of each section (KE greater than 3.0 lb-ft/ft^3)

$KE_1 = 5.6456$ (lb-ft/ft^3)

$KE_2 = 4.9133$ (lb-ft/ft^3)

$KE_3 = 3.9685$ (lb-ft/ft^3)

$KE_4 = 3.7228$ (lb-ft/ft^3)

$KE_5 = 10.1139$ (lb-ft/ft^3)

Important design parameter summary

$q_a = 2400$ air compressor flow rate (acfm)

$p_{in} = 225.2$ required surface air injection pressure (psia)

$p_{bh} = 161.6847$ bottom hole pressure (psia)

$W_{as} = 279.9495$ actual shaft HP required

$HP_{alt} = 684$ derated HP available

$q_{ftotal} = 56.8639$ total volume per hour of fuel consumed (gal/hr)

Water injection to standpipe to saturate gas flow at bottom hole annulus conditions

$T = 604.41$ bottom hole temperature (°R)

$t_{bh} := T - 459.67$ convert to °F

$t_{bh} = 144.74$ °F

$$p_{sat} := 10^{6.39416 - \left(\frac{1750.286}{217.23 + 0.55 \cdot t_{bh}}\right)}$$ gas-saturation pressure of water (psia)

$p_{sat} = 3.1456$ (psia)

$$q_{winj} := \left(\frac{p_{sat}}{p_{bh} - p_{sat}}\right) \cdot \left(\frac{18.02}{28.96}\right) \cdot \left(\frac{3600}{8.33}\right) \cdot w_a$$

$q_{winj} = 14.4842$ required water injection to saturate gas at bottom hole conditions (gal/hr)

Illustrative Example 8.2: SI version, Minimum Volumetric Flow Rate - Drilling Fluid is Air. Determining the bottom hole and injection pressures while drilling at 3047.9 m of depth for the drilling operations described in Illustrative Examples 8.1.

Minimum volumetric flow rate (trial and error to attain minimum KE)

$q_a := 940.97$ flow rate produced from one compressor (actual liters/sec)

Basic standard and elevation atomospheric conditions

$p_{st} := 10.136$ API standard atmospheric pressure (N/cm^2 abs)

$t_{st} := 15.6$ API standard atmospheric temperature (°C)

$alt := 1219.14$ surface elevation of drilling site (m)

$p_{at} := 8.749$ pressure from table for surface elevation (N/cm^2 abs)

$t_{at} := 7.078$ temperature from table for surface elevation (°C)

$P_{st} := p_{st} \cdot 10000$ convert API standard pressure to N/m^2

$P_{st} = 101360$ (N/m^2 abs)

$T_{st} := t_{st} + 273.15$ convert API standard temp to °C

$T_{st} = 288.75$ (°C)

$P_{at} := p_{at} \cdot 10000$ convert actual atm pressure to N/m^2

$P_{at} = 87490$ (N/m^2 abs)

$T_{at} := t_{at} + 273.15$ convert actual atm temp to °C

Borehole geometry

$H_c := 2133.50$ total depth of cased hole (m)

$H_{oh} := 914.36$ total depth of open hole (m)

$ROP := 18.29$	estimated rate of penetration (m/hr)
$d_{bh} := 200.03$	borehole (drill bit) diameter (mm)
$od_{dc} := 158.75$	outer diameter in drill collar (mm)
$id_{dc} := 71.438$	inner diameter in drill collar (mm)
$od_{dp} := 114.30$	outer diameter in drill pipe (mm)
$id_{dp} := 97.180$	inner diameter in drill pipe (mm)
$od_{tj} := 158.75$	outer diameter in drill pipe tool joint (mm)
$id_{tj} := 69.85$	inside diameter in drill pipe tool joint (mm)
$od_{c} := 219.08$	outer diameter in drill casing (mm)
$id_{c} := 201.19$	inside diameter in drill casing (mm)
$od_{bl} := 219.08$	outer diameter in blooey line (mm)
$id_{bl} := 205.66$	inside diameter in blooey line (mm)
$L_{bl} := 60.957$	length of blooey line (m)
$e_{cs} := 0.000152939$	roughness of commercial steel (m)
$e_{r} := 0.003048$	roughness of rock table (m)
$L_{dc} := 152.39$	total length of drill collars (m)
$L_{tj} := 0.457$	average length of one tool joint (m)
$L_{dp} := 9.144$	average length of one drill pipe section (m)
$d_{n} := 17.78$	nozzle diameter (mm)
$N_{n} := 3$	number of nozzles in the bit

$\beta := 0.01823$ temperature gradient (°C/m)

Section height calculations

$H_1 := H_c - L_{tj} \cdot \left(\frac{H_c}{L_{dp}} \right)$ calculate length of drill pipe in cased section

$H_1 = 2026.8717$ (m)

$H_2 := L_{tj} \cdot \left(\frac{H_c}{L_{dp}} \right)$ calculate length of tool joint in cased section

$H_2 = 106.6283$ (m)

$H_3 := H_{oh} - L_{dc} - L_{tj} \cdot \left(\frac{H_{oh} - L_{dc}}{L_{dp}} \right)$ calculate length of drill pipe in open hole section

$H_3 = 723.8882$ (m)

$H_4 := L_{tj} \cdot \left(\frac{H_{oh} - L_{dc}}{L_{dp}} \right)$ calculate length of tool joints in open hole section

$H_4 = 38.0818$ (m)

$H_5 := L_{dc}$ length of drill collar

$H_5 = 152.39$ (m)

Formation natural gas and water influx

$q_w := 0.0$ injected "mist" water (bbl/hr)

$q_{ng} := 0$ natural gas influx (liters/sec)

$q_{fw} := 0$ formation water influx (bbl/hr)

Constants

$SG_a := 1.0$ specific gravity of air

$SG_w := 1.0$ specific gravity of fresh water

$SG_{ng} := 0.70$ specific gravity of natural gas

$SG_{fw} := 1.07$ specific gravity of formation water

$SG_r := 2.7$ specific gravity of rock

$SG_{df} := 0.8156$ specific gravity of diesel fuel

$k := 1.4$ gas ratio of specific heats

$R := 29.31$ engineering gas constant (N-m/N-°C)

$\rho_w := 1000.0$ density of fresh water (kg/m^3)

$g := 9.81$ gravitational constant (m/sec^2)

$\gamma_w := \rho_w \cdot g$

$\gamma_w = 9810$ specific weight of fresh water (N/m^3)

Preliminary calculations

$D_{bh} := \frac{d_{bh}}{1000}$ convert borehole diameter to m

$D_{bh} = 0.2$ (m)

$rop := \frac{ROP}{60 \cdot 60}$ convert ROP to m/sec

$rop = 0.0051$ (m/sec)

$w_s := \left[\left(\frac{\pi}{4}\right) \cdot D_{bh}^{\ 2} \cdot \gamma_w \cdot SG_r \cdot rop\right]$ weight rate of solid flow out of blooey line (lb/sec)

$w_s = 4.2289$ (N/sec)

$Q_a := \frac{q_a}{1000}$ convert air flow rate to m^3/sec

$Q_a = 0.941$ (m^3/sec)

$$Q_w := \frac{(q_w \cdot 42 \cdot 0.003785)}{60 \cdot 60}$$ convert "mist" water flow to m^3/sec

$Q_w = 0$ (m^3/sec)

$$Q_{ng} := \frac{q_{ng}}{1000}$$ convert natural gas flow rate to m^3/sec

$Q_{ng} = 0$ (m^3/sec)

$Q_{gt} := Q_a + Q_{ng}$ total flow of the gas (m^3/sec)

$Q_{gt} = 0.941$ (m^3/sec)

$$Q_{fw} := \frac{(q_{fw} \cdot 42 \cdot 0.003785)}{60 \cdot 60}$$ convert formation water flow to m^3/sec

$Q_{fw} = 0$ (m^3/sec)

$$\gamma_a := \frac{(P_{at} \cdot SG_a)}{R \cdot T_{at}}$$ specific weight of air (N/m^3)

$\gamma_a = 10.652$ (N/m^3)

$$\gamma_{ng} := \frac{(P_{at} \cdot SG_{ng})}{R \cdot T_{at}}$$ specific weight of natural gas (N/m^3)

$\gamma_{ng} = 7.4564$ (N/m^3)

$w_a := \gamma_a \cdot Q_a$ weight rate of flow of air (N/sec)

$w_a = 10.0232$ (N/sec)

$w_w := SG_w \cdot \gamma_w \cdot Q_w$ weight rate of flow of "mist" water (lb/sec)

$w_w = 0$ (N/sec)

$w_{ng} := \gamma_{ng} \cdot Q_{ng}$ weight rate of flow of natural gas (N/sec)

$w_{ng} = 0$ (N/sec)

$w_{fw} := SG_{fw} \cdot \gamma_w \cdot Q_{fw}$ weight rate of flow of formation water (N/sec)

$w_{fw} = 0$ (N/sec)

$w_{gt} := w_a + w_{ng}$ total weight rate of flow of gas (N/sec)

$w_{gt} = 10.0232$ (N/sec)

Blooey line

$ID := \frac{id_{bl}}{1000}$ convert blooey line inner diameter to m

$ID = 0.2057$ (m)

$A := \frac{\pi}{4} \cdot ID^2$ cross sectional area (m^2)

$A = 0.0332$ (m^2)

$f := \left(\frac{1}{2 \cdot \log\left(\frac{ID}{e_{cs}}\right) + 1.14} \right)^2$ friction factor using von Karman correlation

$f = 0.0183$

$\frac{id_c}{od_{dp}} = 1.7602$ compute D1/D3 ratio for Fig. 8-5

$\frac{id_c}{id_{bl}} = 0.9783$ compute D1/D2 ratio for Fig. 8-5

Figure 8.5 gives the following flow resistance coefficient for the blind Tee:

$K_t := 25$

$K_v := 0.2$ loss factor for gate valves in blooey line

$$P_b := \left[\left[\left[\left(f\cdot\frac{L_{bl}}{ID}\right) + K_t + 2\cdot K_v\right]\cdot\left[\frac{\left(w_{gt}^{\ 2}\cdot R\cdot T_{at}\right)}{g\cdot A^2}\right] + P_{at}^{\ 2}\right]^{0.5}\right]$$

calculate pressure of the gas at the entrance to the blooey line

$P_b = 100017.3774$ (N/m2)

$p_b := \frac{P_b}{10000.0}$ convert to N/cm2

$p_b = 10.0017$ (N/cm2 abs)

Annulus cased hole drill pipe body section

$H := H_1$ (m)

$ID := \frac{id_c}{1000}$ convert inner diameter of casing to m

$ID = 0.2012$ (m)

$OD := \frac{od_{dp}}{1000}$ convert outer diameter of drill pipe to m

$OD = 0.1143$ (m)

$A := \frac{\pi}{4}\cdot\left(ID^2 - OD^2\right)$ annulus area between casing and drill pipe (ft2)

$A = 0.0215$ (m2)

$$a_a := \frac{SG_a}{R} \cdot \left[1 + \frac{(w_s + w_w + w_{fw})}{w_{gt}} \right]$$ calculate constant

$a_a = 0.0485$

$$f := \left[\frac{1}{2 \cdot \log\left[\frac{(ID - OD)}{e_{cs}} \right] + 1.14} \right]^2$$ friction factor using von Karman correlation

$f = 0.0226$

$$b_a := \left[\left[\frac{f}{2 \cdot g \cdot (ID - OD)} \right] \cdot \left(\frac{R}{SG_a} \right)^2 \cdot \frac{w_{gt}^2}{\left(\frac{\pi}{4} \right)^2 \cdot \left(ID^2 - OD^2 \right)^2} \right]$$ calculate constant

$b_a = 2470506.3377$

$T := T_{at} + \beta \cdot H_1$ temperature at bottom of section (K)

$T = 317.1779$ (K)

$$T_{ave1} := \frac{(T_{at} + T)}{2}$$ average temperature of section (K)

$T_{ave1} = 298.7029$ (K)

$T_{ave} := T_{ave1}$

$$P_1 := \left[\left[\left(P_b^2 + b_a \cdot T_{ave}^2 \right) \cdot e^{\left[\frac{(2 \cdot a_a \cdot H)}{T_{ave}} \right]} \right] - b_a \cdot T_{ave}^2 \right]^{0.5}$$ calculate pressure

$P_1 = 474008.0178$ (N/m^2)

$p_1 := \frac{P_1}{10000.0}$ convert to N/cm² abs

$p_1 = 47.4008$ (N/cm² abs)

$\gamma := \frac{P_1 \cdot SG_a}{R \cdot T}$ specific weight of gas (N/m³)

$\gamma = 50.9879$ (N/m³)

$V := \frac{w_{gt}}{\gamma \cdot A}$ velocity (m/sec)

$V = 9.1305$ (m/sec)

$KE_1 := \frac{1}{2} \cdot \frac{\gamma}{g} \cdot V^2$ kinetic energy of gas flow (N-m/m³)

$KE_1 = 216.6492$ (N-m/m³)

Annulus cased hole tool joint section

$H := H_2$

$ID := \frac{id_c}{1000}$ convert inner diameter of casing to m

$ID = 0.2012$ (m)

$OD := \frac{od_{tj}}{1000}$ convert outer diameter of tool joint to m

$OD = 0.1587$ (m)

$OD_t := \frac{od_{dp}}{1000}$ use drill pipe OD for kinetic energy to m

$OD_t = 0.1143$ (m)

$$A := \frac{\pi}{4}\left(ID^2 - OD_t^2\right)$$ annulus area for tool joint KE (m^2)

$A = 0.0215$ (m^2)

$$a_a := \frac{SG_a}{R}\cdot\left[1 + \frac{\left(w_s + w_w + w_{fw}\right)}{w_{gt}}\right]$$ calculate constant

$a_a = 0.0485$

$$f := \left[\frac{1}{2\cdot\log\left[\frac{(ID - OD)}{e_{cs}}\right] + 1.14}\right]^2$$ friction factor using von Karman correlation

$f = 0.0275$

$$b_a := \left[\left[\frac{f}{2\cdot g\cdot(ID - OD)}\right]\cdot\left(\frac{R}{SG_a}\right)^2\cdot\frac{w_{gt}^{\;2}}{\left(\frac{\pi}{4}\right)^2\cdot\left(ID^2 - OD^2\right)^2}\right]$$ calculate constant

$b_a = 19826581.549$

$T_{temp} := T$

$T := T_{temp} + \beta\cdot H_2$ temperature at bottom of section (K)

$T = 319.1217$ (K)

$$T_{ave2} := \frac{\left(T_{temp} + T\right)}{2}$$ average temperature of section (K)

$T_{ave2} = 318.1498$ (K)

$T_{ave} := T_{ave2}$

$$P_2 := \left[\left[\left(P_1^2 + b_a \cdot T_{ave}^2 \right) \cdot e^{\left[\frac{\left(2 \cdot a_a \cdot H \right)}{T_{ave}} \right]} \right] - b_a \cdot T_{ave}^2 \right]^{0.5}$$ calculate pressure

$P_2 = 546297.743$ (N/m^2)

$p_2 := \dfrac{P_2}{10000.0}$ convert to N/cm^2 abs

$p_2 = 54.6298$ N/cm^2 abs

$\gamma := \dfrac{P_2 \cdot SG_a}{R \cdot T}$ specific weight of gas (N/m^3)

$\gamma = 58.406$ (N/m^3)

$V := \dfrac{w_{gt}}{\gamma \cdot A}$ velocity of gas (m/sec)

$V = 7.9708$ (m/sec)

$KE_2 := \dfrac{1}{2} \cdot \dfrac{\gamma}{g} \cdot V^2$ kinetic energy of gas flow (N-m/m^3)

$KE_2 = 189.1328$ (N-m/m^3)

Annulus of open hole drill pipe section

$H := H_3$

$ID := D_{bh}$ open hole diameter of borehole (m)

$ID = 0.2$ (m)

$OD := \dfrac{od_{dp}}{1000}$ convert diameter of drill pipe to m

$OD = 0.1143$ (m)

$$A := \frac{\pi}{4}\left(ID^2 - OD^2\right)$$ area between borehole and drill pipe (m^2)

$$A = 0.0212$$ (m^2)

$$a_a := \frac{SG_a}{R}\cdot\left[1 + \frac{\left(w_s + w_w + w_{fw}\right)}{w_{gt}}\right]$$ calculate constant

$$a_a = 0.0485$$

$$e_{ave} := \frac{\left(e_r ID^2 + e_{cs} OD^2\right)}{ID^2 + OD^2}$$ weighted average roughness (m)

$$e_{ave} = 0.0023$$ (m)

$$f := \left[\frac{1}{2\cdot\log\left[\frac{(ID - OD)}{e_{ave}}\right] + 1.14}\right]^2$$ friction factor using von Karman correlation

$$f = 0.0549$$

$$b_a := \left[\left[\frac{f}{2\cdot g\cdot(ID - OD)}\right]\cdot\left(\frac{R}{SG_a}\right)^2\cdot\frac{w_{gt}^{\,2}}{\left(\frac{\pi}{4}\right)^2\cdot\left(ID^2 - OD^2\right)^2}\right]$$ calculate constant

$$b_a = 6283959.2527$$

$$T_{temp} := T$$

$$T := T_{temp} + \beta\cdot H_3$$ temperature at bottom of section (K)

$$T = 332.3182$$

$$T_{ave3} := \frac{\left(T_{temp} + T\right)}{2}$$ average temperature of section (K)

$T_{ave3} = 325.7199$ (K)

$T_{ave} := T_{ave3}$

$$P_3 := \left[\left[\left(P_2^{\,2} + b_a \cdot T_{ave}^{\,2}\right) \cdot e^{\left[\frac{(2 \cdot a_a \cdot H)}{T_{ave}}\right]}\right] - b_a \cdot T_{ave}^{\,2}\right]^{0.5}$$ calculate pressure

$P_3 = 728488.1033$ (N/m²)

$p_3 := \dfrac{P_3}{10000.0}$ convert to N/cm² abs

$p_3 = 72.8488$ (N/cm² abs)

$\gamma := \dfrac{P_3 \cdot SG_a}{R \cdot T}$ specific weight of gas (N/m³)

$\gamma = 74.7916$ (N/m³)

$V := \dfrac{w_{gt}}{\gamma \cdot A}$ velocity of gas (m/sec)

$V = 6.3321$ (m/sec)

$KE_3 := \dfrac{1}{2} \cdot \dfrac{\gamma}{g} \cdot V^2$ kinetic energy of gas flow (N-m/m³)

$KE_3 = 152.8428$ (N-m/m³)

Annulus open hole tool joint section

$H := H_4$

$ID := D_{bh}$ open hole diameter of borehole (m)

$ID = 0.2$ (m)

$OD := \frac{od_{tj}}{1000}$ convert tool joint diameter to m

$OD = 0.1587$ (m)

$OD_t := \frac{od_{dp}}{1000}$ drill pipe diameter (m)

$OD_t = 0.1143$ (m)

$A := \frac{\pi}{4}\left(ID^2 - OD_t^2\right)$ annulus area for tool joint KE (m^2)

$A = 0.0212$ (m^2)

$a_a := \frac{SG_a}{R} \cdot \left[1 + \frac{\left(w_s + w_w + w_{fw}\right)}{w_{gt}}\right]$ calculate constant

$a_a = 0.0485$

$e_{ave} := \frac{\left(e_r ID^2 + e_{cs} OD^2\right)}{ID^2 + OD^2}$ weighted average roughness (m)

$e_{ave} = 0.0019$ (m)

$f := \left[\frac{1}{2 \cdot \log\left[\frac{(ID - OD)}{e_{ave}}\right] + 1.14}\right]^2$ friction factor using von Karman correlation

$f = 0.0692$

$b_a := \left[\left[\frac{f}{2 \cdot g \cdot (ID - OD)}\right] \cdot \left(\frac{R}{SG_a}\right)^2 \cdot \frac{w_{gt}^{\,2}}{\left(\frac{\pi}{4}\right)^2 \cdot \left(ID^2 - OD^2\right)^2}\right]$ calculate constant

$b_a = 54520466.641$

$T_{temp} := T$

$T := T_{temp} + \beta \cdot H_4$ temperature at bottom of section (K)

$T = 333.0124$

$T_{ave4} := \frac{(T_{temp} + T)}{2}$ average temperature of section (K)

$T_{ave4} = 332.6653$ (K)

$T_{ave} := T_{ave4}$

$$P_4 := \left[\left[\left(P_3^{\,2} + b_a \cdot T_{ave}^{\,2}\right) \cdot e^{\left[\frac{(2 \cdot a_a \cdot H)}{T_{ave}}\right]}\right] - b_a \cdot T_{ave}^{\,2}\right]^{0.5}$$ calculate pressure

$P_4 = 777181.1212$ (N/m^2)

$p_4 := \frac{P_4}{10000.0}$ convert to N/cm^2 abs

$p_4 = 77.7181$ (N/cm^2 abs)

$\gamma := \frac{P_4 \cdot SG_a}{R \cdot T}$ specific weight of gas (N/m^3)

$\gamma = 79.6244$ (N/m^3)

$V := \frac{w_{gt}}{\gamma \cdot A}$ velocity of gas (m/sec)

$V = 5.9477$ (m/sec)

$KE_4 := \frac{1}{2} \cdot \frac{\gamma}{g} \cdot V^2$ kinetic energy of gas flow ($N\text{-}m/m^3$)

$KE_4 = 143.566$ (N-m/m³)

Annulus open hole drill collar section

$H := H_5$

$ID := D_{bh}$ open hole diameter of borehole (m)

$ID = 0.2$ (m)

$$OD := \frac{od_{dc}}{1000}$$ convert drill collar diameter to m

$OD = 0.1587$ (m)

$$A := \frac{\pi}{4}\left(ID^2 - OD^2\right)$$ drill pipe diameter (m²)

$A = 0.0116$ (m²)

$$a_a := \frac{SG_a}{R} \cdot \left[1 + \frac{\left(w_s + w_w + w_{fw}\right)}{w_{gt}}\right]$$ calculate constant

$a_a = 0.0485$

$$e_{ave} := \frac{\left(e_r ID^2 + e_{cs} OD^2\right)}{ID^2 + OD^2}$$ weighted average roughness (m)

$e_{ave} = 0.0019$ (m)

$$f := \left[\frac{1}{2 \cdot \log\left[\frac{(ID - OD)}{e_{ave}}\right] + 1.14}\right]^2$$ friction factor using von Karman correlation

$f = 0.0692$

$$b_a := \left[\left[\frac{f}{2\cdot g\cdot(ID - OD)}\right]\cdot\left(\frac{R}{SG_a}\right)^2\cdot\frac{w_{gt}^2}{\left(\frac{\pi}{4}\right)^2\cdot\left(ID^2 - OD^2\right)^2}\right]$$ calculate constant

$b_a = 54520466.641$

$T_{temp} := T$

$T := T_{temp} + \beta\cdot H_5$ temperature at bottom of section (K)

$T = 335.7905$

$$T_{ave5} := \frac{(T_{temp} + T)}{2}$$ average temperature of section (K)

$T_{ave5} = 334.4015$ (K)

$T_{ave} := T_{ave5}$

$$P_5 := \left[\left[\left(P_4^2 + b_a\cdot T_{ave}^2\right)\cdot e^{\left[\frac{(2\cdot a_a\cdot H)}{T_{ave}}\right]}\right] - b_a\cdot T_{ave}^2\right]^{0.5}$$

$P_5 = 952330.575$ (N/m^2)

$$p_5 := \frac{P_5}{1000}$$

$p_5 = 952.3306$ (N/cm^2 abs)

$p_{bh} := p_5$ annulus bottom hole pressure (N/cm^2 abs)

$$\gamma := \frac{P_5 \cdot SG_a}{R \cdot T}$$ specific weight of gas (N/m³)

$\gamma = 96.7617$ (N/m³)

$$V := \frac{w_{gt}}{\gamma \cdot A}$$ velocity of gas (m/sec)

$V = 8.9052$ (m/sec)

$$KE_5 := \frac{1}{2} \cdot \frac{\gamma}{g} \cdot V^2$$ kinetic energy of gas flow (N-m/m³)

$KE_5 = 391.107$ (N-m/m³)

Bottom hole through the bit nozzles

$$D_n := \frac{d_n}{1000}$$ convert nozzle diameter to m

$D_n = 0.0178$ (m)

$$A_n := N_n \cdot \left(\frac{\pi}{4}\right) \cdot D_n^{\,2}$$ total flow area of nozzles in bit (m²)

$A_n = 0.0007$ (m²)

$$\gamma := \frac{(P_5 \cdot SG_a)}{R \cdot T}$$ specific weight of air (N/m³)

$\gamma = 96.7617$ (N/m³)

$$P := P_5 \cdot \left[1 + \frac{\left(\frac{w_{gt}}{A_n}\right)^2}{2 \cdot g \cdot \left(\frac{k}{k-1}\right) \cdot P_5 \cdot \gamma} \right]^{\frac{k}{k-1}}$$ calculate pressure

$P = 1051172.3212$ (N/m²)

$$p := \frac{P}{10000.0}$$ convert to N/cm^2 abs

$$p = 105.1172$$ (N/cm^2 abs)

Open hole inside drill string drill collar section

$$D := \frac{id_{dc}}{1000}$$ convert diameter to m

$$D = 0.0714$$ (m)

$$a_i := \left(\frac{SG_a}{R}\right)\cdot\left(1 + \frac{w_w}{w_{gt}}\right)$$ calculate constant

$$a_i = 0.0341$$

$$f := \left(\frac{1}{2\cdot\log\left(\frac{D}{e_{cs}}\right) + 1.14}\right)^2$$ friction factor using von Karman correlation

$$f = 0.0238$$

$$b_i := \left(\frac{f}{2\cdot g\cdot D}\right)\cdot\left(\frac{R}{SG_a}\right)^2\cdot\frac{w_{gt}^{\;2}}{\left(\frac{\pi}{4}\right)^2\cdot D^4}$$ calculate constant

$$b_i = 91312035.3174$$

$$P := \left[\frac{\left[P^2 + b_i\cdot T_{ave5}^{\;2}\cdot\left[e^{\frac{(2\cdot a_i\cdot H_5)}{T_{ave5}}} - 1\right]\right]}{e^{\frac{(2\cdot a_i\cdot H_5)}{T_{ave5}}}}\right]^{(0.5)}$$ calculate pressure

$$P = 1176334.6175$$ (N/m^2)

$$p := \frac{P}{10000.0}$$ convert to N/cm^2 abs

$p = 117.6335$ (N/cm² abs)

Open hole inside drill string tool joint section

$$D := \frac{id_{tj}}{1000}$$ convert diameter to m

$D = 0.0698$ (m)

$$a_i := \left(\frac{SG_a}{R}\right)\cdot\left(1 + \frac{w_w}{w_{gt}}\right)$$ calculate constant

$a_i = 0.0341$

$$f := \left(\frac{1}{2\cdot\log\left(\frac{D}{e_{cs}}\right) + 1.14}\right)^2$$ friction factor using von Karman correlation

$f = 0.024$

$$b_i := \left(\frac{f}{2\cdot g\cdot D}\right)\cdot\left(\frac{R}{SG_a}\right)^2\cdot\frac{w_{gt}^2}{\left(\frac{\pi}{4}\right)^2\cdot D^4}$$ calculate constant

$b_i = 102793131.7931$

$$P := \left[\frac{\left[P^2 + b_i\cdot T_{ave4}^2\cdot\left[e^{\frac{(2\cdot a_i\cdot H_4)}{T_{ave4}}} - 1\right]\right]}{e^{\frac{(2\cdot a_i\cdot H_4)}{T_{ave4}}}}\right]^{(0.5)}$$ calculate pressure

$P = 1208929.0798$ (N/m²)

$$p := \frac{P}{10000.0}$$ convert to N/cm² abs

$p = 120.8929$ (N/cm² abs)

Open hole inside drill string drill pipe body section

$$D := \frac{id_{dp}}{1000}$$ convert diameter to m

$D = 0.0972$ (m)

$$a_i := \left(\frac{SG_a}{R}\right)\cdot\left(1 + \frac{w_w}{w_{gt}}\right)$$ calculate constant

$a_i = 0.0341$

$$f := \left(\frac{1}{2\cdot\log\left(\frac{D}{e_{cs}}\right) + 1.14}\right)^2$$ friction factor using von Karman correlation

$f = 0.022$

$$b_i := \left(\frac{f}{2\cdot g\cdot D}\right)\cdot\left(\frac{R}{SG_a}\right)^2\cdot\frac{w_{gt}^2}{\left(\frac{\pi}{4}\right)^2\cdot D^4}$$ calculate constant

$b_i = 18078996.3818$

$$P := \left[\frac{\left[P^2 + b_i\cdot T_{ave3}^{\,2}\cdot\left[e^{\frac{(2\cdot a_i\cdot H_3)}{T_{ave3}}} - 1\right]\right]}{e^{\frac{(2\cdot a_i\cdot H_3)}{T_{ave3}}}}\right]^{(0.5)}$$ calculate pressure

$P = 1235213.1684$ (N/m2)

$$p := \frac{P}{10000.0}$$ convert to N/cm2 abs

$p = 123.5213$ (N/cm2 abs)

Cased hole inside drill string tool joint section

$$D := \frac{id_{tj}}{1000}$$ convert diameter to m

$D = 0.0698$ (m)

$$a_i := \left(\frac{SG_a}{R}\right)\cdot\left(1 + \frac{w_w}{w_{gt}}\right)$$ calculate constant

$a_i = 0.0341$

$$f := \left(\frac{1}{2\cdot\log\left(\frac{D}{e_{cs}}\right) + 1.14}\right)^2$$ friction factor using von Karman correlation

$f = 0.024$

$$b_i := \left(\frac{f}{2\cdot g\cdot D}\right)\cdot\left(\frac{R}{SG_a}\right)^2\cdot\frac{w_{gt}^{\,2}}{\left(\frac{\pi}{4}\right)^2\cdot D^4}$$ calculate constant

$b_i = 102793131.7931$

$$P := \left[\frac{\left[p^2 + b_i\cdot T_{ave2}^{\,2}\cdot\left[e^{\frac{(2\cdot a_i\cdot H_2)}{T_{ave2}}} - 1\right]\right]}{e^{\frac{(2\cdot a_i\cdot H_2)}{T_{ave2}}}}\right]^{(0.5)}$$ calculate pressure

$P = 1313964.4224$ (N/m^2)

$$p := \frac{P}{10000.0}$$ convert to N/cm^2 abs

$p = 131.3964$ (N/cm^2 abs)

Cased hole inside drill string pipe body section

$D := \frac{id_{dp}}{1000}$ convert diameter to m

$D = 0.0972$ (m)

$a_i := \left(\frac{SG_a}{R}\right)\cdot\left(1 + \frac{w_w}{w_{gt}}\right)$ calculate constant

$a_i = 0.0341$

$f := \left(\frac{1}{2\cdot\log\left(\frac{D}{e_{cs}}\right) + 1.14}\right)^2$ friction factor using von Karman correlation

$f = 0.022$

$b_i := \left(\frac{f}{2\cdot g\cdot D}\right)\cdot\left(\frac{R}{SG_a}\right)^2\cdot\frac{w_{gt}^2}{\left(\frac{\pi}{4}\right)^2\cdot D^4}$ calculate constant

$b_i = 18078996.3818$

$P := \left[\frac{\left[P^2 + b_i\cdot T_{ave1}^2\cdot\left[e^{\frac{(2\cdot a_i\cdot H_1)}{T_{ave1}}} - 1\right]\right]}{e^{\frac{(2\cdot a_i\cdot H_1)}{T_{ave1}}}}\right]^{0.5}$ calculate pressure

$P = 1297868.3213$ (N/m2)

$p := \frac{P}{10000.0}$ convert to N/cm2 abs

$p = 129.7868$ (N/cm2 abs)

$p_{in} := p$

$p_{in} = 129.7868$ drilling string injection pressure (N/cm^2 abs)

Kinetic energy summary of each section (KE greater than 3.0 $lb\text{-}ft/ft^3$)

$KE_1 = 216.6492$ ($N\text{-}m/m^3$)

$KE_2 = 189.1328$ ($N\text{-}m/m^3$)

$KE_3 = 152.8428$ ($N\text{-}m/m^3$)

$KE_4 = 143.566$ ($N\text{-}m/m^3$)

$KE_5 = 391.107$ ($N\text{-}m/m^3$)

Illustrative Example 8.3: SI version, Two Dresser Clark CFB-4 Compressors - Drilling Fluid is Air. Determining the bottom hole and injection pressures while drilling at 3047.9 m of depth for the drilling operations described in Illustrative Examples 8.1 and 8.2.

Compressor data - Dresser Clark Model CFB-4 with Caterpillar Model D398 prime mover

$q_c := 566.254$ flow rate produced from one compressor (actual liters/sec)

$num_c := 2$ number of compressors (required for SF above min KE)

$\varepsilon_m := 0.8$ mechanical efficiency

$n_s := 4$ number of stages

$PW_{max} := 566.7$ max sea level power rating of compressor prime mover (kW)

$c := 0.02$ clearance volume ratio for compressor

Basic standard and elevation atomospheric conditions

$p_{st} := 10.136$ API standard atmospheric pressure (N/cm^2 abs)

$t_{st} := 15.6$ API standard atmospheric temperature (°C)

$alt := 1219$ surface elevation of drilling site (m)

$p_{at} := 8.749$ pressure from table for surface elevation (N/cm^2 abs)

$t_{at} := 7.078$ temperature from table for surface elevation (°C)

$P_{st} := p_{st} \cdot 10000.0$ convert API standard pressure to N/m^2

$P_{st} = 101360$ (N/m^2 abs)

$T_{st} := t_{st} + 273.15$ convert API standard temp to K

$T_{st} = 288.75$ (K)

$P_{at} := p_{at} \cdot 10000.0$ convert actual atm pressure to N/m^2

$P_{at} = 87490$ (N/m^2 abs)

$T_{at} := t_{at} + 273.15$ convert actual atm temp to K

Borehole geometry

$H_c := 2133.5$ total depth of cased hole (m)

$H_{oh} := 914.36$ total depth of open hole (m)

$ROP := 18.29$ estimated rate of penetration (m/hr)

$d_{bh} := 200.03$ borehole (drill bit) diameter (mm)

$od_{dc} := 158.75$ outer diameter in drill collar (mm)

$id_{dc} := 71.438$ inner diameter in drill collar (mm)

$od_{dp} := 114.30$ outer diameter in drill pipe (mm)

$id_{dp} := 97.18$ inner diameter in drill pipe (mm)

$od_{tj} := 158.75$ outer diameter in drill pipe tool joint (mm)

$id_{tj} := 69.85$ inside diameter in drill pipe tool joint (mm)

$od_c := 219.08$ outer diameter in drill casing (mm)

$id_c := 201.19$ inside diameter in drill casing (mm)

$od_{bl} := 219.08$ outer diameter in blooey line (mm)

$id_{bl} := 205.66$ inside diameter in blooey line (mm)

$L_{bl} := 60.957$ length of blooey line (m)

$e_{cs} := 0.000152939$ roughness of commercial steel (m)

$e_r := 0.003048$ roughness of rock table (m)

$L_{dc} := 152.39$ total length of drill collars (m)

$L_{tj} := 0.457$ average length of one tool joint (m)

$L_{dp} := 9.144$ average length of one drill pipe section (m)

$d_n := 17.78$ nozzle diameter (mm)

$N_n := 3$ number of nozzles in the bit

$\beta := 0.01823$ temperature gradient (°C/m)

Section height calculations

$$H_1 := H_c - L_{tj} \cdot \left(\frac{H_c}{L_{dp}} \right)$$ calculate length of drill pipe in cased section

$H_1 = 2026.87166$ (m)

$$H_2 := L_{tj} \cdot \left(\frac{H_c}{L_{dp}} \right)$$ calculate length of tool joint in cased section

$H_2 = 106.62834$ (m)

$$H_3 := H_{oh} - L_{dc} - L_{tj} \cdot \left(\frac{H_{oh} - L_{dc}}{L_{dp}} \right)$$ calculate length of drill pipe in open hole section

$H_3 = 723.88817$ (m)

$$H_4 := L_{tj} \cdot \left(\frac{H_{oh} - L_{dc}}{L_{dp}} \right)$$ calculate length of tool joints in open hole section

$H_4 = 38.08183$ (m)

$H_5 := L_{dc}$ length of drill collar

$H_5 = 152.39$ (m)

Formation natural gas and water influx

$q_w := 0.5$	injected "mist" water (bbl/hr)
$q_{ng} := 0.0$	natural gas influx (liters/sec)
$q_{fw} := 0.00$	formation water influx (bbl/hr)

Constants

$SG_a := 1.0$	specific gravity of air
$SG_w := 1.0$	specific gravity of fresh water
$SG_{ng} := 0.70$	specific gravity of natural gas
$SG_{fw} := 1.07$	specific gravity of formation water
$SG_r := 2.7$	specific gravity of rock
$SG_{df} := 0.8156$	specific gravity of diesel fuel
$k := 1.4$	gas ratio of specific heats
$R := 29.31$	engineering gas constant (N-m/lN-K)
$\rho_w := 1000.0$	density of fresh water (kg/m^3)
$g := 9.81$	gravitational constant (m/sec^2)
$\gamma_w := \rho_w \cdot g$	specific weight of fresh water (N/m^3)
$\gamma_w = 9810$	(N/m^3)
$\gamma_{wg} := 9.81$	specific weight of water (N/liter)

Preliminary calculations

$D_{bh} := \frac{d_{bh}}{1000}$ convert borehole diameter to m

$D_{bh} = 0.20003$ (m)

$rop := \frac{ROP}{60 \cdot 60}$ convert ROP to m/sec

$rop = 0.00508$ (m/sec)

$w_s := \left[\left(\frac{\pi}{4} \right) \cdot D_{bh}^{\ 2} \cdot \gamma_w \cdot SG_r \cdot rop \right]$ weight rate of solid flow out of blooey line (N/sec)

$w_s = 4.22887$ (N/sec)

$q_a := q_c \cdot num_c$ air flow rate from all compressors (actual liters/sec)

$Q_a := \frac{q_a}{1000}$ convert air flow rate to m^3/ sec

$Q_a = 1.13251$ (m^3/sec)

$Q_w := \frac{(q_w \cdot 42 \cdot 0.003785)}{60 \cdot 60}$ convert "mist" water flow to m^3/sec

$Q_w = 0.00002$ (m^3/sec)

$Q_{ng} := \frac{q_{ng}}{1000}$ convert natural gas flow rate to m^3/sec

$Q_{ng} = 0$ (m^3/sec)

$Q_{gt} := Q_a + Q_{ng}$ total flow of the gas (m^3/sec)

$Q_{gt} = 1.13251$ (m^3/sec)

$Q_{fw} := \frac{(q_{fw} \cdot 42 \cdot 0.003785)}{60 \cdot 60}$ convert formation water flow to m^3/sec

$Q_{fw} = 0$ (m³/sec)

$\gamma_a := \frac{(P_{at} \cdot SG_a)}{R \cdot T_{at}}$ specific weight of air (N/m³)

$\gamma_a = 10.652$ (N/m³)

$\gamma_{ng} := \frac{(P_{at} \cdot SG_{ng})}{R \cdot T_{at}}$ specific weight of natural gas (N/m³)

$\gamma_{ng} = 7.4564$ (N/m³)

$w_a := \gamma_a \cdot Q_a$ weight rate of flow of air (N/sec)

$w_a = 12.06347$ (N/sec)

$w_w := SG_w \cdot \gamma_w \cdot Q_w$ weight rate of flow of "mist" water (N/sec)

$w_w = 0.2166$ (N/sec)

$w_{ng} := \gamma_{ng} \cdot Q_{ng}$ weight rate of flow of natural gas (N/sec)

$w_{ng} = 0$ (N/sec)

$w_{fw} := SG_{fw} \cdot \gamma_w \cdot Q_{fw}$ weight rate of flow of formation water (N/sec)

$w_{fw} = 0$ (N/sec)

$w_{gt} := w_a + w_{ng}$ total weight rate of flow of gas (N/sec)

$w_{gt} = 12.06347$ (N/sec)

Blooey line

$ID := \frac{id_{bl}}{1000}$ convert blooey line inner diameter to m

$ID = 0.20566$ (m)

$A := \frac{\pi}{4} \cdot ID^2$ cross sectional area (m^2)

$A = 0.03322$ (m^2)

$$f := \left(\frac{1}{2 \cdot \log\left(\frac{ID}{e_{cs}}\right) + 1.14} \right)^2$$

friction factor using von Karman correlation

$f = 0.01828$

$\frac{id_c}{od_{dp}} = 1.76019$ compute D1/D3 ratio for Fig. 8-5

$\frac{id_c}{id_{bl}} = 0.97827$ compute D1/D2 ratio for Fig. 8-5

Figure 8.5 gives the following flow resistance coefficient for the blind Tee:

$K_t := 25$

$K_v := 0.2$ loss factor for gate valves in blooey line

$$P_b := \left[\left[\left[\left(f \cdot \frac{L_{bl}}{ID} \right) + K_t + 2 \cdot K_v \right] \cdot \left[\frac{\left(w_{gt}^2 \cdot R \cdot T_{at} \right)}{g \cdot A^2} \right] + P_{at}^2 \right]^{0.5} \right]$$

calculate pressure of the gas at the entrance to the blooey line

$P_b = 105152.69527$ (N/m^2)

$p_b := \frac{P_b}{10000.0}$ convert to N/cm^2

$p_b = 10.51527$ (N/cm^2 abs)

Annulus cased hole drill pipe body section

$H := H_1$ (m)

$ID := \frac{id_c}{1000}$ convert inner diameter of casing to m

$ID = 0.20119$ (m)

$OD := \frac{od_{dp}}{1000}$ convert outer diameter of drill pipe to m

$OD = 0.1143$ (m)

$A := \frac{\pi}{4}\cdot\left(ID^2 - OD^2\right)$ annulus area between casing and drill pipe (m^2)

$A = 0.02153$ (m^2)

$a_a := \frac{SG_a}{R}\cdot\left[1 + \frac{\left(w_s + w_w + w_{fw}\right)}{w_{gt}}\right]$ calculate constant

$a_a = 0.04669$

$f := \left[\frac{1}{2\cdot\log\left[\frac{(ID - OD)}{e_{cs}}\right] + 1.14}\right]^2$ friction factor using von Karman correlation

$f = 0.02262$

$b_a := \left[\left[\frac{f}{2\cdot g\cdot(ID - OD)}\right]\cdot\left(\frac{R}{SG_a}\right)^2\cdot\frac{w_{gt}^{\,2}}{\left(\frac{\pi}{4}\right)^2\cdot\left(ID^2 - OD^2\right)^2}\right]$ calculate constant

$b_a = 3578631.45129$

$T := T_{at} + \beta\cdot H_1$ temperature at bottom of section (K)

$T = 317.17787$ (K)

$T_{ave1} := \frac{(T_{at} + T)}{2}$ average temperature of section (K)

$T_{ave1} = 298.70294$ (K)

$T_{ave} := T_{ave1}$

$$P_1 := \left[\left[\left(P_b^{\,2} + b_a \cdot T_{ave}^{\,2}\right) \cdot e^{\left[\frac{(2 \cdot a_a \cdot H)}{T_{ave}}\right]}\right] - b_a \cdot T_{ave}^{\,2}\right]^{0.5}$$ calculate pressure

$P_1 = 550678.33987$ (N/m^2)

$p_1 := \frac{P_1}{10000.0}$ convert to N/cm^2 abs

$p_1 = 55.06783$ (N/cm^2 abs)

$\gamma := \frac{P_1 \cdot SG_a}{R \cdot T}$ specific weight of gas (N/m^3)

$\gamma = 59.23512$ (N/m^3)

$V := \frac{w_{gt}}{\gamma \cdot A}$ velocity (m/sec)

$V = 9.45906$ (m/sec)

$KE_1 := \frac{1}{2} \cdot \frac{\gamma}{g} \cdot V^2$ kinetic energy of gas flow ($N\text{-}m/m^3$)

$KE_1 = 270.13188$ ($N\text{-}m/m^3$)

Annulus cased hole tool joint section

$H := H_2$

$ID := \frac{id_c}{1000}$ convert inner diameter of casing to m

$ID = 0.20119$ (m)

$OD := \frac{od_{tj}}{1000}$ convert outer diameter of tool joint to m

$OD = 0.15875$ (m)

$OD_t := \frac{od_{dp}}{1000}$ use drill pipe OD for kinetic energy to m

$OD_t = 0.1143$ (m)

$A := \frac{\pi}{4}\left(ID^2 - OD_t^2\right)$ annulus area for tool joint KE (m^2)

$A = 0.02153$ (m^2)

$a_a := \frac{SG_a}{R}\cdot\left[1 + \frac{\left(w_s + w_w + w_{fw}\right)}{w_{gt}}\right]$ calculate constant

$a_a = 0.04669$

$f := \left[\frac{1}{2\cdot\log\left[\frac{(ID - OD)}{e_{cs}}\right] + 1.14}\right]^2$ friction factor using von Karman correlation

$f = 0.02753$

$b_a := \left[\left[\frac{f}{2\cdot g\cdot(ID - OD)}\right]\cdot\left(\frac{R}{SG_a}\right)^2\cdot\frac{w_{gt}^2}{\left(\frac{\pi}{4}\right)^2\cdot\left(ID^2 - OD^2\right)^2}\right]$ calculate constant

$b_a = 28719630.1504$

$T_{temp} := T$

$T := T_{temp} + \beta \cdot H_2$ temperature at bottom of section (K)

$T = 319.1217$ (K)

$T_{ave2} := \frac{(T_{temp} + T)}{2}$ average temperature of section (K)

$T_{ave2} = 318.14979$ (K)

$T_{ave} := T_{ave2}$

$$P_2 := \left[\left[\left(P_1^{\,2} + b_a \cdot T_{ave}^{\,2}\right) \cdot e^{\left[\frac{(2 \cdot a_a \cdot H)}{T_{ave}}\right]}\right] - b_a \cdot T_{ave}^{\,2}\right]^{0.5}$$ calculate pressure

$P_2 = 636636.21661$ (N/m^2)

$p_2 := \frac{P_2}{10000.0}$ convert to N/cm^2 abs

$p_2 = 63.66362$ N/cm^2 abs

$\gamma := \frac{P_2 \cdot SG_a}{R \cdot T}$ specific weight of gas (N/m^3)

$\gamma = 68.06427$ (N/m^3)

$V := \frac{w_{gt}}{\gamma \cdot A}$ velocity of gas (m/sec)

$V = 8.23205$ (m/sec)

$KE_2 := \frac{1}{2} \cdot \frac{\gamma}{g} \cdot V^2$ kinetic energy of gas flow ($N\text{-}m/m^3$)

$KE_2 = 235.09098$ ($N\text{-}m/m^3$)

Annulus of open hole drill pipe section

$H := H_3$

$ID := D_{bh}$ open hole diameter of borehole (m)

$ID = 0.20003$ (m)

$OD := \frac{od_{dp}}{1000}$ convert diameter of drill pipe to m

$OD = 0.1143$ (m)

$A := \frac{\pi}{4}\left(ID^2 - OD^2\right)$ area between borehole and drill pipe (m^2)

$A = 0.02116$ (m^2)

$a_a := \frac{SG_a}{R}\cdot\left[1 + \frac{\left(w_s + w_w + w_{fw}\right)}{w_{gt}}\right]$ calculate constant

$a_a = 0.04669$

$e_{ave} := \frac{\left(e_r ID^2 + e_{cs} OD^2\right)}{ID^2 + OD^2}$ weighted average roughness (m)

$e_{ave} = 0.00234$ (m)

$f := \left[\frac{1}{2\cdot\log\left[\frac{(ID - OD)}{e_{ave}}\right] + 1.14}\right]^2$ friction factor using von Karman correlation

$f = 0.05486$

$$b_a := \left[\left[\frac{f}{2\cdot g\cdot(ID - OD)}\right]\cdot\left(\frac{R}{SG_a}\right)^2\cdot\frac{w_{gt}^{\,2}}{\left(\frac{\pi}{4}\right)^2\cdot\left(ID^2 - OD^2\right)^2}\right]$$ calculate constant

$b_a = 9102577.02118$

$T_{temp} := T$

$T := T_{temp} + \beta\cdot H_3$ temperature at bottom of section (K)

$T = 332.31819$

$$T_{ave3} := \frac{\left(T_{temp} + T\right)}{2}$$ average temperature of section (K)

$T_{ave3} = 325.71995$ (K)

$T_{ave} := T_{ave3}$

$$P_3 := \left[\left[\left(P_2^{\,2} + b_a\cdot T_{ave}^{\,2}\right)\cdot e^{\left[\frac{(2\cdot a_a\cdot H)}{T_{ave}}\right]}\right] - b_a\cdot T_{ave}^{\,2}\right]^{0.5}$$ calculate pressure

$P_3 = 849422.40244$ (N/m^2)

$$p_3 := \frac{P_3}{10000.0}$$ convert to N/cm^2 abs

$p_3 = 84.94224$ (N/cm^2 abs)

$$\gamma := \frac{P_3\cdot SG_a}{R\cdot T}$$ specific weight of gas (N/m^3)

$\gamma = 87.20749$ (N/m^3)

$$V := \frac{w_{gt}}{\gamma\cdot A}$$ velocity of gas (ft/sec)

$V = 6.53597$ (m/sec)

$KE_3 := \frac{1}{2} \cdot \frac{\gamma}{g} \cdot V^2$ kinetic energy of gas flow (N-m/m^3)

$KE_3 = 189.87808$ (N-m/m^3)

Annulus open hole tool joint section

$H := H_4$

$ID := D_{bh}$ open hole diameter of borehole (m)

$ID = 0.20003$ (m)

$OD := \frac{od_{tj}}{1000}$ convert tool joint diameter to m

$OD = 0.15875$ (m)

$OD_t := \frac{od_{dp}}{1000}$ drill pipe diameter (m)

$OD_t = 0.1143$ (m)

$A := \frac{\pi}{4}\left(ID^2 - OD_t^2\right)$ annulus area for tool joint KE (m^2)

$A = 0.02116$ (m^2)

$a_a := \frac{SG_a}{R} \cdot \left[1 + \frac{\left(w_s + w_w + w_{fw}\right)}{w_{gt}}\right]$ calculate constant

$a_a = 0.04669$

$e_{ave} := \frac{\left(e_r ID^2 + e_{cs} OD^2\right)}{ID^2 + OD^2}$ weighted average roughness (m)

$e_{ave} = 0.00193$ (m)

$$f := \left[\frac{1}{2\cdot\log\left[\frac{(ID - OD)}{e_{ave}}\right] + 1.14}\right]^2$$ friction factor using von Karman correlation

$f = 0.06923$

$$b_a := \left[\left[\frac{f}{2\cdot g\cdot(ID - OD)}\right]\cdot\left(\frac{R}{SG_a}\right)^2\cdot\frac{w_{gt}^{\,2}}{\left(\frac{\pi}{4}\right)^2\cdot\left(ID^2 - OD^2\right)^2}\right]$$ calculate constant

$b_a = 78975169.45561$

$T_{temp} := T$

$T := T_{temp} + \beta\cdot H_4$ temperature at bottom of section (K)

$T = 333.01242$

$$T_{ave4} := \frac{(T_{temp} + T)}{2}$$ average temperature of section (K)

$T_{ave4} = 332.6653$ (K)

$T_{ave} := T_{ave4}$

$$P_4 := \left[\left[\left(P_3^{\,2} + b_a\cdot T_{ave}^{\,2}\right)\cdot e^{\left[\frac{(2\cdot a_a\cdot H)}{T_{ave}}\right]}\right] - b_a\cdot T_{ave}^{\,2}\right]^{0.5}$$ calculate pressure

$P_4 = 907304.70045$ (N/m^2)

$$p_4 := \frac{P_4}{10000.0}$$ convert to N/cm^2 abs

$p_4 = 90.73047$ (N/cm2 abs)

$\gamma := \frac{P_4 \cdot SG_a}{R \cdot T}$ specific weight of gas (N/m³)

$\gamma = 92.95589$ (N/m³)

$V := \frac{w_{gt}}{\gamma \cdot A}$ velocity of gas (m/sec)

$V = 6.13178$ (m/sec)

$KE_4 := \frac{1}{2} \cdot \frac{\gamma}{g} \cdot V^2$ kinetic energy of gas flow (N-m/m³)

$KE_4 = 178.136$ (N-m/m³)

Annulus open hole drill collar section

$H := H_5$

$ID := D_{bh}$ open hole diameter of borehole (m)

$ID = 0.20003$ (m)

$OD := \frac{od_{dc}}{1000}$ convert drill collar diameter to m

$OD = 0.15875$ (m)

$A := \frac{\pi}{4}\left(ID^2 - OD^2\right)$ drill pipe diameter (m)

$A = 0.01163$ (m)

$a := \frac{SG_a}{R} \cdot \left[1 + \frac{(w_s + w_w + w_{fw})}{w_{gt}}\right]$ calculate constant

$a_a = 0.04669$

$$e_{ave} := \frac{\left(e_r ID^2 + e_{cs} OD^2\right)}{ID^2 + OD^2}$$ weighted average roughness (m)

$e_{ave} = 0.00193$ (m)

$$f := \left[\frac{1}{2 \cdot \log\left[\frac{(ID - OD)}{e_{ave}}\right] + 1.14}\right]^2$$ friction factor using von Karman correlation

$f = 0.06923$

$$b_a := \left[\left[\frac{f}{2 \cdot g \cdot (ID - OD)}\right] \cdot \left(\frac{R}{SG_a}\right)^2 \cdot \frac{w_{gt}^2}{\left(\frac{\pi}{4}\right)^2 \cdot \left(ID^2 - OD^2\right)^2}\right]$$ calculate constant

$b_a = 78975169.45561$

$T_{temp} := T$

$T := T_{temp} + \beta \cdot H_5$ temperature at bottom of section (K)

$T = 335.79049$

$$T_{ave5} := \frac{\left(T_{temp} + T\right)}{2}$$ average temperature of section (K)

$T_{ave5} = 334.40145$ (K)

$T_{ave} := T_{ave5}$

$$P_5 := \left[\left[\left(P_4^2 + b_a \cdot T_{ave}^2\right) \cdot e^{\left[\frac{(2 \cdot a_a \cdot H)}{T_{ave}}\right]}\right] - b_a \cdot T_{ave}^2\right]^{0.5}$$

$P_5 = 1114861.8506$ (N/m^2)

$P_{bh} := P_5$

$p_5 := \frac{P_5}{10000.0}$

$p_5 = 111.48619$ (N/cm^2 abs)

$p_{bh} := p_5$ annulus bottom hole pressure (N/cm^2 abs)

$\gamma := \frac{P_5 \cdot SG_a}{R \cdot T}$ specific weight of gas (N/m^3)

$\gamma = 113.27572$ (N/m^3)

$V := \frac{w_{gt}}{\gamma \cdot A}$ velocity of gas (m/sec)

$V = 9.15541$ (m/sec)

$KE_5 := \frac{1}{2} \cdot \frac{\gamma}{g} \cdot V^2$ kinetic energy of gas flow (N-m/m^3)

$KE_5 = 483.94194$ (N-m/m^3)

Bottom hole through the bit nozzles

$D_n := \frac{d_n}{1000}$ convert nozzle diameter to m

$D_n = 0.01778$ (m)

$A_n := N_n \cdot \left(\frac{\pi}{4}\right) \cdot D_n^2$ total flow area of nozzles in bit (m^2)

$A_n = 0.00074$ (m^2)

$$\gamma := \frac{(P_5 \cdot SG_a)}{R \cdot T}$$ specific weight of air (N/m^3)

$\gamma = 113.27572$ (N/m^3)

$$P := P_5 \cdot \left[1 + \frac{\left(\frac{w_{gt}}{A_n}\right)^2}{2 \cdot g \cdot \left(\frac{k}{k-1}\right) \cdot P_5 \cdot \gamma} \right]^{\frac{k}{k-1}}$$ calculate pressure

$P = 1237412.76384$ (N/m^2)

$$P_{sonic} := \frac{P_{bh}}{0.528}$$ verify subsonic flow

$P_{sonic} = 2111480.77765$ (N/m^2) - flow is sonic since $P < P_{sonic}$

$$p := \frac{P}{10000.0}$$ convert to N/cm^2 abs

$p = 123.74128$ (N/cm^2 abs)

Open hole inside drill string drill collar section

$$D := \frac{id_{dc}}{1000}$$ convert diameter to m

$D = 0.07144$ (m)

$$a_i := \left(\frac{SG_a}{R}\right) \cdot \left(1 + \frac{w_w}{w_{gt}}\right)$$ calculate constant

$a_i = 0.03473$

$$f := \left(\frac{1}{2 \cdot \log\left(\frac{D}{e_{cs}}\right) + 1.14} \right)^2$$ friction factor using von Karman correlation

$f = 0.02382$

$$b_i := \left(\frac{f}{2\cdot g\cdot D}\right)\cdot\left(\frac{R}{SG_a}\right)^2\cdot\frac{w_{gt}^{\ 2}}{\left(\frac{\pi}{4}\right)^2\cdot D^4}$$ calculate constant

$b_i = 132269290.90696$

$$P := \left[\frac{\left[P^2 + b_i\cdot T_{ave5}^{\ 2}\cdot\left[e^{\frac{(2\cdot a_i\cdot H_5)}{T_{ave5}}} - 1\right]\right]}{e^{\frac{(2\cdot a_i\cdot H_5)}{T_{ave5}}}}\right]^{(0.5)}$$ calculate pressure

$P = 1394396.92355$ (N/m^2)

$$p := \frac{P}{10000.0}$$ convert to N/cm^2 abs

$p = 139.43969$ (N/cm^2 abs)

Open hole inside drill string tool joint section

$$D := \frac{id_{tj}}{1000}$$ convert diameter to m

$D = 0.06985$ (m)

$$a_i := \left(\frac{SG_a}{R}\right)\cdot\left(1 + \frac{w_w}{w_{gt}}\right)$$ calculate constant

$a_i = 0.03473$

$$f := \left(\frac{1}{2\cdot\log\left(\frac{D}{e_{cs}}\right) + 1.14}\right)^2$$ friction factor using von Karman correlation

$f = 0.02397$

$$b_i := \left(\frac{f}{2 \cdot g \cdot D}\right) \cdot \left(\frac{R}{SG_a}\right)^2 \cdot \frac{w_{gt}^2}{\left(\frac{\pi}{4}\right)^2 \cdot D^4}$$ calculate constant

$b_i = 148900137.91852$

$$P := \left[\frac{\left[P^2 + b_i \cdot T_{ave4}^2 \cdot \left[e^{\frac{(2 \cdot a_i \cdot H_4)}{T_{ave4}}} - 1\right]\right]}{e^{\frac{(2 \cdot a_i \cdot H_4)}{T_{ave4}}}}\right]^{(0.5)}$$ calculate pressure

$P = 1435078.9671$ (N/m²)

$$p := \frac{P}{10000.0}$$ convert to N/cm² abs

$p = 143.5079$ (N/cm² abs)

Open hole inside drill string drill pipe body section

$$D := \frac{id_{dp}}{1000}$$ convert diameter to m

$D = 0.09718$ (m)

$$a_i := \left(\frac{SG_a}{R}\right) \cdot \left(1 + \frac{w_w}{w_{gt}}\right)$$ calculate constant

$a_i = 0.03473$

$$f := \left(\frac{1}{2 \cdot \log\left(\frac{D}{e_{cs}}\right) + 1.14}\right)^2$$ friction factor using von Karman correlation

$f = 0.02197$

$$b_i := \left(\frac{f}{2\cdot g\cdot D}\right)\cdot\left(\frac{R}{SG_a}\right)^2\cdot\frac{w_{gt}^{\;2}}{\left(\frac{\pi}{4}\right)^2\cdot D^4}$$ calculate constant

$b_i = 26188180.16067$

$$P := \left[\frac{\left[p^2 + b_i\cdot T_{ave3}^{\;2}\cdot\left[e^{\frac{(2\cdot a_i\cdot H_3)}{T_{ave3}}} - 1\right]\right]}{e^{\frac{(2\cdot a_i\cdot H_3)}{T_{ave3}}}}\right]^{(0.5)}$$ calculate pressure

$P = 1470474.24792$ (N/m2)

$$p := \frac{P}{10000.0}$$ convert to N/cm2 abs

$p = 147.04742$ (N/cm2 abs)

Cased hole inside drill string tool joint section

$$D := \frac{id_{tj}}{1000}$$ convert diameter to m

$D = 0.06985$ (m)

$$a_i := \left(\frac{SG_a}{R}\right)\cdot\left(1 + \frac{w_w}{w_{gt}}\right)$$ calculate constant

$a_i = 0.03473$

$$f := \left(\frac{1}{2\cdot\log\left(\frac{D}{e_{cs}}\right) + 1.14}\right)^2$$ friction factor using von Karman correlation

$f = 0.02397$

$$b_i := \left(\frac{f}{2\cdot g\cdot D}\right)\cdot\left(\frac{R}{SG_a}\right)^2\cdot\frac{w_{gt}^{\;2}}{\left(\frac{\pi}{4}\right)^2\cdot D^4}$$ calculate constant

$b_i = 148900137.91852$

$$P := \left[\frac{\left[p^2 + b_i\cdot T_{ave2}^{\;2}\cdot\left[e^{\frac{(2\cdot a_i\cdot H_2)}{T_{ave2}}} - 1\right]\right]}{e^{\frac{(2\cdot a_i\cdot H_2)}{T_{ave2}}}}\right]^{(0.5)}$$ calculate pressure

$P = 1568232.01259$ (N/m²)

$$p := \frac{P}{10000.0}$$ convert to N/cm² abs

$p = 156.8232$ (N/cm² abs)

Cased hole inside drill string pipe body section

$$D := \frac{id_{dp}}{1000}$$ convert diameter to m

$D = 0.09718$ (m)

$$a_i := \left(\frac{SG_a}{R}\right)\cdot\left(1 + \frac{w_w}{w_{gt}}\right)$$ calculate constant

$a_i = 0.03473$

$$f := \left(\frac{1}{2\cdot\log\left(\frac{D}{e_{cs}}\right) + 1.14}\right)^2$$ friction factor using von Karman correlation

$f = 0.02197$

$$b_i := \left(\frac{f}{2 \cdot g \cdot D}\right) \cdot \left(\frac{R}{SG_a}\right)^2 \cdot \frac{w_{gt}^{\;2}}{\left(\frac{\pi}{4}\right)^2 \cdot D^4}$$ calculate constant

$b_i = 26188180.16067$

$$P := \left[\frac{\left[P^2 + b_i \cdot T_{avel}^{\;2} \cdot \left[e^{\frac{(2 \cdot a_i \cdot H_1)}{T_{avel}}} - 1\right]\right]}{e^{\frac{(2 \cdot a_i \cdot H_1)}{T_{avel}}}}\right]^{0.5}$$ calculate pressure

$P = 1553453.43724$ (N/m2)

$p := \frac{P}{10000.0}$ convert to N/cm2 abs

$p = 155.34534$ (N/cm2 abs)

$p_{in} := p$

$p_{in} = 155.34534$ drilling string injection pressure (N/cm2 abs)

Determine compressor system fuel composition rate

$p_i := p_{at}$ input compressor pressure (N/cm2 abs)

$P_i := p_i \cdot 10000.0$ convert to N/m2 abs

$P_i = 87490$ (N/m2 abs)

$p_o := p_{in}$ output compressor pressure (N/cm2 abs)

$P_o := p_o \cdot 10000.0$ convert to N/cm2abs

$P_o = 1553453.43724$ (N/m2 abs)

$q_i := q_c$ input flow rate for one compressor

$Q_i := \frac{q_c}{1000}$ convert to m^3/sec

$Q_i = 0.56625$ (m^3/sec)

$r_s := \left(\frac{P_o}{P_i}\right)^{\frac{1}{n_s}}$ compression ratio for each stage

$r_s = 2.05274$

$\varepsilon_v := 0.96\left[1 - c\cdot\left(r_s^{\frac{1}{k}} - 1\right)\right]$ volumetric efficiency (recip only)

$\varepsilon_v = 0.94711$

$W_s := \left[\frac{(n_s\cdot k)}{k-1}\right]\cdot(P_i\cdot Q_i)\cdot\left[\left(\frac{P_o}{P_i}\right)^{\frac{(k-1)}{n_s\cdot k}} - 1\right]$ theoretical shaft horsepower (Watts)

$W_s = 158215.49382$ (Watts)

$W_{as} := \frac{W_s}{\varepsilon_m\cdot\varepsilon_v}$ actual shaft power (Watts)

$W_{as} = 208813.9687$ (Watts)

$PW_{as} := \frac{W_{as}}{1000}$ convert to kWatts

Derate turbocharged prime mover for surface elevation of 1219 m above sea

$\%R := 0.10$ % reduction required due to altitude

$PW_{alt} := PW_{max} - \%R \cdot PW_{max}$ maximum power available prime mover (kWatts)

$PW_{alt} = 510.03$ (kWatts)

Actual shaft power is less than prime mover maximum avialable, so OK

$PR := \left(\frac{PW_{as}}{PW_{alt}}\right) \cdot 100$ power ratio of prime mover

$PR = 40.94151$

Use above PR with Fig. 5-16 (extrapolate) and obtain approximate rate of fuel consumption rate

$FC := 4.15$ fuel consumption (N/kW-hr)

$w_f := FC \cdot PW_{as}$ weight rate of fuel consumption (N/hr)

$w_f = 866.57797$ (N/hr)

$q_f := \frac{w_f}{SG_{df} \cdot \gamma_{wg}}$ fuel volume per hour for one compressor (liters/hr)

$q_f = 108.30822$ (liters/hr)

$q_{ftotal} := q_f \cdot num_c$ total fuel consumed for all compressors (liters/hr)

$q_{ftotal} = 216.61644$ diesel fuel rate (liters/hr)

Kinetic energy summary of each section (KE greater than 143.5 N-m/m^3)

$KE_1 = 270.13188$ (N-m/m^3)

$KE_2 = 235.09098$ (N-m/m^3)

$KE_3 = 189.87808$ (N-m/m^3)

$KE_4 = 178.136$ (N-m/m^3)

$KE_5 = 483.94194$ (N-m/m^3)

Important design parameter summary

$q_a = 1132.508$ air compressor flow rate (liters/sec)

$p_{in} = 155.34534$ required surface air injection pressure (N/cm^2 abs)

$p_{bh} = 111.48619$ bottom hole pressure (N/cm^2 abs)

$PW_{as} = 208.81397$ actual shaft power (kWatts) required

$PW_{alt} = 510.03$ derated power (kWatts) available

$q_{ftotal} = 216.61644$ total volume per hour of fuel consumed (liters/hr)

Water injection to standpipe to saturate gas flow at bottom hole annulus conditions

$T = 335.79049$ bottom hole temperature (K)

$t_{bh} := T - 273.15$ convert bottom hole temp to (°C)

$t_{bh} = 62.64049$ (°C)

$t_{bhF} := \left(\frac{9}{5}\right) \cdot t_{bh} + 32$ convert °C to °F

$t_{bhF} = 144.75288$ (°F)

$p_{bhE} := \frac{p_{bh}}{0.6897}$ convert N/cm^2 abs to psia

$p_{bhE} = 161.64446$ (psia)

$w_{aE} := \frac{w_a}{4.4485}$ convert N/sec to lb/sec

$w_{aE} = 2.71181$

$$p_{sat} := 10^{6.39416-\left(\frac{1750.286}{217.23+0.55\cdot t_{bhF}}\right)}$$

gas-saturation pressure of water (psia)

$p_{sat} = 3.14663$ (psia)

$$q_{winj} := \left(\frac{p_{sat}}{p_{bhE} - p_{sat}}\right)\cdot\left(\frac{18.02}{28.96}\right)\cdot\left(\frac{3600}{8.33}\right)\cdot w_{aE}$$

$q_{winj} = 14.47754$ required water injection to saturate gas at bottom hole conditions (gal/hr)

APPENDIX

Chapter 9 Illustrative Example MathCad™ Solutions D

This appendix presents the detailed MathCad solutions for the illustrative examples in Chapter 9.

Illistrative Example 9.3: USCS version, Drilling Fluid is Aerated Fluid.
This program calculates the required gas injection volumetric flow rate in order to obtain a bottom hole annulus pressure of 3360 psig while drilling at 10,000 ft of depth. Assumes homogeneous multiphase flow.

Given input conditions

$q_a := 1263$	air volumetric flow rate trial and error value (acfm)
$q_f := 191$	mud flow rate (gpm)
$\gamma_f := 10$	specific weight of mud (ppg)
$\mu_{air} := 0.012$	viscosity of air (centipoises)
$\mu_{mud} := 30$	viscosity of mud (centipoises)
$C := 0.81$	aerated fluid flow loss coefficient through nozzles
$D_c := 7000$	total depth of cased hole (ft)
$D_{oh} := 3000$	total length of open hole (ft)
$alt := 4000$	elevation of drilling site (ft)
$p_o := 12.685$	pressure from table for elevation (psia)
$t_o := 60$	temperature from table for elevation (°F)
$p_{at} := 12.685$	actual given atmospheric surface pressure (psia)
$t_{at} := 60$	actual given atmospheric surface temperature (°F)
$ROP := 60$	estimated rate of penetration (ft/hr)
$d_{bh} := 7.875$	borehole (drill bit) diameter (in)
$od_{dc} := 6.25$	outer diameter of drill collar (in)
$id_{dc} := 2.8125$	inner diameter of drill collar (in)
$od_{dp} := 4.5$	outer diameter of drill pipe (in)
$id_{dp} := 3.826$	inner diameter of drill pipe (in)
$od_{tj} := 6.25$	outer diameter of drill pipe tool joint (in)
$id_{tj} := 2.75$	inner diameter of drill pipe tool joint (in)
$od_c := 8.625$	outer diameter of casing (in)
$id_c := 7.921$	inner diameter of casing (in)
$L_r := 100$	length of return line (ft)
$id_r := 5.625$	inner diameter of return line (in)
$SG_r := 2.7$	specific gravity of rock
$e_r := 0.01$	roughness of rock (ft)
$L_{dc} := 500$	total length of drill collars (ft)

$L_{tj} := 1.5$	average length of one tool joint (ft)
$L_{dp} := 30$	average length of one drill pipe section (ft)
$d_n := \frac{11}{32}$	nozzle diameter (in)
$N_n := 3$	number of nozzles in the bit
$\beta := 0.01$	geothermal gradient (°F/ft)

Constant

$p_{st} := 14.695$	API atmospheric pressure (psia)
$t_{st} := 60$	API standard temperature (°F)
$SG_a := 1$	specific gravity of air
$R := 53.36$	specific gas constant (lb-ft/lb-°R)
$w_w := 8.33$	specific weight of water (lb/gal)
$\rho_w := 62.4$	specific weight of water (lb/ft³)
$e_{cs} := 0.0005$	roughness of commercial steel (ft)
$g := 32.2$	gravitational constant (ft/sec²)

Preliminary calculations

$P_{at} := p_{at} \cdot 144$	convert actual atm pressure to lb/ft²
$P_{at} = 1826.640$	(lb/ft²)
$T_{at} := t_{at} + 459.67$	convert actual atm temp to °R
$T_{at} = 519.670$	(°R)
$P_o := p_o \cdot 144$	convert table pressure to lb/ft^2
$P_o = 1826.640$	(lb/ft²)
$T_o := t_o + 459.67$	convert table temperature to °R
$T_o = 519.670$	(°R)
$D_{bh} := \frac{d_{bh}}{12}$	convert borehole diameter to ft
$D_{bh} = 0.656$	(ft)
$ID_r := \frac{id_r}{12}$	convert return line diameter to ft
$ID_r = 0.469$	(ft)

$OD_{dp} := \frac{od_{dp}}{12}$ convert drill pipe outer diameter to ft

$OD_{dp} = 0.375$ (ft)

$ID_{dp} := \frac{id_{dp}}{12}$ convert drill pipe inner diameter to ft

$ID_{dp} = 0.319$ (ft)

$ID_c := \frac{id_c}{12}$ convert casing inner diameter to ft

$ID_c = 0.660$ (ft)

$OD_{dc} := \frac{od_{dc}}{12}$ convert drill collar outer diameter to ft

$OD_{dc} = 0.521$ (ft)

$ID_{dc} := \frac{id_{dc}}{12}$ convert drill collar inner diameter to ft

$ID_{dc} = 0.234$ (ft)

$OD_{tj} := \frac{od_{tj}}{12}$ convert tool joint outer diameter to ft

$OD_{tj} = 0.521$ (ft)

$ID_{tj} := \frac{id_{tj}}{12}$ convert tool joint inner diameter to ft

$ID_{tj} = 0.229$ (ft)

$d_n = 0.344$ nozzle diameter (in)

$D_n := \frac{d_n}{12}$ nozzle diameter converted to ft

$D_n = 0.029$ (ft)

$\mu_g := \mu_{air} \cdot 2.089 \cdot 10^{-5}$ convert viscosity to lb-sec/ft^2

$\mu_g = 2.507 \times 10^{-7}$ (lb-sec/ft^2)

$\mu_f := \mu_{mud} \cdot 2.089 \cdot 10^{-5}$ convert viscosity to lb-sec/ft^2

$\mu_f = 6.267 \times 10^{-4}$ (lb-sec/ft^2)

$Q_a := \frac{q_a}{60}$ convert air flow rate to ft^3/sec

$Q_a = 21.050$ (ft^3/sec)

$Q_f := \frac{q_f \cdot 231}{60 \cdot 12^3}$ convert mud flow to (ft^3/sec)

$Q_f = 0.426$ (ft^3/sec)

$\gamma_f := \gamma_f \cdot \frac{12^3}{231}$ convert spec. wt. of mud from ppg to lb/ft^3

$\gamma_f = 74.805$ (lb/ft^3)

$\rho_f := \frac{\gamma_f}{g}$ mud density ($lb\text{-}sec^2/ft^4$)

$\rho_f = 2.3231$ ($lb\text{-}sec^2/ft^4$)

$\nu_f := \frac{\mu_f}{\rho_f}$ mud kinematic viscosity (ft^2/sec)

$\nu_f = 2.698 \times 10^{-4}$ (ft^2/sec)

$\gamma_g := \frac{(P_{at} \cdot SG_a)}{R \cdot T_{at}}$ specific weight of air (lb/ft^3)

$\gamma_g = 0.0659$ (lb/ft^3)

$\rho_g := \frac{\gamma_g}{g}$ gas (air) density ($lb\text{-}sec^2/ft^4$)

$\rho_g = 2.046 \times 10^{-3}$ ($lb\text{-}sec^2/ft^4$)

$\nu_g := \frac{\mu_g}{\rho_g}$ gas (air) kinematic viscosity (ft^2/sec)

$\nu_g = 1.225 \times 10^{-4}$ (ft^2/sec)

$rop := \frac{ROP}{60 \cdot 60}$ convert ROP to ft/sec

$rop = 0.0167$ (ft/sec)

$w_s := \left[\left(\frac{\pi}{4}\right) \cdot D_{bh}^{\,2} \cdot 62.4 \cdot SG_r \cdot rop\right]$ weight rate of solid flow out of blooey line (lb/sec)

$w_s = 0.950$ (lb/sec)

$w_g := \gamma_g \cdot Q_a$ weight rate of flow of air (lb/sec)

$w_g = 1.387$ (lb/sec)

$w_f := \gamma_f \cdot Q_f$ weight rate of flow of mud (lb/sec)

$w_f = 31.833$ (lb/sec)

$w_t := w_g + w_f + w_s$ total weight rate of flow (lb/sec)

$w_t = 34.170$ (lb/sec)

Section height calculations

$H_1 := D_c - L_{tj} \cdot \left(\frac{D_c}{L_{dp}} \right)$ calculate length of drill pipe in cased section

$H_1 = 6650.000$ (ft)

$H_2 := L_{tj} \cdot \left(\frac{D_c}{L_{dp}} \right)$ calculate length of tool joint in cased section

$H_2 = 350.000$ (ft)

$H_3 := D_{oh} - L_{dc} - L_{tj} \cdot \left(\frac{D_{oh} - L_{dc}}{L_{dp}} \right)$ calculate length of drill pipe in open hole section

$H_3 = 2375.000$ (ft)

$H_4 := L_{tj} \cdot \left(\frac{D_{oh} - L_{dc}}{L_{dp}} \right)$ calculate length of tool joints in open hole section

$H_4 = 125.000$ (ft)

$H_5 := L_{dc}$ length of drill collar

$H_5 = 500.000$ (ft)

Return line

$ID := ID_r$ set variable ID equal to return line's ID

$ID = 0.469$ (ft)

$A := \frac{\pi}{4} \cdot ID^2$ cross sectional area (ft^2)

$A = 0.173$ (ft^2)

$V := \frac{(Q_a + Q_f)}{A}$ velocity of flow through the line (ft/sec)

$V = 124.443$ (ft/sec)

$\nu_{ave} := \dfrac{(w_g \cdot \nu_g + w_f \cdot \nu_f)}{w_g + w_f}$ average kinematic viscosity of gas & mud

$\nu_{ave} = 2.636 \times 10^{-4}$ (ft^2/sec)

$N_R := \dfrac{(V \cdot ID)}{\nu_{ave}}$ Reynolds number calculations

$N_R = 2.213 \times 10^5$

$f := \left[\dfrac{1}{-1.8 \cdot \log\left[\left(\dfrac{e_{cs}}{3.7 \cdot ID}\right)^{1.11} + \dfrac{6.9}{N_R}\right]}\right]^2$ calculate friction factor using Haaland equation

$f = 0.0211$ Haaland friction factor

$Q(P) := \dfrac{P_{at}}{P} \cdot \dfrac{T_o}{T_{at}} \cdot Q_a$ flow rate as a function of pressure

$\gamma_{mix}(P) := \dfrac{w_t}{Q(P) + Q_f}$ specific weight as a function of pressure

$V(P) := \dfrac{(Q(P) + Q_f)}{A}$ velocity as a function of pressure

$p_r := 21.597$ trial pressure

$P_r := p_r \cdot 144$ convert pressure to lb/ft^2

$P_r = 3109.968$ (lb/ft^2)

$\displaystyle\int_{P_{at}}^{P_r} \frac{1}{\gamma_{mix}(P) \cdot \left(V(P)^2 \cdot \dfrac{f}{2 \cdot g \cdot ID}\right)} \, dP = 100.085$ left hand side

$\displaystyle\int_0^{L_r} 1 \, dh = 100.000$ right hand side

$K_t := 30$ resistance coefficient for the blind Tee

$K_v := 0.2$ resistance coefficient for gate valves

$$\gamma_{mix} := \frac{w_t}{\left[\left(\frac{P_{at}}{P_r}\right)\cdot\left(\frac{T_o}{T_{at}}\right)\cdot Q_a + Q_f\right]}$$ specific weight (lb/ft³)

$\gamma_{mix} = 2.672$ (lb/ft³)

$$V := \left[\frac{\left[\left(\frac{P_{at}}{P_r}\right)\cdot\left(\frac{T_o}{T_{at}}\right)\cdot Q_a + Q_f\right]}{A}\right]$$ velocity (ft/sec)

$V = 74.109$ (ft/sec)

$$\Delta P_T := \gamma_{mix}\cdot(2\cdot K_v + K_t)\cdot\frac{V^2}{2\cdot g}$$ caclulate pressure loss due to Tee and valves

$\Delta P_T = 6926.764$ (lb/ft²)

$P_e := P_r + \Delta P_T$ sum pressure losses

$P_e = 10036.732$ (lb/ft²)

$$p_e := \frac{P_e}{144}$$ convert from lb/ft² abs to psia

$p_e = 69.700$ (psia)

Annulus cased hole drill pipe section

$ID := ID_c$ set inner diameter to casing ID

$ID = 0.660$ (ft)

$OD := OD_{dp}$ set outer diameter to drill pipe OD

$OD = 0.375$ (ft)

$P := P_e$ set pressure equal to previous section pressure

$P = 10036.732$ (lb/ft²)

$$A := \frac{\pi}{4}\cdot\left(ID^2 - OD^2\right)$$ cross sectional area (ft²)

$A = 0.232$ (ft²)

$T := T_o + 0.01 \cdot H_1$ — calculate temperature at bottom of section

$T = 586.170$ — (°R)

$T_{ave1} := \frac{T_o + T}{2}$ — calculate average temperature of section

$T_{ave1} = 552.920$ — (°R)

$T_{ave} := T_{ave1}$ — set average temp to section average temp

$\gamma_g := \frac{P \cdot SG_a}{R \cdot T_{ave1}}$ — specific weight of the gas (lb/ft^3)

$\gamma_g = 0.340$ — (lb/ft^3)

$\rho_g := \frac{\gamma_g}{g}$ — gas (air) density (lb-sec^2/ft^4)

$\rho_g = 0.011$ — (lb-sec^2/ft^4)

$\nu_g := \frac{\mu_g}{\rho_g}$ — gas (air) kinematic viscosity (ft^2/sec)

$\nu_g = 2.373 \times 10^{-5}$ — (ft^2/sec)

$\nu_{ave} := \frac{(w_g \cdot \nu_g + w_f \cdot \nu_f)}{w_g + w_f}$ — average kinematic viscosity of gas & mud

$\nu_{ave} = 2.595 \times 10^{-4}$ — (ft^2/sec)

$$V := \left[\frac{\left[\left(\frac{P_{at}}{P} \right) \cdot \left(\frac{T_{ave}}{T_{at}} \right) \cdot Q_a + Q_f \right]}{A} \right]$$

velocity of flow in annulus (ft/sec)

$V = 19.424$ — (ft/sec)

$N_R := \frac{V \cdot (ID - OD)}{\nu_{ave}}$ — Reynolds number

$N_R = 2.134 \times 10^4$

$$f := \left[\frac{1}{-1.8 \cdot \log\left[\left[\frac{e_{cs}}{3.7 \cdot (ID - OD)} \right]^{1.11} + \frac{6.9}{N_R} \right]} \right]^2$$

calculate friction factor using Haaland equation

$f = 0.0287$ — Haaland friction factor

$Q(P) := \frac{P_{at}}{P} \cdot \frac{T_{ave}}{T_{at}} \cdot Q_a$ flow rate as a function of pressure

$\gamma_{mix}(P) := \frac{w_t}{Q(P) + Q_f}$ specific weight as a function of pressure

$V(P) := \frac{(Q(P) + Q_f)}{A}$ velocity as a function of pressure

$p_1 := 1839.06$ trial pressure (psia)

$P_1 := p_1 \cdot 144$ convert pressure to lb/ft^2

$P_1 = 2.648 \times 10^5$ (lb/ft^2)

$$\int_{P_e}^{P_1} \frac{1}{\gamma_{mix}(P) \cdot \left[1 + (V(P))^2 \cdot \frac{f}{2 \cdot g \cdot (ID - OD)}\right]} dP = 6650.072$$ left hand side

$$\int_0^{H_1} 1\, dh = 6650.000$$ right hand side

Annulus cased hole tool joint section

$ID := ID_c$ set inner diameter to casing ID

$OD := OD_{tj}$ set outer diameter to drill pipe OD

$P := P_1$ set pressure equal to previous section pressure

$A := \frac{\pi}{4} \cdot \left(ID^2 - OD^2\right)$ cross sectional area (ft^2)

$T_{temp} := T$ keep previous section temperature

$T := T_{temp} + 0.01 \cdot H_2$ calculate temperature at bottom of section

$T = 589.670$ (°R)

$T_{ave2} := \frac{T_{temp} + T}{2}$ calculate average temperature of section

$T_{ave2} = 587.920$ (°R)

$T_{ave} := T_{ave2}$ set average temp to section average temp

$\gamma_g := \frac{P \cdot SG_a}{R \cdot T_{ave}}$ specific weight of the gas (lb/ft^3)

$\gamma_g = 8.442$ (lbs/ft^3)

$\rho_g := \frac{\gamma_g}{g}$ gas (air) density (lb-sec^2/ft^4)

$\rho_g = 0.262$ (lb-sec^2/ft^4)

$\nu_g := \frac{\mu_g}{\rho_g}$ gas (air) kinematic viscosity (ft^2/sec)

$\nu_g = 9.562 \times 10^{-7}$ (ft^2/sec)

$\nu_{ave} := \frac{(w_g \cdot \nu_g + w_f \cdot \nu_f)}{w_g + w_f}$ average kinematic viscosity of gas & mud

$\nu_{ave} = 2.585 \times 10^{-4}$ (ft^2/sec)

$V := \left[\frac{\left[\left(\frac{P_{at}}{P}\right) \cdot \left(\frac{T_{ave}}{T_{at}}\right) \cdot Q_a + Q_f\right]}{A}\right]$ velocity of flow in annulus (ft/sec)

$V = 4.567$ (ft/sec)

$N_R := \frac{V \cdot (ID - OD)}{\nu_{ave}}$ Reynolds number calculations

$N_R = 2.460 \times 10^3$

$f := \left[\frac{1}{-1.8 \cdot \log\left[\left[\frac{e_{cs}}{3.7 \cdot (ID - OD)}\right]^{1.11} + \frac{6.9}{N_R}\right]}\right]^2$ calculate friction factor using Haaland equation

$f = 0.0499$ Haaland friction factor

$Q(P) := \frac{P_{at}}{P} \cdot \frac{T_{ave}}{T_{at}} \cdot Q_a$ flow rate as a function of pressure

$\gamma_{mix}(P) := \frac{w_t}{Q(P) + Q_f}$ specific weight as a function of pressure

$V(P) := \frac{(Q(P) + Q_f)}{A}$ velocity as a function of pressure

$p_2 := 1997.680$ trial pressure (psia)

$P_2 := p_2 \cdot 144$ convert pressure to lb/ft^2

$P_2 = 2.877 \times 10^5$ (lb/ft^2)

$$\int_{P_1}^{P_2} \frac{1}{\gamma_{mix}(P)\cdot\left[1 + (V(P))^2 \cdot \frac{f}{2\cdot g\cdot(ID - OD)}\right]}\, dP = 350.079$$ left hand side

$$\int_{H_1}^{H_1+H_2} 1\, dh = 350.000$$ right hand side

Annulus open hole drill pipe section

$ID := D_{bh}$ set inner diameter to casing ID

$OD := OD_{dp}$ set outer diameter to drill pipe OD

$P := P_2$ set pressure equal to previous section pressure

$A := \frac{\pi}{4}\cdot\left(ID^2 - OD^2\right)$ cross sectional area (ft^2)

$T_{temp} := T$ keep previous section temperature

$T := T_{temp} + 0.01\cdot H_3$ calculate temperature at bottom of section

$T = 613.420$ (°R)

$T_{ave3} := \frac{T_{temp} + T}{2}$ calculate average temperature of section

$T_{ave3} = 601.545$ (°R)

$T_{ave} := T_{ave3}$ set average temp to section average temp

$\gamma_g := \frac{P\cdot SG_a}{R\cdot T_{ave}}$ specific weight of the gas (lb/ft^3)

$\gamma_g = 8.962$ (lbs/ft^3)

$\rho_g := \frac{\gamma_g}{g}$ gas (air) density (lb-sec^2/ft^4)

$\rho_g = 0.278$ (lb-sec^2/ft^4)

$$\nu_g := \frac{\mu_g}{\rho_g}$$ gas (air) kinematic viscosity (ft²/sec)

$$\nu_g = 9.007 \times 10^{-7}$$ (ft²/sec)

$$\nu_{ave} := \frac{(w_g \cdot \nu_g + w_f \cdot \nu_f)}{w_g + w_f}$$ average kinematic viscosity of gas & mud

$$\nu_{ave} = 2.585 \times 10^{-4}$$ (ft²/sec)

$$V := \left[\frac{\left[\left(\frac{P_{at}}{P}\right)\cdot\left(\frac{T_{ave}}{T_{at}}\right)\cdot Q_a + Q_f\right]}{A}\right]$$ velocity of flow in annulus (ft/sec)

$$V = 2.547$$ (ft/sec)

$$N_R := \frac{V\cdot(ID - OD)}{\nu_{ave}}$$ Reynolds number calculations

$$N_R = 2.771 \times 10^{3}$$

$$e_{ave} := \frac{\left[e_r\cdot\left(\frac{\pi}{4}\right)\cdot ID^2 + e_{cs}\cdot\left(\frac{\pi}{4}\right)\cdot OD^2\right]}{\left[\left(\frac{\pi}{4}\right)\cdot ID^2 + \left(\frac{\pi}{4}\right)\cdot OD^2\right]}$$ calculate weighted average of roughness between the steel and the rock

$$e_{ave} = 7.662 \times 10^{-3}$$ weighted average of roughness (ft)

$$f := \left[\frac{1}{-1.8\cdot\log\left[\left[\frac{e_{ave}}{3.7\cdot(ID - OD)}\right]^{1.11} + \frac{6.9}{N_R}\right]}\right]^2$$ calculate friction factor using Haaland equation

$$f = 0.066$$ Haaland friction factor

$$Q(P) := \frac{P_{at}}{P}\cdot\frac{T_{ave}}{T_{at}}\cdot Q_a$$ flow rate as a function of pressure

$$\gamma_{mix}(P) := \frac{w_t}{Q(P) + Q_f}$$ specific weight as a function of pressure

$$V(P) := \frac{(Q(P) + Q_f)}{A}$$ velocity as a function of pressure

$$p_3 := 3044.1$$ trial pressure (psia)

$$P_3 := p_3\cdot 144$$ convert pressure to lb/ft²

$P_3 = 4.384 \times 10^5$ (lb/ft²)

$$\int_{P_2}^{P_3} \frac{1}{\gamma_{mix}(P)\cdot\left[1 + (V(P))^2\cdot\frac{f}{2\cdot g\cdot(ID - OD)}\right]}\,dP = 2375.030$$ left hand side

$$\int_{H_1+H_2}^{H_1+H_2+H_3} 1\,dh = 2375.000$$ right hand side

Annulus open hole tool joint section

$ID := D_{bh}$ set inner diameter to casing ID

$OD := OD_{tj}$ set outer diameter to drill pipe OD

$P := P_3$ set pressure equal to previous section pressure

$A := \frac{\pi}{4}\cdot\left(ID^2 - OD^2\right)$ cross sectional area (ft²)

$T_{temp} := T$ keep previous section temperature

$T := T_{temp} + 0.01\cdot H_4$ calculate temperature at bottom of section

$T = 614.670$ (°R)

$T_{ave4} := \frac{T_{temp} + T}{2}$ calculate average temperature of section

$T_{ave4} = 614.045$ (°R)

$T_{ave} := T_{ave4}$ set average temp to section average temp

$\gamma_g := \frac{P\cdot SG_a}{R\cdot T_{ave}}$ specific weight of the gas (lb/ft³)

$\gamma_g = 13.378$ (lb/ft³)

$\rho_g := \frac{\gamma_g}{g}$ gas (air) density (lb-sec²/ft⁴)

$\rho_g = 0.415$ (lb-sec²/ft⁴)

$\nu_g := \frac{\mu_g}{\rho_g}$ gas (air) kinematic viscosity (ft²/sec)

$\nu_g = 6.034 \times 10^{-7}$ (ft²/sec)

$\nu_{ave} := \dfrac{(w_g \cdot \nu_g + w_f \cdot \nu_f)}{w_g + w_f}$ average kinematic viscosity of gas & mud

$\nu_{ave} = 2.585 \times 10^{-4}$ (ft²/sec)

$V := \left[\dfrac{\left[\left(\frac{P_{at}}{P}\right)\cdot\left(\frac{T_{ave}}{T_{at}}\right)\cdot Q_a + Q_f\right]}{A}\right]$ velocity of flow in annulus (ft/sec)

$V = 4.227$ (ft/sec)

$N_R := \dfrac{V \cdot (ID - OD)}{\nu_{ave}}$ Reynolds number calculations

$N_R = 2.214 \times 10^3$

$e_{ave} := \dfrac{\left[e_r \cdot \left(\frac{\pi}{4}\right)\cdot ID^2 + e_{cs}\cdot\left(\frac{\pi}{4}\right)\cdot OD^2\right]}{\left[\left(\frac{\pi}{4}\right)\cdot ID^2 + \left(\frac{\pi}{4}\right)\cdot OD^2\right]}$ calculate weighted average of roughness between the steel and the rock

$e_{ave} = 6.329 \times 10^{-3}$ weighted average of roughness (ft)

$f := \left[\dfrac{1}{-1.8\cdot\log\left[\left[\dfrac{e_{ave}}{3.7\cdot(ID - OD)}\right]^{1.11} + \dfrac{6.9}{N_R}\right]}\right]^2$ calculate friction factor using Haaland equation

$f = 0.0802$ Haaland friction factor

$Q(P) := \dfrac{P_{at}}{P}\cdot\dfrac{T_{ave}}{T_{at}}\cdot Q_a$ flow rate as a function of pressure

$\gamma_{mix}(P) := \dfrac{w_t}{Q(P) + Q_f}$ specific weight as a function of pressure

$V(P) := \dfrac{(Q(P) + Q_f)}{A}$ velocity as a function of pressure

$p_4 := 3109.50$ trial pressure (psia)

$P_4 := p_4 \cdot 144$ convert pressure to lb/ft²

$P_4 = 4.478 \times 10^5$ (lb/ft²)

$$\int_{P_3}^{P_4} \frac{1}{\gamma_{mix}(P)\cdot\left[1+(V(P))^2\cdot\frac{f}{2\cdot g\cdot(ID-OD)}\right]}\,dP = 125.078$$ left hand side

$$\int_{H_1+H_2+H_3}^{H_1+H_2+H_3+H_4} 1\,dh = 125.000$$ right hand side

Annulus open hole drill collar section

$ID := D_{bh}$	set inner diameter to casing ID
$OD := OD_{dc}$	set outer diameter to drill pipe OD
$P := P_4$	set pressure equal to previous section pressure
$A := \frac{\pi}{4}\cdot\left(ID^2 - OD^2\right)$	cross sectional area (ft^2)
$T_{temp} := T$	keep previous section temperature
$T := T_{temp} + 0.01\cdot H_5$	calculate temperature at bottom of section
$T = 619.670$	(°R)
$T_{ave5} := \frac{T_{temp} + T}{2}$	calculate average temperature of section
$T_{ave5} = 617.170$	(°R)
$T_{ave} := T_{ave5}$	set average temp to section average temp
$\gamma_g := \frac{P\cdot SG_a}{R\cdot T_{ave}}$	specific weight of the gas (lb/ft^3)
$\gamma_g = 13.597$	(lb/ft^3)
$\rho_g := \frac{\gamma_g}{g}$	gas (air) density ($lb\text{-}sec^2/ft^4$)
$\rho_g = 0.422$	($lb\text{-}sec^2/ft^4$)
$\nu_g := \frac{\mu_g}{\rho_g}$	gas (air) kinematic viscosity (ft^2/sec)
$\nu_g = 5.937 \times 10^{-7}$	(ft^2/sec)

$$\nu_{ave} := \frac{(w_g \cdot \nu_g + w_f \cdot \nu_f)}{w_g + w_f}$$ average kinematic viscosity of gas & mud

$\nu_{ave} = 2.585 \times 10^{-4}$ (ft^2/sec)

$$V := \left[\frac{\left[\left(\frac{P_{at}}{P}\right)\cdot\left(\frac{T_{ave}}{T_{at}}\right)\cdot Q_a + Q_f\right]}{A}\right]$$ velocity of flow in annulus (ft/sec)

$V = 4.214$ (ft/sec)

$$N_R := \frac{V\cdot(ID - OD)}{\nu_{ave}}$$ Reynolds number calculations

$N_R = 2.207 \times 10^3$

$$e_{ave} := \frac{\left[e_r\cdot\left(\frac{\pi}{4}\right)\cdot ID^2 + e_{cs}\cdot\left(\frac{\pi}{4}\right)\cdot OD^2\right]}{\left[\left(\frac{\pi}{4}\right)\cdot ID^2 + \left(\frac{\pi}{4}\right)\cdot OD^2\right]}$$ calculate weighted average of roughness between the steel and the rock

$e_{ave} = 6.329 \times 10^{-3}$ weighted average of roughness (ft)

$$f := \left[\frac{1}{-1.8\cdot\log\left[\left[\frac{e_{ave}}{3.7\cdot(ID - OD)}\right]^{1.11} + \frac{6.9}{N_R}\right]}\right]^2$$ calculate friction factor using Haaland equation

$f = 0.0802$ Haaland friction factor

$$Q(P) := \frac{P_{at}}{P}\cdot\frac{T_{ave}}{T_{at}}\cdot Q_a$$ flow rate as a function of pressure

$$\gamma_{mix}(P) := \frac{w_t}{Q(P) + Q_f}$$ specific weight as a function of pressure

$$V(P) := \frac{(Q(P) + Q_f)}{A}$$ velocity as a function of pressure

$p_5 := 3360.00 + p_{at}$ trial pressure (psia)

$p_5 = 3372.685$ (psia)

$P_5 := p_5 \cdot 144$ convert pressure to lb/ft²

$P_5 = 4.857 \times 10^5$ (lb/ft²)

$$\int_{P_4}^{P_5} \frac{1}{\gamma_{mix}(P) \cdot \left[1 + (V(P))^2 \cdot \frac{f}{2 \cdot g \cdot (ID - OD)}\right]} \, dP = 500.113$$ left hand side

$$\int_{H_1+H_2+H_3+H_4}^{H_1+H_2+H_3+H_4+H_5} 1 \, dh = 500.000$$ right hand side

$p_{bh} := p_5 - p_{at}$ bottom hole pressure (pisg)

$p_{bh} = 3360.000$ (psig)

Bottom hole through the bit nozzles

$D_e := \left[N_n \cdot (D_n)^2\right]^{0.5}$ equivalent single diameter (ft)

$D_e = 0.050$ (ft)

$$\gamma_{mix} := \frac{(w_g + w_f)}{\left[Q_f + Q_a \cdot \left(\frac{P_{at}}{P_5}\right) \cdot \frac{T}{T_{at}}\right]}$$ specific weight of mixture (lb/ft³)

$\gamma_{mix} = 63.890$ (lb/ft³)

$$\Delta P := \frac{(w_g + w_f)^2}{2 \cdot C^2 \cdot g \cdot \gamma_{mix} \cdot \left(\frac{\pi}{4}\right)^2 \cdot D_e^4}$$ approximation of pressure change of the aerated fluid through the drill bit

$\Delta P = 109355.738$ (lb/ft²)

$P := P_5 + \Delta P$ sum pressures to get pressure above drill bit

$P = 595022.378$ (lb/ft²)

$p := \frac{P}{144}$ convert prssure to lb/in²

$p = 4132.100$ (psia)

Open hole inside drill string drill collar section

$ID := ID_{dc}$ set inner diameter to casing ID

$A := \frac{\pi}{4} \cdot \left(ID^2\right)$ cross sectional area (ft^2)

$T_{ave} := T_{ave5}$ set ave temp to section ave temp

$\gamma_g := \frac{P \cdot SG_a}{R \cdot T_{ave}}$ specific weight of the gas (lb/ft^3)

$\gamma_g = 18.068$ (lb/ft^3)

$\rho_g := \frac{\gamma_g}{g}$ gas (air) density ($lb\text{-}sec^2/ft^4$)

$\rho_g = 0.561$ ($lb\text{-}sec^2/ft^4$)

$\nu_g := \frac{\mu_g}{\rho_g}$ gas (air) kinematic viscosity (ft^2/sec)

$\nu_g = 4.467 \times 10^{-7}$ (ft^2/sec)

$\nu_{ave} := \frac{\left(w_g \cdot \nu_g + w_f \cdot \nu_f\right)}{w_g + w_f}$ average kinematic viscosity of gas & mud

$\nu_{ave} = 2.585 \times 10^{-4}$ (ft^2/sec)

$V := \left[\frac{\left[\left(\frac{P_{at}}{P}\right) \cdot \left(\frac{T_{ave}}{T_{at}}\right) \cdot Q_a + Q_f\right]}{A}\right]$ velocity of flow in annulus (ft/sec)

$V = 11.642$ (ft/sec)

$N_R := \frac{V \cdot ID}{\nu_{ave}}$ Reynolds number calculations

$N_R = 1.056 \times 10^4$

$$f := \left[\frac{1}{-1.8 \cdot \log\left[\left(\frac{e_{cs}}{3.7 \cdot ID}\right)^{1.11} + \frac{6.9}{N_R}\right]}\right]^2$$ calculate friction factor using Haaland equation

$f = 0.0334$ Haaland friction factor

$$Q(P) := \frac{P_{at}}{P} \cdot \frac{T_{ave}}{T_{at}} \cdot Q_a$$ flow rate as a function of pressure

$$\gamma_{mix}(P) := \frac{w_g + w_f}{Q(P) + Q_f}$$ specific weight as a function of pressure

$$V(P) := \frac{(Q(P) + Q_f)}{A}$$ velocity as a function of pressure

$p_4 := 3972.12$ trial pressure (psia)

$P_4 := p_4 \cdot 144$ convert pressure to lb/ft^2

$P_4 = 5.720 \times 10^5$ (lb/ft^2)

$$\int_{P_4}^{P} \frac{1}{\gamma_{mix}(P) \cdot \left[1 - (V(P))^2 \cdot \frac{f}{2 \cdot g \cdot (ID)}\right]} \, dP = 500.067$$ left hand side

$$\int_{H_1+H_2+H_3+H_4}^{H_1+H_2+H_3+H_4+H_5} 1 \, dh = 500.000$$ right hand side .

Open hole inside drill string tool joint section

$ID := ID_{tj}$ set inner diameter to casing ID

$$A := \frac{\pi}{4} \cdot \left(ID^2\right)$$ cross sectional area (ft^2)

$T_{ave} := T_{ave4}$ set ave temp to section ave temp

$P := P_4$ set pressure to previous section pressure

$$\gamma_g := \frac{P \cdot SG_a}{R \cdot T_{ave}}$$ specific weight of the gas (lb/ft^3)

$\gamma_g = 17.457$ (lb/ft³)

$\rho_g := \frac{\gamma_g}{g}$ gas (air) density (lb-sec²/ft⁴)

$\rho_g = 0.542$ (lb-sec²/ft⁴)

$\nu_g := \frac{\mu_g}{\rho_g}$ gas (air) kinematic viscosity (ft²/sec)

$\nu_g = 4.624 \times 10^{-7}$ (ft²/sec)

$\nu_{ave} := \frac{(w_g \cdot \nu_g + w_f \cdot \nu_f)}{w_g + w_f}$ average kinematic viscosity of gas & mud

$\nu_{ave} = 2.585 \times 10^{-4}$ (ft²/sec)

$V := \left[\frac{\left[\left(\frac{P_{at}}{P}\right) \cdot \left(\frac{T_{ave}}{T_{at}}\right) \cdot Q_a + Q_f\right]}{A}\right]$ velocity of flow in annulus (ft/sec)

$V = 12.243$ (ft/sec)

$N_R := \frac{V \cdot ID}{\nu_{ave}}$ Reynolds number calculations

$N_R = 1.085 \times 10^4$

$f := \left[\frac{1}{-1.8 \cdot \log\left[\left(\frac{e_{cs}}{3.7 \cdot ID}\right)^{1.11} + \frac{6.9}{N_R}\right]}\right]^2$ calculate friction factor using Haaland equation

$f = 0.0332$ Haaland friction factor

$Q(P) := \frac{P_{at}}{P} \cdot \frac{T_{ave}}{T_{at}} \cdot Q_a$ flow rate as a function of pressure

$\gamma_{mix}(P) := \frac{w_g + w_f}{Q(P) + Q_f}$ specific weight as a function of pressure

$V(P) := \frac{(Q(P) + Q_f)}{A}$ velocity as a function of pressure

$p_3 := 3934.34$ trial pressure (psia)

$P_3 := p_3 \cdot 144$ convert pressure to lb/ft²

$P_3 = 5.665 \times 10^5$ (lb/ft²)

$$\int_{P_3}^{P_4} \frac{1}{\gamma_{mix}(P)\cdot\left[1 - (V(P))^2 \cdot \frac{f}{2\cdot g\cdot(ID)}\right]}\, dP = 125.015$$ left hand side

$$\int_{H_1+H_2+H_3}^{H_1+H_2+H_3+H_4} 1\, dh = 125.000$$ right hand side

Open hole inside drill string drill pipe section

$ID := ID_{dp}$ set inner diameter to casing ID

$A := \frac{\pi}{4}\cdot\left(ID^2\right)$ cross sectional area (ft²)

$T_{ave} := T_{ave3}$ set ave temp to section ave temp

$P := P_3$ set pressure to previous section pressure

$\gamma_g := \frac{P\cdot SG_a}{R\cdot T_{ave}}$ specific weight of the gas (lb/ft³)

$\gamma_g = 17.650$ (lb/ft³)

$\rho_g := \frac{\gamma_g}{g}$ gas (air) density (lb-sec²/ft⁴)

$\rho_g = 0.548$ (lb-sec²/ft⁴)

$\nu_g := \frac{\mu_g}{\rho_g}$ gas (air) kinematic viscosity (ft²/sec)

$\nu_g = 4.573 \times 10^{-7}$ (ft²/sec)

$\nu_{ave} := \frac{\left(w_g\cdot\nu_g + w_f\cdot\nu_f\right)}{w_g + w_f}$ average kinematic viscosity of gas & mud

$\nu_{ave} = 2.585 \times 10^{-4}$ (ft²/sec)

$$V := \left[\frac{\left[\left(\frac{P_{at}}{P}\right)\cdot\left(\frac{T_{ave}}{T_{at}}\right)\cdot Q_a + Q_f\right]}{A}\right]$$ velocity of flow in annulus (ft/sec)

$V = 6.314$ (ft/sec)

$$N_R := \frac{V\cdot ID}{\nu_{ave}}$$ Reynolds number calculations

$N_R = 7.787 \times 10^3$

$$f := \left[\frac{1}{-1.8\cdot\log\left[\left(\frac{e_{cs}}{3.7\cdot ID}\right)^{1.11} + \frac{6.9}{N_R}\right]}\right]^2$$ calculate friction factor using Haaland equation

$f = 0.0349$ Haaland friction factor

$$Q(P) := \frac{P_{at}}{P}\cdot\frac{T_{ave}}{T_{at}}\cdot Q_a$$ flow rate as a function of pressure

$$\gamma_{mix}(P) := \frac{w_g + w_f}{Q(P) + Q_f}$$ specific weight as a function of pressure

$$V(P) := \frac{(Q(P) + Q_f)}{A}$$ velocity as a function of pressure

$p_2 := 2948.0$ trial pressure (psia)

$P_2 := p_2\cdot 144$ convert pressure to lb/ft²

$P_2 = 4.245 \times 10^5$ (lb/ft²)

$$\int_{P_2}^{P_3} \frac{1}{\gamma_{mix}(P)\cdot\left[1 - (V(P))^2\cdot\frac{f}{2\cdot g\cdot(ID)}\right]}\,dP = 2375.084$$ left hand side

$$\int_{H_1+H_2}^{H_1+H_2+H_3} 1\, dh = 2375.000$$ right hand side

Cased hole inside drill string tool joint section

$ID := ID_{tj}$ set inner diameter to casing ID

$A := \frac{\pi}{4} \cdot \left(ID^2\right)$ cross sectional area (ft^2)

$T_{ave} := T_{ave2}$ set ave temp to section ave temp

$P := P_2$ set pressure to previous section pressure

$\gamma_g := \frac{P \cdot SG_a}{R \cdot T_{ave}}$ specific weight of the gas (lb/ft^3)

$\gamma_g = 13.532$ (lb/ft^3)

$\rho_g := \frac{\gamma_g}{g}$ gas (air) density ($lb\text{-}sec^2/ft^4$)

$\rho_g = 0.420$ ($lb\text{-}sec^2/ft^4$)

$\nu_g := \frac{\mu_g}{\rho_g}$ gas (air) kinematic viscosity (ft^2/sec)

$\nu_g = 5.965 \times 10^{-7}$ (ft^2/sec)

$\nu_{ave} := \frac{\left(w_g \cdot \nu_g + w_f \cdot \nu_f\right)}{w_g + w_f}$ average kinematic viscosity of gas & mud

$\nu_{ave} = 2.585 \times 10^{-4}$ (ft^2/sec)

$V := \left[\frac{\left[\left(\frac{P_{at}}{P}\right) \cdot \left(\frac{T_{ave}}{T_{at}}\right) \cdot Q_a + Q_f\right]}{A}\right]$ velocity of flow in annulus (ft/sec)

$V = 12.801$ (ft/sec)

$N_R := \frac{V \cdot ID}{\nu_{ave}}$ Reynolds number calculations

$N_R = 1.135 \times 10^4$

$$f := \left[\frac{1}{-1.8\cdot\log\left[\left(\frac{e_{cs}}{3.7\cdot ID}\right)^{1.11} + \frac{6.9}{N_R}\right]}\right]^2$$ calculate friction factor using Haaland equation

$f = 0.0329$ Haaland friction factor

$Q(P) := \frac{P_{at}}{P}\cdot\frac{T_{ave}}{T_{at}}\cdot Q_a$ flow rate as a function of pressure

$\gamma_{mix}(P) := \frac{w_g + w_f}{Q(P) + Q_f}$ specific weight as a function of pressure

$V(P) := \frac{(Q(P) + Q_f)}{A}$ velocity as a function of pressure

$p_1 := 2851.67$ trail pressure (psia)

$P_1 := p_1\cdot 144$ convert pressure to lb/ft^2

$P_1 = 4.106\times 10^5$ (lb/ft^2)

$$\int_{P_1}^{P_2} \frac{1}{\gamma_{mix}(P)\cdot\left[1 - (V(P))^2\cdot\frac{f}{2\cdot g\cdot(ID)}\right]}\, dP = 350.039$$ left hand side

$$\int_{H_1}^{H_1+H_2} 1\, dh = 350.000$$ right hand side of Eqn.

Cased hole inside drill string drill pipe section

$ID := ID_{dp}$ set inner diameter to casing ID

$A := \frac{\pi}{4}\cdot\left(ID^2\right)$ cross sectional area (ft^2)

$T_{ave} := T_{ave1}$ set ave temp to section ave temp

$P := P_1$ set pressure to previous section pressure

$$\gamma_g := \frac{P \cdot SG_a}{R \cdot T_{ave}}$$ specific weight of the gas (lb/ft^3)

$\gamma_g = 13.918$ (lb/ft^3)

$$\rho_g := \frac{\gamma_g}{g}$$ gas (air) density (lb-sec^2/ft^4)

$\rho_g = 0.432$ (lb-sec^2/ft^4)

$$\nu_g := \frac{\mu_g}{\rho_g}$$ gas (air) kinematic viscosity (ft^2/sec)

$\nu_g = 5.800 \times 10^{-7}$ (ft^2/sec)

$$\nu_{ave} := \frac{(w_g \cdot \nu_g + w_f \cdot \nu_f)}{w_g + w_f}$$ average kinematic viscosity of gas & mud

$\nu_{ave} = 2.585 \times 10^{-4}$ (ft^2/sec)

$$V := \left[\frac{\left[\left(\frac{P_{at}}{P}\right) \cdot \left(\frac{T_{ave}}{T_{at}}\right) \cdot Q_a + Q_f\right]}{A}\right]$$ velocity of flow in annulus (ft/sec)

$V = 6.578$ (ft/sec)

$$N_R := \frac{V \cdot ID}{\nu_{ave}}$$ Reynolds number calculations

$N_R = 8.112 \times 10^3$

$$f := \left[\frac{1}{-1.8 \cdot \log\left[\left(\frac{e_{cs}}{3.7 \cdot ID}\right)^{1.11} + \frac{6.9}{N_R}\right]}\right]^2$$ calculate friction factor using Haaland equation

$f = 0.0346$ Haaland friction factor

$$Q(P) := \frac{P_{at}}{P} \cdot \frac{T_{ave}}{T_{at}} \cdot Q_a$$ flow rate as a function of pressure

$$\gamma_{mix}(P) := \frac{w_g + w_f}{Q(P) + Q_f}$$ specific weight as a function of pressure

$$V(P) := \frac{(Q(P) + Q_f)}{A}$$ velocity as a function of pressure

$p_{in} := 637.39$ trial pressure (psia)

$P_{in} := p_{in} \cdot 144$ convert pressure to lb/ft^2

$P_{in} = 9.178 \times 10^4$ (lb/ft^2)

$$\int_{P_{in}}^{P_1} \frac{1}{\gamma_{mix}(P) \cdot \left[1 - (V(P))^2 \cdot \frac{f}{2 \cdot g \cdot (ID)}\right]} dP = 6650.282$$ left hand side

$$\int_0^{H_1} 1\, dh = 6650.000$$ right hand side

Parameter summary

$q_a = 1263.000$ air compressor flow rate (scfm)

$p_{in} = 637.390$ required surface air injection pressure (psia)

$p_{bh} = 3360.000$ bottom hole pressure (psig)

Illustrative Example 9.4: USCS version, Drilling Fluid is Aerated Fluid. H&B solution.
This program calculates the required gas injection volumetric flow rate in order to obtain a bottom hole annulus pressure of 3360 psig while drilling at 10,000 ft of depth. Utilizes the Hegendorn and Brown liquid hold-up correlations.

Given input conditions

$q_a := 301.26$	air compressor flow rate (acfm) iterate value
$q_f := 191$	mud flow rate (gpm)
$\gamma_f := 10$	specific weight of mud (ppg)
$\mu_{air} := 0.012$	viscosity of air (centipoises)
$\mu_{mud} := 30$	viscosity of mud (centipoises)
$C := 0.81$	aerated fluid flow loss coefficient through nozzles
$D_c := 7000$	total depth of cased hole (ft)
$D_{oh} := 3000$	total length of open hole (ft)
$alt := 4000$	elevation of drilling site (ft)
$p_o := 12.685$	pressure from table for surface elevation (psia)
$t_o := 60$	temperature from table for elev. (°F)
$p_{at} := 12.685$	actual given atmospheric surface pressure (psia)
$t_{at} := 60$	actual given atmospheric surface temperature (°F)
$ROP := 60$	estimated rate of penetration (ft/hr)
$d_{bh} := 7.875$	borehole (drill bit) diameter (in)
$od_{dc} := 6.25$	outer diameter of drill collar (in)
$id_{dc} := 2.8125$	inner diameter of drill collar (in)
$od_{dp} := 4.5$	outer diameter of drill pipe (in)
$id_{dp} := 3.826$	inner diameter of drill pipe (in)
$od_{tj} := 6.25$	outer diameter of drill pipe tool joint (in)
$id_{tj} := 2.750$	inner diameter of drill pipe tool joint (in)
$od_c := 8.625$	outer diameter of casing (in)

$id_c := 7.921$	inner diameter of casing (in)
$L_r := 100$	length of return line (ft)
$id_r := 5.625$	inner diameter of return line (in)
$SG_r := 2.7$	specific gravity of rock
$e_r := 0.01$	roughness of rock (ft)
$L_{dc} := 500$	total length of drill collars (ft)
$L_{tj} := 1.5$	ave length of one tool joint (ft)
$L_{dp} := 30$	ave length of one drill pipe section (ft)
$d_n := \frac{11}{32}$	nozzle diameter (in)
$N_n := 3$	number of nozzles in the bit
$\beta := 0.01$	geothermal gradient (deg F/ft)

Constants

$p_{st} := 14.695$	API atmospheric pressure (psia)
$t_{st} := 60$	API standard temperature (°F)
$SG_a := 1$	specific gravity of air
$SG_{df} := 0.8156$	diesel fuel specific gravity
$k := 1.4$	air specific heat constant
$R := 53.36$	specific gas constant (ft-lb/lb-°R)
$w_w := 8.33$	specific weight of water (lb/gal)
$\rho_w := 62.4$	specific weight of water (lb/ft^3)
$e_{cs} := 0.0005$	roughness of commercial steel (ft)
$g := 32.2$	gravitational constant (ft/sec^2)

Preliminary calculations

$P_{at} := p_{at} \cdot 144$	convert actual atm pressure to lb/ft^2
$P_{at} = 1826.640$	(lb/ft^2)

$T_{at} := t_{at} + 459.67$ convert actual atm temp to °R

$T_{at} = 519.670$ (°R)

$P_o := p_o \cdot 144$ convert table pressure to lb/ft²

$P_o = 1826.640$ (lb/ft²)

$T_o := t_o + 459.67$ convert table temperature to °R

$T_o = 519.670$ (°R)

$D_{bh} := \frac{d_{bh}}{12}$ convert borehole diameter to ft

$D_{bh} = 0.656$ (ft)

$ID_r := \frac{id_r}{12}$ convert return line diameter to ft

$ID_r = 0.469$ (ft)

$OD_{dp} := \frac{od_{dp}}{12}$ convert drill pipe outer diameter to ft

$OD_{dp} = 0.375$ (ft)

$ID_{dp} := \frac{id_{dp}}{12}$ convert drill pipe inner diameter to ft

$ID_{dp} = 0.319$ (ft)

$ID_c := \frac{id_c}{12}$ convert casing inner diameter to ft

$ID_c = 0.660$ (ft)

$OD_{dc} := \frac{od_{dc}}{12}$ convert drill collar outer diameter to ft

$OD_{dc} = 0.521$ (ft)

$ID_{dc} := \frac{id_{dc}}{12}$ convert drill collar inner diameter to ft

$ID_{dc} = 0.234$ (ft)

$OD_{tj} := \frac{od_{tj}}{12}$ convert tool joint outer diameter to ft

$OD_{tj} = 0.521$ (ft)

$ID_{tj} := \frac{id_{tj}}{12}$ convert tool joint inner diameter to ft

$ID_{tj} = 0.229$ (ft)

$d_n = 0.344$ nozzle diameter (in)

$D_n := \frac{d_n}{12}$ nozzle diameter converted to ft

$D_n = 0.029$ (ft)

$\mu_g := \mu_{air} \cdot 2.089 \cdot 10^{-5}$ convert viscosity to lbs-sec/ft^2

$\mu_g = 2.507 \times 10^{-7}$ (lbs-sec/ft^2)

$\mu_f := \mu_{mud} \cdot 2.089 \cdot 10^{-5}$ convert viscosity to lbs-sec/ft^2

$\mu_f = 6.267 \times 10^{-4}$ (lbs-sec/ft^2)

$Q_a := \frac{q_a}{60}$ convert air flow rate to ft^3/sec

$Q_a = 5.021$ (ft^3/sec)

$Q_f := \frac{q_f \cdot 231}{60 \cdot 12^3}$ convert mud flow to (ft^3/sec)

$Q_f = 0.426$ (ft^3/sec)

$\gamma_f := \gamma_f \cdot \frac{12^3}{231}$ convert spec. wt. of mud from ppg to lb/ft^3

$\gamma_f = 74.805$ (lb/ft^3)

$\rho_f := \frac{\gamma_f}{g}$ mud density (lb-sec^2/ft^4)

$\rho_f = 2.3231$ (lb-sec^2/ft^4)

$\nu_f := \frac{\mu_f}{\rho_f}$ mud kinematic viscosity (ft^2/sec)

$\nu_f = 2.698 \times 10^{-4}$ (ft^2/sec)

$\gamma_g := \frac{(P_{at} \cdot SG_a)}{R \cdot T_{at}}$ specific weight of air (lb/ft^3)

$\gamma_g = 0.0659$ (lbs/ft^3)

$\rho_g := \frac{\gamma_g}{g}$ gas (air) density (lb-sec^2/ft^4)

$\rho_g = 2.046 \times 10^{-3}$ (lb-sec^2/ft^4)

$\nu_g := \frac{\mu_g}{\rho_g}$ gas (air) kinematic viscosity (ft^2/sec)

$\nu_g = 1.225 \times 10^{-4}$ (ft^2/sec)

$rop := \frac{ROP}{60 \cdot 60}$ convet ROP to ft/sec

$rop = 0.0167$ (ft/sec)

$w_s := \left[\left(\frac{\pi}{4}\right) \cdot D_{bh}{}^2 \cdot 62.4 \cdot SG_r \cdot rop\right]$ weight rate of solid flow out of blooey line (lb/sec)

$w_s = 0.950$ (lb/sec)

$w_g := \gamma_g \cdot Q_a$ weight rate of flow of air (lb/sec)

$w_g = 0.331$ (lb/sec)

$w_f := \gamma_f \cdot Q_f$ weight rate of flow of mud (lb/sec)

$w_f = 31.833$ (lb/sec)

$w_t := w_g + w_f + w_s$ total weight rate of flow

$w_t = 33.114$

Section height calculations

$H_1 := D_c - L_{tj} \cdot \left(\frac{D_c}{L_{dp}}\right)$ calculate length of drill pipe in cased section

$H_1 = 6650.000$ (ft)

$H_2 := L_{tj} \cdot \left(\frac{D_c}{L_{dp}}\right)$ calculate length of tool joint in cased section

$H_2 = 350.000$ (ft)

$H_3 := D_{oh} - L_{dc} - L_{tj} \cdot \left(\frac{D_{oh} - L_{dc}}{L_{dp}}\right)$ calculate length of drill pipe in open hole section

$H_3 = 2375.000$ (ft)

$H_4 := L_{tj} \cdot \left(\frac{D_{oh} - L_{dc}}{L_{dp}} \right)$ calculate length of tool joints in open hole section

$H_4 = 125.000$ (ft)

$H_5 := L_{dc}$ length of drill collar

$H_5 = 500.000$ (ft)

Return line

$ID := ID_r$ set variable ID equal to return line's ID

$ID = 0.469$ (ft)

$A := \frac{\pi}{4} \cdot ID^2$ cross sectional area (ft^2)

$A = 0.173$ (ft^2)

$V := \frac{(Q_a + Q_f)}{A}$ velocity of flow through the line (ft/sec)

$V = 31.561$ (ft/sec)

$\nu_{ave} := \frac{(w_g \cdot \nu_g + w_f \cdot \nu_f)}{w_g + w_f}$ average kinematic viscosity of gas & mud

$\nu_{ave} = 2.682 \times 10^{-4}$ (ft^2/sec)

$N_R := \frac{(V \cdot ID)}{\nu_{ave}}$ Reynolds number calculations

$N_R = 5.515 \times 10^4$

$$f := \left[\frac{1}{-1.8 \cdot \log\left[\left(\frac{e_{cs}}{3.7 \cdot ID} \right)^{1.11} + \frac{6.9}{N_R} \right]} \right]^2$$

calculate friction factor using Haaland equation

$f = 0.0236$ Haaland friction factor

$P := P_o$

$P = 1826.640$

$T := T_o$

$T = 519.670$

$Q(P) := \frac{P_{at}}{P} \cdot \frac{T}{T_{at}} \cdot Q_a$ flow rate as a function of pressure

$\gamma_{mix}(P) := \frac{w_t}{Q(P) + Q_f}$ specific weight as a function of pressure

$V(P) := \frac{(Q(P) + Q_f)}{A}$ velocity as a function of pressure

$p_r := 15.66$ trial pressure (psia)

$P_r := p_r \cdot 144$ convert pressure to lb/ft²

$P_r = 2255.040$ (lb/ft²)

$\int_{P_{at}}^{P_r} \frac{1}{\gamma_{mix}(P) \cdot \left(V(P)^2 \cdot \frac{f}{2 \cdot g \cdot ID} \right)} \, dP = 100.069$ left hand side

$\int_0^{L_r} 1 \, dh = 100.000$ right hand side

$K_t := 25$ resistance coefficient for the blind Tee

$K_v := 0.2$ resistance coefficient for gate valves in blooey line

$\gamma_{mix} := \frac{w_t}{\left[\left(\frac{P_{at}}{P_r} \right) \cdot \left(\frac{T_o}{T_{at}} \right) \cdot Q_a + Q_f \right]}$ specific weight (lbs/ft³)

$\gamma_{mix} = 7.371$ (lb/ft³)

$V := \left[\frac{\left[\left(\frac{P_{at}}{P_r} \right) \cdot \left(\frac{T_o}{T_{at}} \right) \cdot Q_a + Q_f \right]}{A} \right]$ velocity (ft/sec)

$V = 26.034$ (ft/sec)

$\Delta P_T := \gamma_{mix} \cdot (2 \cdot K_v + K_t) \cdot \frac{V^2}{2 \cdot g}$ caclulate pressure loss due to Tee and valves

$\Delta P_T = 1970.240$ (lb/ft²)

$P_e := P_r + \Delta P_T$ sum pressure losses

$P_e = 4225.280$ (lb/ft²)

$p_e := \frac{P_e}{144}$ convert from lb/ft² to psia

$p_e = 29.342$ (psia)

Annulus cased hole drill pipe section

$ID := ID_c$ set inner diameter to casing ID

$ID = 0.660$ (ft)

$OD := OD_{dp}$ set outer diameter to drill pipe OD

$OD = 0.375$ (ft)

$P := P_e$ set pressure equal to previous section pressure

$P = 4225.280$ (lb/ft²)

$A := \frac{\pi}{4} \cdot \left(ID^2 - OD^2\right)$ cross sectional area (ft²)

$A = 0.232$ (ft²)

$T := T_o + 0.01 \cdot H_1$ calculate temperature at bottom of section

$T = 586.170$ (°R)

$T_{ave1} := \frac{T_o + T}{2}$ calculate average temperature of section

$T_{ave1} = 552.920$ (°R)

$T_{ave} := T_{ave1}$ set average temp to section averate temp

$\gamma_g(P) := \frac{P \cdot SG_a}{R \cdot T_{ave1}}$ specific weight of the gas (lb/ft³)

$\gamma_g(P) = 0.143211$ (lb/ft³)

$$\rho_g(P) := \frac{\gamma_g(P)}{g}$$ gas density (lb-sec²/ft⁴)

$$\rho_g(P) = 0.004448$$ (lb-sec²/ft⁴)

$$\nu_g(P) := \frac{\mu_g}{\rho_g(P)}$$ gas kinematic viscosity (ft²/sec)

$$\nu_g(P) = 0.00005636$$ (ft²/sec)

$$\nu_{ave}(P) := \frac{\left(w_g \cdot \nu_g(P) + w_f \cdot \nu_f\right)}{w_g + w_f}$$ average kinematic viscosity of gas & mud

$$\nu_{ave}(P) = 0.00026757$$ (ft²/sec)

$$V(P) := \left[\frac{\left[\left[\left(\frac{P_{at}}{P}\right) \cdot \left(\frac{T_{ave}}{T_{at}}\right) \cdot Q_a + Q_f\right]\right]}{A}\right]$$ velocity of flow in annulus (ft/sec)

$$V(P) = 11.801$$ (ft/sec)

$$N_R(P) := \frac{V(P) \cdot (ID - OD)}{\nu_{ave}(P)}$$ Reynolds number calculations

$$N_R(P) = 12573.819$$

$$f(P) := \left[\frac{1}{-1.8 \cdot \log\left[\left[\frac{e_{cs}}{3.7 \cdot (ID - OD)}\right]^{1.11} + \frac{6.9}{N_R(P)}\right]}\right]^2$$ calculate friction factor using Haaland equation

$$f(P) = 0.03164$$ Haaland friction factor

$$t := T - 459.67$$

$$p(P) := \frac{P}{144}$$

H&B correlations

$$\sigma_{w74}(P) := 75 - 1.108 \cdot p(P)^{0.349}$$

$$\sigma_{w280}(P) := 53 - 0.1048 \cdot p(P)^{0.637}$$

$$\sigma_{wt}(P) := \sigma_{w74}(P) - \frac{(t - 74)\cdot(\sigma_{w74}(P) - \sigma_{w280}(P))}{206}$$

$\sigma_{wt}(P) = 66.478$

$$\upsilon_{sL} := \frac{Q_f}{A}$$

$\upsilon_{sL} = 1.836$

$$\upsilon_{sg}(P) := \frac{\left[\left(\frac{P_{at}}{P}\right)\cdot\left(\frac{T_{ave}}{T_{at}}\right)\cdot Q_a\right]}{A}$$

$\upsilon_{sg}(P) = 9.965$

$d := ID$

$\rho_L := \gamma_f$

$\rho_L = 74.805$

$\mu_L := \mu_{mud}$

$\mu_L = 30.000$

$$N_{L\upsilon}(P) := 1.938\cdot\upsilon_{sL}\cdot\left(\frac{\rho_L}{\sigma_{wt}(P)}\right)^{0.25}$$

$N_{L\upsilon}(P) = 3.665$

$$N_{g\upsilon}(P) := 1.938\cdot\upsilon_{sg}(P)\cdot\left(\frac{\rho_L}{\sigma_{wt}(P)}\right)^{0.25}$$

$N_{g\upsilon}(P) = 19.891$

$$N_d(P) := 120.872\cdot d\cdot\left(\frac{\rho_L}{\sigma_{wt}(P)}\right)^{0.5}$$

$N_d(P) = 84.635$

$$N_L(P) := 0.15726 \cdot \mu_L \cdot \left(\frac{1}{\rho_L \cdot \sigma_{wt}(P)^3} \right)^{0.25}$$

$N_L(P) = 0.069$

$X1(P) := \log\left(N_L(P)\right) + 3$

$Y(P) := -2.69851 + 0.15841 \cdot X1(P) - 0.551 \cdot X1(P)^2 + 0.54785 \cdot X1(P)^3 - 0.12195 \cdot X1(P)^4$

$CN_L(P) := 10^{Y(P)}$

$p_a := p_o$

$$X_{HL}(P) := \frac{N_{L\upsilon}(P) \cdot \left(CN_L(P)\right) \cdot p(P)^{0.1}}{N_d(P) \cdot N_{g\upsilon}(P)^{0.575} \cdot p_a^{0.1}}$$

$FXHL(P) := \log\left(X_{HL}(P)\right) + 6$

$HLoFy(P) := -0.10307 + 0.61777 \cdot FXHL(P) - 0.63295 \cdot FXHL(P)^2 + 0.29598 \cdot FXHL(P)^3 - 0.0401 \cdot FXHL(P)^4$

$$X_\psi(P) := \frac{N_{g\upsilon}(P) \cdot N_L(P)^{0.38}}{N_d(P)^{2.14}}$$

$\psi(P) := 0.91163 - 4.82176 \cdot X_\psi(P) + 1232.25 \cdot X_\psi(P)^2 - 22253.6 \cdot X_\psi(P)^3 + 116174.3 \cdot X_\psi(P)^4$

$H_L(P) := \psi(P) \cdot HLoFy(P)$

$H_L(P) = 0.209$ liquid holdup fraction

$\upsilon_m(P) := \upsilon_{sL} + \upsilon_{sg}(P)$

$\upsilon_m(P) = 11.801$

$\gamma_{mix}(P) := \gamma_f \cdot H_L(P) + \gamma_g(P) \cdot \left(1 - H_L(P)\right)$

$\gamma_{mix}(P) = 15.721$ mixed specific weight (lb/ft^3)

$$Q(P) := \frac{P_{at}}{P} \cdot \frac{T_{ave}}{T_{at}} \cdot Q_a$$ flow rate as a function of pressure

$V(P) := \frac{(Q(P) + Q_f)}{A}$ velocity as a function of pressure

$p_1 := 1976.48$ trial pressure (psia)

$P_1 := p_1 \cdot 144$ convert pressure to lb/ft²

$P_1 = 284613.120$ (lb/ft²)

$$\int_{P_e}^{P_1} \frac{1}{\gamma_{mix}(P) \cdot \left[1 + (V(P))^2 \cdot \frac{f(P)}{2 \cdot g \cdot (ID - OD)}\right]} dP = 6649.943$$ left hand side

$$\int_0^{H_1} 1\, dh = 6650.000$$ right hand side

Annulus cased hole tool joint section

$ID := ID_c$ set inner diameter to casing ID

$OD := OD_{tj}$ set outer diameter to drill pipe OD

$P := P_1$ set pressure equal to previous section pressure

$A := \frac{\pi}{4} \cdot (ID^2 - OD^2)$ cross sectional area (ft²)

$T_{temp} := T$ keep previous section temperature

$T := T_{temp} + 0.01 \cdot H_2$ calculate temperature at bottom of section

$T = 589.670$ (°R)

$T_{ave2} := \frac{T_{temp} + T}{2}$ calculate average temperature of section

$T_{ave2} = 587.920$ (°R)

$T_{ave1} := T_{ave2}$ set average temp to section average temp

$\gamma_g(P) := \frac{P \cdot SG_a}{R \cdot T_{ave1}}$ specific weight of the gas (lb/ft³)

$\gamma_g(P) = 9.646656$ (lb/ft³)

$$\rho_g(P) := \frac{\gamma_g(P)}{g}$$ gas density (lb-sec^2/ft^4)

$$\rho_g(P) = 0.299586$$ (lb-sec^2/ft^4)

$$\nu_g(P) := \frac{\mu_g}{\rho_g(P)}$$ gas kinematic viscosity (ft^2/sec)

$$\nu_g(P) = 0.00000084$$ (ft^2/sec)

$$\nu_{ave}(P) := \frac{\left(w_g \cdot \nu_g(P) + w_f \cdot \nu_f\right)}{w_g + w_f}$$ average kinematic viscosity of gas & mud

$$\nu_{ave}(P) = 0.00026700$$ (ft^2/sec)

$$V(P) := \left[\frac{\left[\left(\frac{P_{at}}{P}\right)\cdot\left(\frac{T_{ave}}{T_{at}}\right)\cdot Q_a + Q_f\right]}{A}\right]$$ velocity of flow in annulus (ft/sec)

$$V(P) = 3.577$$ (ft/sec)

$$N_R(P) := \frac{V(P)\cdot(ID - OD)}{\nu_{ave}(P)}$$ Reynolds number calculations

$$N_R(P) = 1865.651$$

$$f(P) := \left[\frac{1}{-1.8\cdot\log\left[\left[\frac{e_{cs}}{3.7\cdot(ID - OD)}\right]^{1.11} + \frac{6.9}{N_R(P)}\right]}\right]^2$$ calculate friction factor using Haaland equation

Haaland friction factor

$$f(P) = 0.05440$$

$$t := T - 460$$

$$p(P) := \frac{P}{144}$$

H&B correlations

$$\sigma_{w74}(P) := 75 - 1.108 \cdot p(P)^{0.349}$$

$$\sigma_{w280}(P) := 53 - 0.1048 \cdot p(P)^{0.637}$$

$$\sigma_{wt}(P) := \sigma_{w74}(P) - \frac{(t - 74) \cdot (\sigma_{w74}(P) - \sigma_{w280}(P))}{206}$$

$$\sigma_{wt}(P) = 54.065$$

$$\upsilon_{sL} := \frac{Q_f}{A}$$

$$\upsilon_{sL} = 3.295$$

$$\upsilon_{sg}(P) := \frac{\left[\left(\frac{P_{at}}{P}\right) \cdot \left(\frac{T_{ave}}{T_{at}}\right) \cdot Q_a\right]}{A}$$

$$\upsilon_{sg}(P) = 0.282$$

$$d := ID$$

$$\rho_L := \gamma_f$$

$$\rho_L = 74.805$$

$$\mu_L := \mu_{mud}$$

$$\mu_L = 30.000$$

$$N_{L\upsilon}(P) := 1.938 \cdot \upsilon_{sL} \cdot \left(\frac{\rho_L}{\sigma_{wt}(P)}\right)^{0.25}$$

$$N_{L\upsilon}(P) = 6.926$$

$$N_{g\upsilon}(P) := 1.938 \cdot \upsilon_{sg}(P) \cdot \left(\frac{\rho_L}{\sigma_{wt}(P)}\right)^{0.25}$$

$$N_{g\upsilon}(P) = 0.593$$

$N_d(P) := 120.872 \cdot d \cdot \left(\frac{\rho_L}{\sigma_{wt}(P)}\right)^{0.5}$

$N_d(P) = 93.850$

$N_L(P) := 0.15726 \cdot \mu_L \cdot \left(\frac{1}{\rho_L \cdot \sigma_{wt}(P)^3}\right)^{0.25}$

$N_L(P) = 0.080$

$X1(P) := \log(N_L(P)) + 3$

$Y(P) := -2.69851 + 0.15841 \cdot X1(P) - 0.551 \cdot X1(P)^2 + 0.54785 \cdot X1(P)^3 - 0.12195 \cdot X1(P)^4$

$CN_L(P) := 10^{Y(P)}$

$p_a := p_o$

$X_{HL}(P) := \frac{N_{L\upsilon}(P) \cdot (CN_L(P)) \cdot p(P)^{0.1}}{N_d(P) \cdot N_{g\upsilon}(P)^{0.575} \cdot p_a^{0.1}}$

$FXHL(P) := \log(X_{HL}(P)) + 6$

$HLoFy(P) := -0.10307 + 0.61777 \cdot FXHL(P) - 0.63295 \cdot FXHL(P)^2 + 0.29598 \cdot FXHL(P)^3 - 0.0401 \cdot FX$

$X_\psi(P) := \frac{N_{g\upsilon}(P) \cdot N_L(P)^{0.38}}{N_d(P)^{2.14}}$

$\psi(P) := 0.91163 - 4.82176 \cdot X_\psi(P) + 1232.25 \cdot X_\psi(P)^2 - 22253.6 \cdot X_\psi(P)^3 + 116174.3 \cdot X_\psi(P)^4$

$H_L(P) := \psi(P) \cdot HLoFy(P)$

$H_L(P) = 0.728$ liquid holdup fraction

$\upsilon_m(P) := \upsilon_{sL} + \upsilon_{sg}(P)$

$\upsilon_m(P) = 3.577$

$\gamma_{mix}(P) := \gamma_f \cdot H_L(P) + \gamma_g(P) \cdot (1 - H_L(P))$

$\gamma_{mix}(P) = 57.076$ mixture specific weight (lb/ft^3)

$Q(P) := \frac{P_{at}}{P} \cdot \frac{T_{ave}}{T_{at}} \cdot Q_a$ flow rate as a function of pressure

$V(P) := \frac{(Q(P) + Q_f)}{A}$ velocity as a function of pressure

$p_2 := 2127.0$ trial pressure (psia)

$P_2 := p_2 \cdot 144$ convert pressure to lb/ft^2

$P_2 = 3.063 \times 10^5$ (lb/ft^2)

$$\int_{P_1}^{P_2} \frac{1}{\gamma_{mix}(P) \cdot \left[1 + (V(P))^2 \cdot \frac{f(P)}{2 \cdot g \cdot (ID - OD)}\right]} dP = 350.000$$ left hand side

$$\int_{H_1}^{H_1+H_2} 1 \, dh = 350.000$$ right hand side

Annulus of open hole drill pipe section

$ID := D_{bh}$ set inner diameter to casing ID

$OD := OD_{dp}$ set outer diameter to drill pipe OD

$P := P_2$ set pressure equal to previous section pressure

$A := \frac{\pi}{4} \cdot (ID^2 - OD^2)$ cross sectional area (ft^2)

$T_{temp} := T$ keep previous section temperature

$T := T_{temp} + 0.01 \cdot H_3$ calculate temperature at bottom of section

$T = 613.420$ (°R)

$T_{ave3} := \frac{T_{temp} + T}{2}$ calculate average temperature of section

$T_{ave3} = 601.545$ (°R)

$T_{ave} := T_{ave3}$ set average temp to section average temp

$$\gamma_g(P) := \frac{P \cdot SG_a}{R \cdot T_{ave1}}$$ specific weight of the gas (lb/ft³)

$\gamma_g(P) = 10.381303$ (lb/ft³)

$$\rho_g(P) := \frac{\gamma_g(P)}{g}$$ gas density (lb-sec²/ft⁴)

$\rho_g(P) = 0.322401$ (lb/ft³)

$$\nu_g(P) := \frac{\mu_g}{\rho_g(P)}$$ gas kinematic viscosity (ft²/sec)

$\nu_g(P) = 0.00000078$ (ft²/sec)

$$\nu_{ave}(P) := \frac{\left(w_g \cdot \nu_g(P) + w_f \cdot \nu_f\right)}{w_g + w_f}$$ average kinematic viscosity of gas & mud

$\nu_{ave}(P) = 0.00026700$ (ft²/sec)

$$V(P) := \left[\frac{\left[\left(\frac{P_{at}}{P}\right) \cdot \left(\frac{T_{ave}}{T_{at}}\right) \cdot Q_a + Q_f\right]}{A}\right]$$ velocity of flow in annulus (ft/sec)

$V(P) = 2.020$ (ft/sec)

$$N_R(P) := \frac{V(P) \cdot (ID - OD)}{\nu_{ave}(P)}$$ Reynolds number calculations

$N_R(P) = 2128.120$

$$f(P) := \left[\frac{1}{-1.8 \cdot \log\left[\left[\frac{e_{cs}}{3.7 \cdot (ID - OD)}\right]^{1.11} + \frac{6.9}{N_R(P)}\right]}\right]^2$$ calculate friction factor using Haaland equation

$f(P) = 0.05091$ Haaland friction factor

$t := T - 460$

$p(P) := \frac{P}{144}$

H&B correlations

$\sigma_{w74}(P) := 75 - 1.108 \cdot p(P)^{0.349}$

$\sigma_{w280}(P) := 53 - 0.1048 \cdot p(P)^{0.637}$

$\sigma_{wt}(P) := \sigma_{w74}(P) - \frac{(t - 74) \cdot \left(\sigma_{w74}(P) - \sigma_{w280}(P)\right)}{206}$

$\sigma_{wt}(P) = 51.322$

$\upsilon_{sL} := \frac{Q_f}{A}$

$\upsilon_{sL} = 1.868$

$\upsilon_{sg}(P) := \frac{\left[\left(\frac{P_{at}}{P}\right) \cdot \left(\frac{T_{ave}}{T_{at}}\right) \cdot Q_a\right]}{A}$

$\upsilon_{sg}(P) = 0.152$

$d := ID$

$\rho_L := \gamma_f$

$\rho_L = 74.805$

$\mu_L := \mu_{mud}$

$\mu_L = 30.000$

$N_{L\upsilon}(P) := 1.938 \cdot \upsilon_{sL} \cdot \left(\frac{\rho_L}{\sigma_{wt}(P)}\right)^{0.25}$

$N_{L\upsilon}(P) = 3.978$

$$N_{gv}(P) := 1.938 \cdot \upsilon_{sg}(P) \cdot \left(\frac{\rho_L}{\sigma_{wt}(P)} \right)^{0.25}$$

$$N_{gv}(P) = 0.324$$

$$N_d(P) := 120.872 \cdot d \cdot \left(\frac{\rho_L}{\sigma_{wt}(P)} \right)^{0.5}$$

$$N_d(P) = 95.765$$

$$N_L(P) := 0.15726 \cdot \mu_L \cdot \left(\frac{1}{\rho_L \cdot \sigma_{wt}(P)^3} \right)^{0.25}$$

$$N_L(P) = 0.084$$

$$X1(P) := \log(N_L(P)) + 3$$

$$Y(P) := -2.69851 + 0.15841 \cdot X1(P) - 0.551 \cdot X1(P)^2 + 0.54785 \cdot X1(P)^3 - 0.12195 \cdot X1(P)^4$$

$$CN_L(P) := 10^{Y(P)}$$

$$p_a := p_o$$

$$X_{HL}(P) := \frac{N_{L\upsilon}(P) \cdot (CN_L(P)) \cdot p(P)^{0.1}}{N_d(P) \cdot N_{gv}(P)^{0.575} \cdot p_a^{0.1}}$$

$$FXHL(P) := \log(X_{HL}(P)) + 6$$

$$HLoFy(P) := -0.10307 + 0.61777 \cdot FXHL(P) - 0.63295 \cdot FXHL(P)^2 + 0.29598 \cdot FXHL(P)^3 - 0.0401 \cdot FX$$

$$X_\psi(P) := \frac{N_{gv}(P) \cdot N_L(P)^{0.38}}{N_d(P)^{2.14}}$$

$$\psi(P) := 0.91163 - 4.82176 \cdot X_\psi(P) + 1232.25 \cdot X_\psi(P)^2 - 22253.6 \cdot X_\psi(P)^3 + 116174.3 \cdot X_\psi(P)^4$$

$$H_L(P) := \psi(P) \cdot HLoFy(P)$$ liquid holdup fraction

$v_m(P) := v_{sL} + v_{sg}(P)$

$v_m(P) = 2.020$

$\gamma_{mix}(P) := \gamma_f \cdot H_L(P) + \gamma_g(P) \cdot (1 - H_L(P))$ mixture specific weight (lb/ft^3)

$\gamma_{mix}(P) = 54.851$ (lb/ft^3)

$$Q(P) := \frac{P_{at}}{P} \cdot \frac{T_{ave}}{T_{at}} \cdot Q_a$$

$$V(P) := \frac{(Q(P) + Q_f)}{A}$$ velocity as a function of pressure

$p_3 := 3080.3$ trial pressure (psia)

$P_3 := p_3 \cdot 144$ convert pressure to lb/ft^2

$P_3 = 4.436 \times 10^5$ (lb/ft^2)

$$\int_{P_2}^{P_3} \frac{1}{\gamma_{mix}(P) \cdot \left[1 + (V(P))^2 \cdot \frac{f(P)}{2 \cdot g \cdot (ID - OD)}\right]} dP = 2375.014$$ left hand side

$$\int_{H_1+H_2}^{H_1+H_2+H_3} 1 \, dh = 2375.000$$ right hand side

Annulus of open hole tool joint section

$ID := D_{bh}$ set inner diameter to casing ID

$OD := OD_{tj}$ set outer diameter to drill pipe OD

$P := P_3$ set pressure equal to previous section pressure

$$A := \frac{\pi}{4} \cdot \left(ID^2 - OD^2\right)$$ cross sectional area (ft^2)

$T_{temp} := T$ keep previous section temperature

$T := T_{temp} + 0.01 \cdot H_4$ calculate temperature at bottom of section

$T = 614.670$ (°R)

$T_{ave4} := \dfrac{T_{temp} + T}{2}$ calculate average temperature of section

$T_{ave4} = 614.045$ (°R)

$T_{ave} := T_{ave4}$ set average temp to section average temp

$\gamma_g(P) := \dfrac{P \cdot SG_a}{R \cdot T_{ave1}}$ specific weight of the gas (lb/ft³)

$\gamma_g(P) = 15.034098$ (lb/ft³)

$\rho_g(P) := \dfrac{\gamma_g(P)}{g}$ gas density (lb-sec²/ft⁴)

$\rho_g(P) = 0.466897$ (lb-sec²/ft⁴)

$\nu_g(P) := \dfrac{\mu_g}{\rho_g(P)}$ gas kinematic viscosity (ft²/sec)

$\nu_g(P) = 0.00000054$ (ft²/sec)

$\nu_{ave}(P) := \dfrac{\left(w_g \cdot \nu_g(P) + w_f \cdot \nu_f\right)}{w_g + w_f}$ average kinematic viscosity of gas & mud

$\nu_{ave}(P) = 0.00026700$ (ft²/sec)

$V(P) := \left[\dfrac{\left[\left(\dfrac{P_{at}}{P}\right) \cdot \left(\dfrac{T_{ave}}{T_{at}}\right) \cdot Q_a + Q_f\right]}{A}\right]$ velocity of flow in annulus (ft/sec)

$V(P) = 3.594$ (ft/sec)

$N_R(P) := \dfrac{V(P) \cdot (ID - OD)}{\nu_{ave}(P)}$ Reynolds number calculations

$N_R(P) = 1823.031$

$$f(P) := \left[\frac{1}{-1.8\cdot\log\left[\left[\frac{e_{cs}}{3.7\cdot(ID - OD)}\right]^{1.11} + \frac{6.9}{N_R(P)}\right]}\right]^2$$ calculate friction factor using Haaland equation

$f(P) = 0.05488$ Haaland friction factor

$t := T - 460$

$$p(P) := \frac{P}{144}$$

H&B correlations

$\sigma_{w74}(P) := 75 - 1.108\cdot p(P)^{0.349}$

$\sigma_{w280}(P) := 53 - 0.1048\cdot p(P)^{0.637}$

$$\sigma_{wt}(P) := \sigma_{w74}(P) - \frac{(t - 74)\cdot(\sigma_{w74}(P) - \sigma_{w280}(P))}{206}$$

$\sigma_{wt}(P) = 48.415$

$$\upsilon_{sL} := \frac{Q_f}{A}$$

$\upsilon_{sL} = 3.399$

$$\upsilon_{sg}(P) := \frac{\left[\left(\frac{P_{at}}{P}\right)\cdot\left(\frac{T_{ave}}{T_{at}}\right)\cdot Q_a\right]}{A}$$

$\upsilon_{sg}(P) = 0.195$

$d := ID$

$\rho_L := \gamma_f$

$\rho_L = 74.805$

$\mu_L := \mu_{mud}$

$\mu_L = 30.000$

$$N_{L\upsilon}(P) := 1.938 \cdot \upsilon_{sL} \cdot \left(\frac{\rho_L}{\sigma_{wt}(P)}\right)^{0.25}$$

$N_{L\upsilon}(P) = 7.345$

$$N_{g\upsilon}(P) := 1.938 \cdot \upsilon_{sg}(P) \cdot \left(\frac{\rho_L}{\sigma_{wt}(P)}\right)^{0.25}$$

$N_{g\upsilon}(P) = 0.422$

$$N_d(P) := 120.872 \cdot d \cdot \left(\frac{\rho_L}{\sigma_{wt}(P)}\right)^{0.5}$$

$N_d(P) = 98.599$

$$N_L(P) := 0.15726 \cdot \mu_L \cdot \left(\frac{1}{\rho_L \cdot \sigma_{wt}(P)^3}\right)^{0.25}$$

$N_L(P) = 0.087$

$$X1(P) := \log\left(N_L(P)\right) + 3$$

$$Y(P) := -2.69851 + 0.15841 \cdot X1(P) - 0.551 \cdot X1(P)^2 + 0.54785 \cdot X1(P)^3 - 0.12195 \cdot X1(P)^4$$

$$CN_L(P) := 10^{Y(P)}$$

$$p_a := p_o$$

$$X_{HL}(P) := \frac{N_{L\upsilon}(P) \cdot \left(CN_L(P)\right) \cdot p(P)^{0.1}}{N_d(P) \cdot N_{g\upsilon}(P)^{0.575} \cdot p_a{}^{0.1}}$$

$$FXHL(P) := \log\left(X_{HL}(P)\right) + 6$$

$$HLoFy(P) := -0.10307 + 0.61777 \cdot FXHL(P) - 0.63295 \cdot FXHL(P)^2 + 0.29598 \cdot FXHL(P)^3 - 0.0401 \cdot FXHL$$

$$X_\psi(P) := \frac{N_{g\upsilon}(P) \cdot N_L(P)^{0.38}}{N_d(P)^{2.14}}$$

$$\psi(P) := 0.91163 - 4.82176 \cdot X_\psi(P) + 1232.25 \cdot X_\psi(P)^2 - 22253.6 \cdot X_\psi(P)^3 + 116174.3 \cdot X_\psi(P)^4$$

$H_L(P) := \psi(P)\cdot HLoFy(P)$

$H_L(P) = 0.783$ liquid holdup fraction

$\upsilon_m(P) := \upsilon_{sL} + \upsilon_{sg}(P)$

$\upsilon_m(P) = 3.594$

$\gamma_{mix}(P) := \gamma_f\cdot H_L(P) + \gamma_g(P)\cdot(1 - H_L(P))$

$\gamma_{mix}(P) = 61.839$ mixture specific weight (lb/ft^3)

$Q(P) := \frac{P_{at}}{P}\cdot\frac{T_{ave}}{T_{at}}\cdot Q_a$ flow rate as a function of pressure

$V(P) := \frac{(Q(P) + Q_f)}{A}$ velocity as a function of pressure

$p_4 := 3138.470$ trial pressure (psia)

$P_4 := p_4\cdot 144$ convert pressure to lb/ft^2

$P_4 = 4.519 \times 10^5$ (lb/ft^2)

$$\int_{P_3}^{P_4} \frac{1}{\gamma_{mix}(P)\cdot\left[1 + (V(P))^2\cdot\frac{f(P)}{2\cdot g\cdot(ID - OD)}\right]}\,dP = 125.082$$ left hand side

$$\int_{H_1+H_2+H_3}^{H_1+H_2+H_3+H_4} 1\,dh = 125.000$$ right hand side

Annulus open hole drill of drill collar section

$ID := D_{bh}$ set inner diameter to casing ID

$OD := OD_{dc}$	set outer diameter to drill pipe OD
$P := P_4$	set pressure equal to previous section pressure
$A := \frac{\pi}{4} \cdot \left(ID^2 - OD^2\right)$	cross sectional area (ft²)
$T_{temp} := T$	keep previous section temperature
$T := T_{temp} + 0.01 \cdot H_5$	calculate temperature at bottom of section
$T = 619.670$	(°R)
$T_{ave5} := \frac{T_{temp} + T}{2}$	calculate average temperature of section
$T_{ave5} = 617.170$	(°R)
$T_{ave} := T_{ave5}$	set average temp to section average temp
$\gamma_g(P) := \frac{P \cdot SG_a}{R \cdot T_{ave1}}$	specific weight of the gas (lb/ft³)
$\gamma_g(P) = 15.318010$	(lb/ft³)
$\rho_g(P) := \frac{\gamma_g(P)}{g}$	gas density (lb-sec²/ft⁴)
$\rho_g(P) = 0.475715$	(lb-sec²/ft⁴)
$\nu_g(P) := \frac{\mu_g}{\rho_g(P)}$	gas kinematic viscosity (ft²/sec)
$\nu_g(P) = 0.00000053$	(ft²/sec)
$\nu_{ave}(P) := \frac{\left(w_g \cdot \nu_g(P) + w_f \cdot \nu_f\right)}{w_g + w_f}$	average kinematic viscosity of gas & mud
$\nu_{ave}(P) = 0.00026700$	(ft²/sec)
$V(P) := \left[\frac{\left[\left[\left(\frac{P_{at}}{P}\right) \cdot \left(\frac{T_{ave}}{T_{at}}\right) \cdot Q_a + Q_f\right]\right]}{A}\right]$	velocity of flow in annulus (ft/sec)
$V(P) = 3.592$	(ft/sec)

$$N_R(P) := \frac{V(P)\cdot(ID - OD)}{\nu_{ave}(P)}$$ Reynolds number calculations

$N_R(P) = 1821.692$

$$f(P) := \left[\frac{1}{-1.8\cdot\log\left[\left[\frac{e_{cs}}{3.7\cdot(ID - OD)}\right]^{1.11} + \frac{6.9}{N_R(P)}\right]}\right]^2$$ friction factor using Haaland equation

$f(P) = 0.05489$ Haaland friction factor

$t := T - 460$

$$p(P) := \frac{P}{144}$$

H&B correlations

$$\sigma_{w74}(P) := 75 - 1.108\cdot p(P)^{0.349}$$

$$\sigma_{w280}(P) := 53 - 0.1048\cdot p(P)^{0.637}$$

$$\sigma_{wt}(P) := \sigma_{w74}(P) - \frac{(t - 74)\cdot(\sigma_{w74}(P) - \sigma_{w280}(P))}{206}$$

$\sigma_{wt}(P) = 47.743$

$$\upsilon_{sL} := \frac{Q_f}{A}$$

$\upsilon_{sL} = 3.399$

$$\upsilon_{sg}(P) := \frac{\left[\left(\frac{P_{at}}{P}\right)\cdot\left(\frac{T_{ave}}{T_{at}}\right)\cdot Q_a\right]}{A}$$

$\upsilon_{sg}(P) = 0.193$

$d := ID$

$\rho_L := \gamma_f$

$\rho_L = 74.805$

$\mu_L := \mu_{mud}$

$\mu_L = 30.000$

$$N_{L\upsilon}(P) := 1.938 \cdot \upsilon_{sL} \cdot \left(\frac{\rho_L}{\sigma_{wt}(P)}\right)^{0.25}$$

$N_{L\upsilon}(P) = 7.370$

$$N_{g\upsilon}(P) := 1.938 \cdot \upsilon_{sg}(P) \cdot \left(\frac{\rho_L}{\sigma_{wt}(P)}\right)^{0.25}$$

$N_{g\upsilon}(P) = 0.417$

$$N_d(P) := 120.872 \cdot d \cdot \left(\frac{\rho_L}{\sigma_{wt}(P)}\right)^{0.5}$$

$N_d(P) = 99.290$

$$N_L(P) := 0.15726 \cdot \mu_L \cdot \left(\frac{1}{\rho_L \cdot \sigma_{wt}(P)^3}\right)^{0.25}$$

$N_L(P) = 0.088$

$$X1(P) := \log\left(N_L(P)\right) + 3$$

$$Y(P) := -2.69851 + 0.15841 \cdot X1(P) - 0.551 \cdot X1(P)^2 + 0.54785 \cdot X1(P)^3 - 0.12195 \cdot X1(P)^4$$

$$CN_L(P) := 10^{Y(P)}$$

$p_a := p_o$

$$X_{HL}(P) := \frac{N_{L\upsilon}(P) \cdot \left(CN_L(P)\right) \cdot p(P)^{0.1}}{N_d(P) \cdot N_{g\upsilon}(P)^{0.575} \cdot p_a^{\ 0.1}}$$

$$FXHL(P) := \log\left(X_{HL}(P)\right) + 6$$

$$HLoFy(P) := -0.10307 + 0.61777 \cdot FXHL(P) - 0.63295 \cdot FXHL(P)^2 + 0.29598 \cdot FXHL(P)^3 - 0.0401 \cdot FXHL(P)^4$$

$$X_\psi(P) := \frac{N_{g\upsilon}(P)\cdot N_L(P)^{0.38}}{N_d(P)^{2.14}}$$

$$\psi(P) := 0.91163 - 4.82176\cdot X_\psi(P) + 1232.25\cdot X_\psi(P)^2 - 22253.6\cdot X_\psi(P)^3 + 116174.3\cdot X_\psi(P)^4$$

$$H_L(P) := \psi(P)\cdot HLoFy(P)$$

$H_L(P) = 0.785$ liquid holdup fraction

$$\upsilon_m(P) := \upsilon_{sL} + \upsilon_{sg}(P)$$

$\upsilon_m(P) = 3.592$

$\gamma_{mix}(P) := \gamma_f\cdot H_L(P) + \gamma_g(P)\cdot(1 - H_L(P))$ mixture specific weight (lb/ft^3)

$\gamma_{mix}(P) = 62.013$ (lb/ft^3)

$Q(P) := \frac{P_{at}}{P}\cdot\frac{T_{ave}}{T_{at}}\cdot Q_a$ flow rate as a function of pressure

$V(P) := \frac{(Q(P) + Q_f)}{A}$ velocity as a function of pressure

$p_5 := 3360.00 + p_{at}$ bottom hole pressure (psia)

$P_5 := p_5\cdot 144$ convert pressure to lb/ft^2

$P_5 = 4.857 \times 10^5$ (lb/ft^2)

$$\int_{P_4}^{P_5} \frac{1}{\gamma_{mix}(P)\cdot\left[1 + (V(P))^2\cdot\frac{f(P)}{2\cdot g\cdot(ID - OD)}\right]}\,dP = 500.136$$ left hand side

$$\int_{H_1+H_2+H_3+H_4}^{H_1+H_2+H_3+H_4+H_5} 1\,dh = 500.000$$ right hand side

$p_{bh} := p_5 - p_{at}$ $\quad p_5 = 3372.685$ bottom hole pressure (pisg)

$p_{bh} = 3360.000$ (psig)

Bottom hole through the bit nozzles

$$D_e := \left[N_n \cdot (D_n)^2\right]^{0.5}$$ equivalent single diameter (ft)

$D_e = 0.050$ (ft)

$$\gamma_{mix} := \frac{(w_g + w_f)}{\left[Q_f + Q_a \cdot \left(\frac{P_{at}}{P_5}\right) \cdot \frac{T}{T_{at}}\right]}$$ specific weight of mixture (lb/ft^3)

$\gamma_{mix} = 71.784$ (lb/ft^3)

$$\Delta P := \frac{(w_g + w_f)^2}{2 \cdot C^2 \cdot g \cdot \gamma_{mix} \cdot \left(\frac{\pi}{4}\right)^2 \cdot D_e^4}$$ approximate of the pressure change of the aerated fluid through the drill bit

$\Delta P = 91241.257$ (lb/ft^2)

$P := P_5 + \Delta P$ sum pressures to get pressure above drill bit

$P = 576907.897$ (lb/ft^2)

$$p := \frac{P}{144}$$ convert pressure to lb/in^2

$p = 4006.305$ (lb/in^2)

Open hole inside drill string drill collar section

$ID := ID_{dc}$ set inner diameter to casing ID

$$A := \frac{\pi}{4} \cdot (ID^2)$$ cross sectional area (ft^2)

$T_{ave} := T_{ave5}$ set ave temp to section ave temp

$$\gamma_g := \frac{P \cdot SG_a}{R \cdot T_{ave}}$$ specific weight of the gas (lb/ft^3)

$\gamma_g = 17.518$ (lb/ft^3)

$$\rho_g := \frac{\gamma_g}{g}$$ gas density ($lb\text{-}sec^2/ft^4$)

$\rho_g = 0.544$ ($lb\text{-}sec^2/ft^4$)

$$\nu_g := \frac{\mu_g}{\rho_g}$$ gas kinematic viscosity (ft^2/sec)

$\nu_g = 4.608 \times 10^{-7}$ (ft^2/sec)

$$\nu_{ave} := \frac{\left(w_g \cdot \nu_g + w_f \cdot \nu_f\right)}{w_g + w_f}$$ average kinematic viscosity of gas & mud

$\nu_{ave} = 2.670 \times 10^{-4}$ (ft^2/sec)

$$V := \left[\frac{\left[\left(\frac{P_{at}}{P}\right)\cdot\left(\frac{T_{ave}}{T_{at}}\right)\cdot Q_a + Q_f\right]}{A}\right]$$ velocity of flow in annulus (ft/sec)

$V = 10.301$ (ft/sec)

$$N_R := \frac{V \cdot ID}{\nu_{ave}}$$ Reynolds number calculations

$N_R = 9.043 \times 10^3$

$$f := \left[\frac{1}{-1.8 \cdot \log\left[\left(\frac{e_{cs}}{3.7 \cdot ID}\right)^{1.11} + \frac{6.9}{N_R}\right]}\right]^2$$ calculate friction factor using Haaland equation

$f = 0.0345$ Haaland friction factor

$$Q(P) := \frac{P_{at}}{P} \cdot \frac{T_{ave}}{T_{at}} \cdot Q_a$$ flow rate as a function of pressure

$\gamma_{mix}(P) := \dfrac{w_g + w_f}{Q(P) + Q_f}$ specific weight as a function of pressure

$V(P) := \dfrac{(Q(P) + Q_f)}{A}$ velocity as a function of pressure

$p_4 := 3816.2$ trial pressure (psia)

$P_4 := p_4 \cdot 144$ convert pressure to lb/ft^2

$P_4 = 5.495 \times 10^5$ (lb/ft^2)

$$\int_{P_4}^{P} \frac{1}{\gamma_{mix}(P) \cdot \left[1 - (V(P))^2 \cdot \dfrac{f}{2 \cdot g \cdot (ID)}\right]} \, dP = 500.054$$ left hand side

$$\int_{H_1+H_2+H_3+H_4}^{H_1+H_2+H_3+H_4+H_5} 1 \, dh = 500.000$$ right hand side

Open hole inside drill string tool joint section

$ID := ID_{tj}$ set inner diameter to casing ID

$A := \dfrac{\pi}{4} \cdot \left(ID^2\right)$ cross sectional area (ft^2)

$T_{ave} := T_{ave4}$ set ave temp to section ave temp

$P := P_4$ set pressure to previous section pressure

$\gamma_g := \dfrac{P \cdot SG_a}{R \cdot T_{ave}}$ specific weight of the gas (lb/ft^3)

$\gamma_g = 16.772$ (lb/ft^3)

$\rho_g := \dfrac{\gamma_g}{g}$ gas density (lb-sec^2/ft^4)

$\rho_g = 0.521$ (lb-sec^2/ft^4)

$\nu_g := \frac{\mu_g}{\rho_g}$ gas kinematic viscosity (ft^2/sec)

$\nu_g = 4.813 \times 10^{-7}$ (ft^2/sec)

$\nu_{ave} := \frac{(w_g \cdot \nu_g + w_f \cdot \nu_f)}{w_g + w_f}$ average kinematic viscosity of gas & mud

$\nu_{ave} = 2.670 \times 10^{-4}$ (ft^2/sec)

$V := \left[\frac{\left[\left(\frac{P_{at}}{P}\right) \cdot \left(\frac{T_{ave}}{T_{at}}\right) \cdot Q_a + Q_f\right]}{A}\right]$ velocity of flow in annulus (ft/sec)

$V = 10.795$ (ft/sec)

$N_R := \frac{V \cdot ID}{\nu_{ave}}$ Reynolds number calculations

$N_R = 9.266 \times 10^3$

$f := \left[\frac{1}{-1.8 \cdot \log\left[\left(\frac{e_{cs}}{3.7 \cdot ID}\right)^{1.11} + \frac{6.9}{N_R}\right]}\right]^2$ calculate friction factor using Haaland equation

$f = 0.0343$ Haaland friction factor

$Q(P) := \frac{P_{at}}{P} \cdot \frac{T_{ave}}{T_{at}} \cdot Q_a$ flow rate as a function of pressure

$\gamma_{mix}(P) := \frac{w_g + w_f}{Q(P) + Q_f}$ specific weight as a function of pressure

$V(P) := \frac{(Q(P) + Q_f)}{A}$ velocity as a function of pressure

$p_3 := 3770.489$

$P_3 := p_3 \cdot 144$ convert pressure to lb/ft^2

$P_3 = 5.430 \times 10^5$ (lb/ft^2)

$$\int_{P_3}^{P_4} \frac{1}{\gamma_{mix}(P)\cdot\left[1-(V(P))^2\cdot\frac{f}{2\cdot g\cdot(ID)}\right]}\,dP = 125.088$$ left hand side

$$\int_{H_1+H_2+H_3}^{H_1+H_2+H_3+H_4} 1\,dh = 125.000$$ right hand side

Open hole inside drill string drill pipe section

$ID := ID_{dp}$	set inner diameter to casing ID
$A := \frac{\pi}{4}\cdot\left(ID^2\right)$	cross sectional area (ft^2)
$T_{ave} := T_{ave3}$	set ave temp to section ave temp
$P := P_3$	set pressure to previous section pressure
$\gamma_g := \frac{P\cdot SG_a}{R\cdot T_{ave}}$	specific weight of the gas (lb/ft^3)
$\gamma_g = 16.915$	(lb/ft^3)
$\rho_g := \frac{\gamma_g}{g}$	gas density ($lb\text{-}sec^2/ft^4$)
$\rho_g = 0.525$	($lb\text{-}sec^2/ft^4$)
$\nu_g := \frac{\mu_g}{\rho_g}$	gas (air) kinematic viscosity (ft^2/sec)
$\nu_g = 4.772\times10^{-7}$	(ft^2/sec)
$\nu_{ave} := \frac{\left(w_g\cdot\nu_g + w_f\cdot\nu_f\right)}{w_g + w_f}$	average kinematic viscosity of gas & mud
$\nu_{ave} = 2.670\times10^{-4}$	(ft^2/sec)
$V := \left[\frac{\left[\left(\frac{P_{at}}{P}\right)\cdot\left(\frac{T_{ave}}{T_{at}}\right)\cdot Q_a + Q_f\right]}{A}\right]$	velocity of flow in annulus (ft/sec)
$V = 5.575$	(ft/sec)

$$N_R := \frac{V \cdot ID}{\nu_{ave}}$$ Reynolds number calculations

$$N_R = 6.657 \times 10^3$$

$$f := \left[\frac{1}{-1.8 \cdot \log\left[\left(\frac{e_{cs}}{3.7 \cdot ID}\right)^{1.11} + \frac{6.9}{N_R}\right]}\right]^2$$ calculate friction factor using Haaland equation

$$f = 0.0363$$ Haaland friction factor

$$Q(P) := \frac{P_{at}}{P} \cdot \frac{T_{ave}}{T_{at}} \cdot Q_a$$ flow rate as a function of pressure

$$\gamma_{mix}(P) := \frac{w_g + w_f}{Q(P) + Q_f}$$ specific weight as a function of pressure

$$V(P) := \frac{\left(Q(P) + Q_f\right)}{A}$$ velocity as a function of pressure

$$p_2 := 2654.35$$ trial pressure (psia)

$$P_2 := p_2 \cdot 144$$ convert pressure to lb/ft^2

$$P_2 = 3.822 \times 10^5$$ (lb/ft^2)

$$\int_{P_2}^{P_3} \frac{1}{\gamma_{mix}(P) \cdot \left[1 - (V(P))^2 \cdot \frac{f}{2 \cdot g \cdot (ID)}\right]} \, dP = 2375.089$$ left hand side

$$\int_{H_1+H_2}^{H_1+H_2+H_3} 1 \, dh = 2375.000$$ right hand side

Casing inside drill string tool joint section

$$ID := ID_{tj}$$ set inner diameter to casing ID

$$A := \frac{\pi}{4} \cdot \left(ID^2\right)$$ cross sectional area (ft^2)

$$T_{ave} := T_{ave2}$$ set ave temp to section ave temp

$P := P_2$ set pressure to previous section pressure

$\gamma_g := \frac{P \cdot SG_a}{R \cdot T_{ave}}$ specific weight of the gas (lb/ft^3)

$\gamma_g = 12.184$ (lb/ft^3)

$\rho_g := \frac{\gamma_g}{g}$ gas (air) density (lb-sec^2/ft^4)

$\rho_g = 0.378$ (lb-sec^2/ft^4)

$\nu_g := \frac{\mu_g}{\rho_g}$ gas (air) kinematic viscosity (ft^2/sec)

$\nu_g = 6.625 \times 10^{-7}$ (ft^2/sec)

$\nu_{ave} := \frac{(w_g \cdot \nu_g + w_f \cdot \nu_f)}{w_g + w_f}$ average kinematic viscosity of gas & mud

$\nu_{ave} = 2.670 \times 10^{-4}$ (ft^2/sec)

$V := \left[\frac{\left[\left(\frac{P_{at}}{P}\right) \cdot \left(\frac{T_{ave}}{T_{at}}\right) \cdot Q_a + Q_f\right]}{A}\right]$ velocity of flow in annulus (ft/sec)

$V = 10.975$ (ft/sec)

$N_R := \frac{V \cdot ID}{\nu_{ave}}$ Reynolds number calculations

$N_R = 9.420 \times 10^3$

$f := \left[\frac{1}{-1.8 \cdot \log\left[\left(\frac{e_{cs}}{3.7 \cdot ID}\right)^{1.11} + \frac{6.9}{N_R}\right]}\right]^2$ calculate friction factor using Haaland equation

$f = 0.0342$ Haaland friction factor

$Q(P) := \frac{P_{at}}{P} \cdot \frac{T_{ave}}{T_{at}} \cdot Q_a$ flow rate as a function of pressure

$\gamma_{mix}(P) := \dfrac{w_g + w_f}{Q(P) + Q_f}$ specific weight as a function of pressure

$V(P) := \dfrac{(Q(P) + Q_f)}{A}$ velocity as a function of pressure

$p_1 := 2530.2$ trial pressure (psia)

$P_1 := p_1 \cdot 144$ convert pressure to lb/ft^2

$P_1 = 3.643 \times 10^5$ (lb/ft^2)

$$\int_{P_1}^{P_2} \frac{1}{\gamma_{mix}(P) \cdot \left[1 - (V(P))^2 \cdot \dfrac{f}{2 \cdot g \cdot (ID)}\right]} \, dP = 350.036$$ left hand side

$$\int_{H_1}^{H_1+H_2} 1 \, dh = 350.000$$ right hand side

Casing inside drill string drill pipe section

$ID := ID_{dp}$ set inner diameter to casing ID

$A := \dfrac{\pi}{4} \cdot \left(ID^2\right)$ cross sectional area (ft^2)

$T_{ave} := T_{ave1}$ set ave temp to section ave temp

$P := P_1$ set pressure to previous section pressure

$\gamma_g := \dfrac{P \cdot SG_a}{R \cdot T_{ave}}$ specific weight of the gas (lb/ft^3)

$\gamma_g = 12.349$ (lb/ft^3)

$\rho_g := \dfrac{\gamma_g}{g}$ gas density (lb-sec^2/ft^4)

$\rho_g = 0.384$ (lb-sec^2/ft^4)

$\nu_g := \dfrac{\mu_g}{\rho_g}$ gas (air) kinematic viscosity (ft^2/sec)

$\nu_g = 6.536 \times 10^{-7}$ (ft²/sec)

$\nu_{ave} := \dfrac{(w_g \cdot \nu_g + w_f \cdot \nu_f)}{w_g + w_f}$ average kinematic viscosity of gas & mud

$\nu_{ave} = 2.670 \times 10^{-4}$ (ft²/sec)

$V := \left[\dfrac{\left[\left(\dfrac{P_{at}}{P}\right)\cdot\left(\dfrac{T_{ave}}{T_{at}}\right)\cdot Q_a + Q_f\right]}{A}\right]$ velocity of flow in annulus (ft/sec)

$V = 5.666$ (ft/sec)

$N_R := \dfrac{V \cdot ID}{\nu_{ave}}$ Reynolds number calculations

$N_R = 6.765 \times 10^{3}$

$f := \left[\dfrac{1}{-1.8\cdot\log\left[\left(\dfrac{e_{cs}}{3.7\cdot ID}\right)^{1.11} + \dfrac{6.9}{N_R}\right]}\right]^2$ calculate Haaland friction factor

$f = 0.0362$ Haaland friction factor

$Q(P) := \dfrac{P_{at}}{P}\cdot\dfrac{T_{ave}}{T_{at}}\cdot Q_a$ flow rate as a function of pressure

$\gamma_{mix}(P) := \dfrac{w_g + w_f}{Q(P) + Q_f}$ specific weight as a function of pressure

$V(P) := \dfrac{(Q(P) + Q_f)}{A}$ velocity as a function of pressure

$p_{in} := 62.35$ trial pressure (psia)

$P_{in} := p_{in} \cdot 144$ convert pressure to lb/ft²

$P_{in} = 8.978 \times 10^{3}$ (lb/ft²)

$$\int_{P_{in}}^{P_1} \frac{1}{\gamma_{mix}(P)\cdot\left[1-(V(P))^2\cdot\frac{f}{2\cdot g\cdot(ID)}\right]}\,dP = 6550.112$$ left hand side

$$\int_0^{H_1} 1\,dh = 6650.000$$ right hand side

Important design parameter summary

$q_a = 301.260$ air compressor flow rate (scfm)

$p_{in} = 62.350$ required surface air injection pressure (psia)

$p_{bh} = 3360.000$ bottom hole pressure (psig)

Illistrative Example 9.3: SI version, Drilling Fluid is Aerated Fluid.
This program calculates the required gas injection volumetric flow rate in order to obtain a bottom hole annulus pressure of 2317 N/cm² guage while drilling at 3048 m of depth. Assumes homogeneous multiphase flow.

Given input conditions

$q_a := 595.5$	air volumetric flow rate trial and error value (actual liters/sec)
$q_f := 722.9$	mud flow rate (lpm)
$\gamma_f := 11.772$	specific weight of mud (N/liter)
$\mu_{air} := 0.012$	viscosity of air (centipoises)
$\mu_{mud} := 30$	viscosity of mud (centipoises)
$C := 0.81$	aerated fluid flow loss coefficient through nozzles
$D_c := 2133.8$	total depth of cased hole (m)
$D_{oh} := 914.4$	total length of open hole (m)
$alt := 1219.1$	elevation of drilling site (m)
$p_o := 8.749$	pressure from table for elevation (N/cm² abs)
$t_o := 15.6$	temperature from table for elev. (°C)
$p_{at} := 8.749$	actual surface pressure (N/cm² abs)
$t_{at} := 15.6$	actual given atmospheric surface temperature (°C)
$ROP := 18.3$	estimated rate of penetration (m/hr)
$d_{bh} := 200.03$	borehole (drill bit) diameter (mm)
$od_{dc} := 158.75$	outer diameter of drill collar (mm)
$id_{dc} := 71.438$	inner diameter of drill collar (mm)
$od_{dp} := 114.30$	outer diameter of drill pipe (mm)
$id_{dp} := 97.18$	inner diameter of drill pipe (mm)
$od_{tj} := 158.75$	outer diameter of drill pipe tool joint (mm)
$id_{tj} := 69.85$	inner diameter of drill pipe tool joint (mm)
$od_c := 219.08$	outer diameter of casing (mm)
$id_c := 201.19$	inner diameter of casing (mm)
$L_r := 30.49$	length of return line (m)
$id_r := 142.87$	inner diameter of return line (mm)
$SG_r := 2.7$	specific gravity of rock
$e_r := 0.003048$	roughness of rock (m)
$L_{dc} := 152.39$	total length of drill collars (m)

$L_{tj} := 0.457$	ave length of one tool joint (m)
$L_{dp} := 9.144$	ave length of one drill pipe section (m)
$d_n := 8.73$	nozzle diameter (mm)
$N_n := 3$	number of nozzles in the bit
$\beta := 0.01823$	geothermal gradient (°C/m)

Constant

$p_{st} := 10.136$	API atmospheric pressure (N/cm^2 abs)
$t_{st} := 15.6$	API standard temperature (°C)
$SG_a := 1$	specific gravity of air
$R := 29.31$	specific gas constant (N-m/N-K)
$w_w := 9.81$	specific weight of water (N/liter)
$e_{cs} := 0.000152939$	roughness of commercial steel (m)
$g := 9.81$	gravitational constant (m/sec^2)

Preliminary calculations

$P_{at} := p_{at} \cdot 10000.0$	convert actual atm pressure to N/m^2
$P_{at} = 87490.000$	(N/m^2)
$T_{at} := t_{at} + 273.15$	convert actual atm temp to K
$T_{at} = 288.750$	(K)
$P_o := p_o \cdot 100000.0$	convert table pressure to N/m^2
$P_o = 874900.000$	(N/m^2)
$T_o := t_o + 273.15$	convert table temperature to K
$T_o = 288.750$	(K)
$D_{bh} := \frac{d_{bh}}{1000}$	convert borehole diameter to m
$D_{bh} = 0.200$	(m)
$ID_r := \frac{id_r}{1000}$	convert return line diameter to m

$ID_r = 0.143$	(m)
$OD_{dp} := \frac{od_{dp}}{1000}$	convert drill pipe outer diameter to m
$OD_{dp} = 0.114$	(m)
$ID_{dp} := \frac{id_{dp}}{1000}$	convert drill pipe inner diameter to m
$ID_{dp} = 0.097$	(m)
$ID_c := \frac{id_c}{1000}$	convert casing inner diameter to m
$ID_c = 0.201$	(m)
$OD_{dc} := \frac{od_{dc}}{1000}$	convert drill collar outer diameter to m
$OD_{dc} = 0.159$	(m)
$ID_{dc} := \frac{id_{dc}}{1000}$	convert drill collar inner diameter to m
$ID_{dc} = 0.071$	(m)
$OD_{tj} := \frac{od_{tj}}{1000}$	convert tool joint outer diameter to m
$OD_{tj} = 0.159$	(m)
$ID_{tj} := \frac{id_{tj}}{1000}$	convert tool joint inner diameter to m
$ID_{tj} = 0.070$	(m)
$d_n = 8.730$	nozzle diameter (mm)
$D_n := \frac{d_n}{1000}$	nozzle diameter converted to m
$D_n = 0.009$	(m)
$\mu_g := \mu_{air} \cdot 10^{-3}$	convert viscosity to N-sec/m^2
$\mu_g = 1.200 \times 10^{-5}$	(N-sec/m^2)
$\mu_f := \mu_{mud} \cdot 10^{-3}$	convert viscosity to N-sec/m^2
$\mu_f = 3.000 \times 10^{-2}$	(N-sec/m^2)
$Q_a := \frac{q_a}{1000}$	convert air flow rate to m^3/sec

$Q_a = 0.596$ (m^3/sec)

$Q_f := \frac{q_f}{60 \cdot 1000}$ convert mud flow to (m^3/sec)

$Q_f = 0.012$ (m^3/sec)

$\gamma_f := \gamma_f \cdot 1000$ convert spec. wt. of mud from ppg to N/m^3

$\gamma_f = 11772.000$ (N/m^3)

$\rho_f := \frac{\gamma_f}{g}$ mud density (N-sec^2/m^4)

$\rho_f = 1.2000 \times 10^3$ (N-sec^2/m^4)

$\nu_f := \frac{\mu_f}{\rho_f}$ mud kinematic viscosity (m^2/sec)

$\nu_f = 2.500 \times 10^{-5}$ (m^2/sec)

$\gamma_g := \frac{(P_{at} \cdot SG_a)}{R \cdot T_{at}}$ specific weight of air (N/m^3)

$\gamma_g = 10.3376$ (N/m^3)

$\rho_g := \frac{\gamma_g}{g}$ gas (air) density (N-sec^2/m^4)

$\rho_g = 1.054 \times 10^0$ (N-sec^2/m^4)

$\nu_g := \frac{\mu_g}{\rho_g}$ gas kinematic viscosity (m^2/sec)

$\nu_g = 1.139 \times 10^{-5}$ (m^2/sec)

$rop := \frac{ROP}{60 \cdot 60}$ convert ROP to m/sec

$rop = 0.0051$ (m/sec)

$w_s := \left[\left(\frac{\pi}{4}\right) \cdot D_{bh}{}^2 \cdot 9810 \cdot SG_r \cdot rop\right]$ weight rate of solid flow out of blooey line (N/sec)

$w_s = 4.231$ (N/sec)

$w_g := \gamma_g \cdot Q_a$ weight rate of flow of air (N/sec)

$w_g = 6.156$ (N/sec)

$w_f := \gamma_f \cdot Q_f$ weight rate of flow of mud (N/sec)

$w_f = 141.833$ (N/sec)

$w_t := w_g + w_f + w_s$ total weight rate of flow (lb/sec)

$w_t = 152.220$ (N/sec)

Section height calculations

$H_1 := D_c - L_{tj} \cdot \left(\frac{D_c}{L_{dp}}\right)$ calculate length of drill pipe in cased section

$H_1 = 2027.157$ (m)

$H_2 := L_{tj} \cdot \left(\frac{D_c}{L_{dp}}\right)$ calculate length of tool joint in cased section

$H_2 = 106.643$ (m)

$H_3 := D_{oh} - L_{dc} - L_{tj} \cdot \left(\frac{D_{oh} - L_{dc}}{L_{dp}}\right)$ calculate length of drill pipe in open hole section

$H_3 = 723.926$ (m)

$H_4 := L_{tj} \cdot \left(\frac{D_{oh} - L_{dc}}{L_{dp}}\right)$ calculate length of tool joints in open hole section

$H_4 = 38.084$ (m)

$H_5 := L_{dc}$ length of drill collar

$H_5 = 152.390$ (m)

Return line

$ID := ID_r$ set variable ID equal to return line's ID

$ID = 0.143$ (m)

$A := \frac{\pi}{4} \cdot ID^2$ cross sectional area (m^2)

$A = 0.016$ (m^2)

$V := \frac{(Q_a + Q_f)}{A}$ velocity of flow through the line (m/sec)

$V = 37.897$ (m/sec)

$\nu_{ave} := \frac{(w_g \cdot \nu_g + w_f \cdot \nu_f)}{w_g + w_f}$ average kinematic viscosity of gas & mud

$\nu_{ave} = 2.443 \times 10^{-5}$ (m²/sec)

$N_R := \frac{(V \cdot ID)}{\nu_{ave}}$ Reynolds number calculations

$N_R = 2.216 \times 10^5$

$f := \left[\frac{1}{-1.8 \cdot \log\left[\left(\frac{e_{cs}}{3.7 \cdot ID}\right)^{1.11} + \frac{6.9}{N_R}\right]}\right]^2$ calculate friction factor using Haaland equation

$f = 0.0211$ Haaland friction factor

$Q(P) := \frac{P_{at}}{P} \cdot \frac{T_o}{T_{at}} \cdot Q_a$ flow rate as a function of pressure

$\gamma_{mix}(P) := \frac{w_t}{Q(P) + Q_f}$ specific weight as a function of pressure

$V(P) := \frac{(Q(P) + Q_f)}{A}$ velocity as a function of pressure

$p_r := 14.656$ trial pressure

$P_r := p_r \cdot 10000.0$ convert pressure to N/m²

$P_r = 146560.000$ (N/m²)

$\int_{P_{at}}^{P_r} \frac{1}{\gamma_{mix}(P) \cdot \left(V(P)^2 \cdot \frac{f}{2 \cdot g \cdot ID}\right)} dP = 28.985$ left hand side

$\int_0^{L_r} 1 \, dh = 30.490$ right hand side

$K_t := 30$ resistance coefficient for the blind Tee

$K_v := 0.2$ resistance coefficient for gate valves

$\gamma_{mix} := \dfrac{w_t}{\left[\left(\dfrac{P_{at}}{P_r}\right)\cdot\left(\dfrac{T_o}{T_{at}}\right)\cdot Q_a + Q_f\right]}$ specific weight (N/m^3)

$\gamma_{mix} = 414.164$ (N/m^3)

$V := \left[\dfrac{\left[\left(\dfrac{P_{at}}{P_r}\right)\cdot\left(\dfrac{T_o}{T_{at}}\right)\cdot Q_a + Q_f\right]}{A}\right]$ velocity (m/sec)

$V = 22.926$ (m/sec)

$\Delta P_T := \gamma_{mix}\cdot(2\cdot K_v + K_t)\cdot\dfrac{V^2}{2\cdot g}$ caclulate pressure loss due to Tee and valves

$\Delta P_T = 337289.663$ (N/m^2)

$P_e := P_r + \Delta P_T$ sum pressure losses

$P_e = 483849.663$ (N/m^2)

$p_e := \dfrac{P_e}{10000.0}$ convert from N/m^2 abs to N/cm^2 abs

$p_e = 48.385$ (N/cm^2 abs)

Annulus cased hole drill pipe section

$ID := ID_c$ set inner diameter to casing ID

$ID = 0.201$ (m)

$OD := OD_{dp}$ set outer diameter to drill pipe OD

$OD = 0.114$ (m)

$P := P_e$ set pressure equal to previous section pressure

$P = 483849.663$ (N/m^2)

$A := \dfrac{\pi}{4}\cdot\left(ID^2 - OD^2\right)$ cross sectional area (m^2)

$A = 0.022$ (m^2)

$T := T_o + \beta \cdot H_1$ calculate temperature at bottom of section

$T = 325.705$ (K)

$T_{ave1} := \frac{T_o + T}{2}$ calculate average temperature of section

$T_{ave1} = 307.228$ (K)

$T_{ave} := T_{ave1}$ set average temp to section average temp

$\gamma_g := \frac{P \cdot SG_a}{R \cdot T_{ave1}}$ specific weight of the gas (N/m^3)

$\gamma_g = 53.732$ (N/m^3)

$\rho_g := \frac{\gamma_g}{g}$ gas (air) density (N-sec^2/m^4)

$\rho_g = 5.477$ (N-sec^2/m^4)

$\nu_g := \frac{\mu_g}{\rho_g}$ gas kinematic viscosity (m^2/sec)

$\nu_g = 2.191 \times 10^{-6}$ (m^2/sec)

$\nu_{ave} := \frac{(w_g \cdot \nu_g + w_f \cdot \nu_f)}{w_g + w_f}$ average kinematic viscosity of gas & mud

$\nu_{ave} = 2.405 \times 10^{-5}$ (m^2/sec)

$V := \left[\frac{\left[\left(\frac{P_{at}}{P} \right) \cdot \left(\frac{T_{ave}}{T_{at}} \right) \cdot Q_a + Q_f \right]}{A} \right]$ velocity of flow in annulus (m/sec)

$V = 5.881$ (m/sec)

$N_R := \frac{V \cdot (ID - OD)}{\nu_{ave}}$ Reynolds number

$N_R = 2.125 \times 10^4$

$f := \left[\frac{1}{-1.8 \cdot \log \left[\left[\frac{e_{cs}}{3.7 \cdot (ID - OD)} \right]^{1.11} + \frac{6.9}{N_R} \right]} \right]^2$ calculate friction factor using Haaland equation

$f = 0.0288$ Haaland friction factor

$$Q(P) := \frac{P_{at}}{P} \cdot \frac{T_{ave}}{T_{at}} \cdot Q_a$$ flow rate as a function of pressure

$$\gamma_{mix}(P) := \frac{w_t}{Q(P) + Q_f}$$ specific weight as a function of pressure

$$V(P) := \frac{(Q(P) + Q_f)}{A}$$ velocity as a function of pressure

$p_1 := 1273.40$ trial pressure (psia)

$P_1 := p_1 \cdot 10000.0$ convert pressure to N/m²

$P_1 = 1.273 \times 10^7$ (N/m²)

$$\int_{P_e}^{P_1} \frac{1}{\gamma_{mix}(P) \cdot \left[1 + (V(P))^2 \cdot \frac{f}{2 \cdot g \cdot (ID - OD)}\right]} dP = 2027.101$$ left hand side

$$\int_0^{H_1} 1\, dh = 2027.157$$ right hand side

Annulus cased hole tool joint section

$ID := ID_c$ set inner diameter to casing ID

$OD := OD_{tj}$ set outer diameter to drill pipe OD

$P := P_1$ set pressure equal to previous section pressure

$A := \frac{\pi}{4} \cdot \left(ID^2 - OD^2\right)$ cross sectional area (m²)

$T_{temp} := T$ keep previous section temperature

$T := T_{temp} + \beta \cdot H_2$ calculate temperature at bottom of section

$T = 327.649$ (K)

$T_{ave2} := \frac{T_{temp} + T}{2}$ calculate average temperature of section

$T_{ave2} = 326.677$ (K)

$T_{ave} := T_{ave2}$ set average temp to section average temp

$$\gamma_g := \frac{P \cdot SG_a}{R \cdot T_{ave}}$$ specific weight of the gas (N/m^3)

$\gamma_g = 1329.935$ (N/m^3)

$$\rho_g := \frac{\gamma_g}{g}$$ gas (air) density (N-sec^2/m^4)

$\rho_g = 135.569$ (N-sec^2/m^4)

$$\nu_g := \frac{\mu_g}{\rho_g}$$ gas (air) kinematic viscosity (m^2/sec)

$\nu_g = 8.852 \times 10^{-8}$ (m^2/sec)

$$\nu_{ave} := \frac{(w_g \cdot \nu_g + w_f \cdot \nu_f)}{w_g + w_f}$$ average kinematic viscosity of gas & mud

$\nu_{ave} = 2.396 \times 10^{-5}$ (m^2/sec)

$$V := \left[\frac{\left[\left(\frac{P_{at}}{P}\right) \cdot \left(\frac{T_{ave}}{T_{at}}\right) \cdot Q_a + Q_f\right]}{A}\right]$$ velocity of flow in annulus (m/sec)

$V = 1.390$ (m/sec)

$$N_R := \frac{V \cdot (ID - OD)}{\nu_{ave}}$$ Reynolds number calculations

$N_R = 2.462 \times 10^3$

$$f := \left[\frac{1}{-1.8 \cdot \log\left[\left[\frac{e_{cs}}{3.7 \cdot (ID - OD)}\right]^{1.11} + \frac{6.9}{N_R}\right]}\right]^2$$ calculate friction factor using Haaland equation

$f = 0.0499$ Haaland friction factor

$$Q(P) := \frac{P_{at}}{P} \cdot \frac{T_{ave}}{T_{at}} \cdot Q_a$$ flow rate as a function of pressure

$$\gamma_{mix}(P) := \frac{w_t}{Q(P) + Q_f}$$ specific weight as a function of pressure

$$V(P) := \frac{(Q(P) + Q_f)}{A}$$ velocity as a function of pressure

$p_2 := 1382.95$ trial pressure (N/cm^2 abs)

$P_2 := p_2 \cdot 10000.0$ — convert pressure to lb/ft^2

$P_2 = 1.383 \times 10^7$ — (lb/ft^2)

$$\int_{P_1}^{P_2} \frac{1}{\gamma_{mix}(P)\cdot\left[1 + (V(P))^2 \cdot \frac{f}{2\cdot g\cdot(ID - OD)}\right]}\, dP = 106.601$$ left hand side

$$\int_{H_1}^{H_1+H_2} 1\, dh = 106.643$$ right hand side

Annulus open hole drill pipe section

$ID := D_{bh}$ — set inner diameter to casing ID

$OD := OD_{dp}$ — set outer diameter to drill pipe OD

$P := P_2$ — set pressure equal to previous section pressure

$A := \frac{\pi}{4}\cdot\left(ID^2 - OD^2\right)$ — cross sectional area (m^2)

$T_{temp} := T$ — keep previous section temperature

$T := T_{temp} + \beta\cdot H_3$ — calculate temperature at bottom of section

$T = 340.846$ — (K)

$T_{ave3} := \frac{T_{temp} + T}{2}$ — calculate average temperature of section

$T_{ave3} = 334.248$ — (K)

$T_{ave} := T_{ave3}$ — set average temp to section average temp

$\gamma_g := \frac{P\cdot SG_a}{R\cdot T_{ave}}$ — specific weight of the gas (N/m^3)

$\gamma_g = 1411.634$ — (N/m^3)

$\rho_g := \frac{\gamma_g}{g}$ — gas density (N-sec^2/m^4)

$\rho_g = 143.897$ — (N-sec^2/m^4)

$\nu_g := \frac{\mu_g}{\rho_g}$ gas kinematic viscosity (m^2/sec)

$\nu_g = 8.339 \times 10^{-8}$ (m^2/sec)

$\nu_{ave} := \frac{(w_g \cdot \nu_g + w_f \cdot \nu_f)}{w_g + w_f}$ average kinematic viscosity of gas & mud

$\nu_{ave} = 2.396 \times 10^{-5}$ (m^2/sec)

$V := \left[\frac{\left[\left(\frac{P_{at}}{P}\right)\cdot\left(\frac{T_{ave}}{T_{at}}\right)\cdot Q_a + Q_f\right]}{A}\right]$ velocity of flow in annulus (m/sec)

$V = 0.775$ (m/sec)

$N_R := \frac{V\cdot(ID - OD)}{\nu_{ave}}$ Reynolds number calculations

$N_R = 2.774 \times 10^3$

$e_{ave} := \frac{\left[e_r\cdot\left(\frac{\pi}{4}\right)\cdot ID^2 + e_{cs}\cdot\left(\frac{\pi}{4}\right)\cdot OD^2\right]}{\left[\left(\frac{\pi}{4}\right)\cdot ID^2 + \left(\frac{\pi}{4}\right)\cdot OD^2\right]}$ calculate weighted average of roughness between the steel and the rock

$e_{ave} = 2.335 \times 10^{-3}$ weighted average of roughness (m)

$f := \left[\frac{1}{-1.8\cdot\log\left[\left[\frac{e_{ave}}{3.7\cdot(ID - OD)}\right]^{1.11} + \frac{6.9}{N_R}\right]}\right]^2$ calculate friction factor using Haaland equation

$f = 0.066$ Haaland friction factor

$Q(P) := \frac{P_{at}}{P}\cdot\frac{T_{ave}}{T_{at}}\cdot Q_a$ flow rate as a function of pressure

$\gamma_{mix}(P) := \frac{w_t}{Q(P) + Q_f}$ specific weight as a function of pressure

$V(P) := \frac{(Q(P) + Q_f)}{A}$ velocity as a function of pressure

$p_3 := 2106.3$ trial pressure (N/cm^2 abs)

$P_3 := p_3\cdot 10000.0$ convert pressure to N/m^2

$P_3 = 2.106 \times 10^7$ (N/m²)

$$\int_{P_2}^{P_3} \frac{1}{\gamma_{mix}(P)\cdot\left[1 + (V(P))^2\cdot\frac{f}{2\cdot g\cdot(ID - OD)}\right]}\, dP = 723.935$$ left hand side

$$\int_{H_1+H_2}^{H_1+H_2+H_3} 1\, dh = 723.926$$ right hand side

Annulus open hole tool joint section

$ID := D_{bh}$ set inner diameter to casing ID

$OD := OD_{tj}$ set outer diameter to drill pipe OD

$P := P_3$ set pressure equal to previous section pressure

$A := \frac{\pi}{4}\cdot\left(ID^2 - OD^2\right)$ cross sectional area (m²)

$T_{temp} := T$ keep previous section temperature

$T := T_{temp} + \beta\cdot H_4$ calculate temperature at bottom of section

$T = 341.541$ (K)

$T_{ave4} := \frac{T_{temp} + T}{2}$ calculate average temperature of section

$T_{ave4} = 341.193$ (K)

$T_{ave} := T_{ave4}$ set average temp to section average temp

$\gamma_g := \frac{P\cdot SG_a}{R\cdot T_{ave}}$ specific weight of the gas (N/m³)

$\gamma_g = 2106.220$ (N/m³)

$\rho_g := \frac{\gamma_g}{g}$ gas (air) density (N-sec²/m⁴)

$\rho_g = 214.701$ (N-sec²/m⁴)

$\nu_g := \frac{\mu_g}{\rho_g}$ gas (air) kinematic viscosity (m²/sec)

$\nu_g = 5.589 \times 10^{-8}$ (m²/sec)

$\nu_{ave} := \frac{(w_g \cdot \nu_g + w_f \cdot \nu_f)}{w_g + w_f}$ average kinematic viscosity of gas & mud

$\nu_{ave} = 2.396 \times 10^{-5}$ (m²/sec)

$V := \left[\frac{\left[\left(\frac{P_{at}}{P}\right)\cdot\left(\frac{T_{ave}}{T_{at}}\right)\cdot Q_a + Q_f\right]}{A}\right]$ velocity of flow in annulus (m/sec)

$V = 1.287$ (m/sec)

$N_R := \frac{V\cdot(ID - OD)}{\nu_{ave}}$ Reynolds number calculations

$N_R = 2.217 \times 10^3$

$e_{ave} := \frac{\left[e_r\cdot\left(\frac{\pi}{4}\right)\cdot ID^2 + e_{cs}\cdot\left(\frac{\pi}{4}\right)\cdot OD^2\right]}{\left[\left(\frac{\pi}{4}\right)\cdot ID^2 + \left(\frac{\pi}{4}\right)\cdot OD^2\right]}$ calculate weighted average of roughness between the steel and the rock

$e_{ave} = 1.929 \times 10^{-3}$ weighted average of roughness (m)

$f := \left[\frac{1}{-1.8\cdot\log\left[\left[\frac{e_{ave}}{3.7\cdot(ID - OD)}\right]^{1.11} + \frac{6.9}{N_R}\right]}\right]^2$ calculate friction factor using Haaland equation

$f = 0.0802$ Haaland friction factor

$Q(P) := \frac{P_{at}}{P}\cdot\frac{T_{ave}}{T_{at}}\cdot Q_a$ flow rate as a function of pressure

$\gamma_{mix}(P) := \frac{w_t}{Q(P) + Q_f}$ specific weight as a function of pressure

$V(P) := \frac{(Q(P) + Q_f)}{A}$ velocity as a function of pressure

$p_4 := 2151.4$ trial pressure (N/cm² abs)

$P_4 := p_4\cdot 10000.0$ convert pressure to N/m²

$P_4 = 2.151 \times 10^7$ (N/m²)

$$\int_{P_3}^{P_4} \frac{1}{\gamma_{mix}(P)\cdot\left[1 + (V(P))^2\cdot\frac{f}{2\cdot g\cdot(ID - OD)}\right]}\,dP = 38.050$$ left hand side

$$\int_{H_1+H_2+H_3}^{H_1+H_2+H_3+H_4} 1\,dh = 38.084$$ right hand side

Annulus open hole drill collar section

$ID := D_{bh}$ set inner diameter to casing ID

$OD := OD_{dc}$ set outer diameter to drill pipe OD

$P := P_4$ set pressure equal to previous section pressure

$A := \frac{\pi}{4}\cdot\left(ID^2 - OD^2\right)$ cross sectional area (m^2)

$T_{temp} := T$ keep previous section temperature

$T := T_{temp} + \beta\cdot H_5$ calculate temperature at bottom of section

$T = 344.319$ (K)

$T_{ave5} := \frac{T_{temp} + T}{2}$ calculate average temperature of section

$T_{ave5} = 342.930$ (K)

$T_{ave} := T_{ave5}$ set average temp to section average temp

$\gamma_g := \frac{P\cdot SG_a}{R\cdot T_{ave}}$ specific weight of the gas (N/m^3)

$\gamma_g = 2140.426$ (N/m^3)

$\rho_g := \frac{\gamma_g}{g}$ gas (air) density ($N\text{-}sec^2/m^4$)

$\rho_g = 218.188$ ($N\text{-}sec^2/m^4$)

$\nu_g := \frac{\mu_g}{\rho_g}$ gas (air) kinematic viscosity (m^2/sec)

$\nu_g = 5.500 \times 10^{-8}$ (m^2/sec)

$$\nu_{ave} := \frac{(w_g \cdot \nu_g + w_f \cdot \nu_f)}{w_g + w_f}$$ average kinematic viscosity of gas & mud

$\nu_{ave} = 2.396 \times 10^{-5}$ (m²/sec)

$$V := \left[\frac{\left[\left(\frac{P_{at}}{P} \right) \cdot \left(\frac{T_{ave}}{T_{at}} \right) \cdot Q_a + Q_f \right]}{A} \right]$$ velocity of flow in annulus (m/sec)

$V = 1.283$ (m/sec)

$$N_R := \frac{V \cdot (ID - OD)}{\nu_{ave}}$$ Reynolds number calculations

$N_R = 2.210 \times 10^{3}$

$$e_{ave} := \frac{\left[e_r \cdot \left(\frac{\pi}{4} \right) \cdot ID^2 + e_{cs} \cdot \left(\frac{\pi}{4} \right) \cdot OD^2 \right]}{\left[\left(\frac{\pi}{4} \right) \cdot ID^2 + \left(\frac{\pi}{4} \right) \cdot OD^2 \right]}$$ calculate weighted average of roughness between the steel and the rock

$e_{ave} = 1.929 \times 10^{-3}$ weighted average of roughness (m)

$$f := \left[\frac{1}{-1.8 \cdot \log \left[\left[\frac{e_{ave}}{3.7 \cdot (ID - OD)} \right]^{1.11} + \frac{6.9}{N_R} \right]} \right]^2$$ calculate friction factor using Haaland equation

$f = 0.0802$ Haaland friction factor

$$Q(P) := \frac{P_{at}}{P} \cdot \frac{T_{ave}}{T_{at}} \cdot Q_a$$ flow rate as a function of pressure

$$\gamma_{mix}(P) := \frac{w_t}{Q(P) + Q_f}$$ specific weight as a function of pressure

$$V(P) := \frac{(Q(P) + Q_f)}{A}$$ velocity as a function of pressure

$p_5 := 2324.4 + p_{at}$ trial pressure (N/cm² abs)

$p_5 = 2333.149$ (N/cm² abs)

$P_5 := p_5 \cdot 10000.0$ convert pressure to N/m2

$P_5 = 2.333 \times 10^7$ (N/m2)

$$\int_{P_4}^{P_5} \frac{1}{\gamma_{mix}(P)\cdot\left[1 + (V(P))^2 \cdot \frac{f}{2\cdot g\cdot(ID - OD)}\right]}\, dP = 152.357$$ left hand side

$$\int_{H_1+H_2+H_3+H_4}^{H_1+H_2+H_3+H_4+H_5} 1\, dh = 152.390$$ right hand side

$p_{bh} := p_5 - p_{at}$ bottom hole pressure (N/cm2 gauge)

$p_{bh} = 2324.400$ (N/cm2 gauge)

Bottom hole through the bit nozzles

$D_e := \left[N_n \cdot (D_n)^2\right]^{0.5}$ equivalent single diameter (m)

$D_e = 0.015$ (m)

$$\gamma_{mix} := \frac{(w_g + w_f)}{\left[Q_f + Q_a \cdot \left(\frac{P_{at}}{P_5}\right) \cdot \frac{T}{T_{at}}\right]}$$ specific weight of mixture (N/m3)

$\gamma_{mix} = 10059.672$ (N/m3)

$$\Delta P := \frac{(w_g + w_f)^2}{2\cdot C^2 \cdot g \cdot \gamma_{mix} \cdot \left(\frac{\pi}{4}\right)^2 \cdot D_e^4}$$ approximation of pressure change of the aerated fluid through the drill bit

$\Delta P = 5244776.325$ (N/m2)

$P := P_5 + \Delta P$ sum pressures to get pressure above drill bit

$P = 28576266.325$ (N/m2)

$$p := \frac{P}{10000.0}$$ convert prssure to N/cm2

$p = 2857.627$ (N/cm^2 abs)

Open hole inside drill string drill collar section

$ID := ID_{dc}$ set inner diameter to casing ID

$A := \frac{\pi}{4} \cdot (ID^2)$ cross sectional area (m^2)

$T_{ave} := T_{ave5}$ set ave temp to section ave temp

$\gamma_g := \frac{P \cdot SG_a}{R \cdot T_{ave}}$ specific weight of the gas (N/m^3)

$\gamma_g = 2843.051$ (N/m^3)

$\rho_g := \frac{\gamma_g}{g}$ gas density (N-sec^2/m^4)

$\rho_g = 289.812$ (N-sec^2/m^4)

$\nu_g := \frac{\mu_g}{\rho_g}$ gas kinematic viscosity (m^2/sec)

$\nu_g = 4.141 \times 10^{-8}$ (m^2/sec)

$\nu_{ave} := \frac{(w_g \cdot \nu_g + w_f \cdot \nu_f)}{w_g + w_f}$ average kinematic viscosity of gas & mud

$\nu_{ave} = 2.396 \times 10^{-5}$ (m^2/sec)

$V := \left[\frac{\left[\left(\frac{P_{at}}{P} \right) \cdot \left(\frac{T_{ave}}{T_{at}} \right) \cdot Q_a + Q_f \right]}{A} \right]$ velocity of flow in annulus (m/sec)

$V = 3.546$ (m/sec)

$N_R := \frac{V \cdot ID}{\nu_{ave}}$ Reynolds number calculations

$N_R = 1.057 \times 10^4$

$$f := \left[\frac{1}{-1.8\cdot\log\left[\left(\frac{e_{cs}}{3.7\cdot ID}\right)^{1.11} + \frac{6.9}{N_R}\right]}\right]^2$$ calculate friction factor using Haaland equation

$f = 0.0333$ Haaland friction factor

$$Q(P) := \frac{P_{at}}{P}\cdot\frac{T_{ave}}{T_{at}}\cdot Q_a$$ flow rate as a function of pressure

$$\gamma_{mix}(P) := \frac{w_g + w_f}{Q(P) + Q_f}$$ specific weight as a function of pressure

$$V(P) := \frac{\left(Q(P) + Q_f\right)}{A}$$ velocity as a function of pressure

$p_4 := 2747.0$ trial pressure (N/cm² abs)

$P_4 := p_4\cdot 10000.0$ convert pressure to N/m²

$P_4 = 2.747 \times 10^7$ (N/m²)

$$\int_{P_4}^{P} \frac{1}{\gamma_{mix}(P)\cdot\left[1 - (V(P))^2\cdot\frac{f}{2\cdot g\cdot(ID)}\right]}\,dP = 152.471$$ left hand side

$$\int_{H_1+H_2+H_3+H_4}^{H_1+H_2+H_3+H_4+H_5} 1\,dh = 152.390$$ right hand side .

Open hole inside drill string tool joint section

$ID := ID_{tj}$ set inner diameter to casing ID

$A := \frac{\pi}{4}\cdot\left(ID^2\right)$ cross sectional area (m²)

$T_{ave} := T_{ave4}$ set ave temp to section ave temp

$P := P_4$ set pressure to previous section pressure

$\gamma_g := \frac{P \cdot SG_a}{R \cdot T_{ave}}$ specific weight of the gas (N/m³)

$\gamma_g = 2746.895$ (N/m³)

$\rho_g := \frac{\gamma_g}{g}$ gas (air) density (N-sec²/m⁴)

$\rho_g = 280.010$ (N-sec²/m⁴)

$\nu_g := \frac{\mu_g}{\rho_g}$ gas kinematic viscosity (M²/sec)

$\nu_g = 4.286 \times 10^{-8}$ (m²/sec)

$\nu_{ave} := \frac{(w_g \cdot \nu_g + w_f \cdot \nu_f)}{w_g + w_f}$ average kinematic viscosity of gas & mud

$\nu_{ave} = 2.396 \times 10^{-5}$ (m²/sec)

$V := \left[\frac{\left[\left(\frac{P_{at}}{P}\right) \cdot \left(\frac{T_{ave}}{T_{at}}\right) \cdot Q_a + Q_f\right]}{A}\right]$ velocity of flow in annulus (m/sec)

$V = 3.729$ (m/sec)

$N_R := \frac{V \cdot ID}{\nu_{ave}}$ Reynolds number calculations

$N_R = 1.087 \times 10^4$

$f := \left[\frac{1}{-1.8 \cdot \log\left[\left(\frac{e_{cs}}{3.7 \cdot ID}\right)^{1.11} + \frac{6.9}{N_R}\right]}\right]^2$ calculate friction factor using Haaland equation

$f = 0.0332$ Haaland friction factor

$Q(P) := \frac{P_{at}}{P} \cdot \frac{T_{ave}}{T_{at}} \cdot Q_a$ flow rate as a function of pressure

$\gamma_{mix}(P) := \frac{w_g + w_f}{Q(P) + Q_f}$ specific weight as a function of pressure

$V(P) := \frac{(Q(P) + Q_f)}{A}$ velocity as a function of pressure

$p_3 := 2720.9$ trial pressure (N/cm²)

$P_3 := p_3 \cdot 10000.0$ convert pressure to N/m²

$P_3 = 2.721 \times 10^{7}$ (N/m²)

$$\int_{P_3}^{P_4} \frac{1}{\gamma_{mix}(P) \cdot \left[1 - (V(P))^2 \cdot \frac{f}{2 \cdot g \cdot (ID)}\right]} dP = 38.078$$ left hand side

$$\int_{H_1+H_2+H_3}^{H_1+H_2+H_3+H_4} 1 \, dh = 38.084$$ right hand side

Open hole inside drill string drill pipe section

$ID := ID_{dp}$ set inner diameter to casing ID

$A := \frac{\pi}{4} \cdot \left(ID^2\right)$ cross sectional area (m²)

$T_{ave} := T_{ave3}$ set ave temp to section ave temp

$P := P_3$ set pressure to previous section pressure

$\gamma_g := \frac{P \cdot SG_a}{R \cdot T_{ave}}$ specific weight of the gas (N/m³)

$\gamma_g = 2777.335$ (N/m³)

$\rho_g := \frac{\gamma_g}{g}$ gas (air) density (N-sec²/m⁴)

$\rho_g = 283.113$ (N-sec²/m⁴)

$\nu_g := \frac{\mu_g}{\rho_g}$ gas kinematic viscosity (m²/sec)

$\nu_g = 4.239 \times 10^{-8}$ (m²/sec)

$\nu_{ave} := \frac{\left(w_g \cdot \nu_g + w_f \cdot \nu_f\right)}{w_g + w_f}$ average kinematic viscosity of gas & mud

$\nu_{ave} = 2.396 \times 10^{-5}$ (m²/sec)

$$V := \left[\frac{\left[\left(\frac{P_{at}}{P}\right)\cdot\left(\frac{T_{ave}}{T_{at}}\right)\cdot Q_a + Q_f\right]}{A}\right]$$ velocity of flow in annulus (m/sec)

$V = 1.923$ (m/sec)

$$N_R := \frac{V\cdot ID}{\nu_{ave}}$$ Reynolds number calculations

$N_R = 7.800 \times 10^3$

$$f := \left[\frac{1}{-1.8\cdot\log\left[\left(\frac{e_{cs}}{3.7\cdot ID}\right)^{1.11} + \frac{6.9}{N_R}\right]}\right]^2$$ calculate friction factor using Haaland equation

$f = 0.0349$ Haaland friction factor

$$Q(P) := \frac{P_{at}}{P}\cdot\frac{T_{ave}}{T_{at}}\cdot Q_a$$ flow rate as a function of pressure

$$\gamma_{mix}(P) := \frac{w_g + w_f}{Q(P) + Q_f}$$ specific weight as a function of pressure

$$V(P) := \frac{\left(Q(P) + Q_f\right)}{A}$$ velocity as a function of pressure

$p_2 := 2039.3$ trial pressure (N/cm² abs)

$P_2 := p_2\cdot 10000.0$ convert pressure to N/m²

$P_2 = 2.039 \times 10^7$ (N/m²)

$$\int_{P_2}^{P_3} \frac{1}{\gamma_{mix}(P)\cdot\left[1 - (V(P))^2\cdot\frac{f}{2\cdot g\cdot(ID)}\right]}\,dP = 723.849$$ left hand side

$$\int_{H_1+H_2}^{H_1+H_2+H_3} 1\,dh = 723.926$$ right hand side

Cased hole inside drill string tool joint section

$ID := ID_{tj}$ set inner diameter to casing ID

$A := \frac{\pi}{4} \cdot \left(ID^2\right)$ cross sectional area (m^2)

$T_{ave} := T_{ave2}$ set ave temp to section ave temp

$P := P_2$ set pressure to previous section pressure

$\gamma_g := \frac{P \cdot SG_a}{R \cdot T_{ave}}$ specific weight of the gas (N/m^3)

$\gamma_g = 2129.838$ (N/m^3)

$\rho_g := \frac{\gamma_g}{g}$ gas (air) density ($N\text{-}sec^2/m^4$)

$\rho_g = 217.109$ ($N\text{-}sec^2/m^4$)

$\nu_g := \frac{\mu_g}{\rho_g}$ gas kinematic viscosity (m^2/sec)

$\nu_g = 5.527 \times 10^{-8}$ (m^2/sec)

$\nu_{ave} := \frac{\left(w_g \cdot \nu_g + w_f \cdot \nu_f\right)}{w_g + w_f}$ average kinematic viscosity of gas & mud

$\nu_{ave} = 2.396 \times 10^{-5}$ (m^2/sec)

$V := \left[\frac{\left[\left(\frac{P_{at}}{P}\right) \cdot \left(\frac{T_{ave}}{T_{at}}\right) \cdot Q_a + Q_f\right]}{A}\right]$ velocity of flow in annulus (m/sec)

$V = 3.898$ (m/sec)

$N_R := \frac{V \cdot ID}{\nu_{ave}}$ Reynolds number calculations

$N_R = 1.136 \times 10^4$

$f := \left[\frac{1}{-1.8 \cdot \log\left[\left(\frac{e_{cs}}{3.7 \cdot ID}\right)^{1.11} + \frac{6.9}{N_R}\right]}\right]^2$ calculate friction factor using Haaland equation

$f = 0.0329$ Haaland friction factor

$$Q(P) := \frac{P_{at}}{P} \cdot \frac{T_{ave}}{T_{at}} \cdot Q_a$$ flow rate as a function of pressure

$$\gamma_{mix}(P) := \frac{w_g + w_f}{Q(P) + Q_f}$$ specific weight as a function of pressure

$$V(P) := \frac{(Q(P) + Q_f)}{A}$$ velocity as a function of pressure

$p_1 := 1972.730$ trail pressure (N/cm^2 abs)

$P_1 := p_1 \cdot 10000.0$ convert pressure to N/m^2

$P_1 = 1.973 \times 10^7$ (N/m^2)

$$\int_{P_1}^{P_2} \frac{1}{\gamma_{mix}(P) \cdot \left[1 - (V(P))^2 \cdot \frac{f}{2 \cdot g \cdot (ID)}\right]} \, dP = 106.604$$ left hand side

$$\int_{H_1}^{H_1+H_2} 1 \, dh = 106.643$$ right hand side

Cased hole inside drill string drill pipe section

$ID := ID_{dp}$ set inner diameter to casing ID

$A := \frac{\pi}{4} \cdot (ID^2)$ cross sectional area (ft^2)

$T_{ave} := T_{ave1}$ set ave temp to section ave temp

$P := P_1$ set pressure to previous section pressure

$$\gamma_g := \frac{P \cdot SG_a}{R \cdot T_{ave}}$$ specific weight of the gas (N/m^3)

$\gamma_g = 2190.744$ (N/m^3)

$\rho_g := \frac{\gamma_g}{g}$ gas (air) density (N-sec^2/m^4)

$\rho_g = 223.317$ (N-sec²/m⁴)

$\nu_g := \frac{\mu_g}{\rho_g}$ gas (air) kinematic viscosity (m²/sec)

$\nu_g = 5.374 \times 10^{-8}$ (m²/sec)

$\nu_{ave} := \frac{(w_g \cdot \nu_g + w_f \cdot \nu_f)}{w_g + w_f}$ average kinematic viscosity of gas & mud

$\nu_{ave} = 2.396 \times 10^{-5}$ (m²/sec)

$V := \left[\frac{\left[\left(\frac{P_{at}}{P}\right)\cdot\left(\frac{T_{ave}}{T_{at}}\right)\cdot Q_a + Q_f\right]}{A}\right]$ velocity of flow in annulus (m/sec)

$V = 2.003$ (m/sec)

$N_R := \frac{V \cdot ID}{\nu_{ave}}$ Reynolds number calculations

$N_R = 8.124 \times 10^3$

$f := \left[\frac{1}{-1.8 \cdot \log\left[\left(\frac{e_{cs}}{3.7 \cdot ID}\right)^{1.11} + \frac{6.9}{N_R}\right]}\right]^2$ calculate friction factor using Haaland equation

$f = 0.0346$ Haaland friction factor

$Q(P) := \frac{P_{at}}{P} \cdot \frac{T_{ave}}{T_{at}} \cdot Q_a$ flow rate as a function of pressure

$\gamma_{mix}(P) := \frac{w_g + w_f}{Q(P) + Q_f}$ specific weight as a function of pressure

$V(P) := \frac{(Q(P) + Q_f)}{A}$ velocity as a function of pressure

$p_{in} := 441.03$ trial pressure (N/cm² abs)

$P_{in} := p_{in} \cdot 10000.0$ convert pressure to N/m²

$P_{in} = 4.410 \times 10^6$ (N/m²)

$$\int_{P_{in}}^{P_1} \frac{1}{\gamma_{mix}(P)\cdot\left[1-(V(P))^2\cdot\frac{f}{2\cdot g\cdot(ID)}\right]}\,dP = 2027.189$$ left hand side

$$\int_0^{H_1} 1\,dh = 2027.157$$ right hand side

Parameter summary

$q_a = 595.500$ air compressor flow rate (acfm)

$p_{in} = 441.030$ required surface air injection pressure (N/cm^2 abs)

$p_{bh} = 2324.400$ bottom hole pressure (N/cm^2 gauge)

Illustrative Example 9.4: SI version, Drilling Fluid is Aerated Fluid. H&B solution.
This program calculates the required gas injection volumetic flow rate in order to obtain a bottom hole annulus pressure of 2317.4 N/cm^2 gauge while drilling at 3048 m of depth. Utilizes the Hegendorn and Brown liquid hole-up correlation.

Given input conditions

$q_a := 8.5$	air compressor flow rate (scm/min) iterate value
$q_f := 12.05$	mud flow rate (L/sec)
$\gamma_f := 1.2$	specific gravity of mud (water = 1.0)
$\mu_{air} := 0.012$	viscosity of air (centipoises)
$\mu_{mud} := 30$	viscosity of mud (centipoises)
$C := 0.81$	aerated fluid flow loss coefficient through nozzles
$D_c := 2134$	total depth of cased hole (m)
$D_{oh} := 915$	total length of open hole (m)
$alt := 1220$	elevation of drilling site (m)
$p_o := 0.08629$	pressure from table for elevation (MPa)
$t_o := 15.56$	temperature from table for elevation (deg C)
$p_{at} := 0.08629$	actual given atmospheric surface pressure (MPa)
$t_{at} := 15.56$	actual given atmospheric surface temperature (C)
$ROP := 18.29$	estimated rate of penetration (m/hr)
$d_{bh} := 200$	borehole (drill bit) diameter (mm)
$od_{dc} := 171$	outer diameter of drill collar (mm)
$id_{dc} := 71$	inner diameter of drill collar (mm)
$od_{dp} := 114$	outer diameter of drill pipe (mm)
$id_{dp} := 97$	inner diameter of drill pipe (mm)
$od_{tj} := 152$	outer diameter of drill pipe tool joint (mm)
$id_{tj} := 76$	inner diameter of drill pipe tool joint (mm)
$od_c := 219$	outer diameter of casing (mm)
$id_c := 204$	inner diameter of casing (mm)
$L_r := 30.48$	length of return line (m)
$id_r := 143$	inner diameter of return line (mm)
$SG_r := 2.7$	specific gravity of rock (water = 1)
$e_r := 0.003$	roughness of rock (m)
$L_{dc} := 152$	total length of drill collars (m)
$L_{tj} := 0.46$	average length of one tool joint (m)

$L_{dp} := 9.1$	average length of one drill pipe section (m)
$d_n := 12.7$	nozzle diameter (mm)
$N_n := 3$	number of nozzles in the bit
$\beta := 0.018$	geothermal gradient (deg C/m)

Constants

$p_{st} := 0.1$	API atmospheric pressure (MPa)
$t_{st} := 15.56$	API standard temperature (deg C)
$SG_a := 1$	specific gravity of gas (air = 1.0)
$R := 29.31$	specific gas constant (m-N/kgm-K)
$w_w := 9807$	specific weight of water (N/m^3)
$\rho_w := 1000$	density of water (kg/m^3)
$e_{cs} := 0.0000457$	roughness of commercial steel (m)
$g := 9.807$	gravitational constant (m/sec^2)

Solution

$P_{at} := p_{at} \cdot 1000000$	convert actual atm pressure to N/m^2
$P_{at} = 86290.000$	(N/m^2)
$T_{at} := t_{at} + 273$	convert actual atm temp to deg K
$T_{at} = 288.560$	(deg K)
$P_o := p_o \cdot 1000000$	convert table pressure to N/m^2
$P_o = 86290.000$	(N/m^2)
$T_o := t_o + 273$	convert table temperature to deg K
$T_o = 288.560$	(deg K)
$D_{bh} := \frac{d_{bh}}{1000}$	convert borehole diameter to m
$D_{bh} = 0.200$	(m)
$ID_r := \frac{id_r}{1000}$	convert return line diameter to m
$ID_r = 0.143$	(m)
$OD_{dp} := \frac{od_{dp}}{1000}$	convert drill pipe outer diameter to m
$OD_{dp} = 0.114$	(m)
$ID_{dp} := \frac{id_{dp}}{1000}$	convert drill pipe inner diameter to m
$ID_{dp} = 0.097$	(m)

$ID_c := \frac{id_c}{1000}$	convert casing inner diameter to m
$ID_c = 0.204$	(m)
$OD_{dc} := \frac{od_{dc}}{1000}$	convert drill collar outer diameter to m
$OD_{dc} = 0.171$	(m)
$ID_{dc} := \frac{id_{dc}}{1000}$	convert drill collar inner diameter to m
$ID_{dc} = 0.071$	(m)
$OD_{tj} := \frac{od_{tj}}{1000}$	convert tool joint outer diameter to m
$OD_{tj} = 0.152$	(m)
$ID_{tj} := \frac{id_{tj}}{1000}$	convert tool joint inner diameter to m
$ID_{tj} = 0.076$	(m)
$D_n := \frac{d_n}{1000}$	nozzle diameter converted to m
$D_n = 0.013$	(m)
$\mu_g := \frac{\mu_{air}}{1000}$	convert viscosity to N-sec/m^2
$\mu_g = 1.200 \times 10^{-5}$	(N-sec/m^2)
$\mu_f := \frac{\mu_{mud}}{1000}$	convert viscosity to N-sec/m^2
$\mu_f = 3.000 \times 10^{-2}$	(N-sec/m^2)
$Q_a := \frac{q_a}{60}$	convert air flow rate to m^3/sec
$Q_a = 0.142$	(m^3/sec)
$Q_f := \frac{q_f}{1000}$	convert mud flow to (m^3/sec)
$Q_f = 0.012$	(m^3/sec)
$\gamma_f := \gamma_f g \cdot 1000$	convert spec. wt. of mud from S.G. to N/m^3
$\gamma_f = 11768.400$	(N/m^3)
$\rho_f := \frac{\gamma_f}{g}$	mud density (kg*sec^2/m^4)

$\rho_f = 1.2000 \times 10^3$ (kg*sec^2/m^4)

$\nu_f := \dfrac{\mu_f}{\rho_f}$ mud kinematic viscosity (m^2/sec)

$\nu_f = 2.500 \times 10^{-5}$ (m^2/sec)

$\gamma_g := \dfrac{(P_{at} \cdot SG_a)}{R \cdot T_{at}}$ specific weight of air (N/m^3)

$\gamma_g = 10.2025$ (N/m^3)

$\rho_g := \dfrac{\gamma_g}{g}$ gas (air) density (kg/m^3)

$\rho_g = 1.040 \times 10^0$ (kg/m^3)

$\nu_g := \dfrac{\mu_g}{\rho_g}$ gas (air) kinematic viscosity (m^2/sec)

$\nu_g = 1.153 \times 10^{-5}$ (m^2/sec)

$rop := \dfrac{ROP}{60 \cdot 60}$ convet ROP to m/sec

$rop = 0.0051$ (m/sec)

$w_s := \left[\left(\dfrac{\pi}{4}\right) \cdot D_{bh}{}^2 \cdot 9807 \cdot SG_r \cdot rop\right]$ weight rate of solid flow out of blooey line (N/sec)

$w_s = 4.226$ (m/sec)

$w_g := \gamma_g \cdot Q_a$ weight rate of flow of air (N/sec)

$w_g = 1.445$ (N/sec)

$w_f := \gamma_f \cdot Q_f$ weight rate of flow of mud (N/sec)

$w_f = 141.809$ (N/sec)

$w_t := w_g + w_f + w_s$ total weight rate of flow

$w_t = 147.481$ (N/sec)

Section height calculations

$H_1 := D_c - L_{tj} \cdot \left(\dfrac{D_c}{L_{dp}}\right)$ calculate length of drill pipe in cased section

$H_1 = 2026.127$ (m)

$H_2 := L_{tj} \cdot \left(\dfrac{D_c}{L_{dp}}\right)$ calculate length of tool joint in cased section

$H_2 = 107.873$ (m)

$H_3 := D_{oh} - L_{dc} - L_{tj} \cdot \left(\frac{D_{oh} - L_{dc}}{L_{dp}} \right)$ calculate length of drill pipe in open hole section

$H_3 = 724.431$ (m)

$H_4 := L_{tj} \cdot \left(\frac{D_{oh} - L_{dc}}{L_{dp}} \right)$ calculate length of tool joints in open hole section

$H_4 = 38.569$ (m)

$H_5 := L_{dc}$ length of drill collar

$H_5 = 152.000$ (m)

Return line

$ID := ID_r$ set variable ID equal to return line's ID

$ID = 0.143$ (m)

$A := \frac{\pi}{4} \cdot ID^2$ cross sectional area (m^2)

$A = 0.016$ (m^2)

$V := \frac{(Q_a + Q_f)}{A}$ velocity of flow through the line (m/sec)

$V = 9.571$ (m/sec)

$\nu_{ave} := \frac{(w_g \cdot \nu_g + w_f \cdot \nu_f)}{w_g + w_f}$ average kinematic viscosity of gas & mud

$\nu_{ave} = 2.486 \times 10^{-5}$ (m^2/sec)

$N_R := \frac{(V \cdot ID)}{\nu_{ave}}$ Reynolds number calculations

$N_R = 5.505 \times 10^4$ NR > 2000 => Transition or turbulent flow

$f := \left[\frac{1}{-1.8 \cdot \log\left[\left(\frac{e_{cs}}{3.7 \cdot ID} \right)^{1.11} + \frac{6.9}{N_R} \right]} \right]^2$ calculate friction factor using Haaland equation

$f = 0.0213$ Haaland friction factor

$$Q(P) := \frac{P_{at}}{P} \cdot \frac{T_o}{T_{at}} \cdot Q_a$$ flow rate as a function of pressure

$$CQ := P_{at} \cdot \frac{T_o}{T_{at}} \cdot Q_a$$

$$CQ = 12224.417$$

$$\gamma_{mix}(P) := \frac{w_t}{Q(P) + Q_f}$$ specific weight as a function of pressure

$$\gamma_{mix}(P_{at}) = 959.433$$

$$V(P) := \frac{(Q(P) + Q_f)}{A}$$ velocity as a function of pressure

$$V(P_{at}) = 9.571$$

$$p_r := 0.105$$ guess a pressure at the beginning of the return line (MPa) and vary it until matching the two integrals below

$$P_r := p_r \cdot 1000000$$ convert pressure to N/m^2

$$P_r = 105000.000$$ (N/m^2)

$$\int_{P_{at}}^{P_r} \frac{1}{\gamma_{mix}(P) \cdot \left(V(P)^2 \cdot \frac{f}{2 \cdot g \cdot ID} \right)} \, dP = 30.797$$ left hand side of equation

$$\int_0^{L_r} 1 \, dh = 30.480$$ right hand side of equation

$$K_t := 27$$ Figure 8-5 gives the following flow resistance coefficient for the blind Tee

$$K_v := 0.2$$ loss factor for gate valves in blooey line

$$\gamma_{mix} := \frac{w_t}{\left[\left(\frac{P_{at}}{P_r} \right) \cdot \left(\frac{T_o}{T_{at}} \right) \cdot Q_a + Q_f \right]}$$ specific weight (N/m^3)

$$\gamma_{mix} = 1147.952$$ (N/m^3)

$$V := \left[\frac{\left[\left[\left(\frac{P_{at}}{P_r} \right) \cdot \left(\frac{T_o}{T_{at}} \right) \cdot Q_a + Q_f \right] \right]}{A} \right]$$ velocity (m/sec)

$V = 7.999$ (m/sec)

$\Delta P_T := \gamma_{mix} \cdot (2 \cdot K_v + K_t) \cdot \frac{V^2}{2 \cdot g}$ caclulate pressure loss due to Tee and valves

$\Delta P_T = 102614.360$ (N/m^2)

$P_e := P_r + \Delta P_T$ sum pressure losses

$P_e = 207614.360$ (N/m^2)

$p_e := \frac{P_e}{1000000}$ convert from N/m^2 to MPa

$p_e = 0.208$ (MPa)

Cased hole section

Cased hole drill pipe section

$ID := ID_c$ set inner diameter to casing ID

$ID = 0.204$ (m)

$OD := OD_{dp}$ set outer diameter to drill pipe OD

$OD = 0.114$ (m)

$P := P_e$ set pressure equal to previous section pressure

$P = 207614.360$ (N/m^2)

$A := \frac{\pi}{4} \cdot (ID^2 - OD^2)$ cross sectional area (m^2)

$A = 0.022$ (m^2)

$T := T_o + 0.01 \cdot H_1$ calculate temperature at bottom of section

$T = 308.821$ (deg K)

$T_{ave1} := \frac{T_o + T}{2}$ calculate average temperature of section

$T_{ave1} = 298.691$ (deg K)

$T_{ave} := T_{ave1}$ set average temp to section average temp

$\gamma_g(P) := \frac{P \cdot SG_a}{R \cdot T_{ave1}}$ specific weight of the gas (N/m^3)

$\gamma_g(P) = 23.714827$

$\rho_g(P) := \frac{\gamma_g(P)}{g}$ gas (air) density (kg/m^3)

$\rho_g(P) = 2.418153$

$$\nu_g(P) := \frac{\mu_g}{\rho_g(P)}$$ gas (air) kinematic viscosity (m^2/sec)

$\nu_g(P) = 0.00000496$

$$\nu_{ave}(P) := \frac{\left(w_g \cdot \nu_g(P) + w_f \cdot \nu_f\right)}{w_g + w_f}$$ average kinematic viscosity of gas & mud

$\nu_{ave}(P) = 0.00002480$

$$V(P) := \left[\frac{\left[\left(\frac{P_{at}}{P}\right)\cdot\left(\frac{T_{ave}}{T_{at}}\right)\cdot Q_a + Q_f\right]}{A}\right]$$ velocity of flow in annulus (m/sec)

$V(P) = 3.247$

$$N_R(P) := \frac{V(P)\cdot(ID - OD)}{\nu_{ave}(P)}$$ Reynolds number calculations

$N_R(P) = 11786.302$

$$f(P) := \left[\frac{1}{-1.8\cdot\log\left[\left[\frac{e_{cs}}{3.7\cdot(ID - OD)}\right]^{1.11} + \frac{6.9}{N_R(P)}\right]}\right]^2$$ calculate friction factor using Haaland equation

$f(P) = 0.03022$

$t := T - 273$

$$p(P) := \frac{P}{1000000}$$

$\sigma_{w74}(P) := 75 - 6.323\cdot p(P)^{0.349}$ $\frac{dynes}{cm}$

$\sigma_{w280}(P) := 53 - 2.517\cdot p(P)^{0.637}$ $\frac{dynes}{cm}$

$$\sigma_{wt}(P) := \sigma_{w74}(P) - \frac{(1.8t - 42)\cdot(\sigma_{w74}(P) - \sigma_{w280}(P))}{206} \qquad \frac{dynes}{cm}$$

$\sigma_{wt}(P) = 69.244$

$$\upsilon_{sL} := \frac{Q_f}{A}$$ superficial liquid velocity = qL/Ap, m/sec

$\upsilon_{sL} = 0.536$

$$\upsilon_{sg}(P) := \frac{\left[\left(\frac{P_{at}}{P}\right)\cdot\left(\frac{T_{ave}}{T_{at}}\right)\cdot Q_a\right]}{A}$$ superficial gas velocity = qg/Ap, m/sec

$\upsilon_{sg}(P) = 2.711$

$d := ID$

$\rho_L := \gamma_f$ liquid density

$\rho_L = 11768.400$

$\mu_L := \mu_{mud}$

$\mu_L = 30.000$

$$N_{L\upsilon}(P) := 1.7964\cdot\upsilon_{sL}\cdot\left(\frac{\rho_L}{\sigma_{wt}(P)}\right)^{.25}$$

$N_{L\upsilon}(P) = 3.477$

$$N_{g\upsilon}(P) := 1.7964\cdot\upsilon_{sg}(P)\cdot\left(\frac{\rho_L}{\sigma_{wt}(P)}\right)^{.25}$$

$N_{g\upsilon}(P) = 17.587$

$$N_d(P) := 31.664\cdot d\cdot\left(\frac{\rho_L}{\sigma_{wt}(P)}\right)^{.5}$$

$N_d(P) = 84.210$

$$N_L(P) := 0.55646\cdot\mu_L\cdot\left(\frac{1}{\rho_L\cdot\sigma_{wt}(P)^3}\right)^{.25}$$

$N_L(P) = 0.067$

$X1(P) := \log(N_L(P)) + 3$

$Y(P) := -2.69851 + 0.15841 \cdot X1(P) - 0.551 \cdot X1(P)^2 + 0.54785 \cdot X1(P)^3 - 0.12195 \cdot X1(P)^4$

$CN_L(P) := 10^{Y(P)}$

$p_a := p_o$ base pressure (0.1 MPa)

$$X_{HL}(P) := \frac{N_{L\upsilon}(P) \cdot (CN_L(P)) \cdot p(P)^{0.1}}{N_d(P) \cdot N_{g\upsilon}(P)^{0.575} \cdot p_a^{0.1}}$$

$FXHL(P) := \log(X_{HL}(P)) + 6$

$HLoFy(P) := -0.10307 + 0.61777 \cdot FXHL(P) - 0.63295 \cdot FXHL(P)^2 + 0.29598 \cdot FXHL(P)^3 - 0.0401 \cdot FXH$

$$X_\psi(P) := \frac{N_{g\upsilon}(P) \cdot N_L(P)^{0.38}}{N_d(P)^{2.14}}$$

$\psi(P) := 0.91163 - 4.82176 \cdot X_\psi(P) + 1232.25 \cdot X_\psi(P)^2 - 22253.6 \cdot X_\psi(P)^3 + 116174.3 \cdot X_\psi(P)^4$

$H_L(P) := \psi(P) \cdot HLoFy(P)$ liquid holdup (fraction of pipe occupied by liquid)

$H_L(P) = 0.209$

$\upsilon_m(P) := \upsilon_{sL} + \upsilon_{sg}(P)$

$\upsilon_m(P) = 3.247$

$\gamma_{mix}(P) := \gamma_f \cdot H_L(P) + \gamma_g(P) \cdot (1 - H_L(P))$

$\gamma_{mix}(P) = 2481.702$ check a value

$$Q(P) := \frac{P_{at}}{P} \cdot \frac{T_{ave}}{T_{at}} \cdot Q_a$$ flow rate as a function of pressure

$$V(P) := \frac{(Q(P) + Q_f)}{A}$$ velocity as a function of pressure

$p_1 := 13.56$ guess a pressure at the bottom of the section (MPa) and vary it until matching the two integrals below

$P_1 := p_1 \cdot 1000000$ convert pressure to N/m^2

$P_1 = 1.356 \times 10^7$ (N/m^2)

$$\int_{P_e}^{P_1} \frac{1}{\gamma_{mix}(P) \cdot \left[1 + (V(P))^2 \cdot \frac{f(P)}{2 \cdot g \cdot (ID - OD)}\right]} dP = 2026.762$$ left hand side of equation

$$\int_0^{H_1} 1 \, dh = 2026.127$$ right hand side of Eqn. 6-26

Cased hole tool joint section

$ID := ID_c$ set inner diameter to casing ID

$OD := OD_{tj}$ set outer diameter to drill pipe OD

$P := P_1$ set pressure equal to previous section pressure

$A := \frac{\pi}{4} \cdot \left(ID^2 - OD^2\right)$ cross sectional area (m^2)

$T_{temp} := T$ keep previous section temperature

$T := T_{temp} + 0.01 \cdot H_2$ calculate temperature at bottom of section

$T = 309.900$ (deg K)

$T_{ave2} := \frac{T_{temp} + T}{2}$ calculate average temperature of section

$T_{ave2} = 309.361$ (deg K)

$T_{ave1} := T_{ave2}$ set average temp to section averate temp

$\gamma_g(P) := \frac{P \cdot SG_a}{R \cdot T_{ave1}}$ specific weight of the gas (N/m^3)

$\gamma_g(P) = 1548.896012$

$\rho_g(P) := \frac{\gamma_g(P)}{g}$ gas (air) density (kg/m^3)

$\rho_g(P) = 157.937801$

$\nu_g(P) := \frac{\mu_g}{\rho_g(P)}$ gas (air) kinematic viscosity (m^2/sec)

$\nu_g(P) = 0.00000008$

$$\nu_{ave}(P) := \frac{(w_g \cdot \nu_g(P) + w_f \cdot \nu_f)}{w_g + w_f}$$ average kinematic viscosity of gas & mud

$\nu_{ave}(P) = 0.00002475$

$$V(P) := \left[\frac{\left[\left[\left(\frac{P_{at}}{P}\right) \cdot \left(\frac{T_{ave}}{T_{at}}\right) \cdot Q_a + Q_f\right]\right]}{A}\right]$$ velocity of flow in annulus (m/sec)

$V(P) = 0.895$

$$N_R(P) := \frac{V(P) \cdot (ID - OD)}{\nu_{ave}(P)}$$ Reynolds number calculations

$N_R(P) = 1881.068$

$$f(P) := \left[\frac{1}{-1.8 \cdot \log\left[\left[\frac{e_{cs}}{3.7 \cdot (ID - OD)}\right]^{1.11} + \frac{6.9}{N_R(P)}\right]}\right]^2$$ calculate friction factor using Haaland equation

$f(P) = 0.05251$

$t := T - 273$

$$p(P) := \frac{P}{1000000}$$

$$\sigma_{w74}(P) := 75 - 6.323 \cdot p(P)^{0.349} \quad \frac{dynes}{cm}$$

$$\sigma_{w280}(P) := 53 - 2.517 \cdot p(P)^{0.637} \quad \frac{dynes}{cm}$$

$$\sigma_{wt}(P) := \sigma_{w74}(P) - \frac{(1.8t - 42) \cdot (\sigma_{w74}(P) - \sigma_{w280}(P))}{206} \quad \frac{dynes}{cm}$$

$\sigma_{wt}(P) = 56.977$

$$v_{sL} := \frac{Q_f}{A}$$ superficial liquid velocity = qL/Ap, m/sec

$\upsilon_{sL} = 0.829$

$$\upsilon_{sg}(P) := \frac{\left[\left(\frac{P_{at}}{P}\right)\cdot\left(\frac{T_{ave}}{T_{at}}\right)\cdot Q_a\right]}{A}$$

superficial gas velocity = qg/Ap, m/sec

$\upsilon_{sg}(P) = 0.066$

$d := ID$

$\rho_L := \gamma_f$ liquid density

$\rho_L = 11768.400$

$\mu_L := \mu_{mud}$

$\mu_L = 30.000$

$$N_{L\upsilon}(P) := 1.7964\cdot\upsilon_{sL}\cdot\left(\frac{\rho_L}{\sigma_{wt}(P)}\right)^{.25}$$

$N_{L\upsilon}(P) = 5.644$

$$N_{g\upsilon}(P) := 1.7964\cdot\upsilon_{sg}(P)\cdot\left(\frac{\rho_L}{\sigma_{wt}(P)}\right)^{.25}$$

$N_{g\upsilon}(P) = 0.453$

$$N_d(P) := 31.664\cdot d\cdot\left(\frac{\rho_L}{\sigma_{wt}(P)}\right)^{.5}$$

$N_d(P) = 92.834$

$$N_L(P) := 0.55646\cdot\mu_L\cdot\left(\frac{1}{\rho_L\cdot\sigma_{wt}(P)^3}\right)^{.25}$$

$N_L(P) = 0.077$

$$X1(P) := \log(N_L(P)) + 3$$

$$Y(P) := -2.69851 + 0.15841\cdot X1(P) - 0.551\cdot X1(P)^2 + 0.54785\cdot X1(P)^3 - 0.12195\cdot X1(P)^4$$

$$CN_L(P) := 10^{Y(P)}$$

$p_a := p_o$ base pressure (0.1 MPa)

$$X_{HL}(P) := \frac{N_{L\upsilon}(P)\cdot(CN_L(P))\cdot p(P)^{0.1}}{N_d(P)\cdot N_{g\upsilon}(P)^{0.575}\cdot p_a^{\ 0.1}}$$

$$FXHL(P) := \log(X_{HL}(P)) + 6$$

$$HLoFy(P) := -0.10307 + 0.61777\cdot FXHL(P) - 0.63295\cdot FXHL(P)^2 + 0.29598\cdot FXHL(P)^3 - 0.0401\cdot FXHL$$

$$X_\psi(P) := \frac{N_{g\upsilon}(P)\cdot N_L(P)^{0.38}}{N_d(P)^{2.14}}$$

$$\psi(P) := 0.91163 - 4.82176\cdot X_\psi(P) + 1232.25\cdot X_\psi(P)^2 - 22253.6\cdot X_\psi(P)^3 + 116174.3\cdot X_\psi(P)^4$$

$H_L(P) := \psi(P)\cdot HLoFy(P)$ liquid holdup (fraction of pipe occupied by liquid)

$H_L(P) = 0.716$

$\upsilon_m(P) := \upsilon_{sL} + \upsilon_{sg}(P)$

$\upsilon_m(P) = 0.895$

$\gamma_{mix}(P) := \gamma_f\cdot H_L(P) + \gamma_g(P)\cdot(1 - H_L(P))$

$\gamma_{mix}(P) = 8863.518$ check a value

$Q(P) := \frac{P_{at}}{P}\cdot\frac{T_{ave}}{T_{at}}\cdot Q_a$ flow rate as a function of pressure

$V(P) := \frac{(Q(P) + Q_f)}{A}$ velocity as a function of pressure

$p_2 := 14.56$ guess a pressure at the bottom of the section (lMPa) and vary it until matching the two integrals below

$P_2 := p_2\cdot 1000000$ convert pressure to N/m^2

$P_2 = 1.456 \times 10^7$ (N/m^2)

$$\int_{P_1}^{P_2} \frac{1}{\gamma_{mix}(P)\cdot\left[1+(V(P))^2\cdot\frac{f(P)}{2\cdot g\cdot(ID-OD)}\right]}\,dP = 107.591$$ left hand side of equation

$$\int_{H_1}^{H_1+H_2} 1\,dh = 107.873$$ right hand side of equation

Open hole section

Open hole drill pipe section

$ID := D_{bh}$ set inner diameter to casing ID

$OD := OD_{dp}$ set outer diameter to drill pipe OD

$P := P_2$ set pressure equal to previous section pressure

$A := \frac{\pi}{4}\cdot\left(ID^2 - OD^2\right)$ cross sectional area (m^2)

$T_{temp} := T$ keep previous section temperature

$T := T_{temp} + 0.01\cdot H_3$ calculate temperature at bottom of section

$T = 317.144$ (deg K)

$T_{ave3} := \frac{T_{temp} + T}{2}$ calculate average temperature of section

$T_{ave3} = 313.522$ (deg K)

$T_{ave} := T_{ave3}$ set average temp to section average temp

$\gamma_g(P) := \frac{P\cdot SG_a}{R\cdot T_{ave1}}$ specific weight of the gas (N/m^3)

$\gamma_g(P) = 1663.121381$

$\rho_g(P) := \frac{\gamma_g(P)}{g}$ gas (air) density (kg/m^3)

$\rho_g(P) = 169.585131$

$\nu_g(P) := \frac{\mu_g}{\rho_g(P)}$ gas (air) kinematic viscosity (m^2/sec)

$\nu_g(P) = 0.00000007$

$$\nu_{ave}(P) := \frac{(w_g \cdot \nu_g(P) + w_f \cdot \nu_f)}{w_g + w_f}$$ average kinematic viscosity of gas & mud

$$\nu_{ave}(P) = 0.00002475$$

$$V(P) := \left[\frac{\left[\left(\frac{P_{at}}{P}\right) \cdot \left(\frac{T_{ave}}{T_{at}}\right) \cdot Q_a + Q_f\right]}{A}\right]$$ velocity of flow in annulus (m/sec)

$$V(P) = 0.611$$

$$N_R(P) := \frac{V(P) \cdot (ID - OD)}{\nu_{ave}(P)}$$ Reynolds number calculations

$$N_R(P) = 2123.789$$

$$f(P) := \left[\frac{1}{-1.8 \cdot \log\left[\left[\frac{e_{cs}}{3.7 \cdot (ID - OD)}\right]^{1.11} + \frac{6.9}{N_R(P)}\right]}\right]^2$$ calculate friction factor using Haaland equation

$$f(P) = 0.05014$$

$$t := T - 273$$

$$p(P) := \frac{P}{1000000}$$

$$\sigma_{w74}(P) := 75 - 6.323 \cdot p(P)^{0.349} \quad \frac{dynes}{cm}$$

$$\sigma_{w280}(P) := 53 - 2.517 \cdot p(P)^{0.637} \quad \frac{dynes}{cm}$$

$$\sigma_{wt}(P) := \sigma_{w74}(P) - \frac{(1.8t - 42) \cdot (\sigma_{w74}(P) - \sigma_{w280}(P))}{206} \quad \frac{dynes}{cm}$$

$$\sigma_{wt}(P) = 55.305$$

$$\upsilon_{sL} := \frac{Q_f}{A}$$ superficial liquid velocity = qL/Ap, m/sec

$$\upsilon_{sL} = 0.568$$

$$\upsilon_{sg}(P) := \frac{\left[\left(\frac{P_{at}}{P}\right)\cdot\left(\frac{T_{ave}}{T_{at}}\right)\cdot Q_a\right]}{A}$$ superficial gas velocity = qg/Ap, m/sec

$\upsilon_{sg}(P) = 0.043$

$d := ID$

$\rho_L := \gamma_f$ liquid density

$\rho_L = 11768.400$

$\mu_L := \mu_{mud}$

$\mu_L = 30.000$

$$N_{L\upsilon}(P) := 1.7964\cdot\upsilon_{sL}\cdot\left(\frac{\rho_L}{\sigma_{wt}(P)}\right)^{.25}$$

$N_{L\upsilon}(P) = 3.898$

$$N_{g\upsilon}(P) := 1.7964\cdot\upsilon_{sg}(P)\cdot\left(\frac{\rho_L}{\sigma_{wt}(P)}\right)^{.25}$$

$N_{g\upsilon}(P) = 0.295$

$$N_d(P) := 31.664\cdot d\cdot\left(\frac{\rho_L}{\sigma_{wt}(P)}\right)^{.5}$$

$N_d(P) = 92.379$

$$N_L(P) := 0.55646\cdot\mu_L\cdot\left(\frac{1}{\rho_L\cdot\sigma_{wt}(P)^3}\right)^{.25}$$

$N_L(P) = 0.079$

$$X1(P) := \log(N_L(P)) + 3$$

$$Y(P) := -2.69851 + 0.15841\cdot X1(P) - 0.551\cdot X1(P)^2 + 0.54785\cdot X1(P)^3 - 0.12195\cdot X1(P)^4$$

$$CN_L(P) := 10^{Y(P)}$$

$p_a := p_o$ base pressure (0.1 MPa)

$$X_{HL}(P) := \frac{N_{L\upsilon}(P)\cdot(CN_L(P))\cdot p(P)^{0.1}}{N_d(P)\cdot N_{g\upsilon}(P)^{0.575}\cdot p_a^{0.1}}$$

$$FXHL(P) := \log(X_{HL}(P)) + 6$$

$$HLoFy(P) := -0.10307 + 0.61777\cdot FXHL(P) - 0.63295\cdot FXHL(P)^2 + 0.29598\cdot FXHL(P)^3 - 0.0401\cdot FXHL$$

$$X_\psi(P) := \frac{N_{g\upsilon}(P)\cdot N_L(P)^{0.38}}{N_d(P)^{2.14}}$$

$$\psi(P) := 0.91163 - 4.82176\cdot X_\psi(P) + 1232.25\cdot X_\psi(P)^2 - 22253.6\cdot X_\psi(P)^3 + 116174.3\cdot X_\psi(P)^4$$

$H_L(P) := \psi(P)\cdot HLoFy(P)$ liquid holdup (fraction of pipe occupied by liquid)

$H_L(P) = 0.697$

$\upsilon_m(P) := \upsilon_{sL} + \upsilon_{sg}(P)$

$\upsilon_m(P) = 0.611$

$\gamma_{mix}(P) := \gamma_f\cdot H_L(P) + \gamma_g(P)\cdot(1 - H_L(P))$

$\gamma_{mix}(P) = 8704.363$ check a value

$Q(P) := \frac{P_{at}}{P}\cdot\frac{T_{ave}}{T_{at}}\cdot Q_a$ flow rate as a function of pressure

$V(P) := \frac{(Q(P) + Q_f)}{A}$ velocity as a function of pressure

$p_3 := 21.2$ guess a pressure at the bottom of the section (MPa) and vary it until matching the two integrals below

$P_3 := p_3\cdot 1000000$ convert pressure to N/m^2

$P_3 = 2.120 \times 10^7$ (lb/ft^2)

$\int_{P_2}^{P_3} \frac{1}{\gamma_{mix}(P)\cdot\left[1 + (V(P))^2\cdot\frac{f(P)}{2\cdot g\cdot(ID - OD)}\right]}\,dP = 724.275$ left hand side of equation

$\int_{H_1+H_2}^{H_1+H_2+H_3} 1\,dh = 724.431$ right hand side of equation

Open hole tool joint section

$ID := D_{bh}$ set inner diameter to casing ID

$OD := OD_{tj}$ set outer diameter to drill pipe OD

$P := P_3$ set pressure equal to previous section pressure

$A := \frac{\pi}{4}\cdot\left(ID^2 - OD^2\right)$ cross sectional area (m^2)

$T_{temp} := T$ keep previous section temperature

$T := T_{temp} + 0.01\cdot H_4$ calculate temperature at bottom of section

$T = 317.530$ (deg K)

$T_{ave4} := \frac{T_{temp} + T}{2}$ calculate average temperature of section

$T_{ave4} = 317.337$ (deg K)

$T_{ave} := T_{ave4}$ set average temp to section average temp

$\gamma_g(P) := \frac{P\cdot SG_a}{R\cdot T_{ave1}}$ specific weight of the gas (N/m^3)

$\gamma_g(P) = 2421.577835$

$\rho_g(P) := \frac{\gamma_g(P)}{g}$ gas (air) density (kg/m^3)

$\rho_g(P) = 246.923405$

$\nu_g(P) := \frac{\mu_g}{\rho_g(P)}$ gas (air) kinematic viscosity (m^2/sec)

$\nu_g(P) = 0.00000005$

$$\nu_{ave}(P) := \frac{(w_g \cdot \nu_g(P) + w_f \cdot \nu_f)}{w_g + w_f}$$ average kinematic viscosity of gas & mud

$\nu_{ave}(P) = 0.00002475$

$$V(P) := \left[\frac{\left[\left(\frac{P_{at}}{P} \right) \cdot \left(\frac{T_{ave}}{T_{at}} \right) \cdot Q_a + Q_f \right]}{A} \right]$$ velocity of flow in annulus (m/sec)

$V(P) = 0.956$

$$N_R(P) := \frac{V(P) \cdot (ID - OD)}{\nu_{ave}(P)}$$ Reynolds number calculations

$N_R(P) = 1853.888$

$$f(P) := \left[\frac{1}{-1.8 \cdot \log\left[\left[\frac{e_{cs}}{3.7 \cdot (ID - OD)} \right]^{1.11} + \frac{6.9}{N_R(P)} \right]} \right]^2$$ calculate friction factor using Haaland equation

$f(P) = 0.05282$

$t := T - 273$

$$p(P) := \frac{P}{1000000}$$

$$\sigma_{w74}(P) := 75 - 6.323 \cdot p(P)^{0.349} \quad \frac{dynes}{cm}$$

$$\sigma_{w280}(P) := 53 - 2.517 \cdot p(P)^{0.637} \quad \frac{dynes}{cm}$$

$$\sigma_{wt}(P) := \sigma_{w74}(P) - \frac{(1.8t - 42) \cdot (\sigma_{w74}(P) - \sigma_{w280}(P))}{206} \quad \frac{dynes}{cm}$$

$\sigma_{wt}(P) = 52.706$

$$\upsilon_{sL} := \frac{Q_f}{A}$$ superficial liquid velocity = qL/Ap, m/sec

$\upsilon_{sL} = 0.908$

$$\upsilon_{sg}(P) := \frac{\left[\left(\frac{P_{at}}{P}\right)\cdot\left(\frac{T_{ave}}{T_{at}}\right)\cdot Q_a\right]}{A}$$ superficial gas velocity = qg/Ap, m/sec

$\upsilon_{sg}(P) = 0.048$

$d := ID$

$\rho_L := \gamma_f$ liquid density

$\rho_L = 11768.400$

$\mu_L := \mu_{mud}$

$\mu_L = 30.000$

$$N_{L\upsilon}(P) := 1.7964\cdot\upsilon_{sL}\cdot\left(\frac{\rho_L}{\sigma_{wt}(P)}\right)^{.25}$$

$N_{L\upsilon}(P) = 6.306$

$$N_{g\upsilon}(P) := 1.7964\cdot\upsilon_{sg}(P)\cdot\left(\frac{\rho_L}{\sigma_{wt}(P)}\right)^{.25}$$

$N_{g\upsilon}(P) = 0.332$

$$N_d(P) := 31.664\cdot d\cdot\left(\frac{\rho_L}{\sigma_{wt}(P)}\right)^{.5}$$

$N_d(P) = 94.629$

$$N_L(P) := 0.55646\cdot\mu_L\cdot\left(\frac{1}{\rho_L\cdot\sigma_{wt}(P)^3}\right)^{.25}$$

$N_L(P) = 0.082$

$$X1(P) := \log(N_L(P)) + 3$$

$$Y(P) := -2.69851 + 0.15841\cdot X1(P) - 0.551\cdot X1(P)^2 + 0.54785\cdot X1(P)^3 - 0.12195\cdot X1(P)^4$$

$$CN_L(P) := 10^{Y(P)}$$

$p_a := p_o$ base pressure (0.1 MPa)

$$X_{HL}(P) := \frac{N_{L\upsilon}(P)\cdot(CN_L(P))\cdot p(P)^{0.1}}{N_d(P)\cdot N_{g\upsilon}(P)^{0.575}\cdot p_a^{0.1}}$$

$$FXHL(P) := \log(X_{HL}(P)) + 6$$

$$HLoFy(P) := -0.10307 + 0.61777\cdot FXHL(P) - 0.63295\cdot FXHL(P)^2 + 0.29598\cdot FXHL(P)^3 - 0.0401\cdot FXHL$$

$$X_\psi(P) := \frac{N_{g\upsilon}(P)\cdot N_L(P)^{0.38}}{N_d(P)^{2.14}}$$

$$\psi(P) := 0.91163 - 4.82176\cdot X_\psi(P) + 1232.25\cdot X_\psi(P)^2 - 22253.6\cdot X_\psi(P)^3 + 116174.3\cdot X_\psi(P)^4$$

$H_L(P) := \psi(P)\cdot HLoFy(P)$ — liquid holdup (fraction of pipe occupied by liquid)

$H_L(P) = 0.781$

$\upsilon_m(P) := \upsilon_{sL} + \upsilon_{sg}(P)$

$\upsilon_m(P) = 0.956$

$\gamma_{mix}(P) := \gamma_f\cdot H_L(P) + \gamma_g(P)\cdot(1 - H_L(P))$

$\gamma_{mix}(P) = 9717.889$ — check a value

$Q(P) := \frac{P_{at}}{P}\cdot\frac{T_{ave}}{T_{at}}\cdot Q_a$ — flow rate as a function of pressure

$V(P) := \frac{(Q(P) + Q_f)}{A}$ — velocity as a function of pressure

$p_4 := 21.59$ — guess a pressure at the bottom of the section (MPa) and vary it until matching the two integrals below

$P_4 := p_4\cdot 1000000$ — convert pressure to N/m^2

$P_4 = 2.159 \times 10^7$ — (N/m^2)

$$\int_{P_3}^{P_4} \frac{1}{\gamma_{mix}(P)\cdot\left[1 + (V(P))^2\cdot\frac{f(P)}{2\cdot g\cdot(ID - OD)}\right]}\,dP = 38.118$$ left hand side of equation

$$\int_{H_1+H_2+H_3}^{H_1+H_2+H_3+H_4} 1\,dh = 38.569$$ right hand side of equation

Open hole drill collar section

$ID := D_{bh}$ set inner diameter to casing ID

$OD := OD_{dc}$ set outer diameter to drill pipe OD

$P := P_4$ set pressure equal to previous section pressure

$A := \frac{\pi}{4}\cdot\left(ID^2 - OD^2\right)$ cross sectional area (m^2)

$T_{temp} := T$ keep previous section temperature

$T := T_{temp} + 0.01\cdot H_5$ calculate temperature at bottom of section

$T = 319.050$ (deg K)

$T_{ave5} := \frac{T_{temp} + T}{2}$ calculate average temperature of section

$T_{ave5} = 318.290$ (deg K)

$T_{ave} := T_{ave5}$ set average temp to section average temp

$\gamma_g(P) := \frac{P\cdot SG_a}{R\cdot T_{ave1}}$ specific weight of the gas (N/m^3)

$\gamma_g(P) = 2466.125730$

$\rho_g(P) := \frac{\gamma_g(P)}{g}$ gas (air) density (kg/m^3)

$\rho_g(P) = 251.465864$

$\nu_g(P) := \frac{\mu_g}{\rho_g(P)}$ gas (air) kinematic viscosity (m^2/sec)

$\nu_g(P) = 0.00000005$

$$\nu_{ave}(P) := \frac{(w_g \cdot \nu_g(P) + w_f \cdot \nu_f)}{w_g + w_f}$$ average kinematic viscosity of gas & mud

$$\nu_{ave}(P) = 0.00002475$$

$$V(P) := \left[\frac{\left[\left(\frac{P_{at}}{P}\right) \cdot \left(\frac{T_{ave}}{T_{at}}\right) \cdot Q_a + Q_f\right]}{A}\right]$$ velocity of flow in annulus (m/sec)

$$V(P) = 1.500$$

$$N_R(P) := \frac{V(P) \cdot (ID - OD)}{\nu_{ave}(P)}$$ Reynolds number calculations

$$N_R(P) = 1757.616$$

$$f(P) := \left[\frac{1}{-1.8 \cdot \log\left[\left[\frac{e_{cs}}{3.7 \cdot (ID - OD)}\right]^{1.11} + \frac{6.9}{N_R(P)}\right]}\right]^2$$ calculate friction factor using Haaland equation

$$f(P) = 0.05419$$

$$t := T - 273$$

$$p(P) := \frac{P}{1000000}$$

$$\sigma_{w74}(P) := 75 - 6.323 \cdot p(P)^{0.349} \quad \frac{dynes}{cm}$$

$$\sigma_{w280}(P) := 53 - 2.517 \cdot p(P)^{0.637} \quad \frac{dynes}{cm}$$

$$\sigma_{wt}(P) := \sigma_{w74}(P) - \frac{(1.8t - 42) \cdot (\sigma_{w74}(P) - \sigma_{w280}(P))}{206} \quad \frac{dynes}{cm}$$

$$\sigma_{wt}(P) = 52.289$$

$$\upsilon_{sL} := \frac{Q_f}{A}$$ superficial liquid velocity = qL/Ap, m/sec

$$\upsilon_{sL} = 1.426$$

$$\upsilon_{sg}(P) := \frac{\left[\left(\frac{P_{at}}{P}\right)\cdot\left(\frac{T_{ave}}{T_{at}}\right)\cdot Q_a\right]}{A}$$ superficial gas velocity = qg/Ap, m/sec

$\upsilon_{sg}(P) = 0.074$

$d := ID$

$\rho_L := \gamma_f$ liquid density

$\rho_L = 11768.400$

$\mu_L := \mu_{mud}$

$\mu_L = 30.000$

$$N_{L\upsilon}(P) := 1.7964\cdot\upsilon_{sL}\cdot\left(\frac{\rho_L}{\sigma_{wt}(P)}\right)^{.25}$$

$N_{L\upsilon}(P) = 9.922$

$$N_{g\upsilon}(P) := 1.7964\cdot\upsilon_{sg}(P)\cdot\left(\frac{\rho_L}{\sigma_{wt}(P)}\right)^{.25}$$

$N_{g\upsilon}(P) = 0.514$

$$N_d(P) := 31.664\cdot d\cdot\left(\frac{\rho_L}{\sigma_{wt}(P)}\right)^{.5}$$

$N_d(P) = 95.005$

$$N_L(P) := 0.55646\cdot\mu_L\cdot\left(\frac{1}{\rho_L\cdot\sigma_{wt}(P)^3}\right)^{.25}$$

$N_L(P) = 0.082$

$$X1(P) := \log(N_L(P)) + 3$$

$$Y(P) := -2.69851 + 0.15841\cdot X1(P) - 0.551\cdot X1(P)^2 + 0.54785\cdot X1(P)^3 - 0.12195\cdot X1(P)^4$$

$$CN_L(P) := 10^{Y(P)}$$

$p_a := p_o$ base pressure (0.1 MPa)

$$X_{HL}(P) := \frac{N_{L\upsilon}(P)\cdot(CN_L(P))\cdot p(P)^{0.1}}{N_d(P)\cdot N_{g\upsilon}(P)^{0.575}\cdot p_a^{\,0.1}}$$

$$FXHL(P) := \log(X_{HL}(P)) + 6$$

$$HLoFy(P) := -0.10307 + 0.61777\cdot FXHL(P) - 0.63295\cdot FXHL(P)^2 + 0.29598\cdot FXHL(P)^3 - 0.0401\cdot FXHL$$

$$X_{\psi}(P) := \frac{N_{g\upsilon}(P)\cdot N_L(P)^{0.38}}{N_d(P)^{2.14}}$$

$$\psi(P) := 0.91163 - 4.82176\cdot X_{\psi}(P) + 1232.25\cdot X_{\psi}(P)^2 - 22253.6\cdot X_{\psi}(P)^3 + 116174.3\cdot X_{\psi}(P)^4$$

$H_L(P) := \psi(P)\cdot HLoFy(P)$ liquid holdup (fraction of pipe occupied by liquid)

$H_L(P) = 0.814$

$\upsilon_m(P) := \upsilon_{sL} + \upsilon_{sg}(P)$

$\upsilon_m(P) = 1.500$

$\gamma_{mix}(P) := \gamma_f\cdot H_L(P) + \gamma_g(P)\cdot(1 - H_L(P))$

$\gamma_{mix}(P) = 10038.983$ check a value

$Q(P) := \frac{P_{at}}{P}\cdot\frac{T_{ave}}{T_{at}}\cdot Q_a$ flow rate as a function of pressure

$V(P) := \frac{(Q(P) + Q_f)}{A}$ velocity as a function of pressure

5 V(P)

$p_5 := 23.38 + p_{at}$ guess a pressure at the bottom of the section (MPa) and vary it until matching the two integrals below

$P_5 := p_5\cdot 1000000$ convert pressure to N/m^2

$P_5 = 2.347 \times 10^7$ (N/m^2)

$$\int_{P_4}^{P_5} \frac{1}{\gamma_{mix}(P)\cdot\left[1 + (V(P))^2\cdot\frac{f(P)}{2\cdot g\cdot(ID - OD)}\right]}\,dP = 153.071$$ left hand side of equation

$$\int_{H_1+H_2+H_3+H_4}^{H_1+H_2+H_3+H_4+H_5} 1\,dh = 152.000$$ right hand side of equation

$p_{bh} := p_5 - p_{at}$ bottom hole pressure (NPa)

$p_{bh} = 23.380$ (NPa)

Inside drill string

Nozzle

$D_e := \left[N_n\cdot(D_n)^2\right]^{0.5}$ equivalent single diameter (m)

$D_e = 0.022$ (m)

$$\gamma_{mix} := \frac{(w_g + w_f)}{\left[Q_f + Q_a\cdot\left(\frac{P_{at}}{P_5}\right)\cdot\frac{T}{T_{at}}\right]}$$ specific weight of mixture (N/m^3)

$\gamma_{mix} = 11346.018$ (N/m^3)

$$\Delta P := \frac{(w_g + w_f)^2}{2\cdot C^2\cdot g\cdot\gamma_{mix}\cdot\left(\frac{\pi}{4}\right)^2\cdot D_e^4}$$ equation to approximate the pressure change of the aerated fluid through the drill bit

$\Delta P = 973195.743$ (N/m^2)

$P := P_5 + \Delta P$ sum pressures to get pressure above drill bit

$P = 24439485.743$ (N/m^2)

$p := \frac{P}{1000000}$ convert prssure to MPa

$p = 24.439$ (MPa)

Pressure at top of collars

$ID := ID_{dc}$ set inner diameter to casing ID

$A := \frac{\pi}{4}\cdot(ID^2)$ cross sectional area (m^2)

$T_{ave} := T_{ave5}$ set ave temp to section ave temp

$\gamma_g := \frac{P \cdot SG_a}{R \cdot T_{ave}}$ specific weight of the gas (N/m^3)

$\gamma_g = 2619.710$ (N/m^3)

$\rho_g := \frac{\gamma_g}{g}$ gas (air) density (kg/m^3)

$\rho_g = 267.127$ (kg/m^3)

$\nu_g := \frac{\mu_g}{\rho_g}$ gas (air) kinematic viscosity (m^2/sec)

$\nu_g = 4.492 \times 10^{-8}$ (m^2/sec)

$\nu_{ave} := \frac{(w_g \cdot \nu_g + w_f \cdot \nu_f)}{w_g + w_f}$ average kinematic viscosity of gas & mud

$\nu_{ave} = 2.475 \times 10^{-5}$ (m^2/sec)

$V := \left[\frac{\left[\left(\frac{P_{at}}{P} \right) \cdot \left(\frac{T_{ave}}{T_{at}} \right) \cdot Q_a + Q_f \right]}{A} \right]$ velocity of flow in annulus (m/sec)

$V = 3.183$ (m/sec)

$N_R := \frac{V \cdot ID}{\nu_{ave}}$ Reynolds number calculations

$N_R = 9.131 \times 10^3$ NR > 4000 => Turbulent Flow

$f := \left[\frac{1}{-1.8 \cdot \log\left[\left(\frac{e_{cs}}{3.7 \cdot ID} \right)^{1.11} + \frac{6.9}{N_R} \right]} \right]^2$ calculate friction factor using Haaland equation

$f = 0.0324$ Haaland friction factor

$Q(P) := \frac{P_{at}}{P} \cdot \frac{T_{ave}}{T_{at}} \cdot Q_a$ flow rate as a function of pressure

$CQ := P_{at} \cdot \frac{T_{ave}}{T_{at}} \cdot Q_a$

$CQ = 13483.884$

$\gamma_{mix}(P) := \dfrac{w_g + w_f}{Q(P) + Q_f}$ specific weight as a function of pressure

$V(P) := \dfrac{(Q(P) + Q_f)}{A}$ velocity as a function of pressure

$p_4 := 23.12$ guess a pressure at the top of the section (MPa) and vary it until matching the two integrals below

$P_4 := p_4 \cdot 1000000$ convert pressure to N/m^2

$P_4 = 2.312 \times 10^7$ (N/m^2)

$$\int_{P_4}^{P} \frac{1}{\gamma_{mix}(P)\cdot\left[1 - (V(P))^2 \cdot \dfrac{f}{2\cdot g\cdot(ID)}\right]}\, dP = 152.221$$ left hand side of equation

$$\int_{H_1+H_2+H_3+H_4}^{H_1+H_2+H_3+H_4+H_5} 1\, dh = 152.000$$ right hand side of equation

Pressure at top of tool joints

$ID := ID_{tj}$ set inner diameter to casing ID

$A := \dfrac{\pi}{4}\cdot(ID^2)$ cross sectional area (m^2)

$T_{ave} := T_{ave4}$ set ave temp to section ave temp

$P := P_4$ set pressure to previous section pressure

$\gamma_g := \dfrac{P\cdot SG_a}{R\cdot T_{ave}}$ specific weight of the gas (N/m^3)

$\gamma_g = 2485.714$ (N/m^3)

$\rho_g := \dfrac{\gamma_g}{g}$ gas (air) density (kg/m^3)

$\rho_g = 253.463$ (kg/m^3)

$\nu_g := \dfrac{\mu_g}{\rho_g}$ gas (air) kinematic viscosity (m^2/sec)

$\nu_g = 4.734 \times 10^{-8}$ (m^2/sec)

$\nu_{ave} := \dfrac{(w_g\cdot\nu_g + w_f\cdot\nu_f)}{w_g + w_f}$ average kinematic viscosity of gas & mud

$\nu_{ave} = 2.475 \times 10^{-5}$ (m^2/sec)

$$V := \left[\frac{\left[\left(\frac{P_{at}}{P}\right)\cdot\left(\frac{T_{ave}}{T_{at}}\right)\cdot Q_a + Q_f\right]}{A}\right]$$ velocity of flow in annulus (m/sec)

$V = 2.784$ (m/sec)

$$N_R := \frac{V\cdot ID}{\nu_{ave}}$$ Reynolds number calculations

$N_R = 8.551 \times 10^3$ NR > 4000 => Turbulent

$$f := \left[\frac{1}{-1.8\cdot\log\left[\left(\frac{e_{cs}}{3.7\cdot ID}\right)^{1.11} + \frac{6.9}{N_R}\right]}\right]^2$$ calculate friction factor using Haaland equation

$f = 0.0329$ Haaland friction factor

$$Q(P) := \frac{P_{at}}{P}\cdot\frac{T_{ave}}{T_{at}}\cdot Q_a$$ flow rate as a function of pressure

$$\gamma_{mix}(P) := \frac{w_g + w_f}{Q(P) + Q_f}$$ specific weight as a function of pressure

$$V(P) := \frac{\left(Q(P) + Q_f\right)}{A}$$ velocity as a function of pressure

$p_3 := 22.76$ guess a pressure at the top of the section (MPa) and vary it until matching the two integrals below

$P_3 := p_3\cdot 1000000$ convert pressure to N/m^2

$P_3 = 2.276 \times 10^7$ (N/m^2)

$$\int_{P_3}^{P_4} \frac{1}{\gamma_{mix}(P)\cdot\left[1 - (V(P))^2\cdot\frac{f}{2\cdot g\cdot(ID)}\right]}\,dP = 38.326$$ left hand side of equation

$$\int_{H_1+H_2+H_3}^{H_1+H_2+H_3+H_4} 1\,dh = 38.569$$ right hand side of equation

Pressure at top of drill pipe

$ID := ID_{dp}$ set inner diameter to casing ID

$$A := \frac{\pi}{4}\cdot\left(ID^2\right)$$ cross sectional area (m^2)

$T_{ave} := T_{ave3}$ set ave temp to section ave temp

$P := P_3$ set pressure to previous section pressure

$\gamma_g := \frac{P \cdot SG_a}{R \cdot T_{ave}}$ specific weight of the gas (N/m^3)

$\gamma_g = 2476.784$ (N/m^3)

$\rho_g := \frac{\gamma_g}{g}$ gas (air) density (kg/m^3)

$\rho_g = 252.553$ (kg/m^3)

$\nu_g := \frac{\mu_g}{\rho_g}$ gas (air) kinematic viscosity (m^2/sec)

$\nu_g = 4.751 \times 10^{-8}$ (m^2/sec)

$\nu_{ave} := \frac{(w_g \cdot \nu_g + w_f \cdot \nu_f)}{w_g + w_f}$ average kinematic viscosity of gas & mud

$\nu_{ave} = 2.475 \times 10^{-5}$ (m^2/sec)

$V := \left[\frac{\left[\left(\frac{P_{at}}{P}\right)\cdot\left(\frac{T_{ave}}{T_{at}}\right)\cdot Q_a + Q_f\right]}{A}\right]$ velocity of flow in annulus (m/sec)

$V = 1.710$ (m/sec)

$N_R := \frac{V \cdot ID}{\nu_{ave}}$ Reynolds number calculations

$N_R = 6.701 \times 10^3$ NR > 4000 => Turbulent

$f := \left[\frac{1}{-1.8 \cdot \log\left[\left(\frac{e_{cs}}{3.7 \cdot ID}\right)^{1.11} + \frac{6.9}{N_R}\right]}\right]^2$ calculate friction factor using Haaland equation

$f = 0.0350$ Haaland friction factor

$Q(P) := \frac{P_{at}}{P} \cdot \frac{T_{ave}}{T_{at}} \cdot Q_a$ flow rate as a function of pressure

$CQ := P_{at} \cdot \frac{T_{ave}}{T_{at}} \cdot Q_a$

$CQ = 13281.901$

$\gamma_{mix}(P) := \frac{w_g + w_f}{Q(P) + Q_f}$ specific weight as a function of pressure

$V(P) := \frac{(Q(P) + Q_f)}{A}$ velocity as a function of pressure

$p_2 := 15.07$ guess a pressure at the top of the section (MPa) and vary it until matching the two integrals below

$P_2 := p_2 \cdot 1000000$ convert pressure to N/m^2

$P_2 = 1.507 \times 10^{7}$ (N/m^2)

$$\int_{P_2}^{P_3} \frac{1}{\gamma_{mix}(P) \cdot \left[1 - (V(P))^2 \cdot \frac{f}{2 \cdot g \cdot (ID)}\right]} dP = 724.908$$ left hand side of equation

$$\int_{H_1+H_2}^{H_1+H_2+H_3} 1 \, dh = 724.431$$ right hand side of equation

Pressure at top of tool joints

$ID := ID_{tj}$ set inner diameter to casing ID

$A := \frac{\pi}{4} \cdot (ID^2)$ cross sectional area (m^2)

$T_{ave} := T_{ave2}$ set ave temp to section ave temp

$P := P_2$ set pressure to previous section pressure

$\gamma_g := \frac{P \cdot SG_a}{R \cdot T_{ave}}$ specific weight of the gas (N/m^3)

$\gamma_g = 1662.005$ (N/m^3)

$\rho_g := \frac{\gamma_g}{g}$ gas (air) density (kg/m^3)

$\rho_g = 169.471$ (kg/m^3)

$\nu_g := \frac{\mu_g}{\rho_g}$ gas (air) kinematic viscosity (m^2/sec)

$\nu_g = 7.081 \times 10^{-8}$ (m^2/sec)

$\nu_{ave} := \frac{(w_g \cdot \nu_g + w_f \cdot \nu_f)}{w_g + w_f}$ average kinematic viscosity of gas & mud

$\nu_{ave} = 2.475 \times 10^{-5}$ (m^2/sec)

$$V := \left[\frac{\left[\left(\frac{P_{at}}{P}\right)\cdot\left(\frac{T_{ave}}{T_{at}}\right)\cdot Q_a + Q_f\right]}{A}\right]$$ velocity of flow in annulus (m/sec)

$V = 2.848$ (m/sec)

$$N_R := \frac{V\cdot ID}{\nu_{ave}}$$ Reynolds number calculations

$N_R = 8.746 \times 10^3$ NR > 4000 => Turbulent

$$f := \left[\frac{1}{-1.8\cdot\log\left[\left(\frac{e_{cs}}{3.7\cdot ID}\right)^{1.11} + \frac{6.9}{N_R}\right]}\right]^2$$ calculate friction factor using Haaland equation

$f = 0.0327$ Haaland friction factor

$$Q(P) := \frac{P_{at}}{P}\cdot\frac{T_{ave}}{T_{at}}\cdot Q_a$$ flow rate as a function of pressure

$$CQ := P_{at}\cdot\frac{T_{ave}}{T_{at}}\cdot Q_a$$

$CQ = 13105.605$

$$\gamma_{mix}(P) := \frac{w_g + w_f}{Q(P) + Q_f}$$ specific weight as a function of pressure

$$V(P) := \frac{(Q(P) + Q_f)}{A}$$ velocity as a function of pressure

$p_1 := 14.09$ guess a pressure at the top of the section (MPa) and vary it until matching the two integrals below

$P_1 := p_1\cdot 1000000$ convert pressure to N/m^2

$P_1 = 1.409 \times 10^7$ (N/m^2)

$$\int_{P_1}^{P_2} \frac{1}{\gamma_{mix}(P)\cdot\left[1 - (V(P))^2\cdot\frac{f}{2\cdot g\cdot(ID)}\right]}\,dP = 107.900$$ left hand side of equation

$$\int_{H_1}^{H_1+H_2} 1\,dh = 107.873$$ right hand side of equation

Pressrue at top of drill pipe

$ID := ID_{dp}$ set inner diameter to casing ID

$A := \frac{\pi}{4}\cdot\left(ID^2\right)$ cross sectional area (m^2)

$T_{ave} := T_{ave1}$ set ave temp to section ave temp

$P := P_1$ set pressure to previous section pressure

$\gamma_g := \frac{P\cdot SG_a}{R\cdot T_{ave}}$ specific weight of the gas (N/m^3)

$\gamma_g = 1609.435$ (N/m^3)

$\rho_g := \frac{\gamma_g}{g}$ gas (air) density (kg/m^3)

$\rho_g = 164.111$ (kg/m^3)

$\nu_g := \frac{\mu_g}{\rho_g}$ gas (air) kinematic viscosity (m^2/sec)

$\nu_g = 7.312\times 10^{-8}$ (m^2/sec)

$\nu_{ave} := \frac{\left(w_g\cdot\nu_g + w_f\cdot\nu_f\right)}{w_g + w_f}$ average kinematic viscosity of gas & mud

$\nu_{ave} = 2.475\times 10^{-5}$ (m^2/sec)

$V := \left[\frac{\left[\left(\frac{P_{at}}{P}\right)\cdot\left(\frac{T_{ave}}{T_{at}}\right)\cdot Q_a + Q_f\right]}{A}\right]$ velocity of flow in annulus (m/sec)

$V = 1.752$ (m/sec)

$N_R := \frac{V\cdot ID}{\nu_{ave}}$ Reynolds number calculations

$N_R = 6.867\times 10^{3}$ NR > 4000 => Turbulent

$f := \left[\frac{1}{-1.8\cdot\log\left[\left(\frac{e_{cs}}{3.7\cdot ID}\right)^{1.11} + \frac{6.9}{N_R}\right]}\right]^2$ calculate friction factor using Haaland equation

$f = 0.0348$ Haaland friction factor

$Q(P) := \frac{P_{at}}{P}\cdot\frac{T_{ave}}{T_{at}}\cdot Q_a$ flow rate as a function of pressure

$$CQ := P_{at} \cdot \frac{T_{ave}}{T_{at}} \cdot Q_a$$

$CQ = 12653.586$

$$\gamma_{mix}(P) := \frac{w_g + w_f}{Q(P) + Q_f}$$ specific weight as a function of pressure

$$V(P) := \frac{(Q(P) + Q_f)}{A}$$ velocity as a function of pressure

$p_{in} := 0.30$ guess a pressure at the top of the section (MPa) and vary it until matching the two integrals below

$P_{in} := p_{in} \cdot 1000000$ convert pressure to N/m^2

$P_{in} = 3.000 \times 10^5$ (N/m^2)

$$\int_{P_{in}}^{P_1} \frac{1}{\gamma_{mix}(P) \cdot \left[1 - (V(P))^2 \cdot \frac{f}{2 \cdot g \cdot (ID)}\right]} \, dP = 1935.407$$ left hand side of equation

$$\int_0^{H_1} 1 \, dh = 2026.127$$ right hand side of equation

Important design parameter summary

$q_a = 8.500$ air compressor flow rate (scm/min)

$p_{in} = 0.300$ required surface air injection pressure (MPa)

$p_{bh} = 23.380$ bottom hole pressure (MPa)

APPENDIX

E

Chapter 10 Illustrative Example MathCad™ Solutions

This appendix presents the detailed MathCad solutions for the illustrative examples in Chapter 10.

Illustrative Example 10.1: USCS version, Non-Friction Solution - Drilling Fluid is Stable Foam. This is a closed form solution. However, it is solved by simple trial and error to match the left and right sides of the equation.

Given input conditions

$q_f := 50$	mud flow rate (gal/min)
$q_c := 2120.0$	flow rate single compressors (scfm)
$p_{bp} := 80.0$	back pressure at top of annulus (psig)
$Q_f := \frac{q_f \cdot 231}{12^3 \cdot 60}$	convert water gpm to ft³/sec
$Q_f = 0.111$	volumetric flow rate of water (ft³/sec)
$Q_c := \frac{q_c}{60}$	convert scfm to ft³/sec
$Q_c = 35.333$	air volumetric flow rate from compressor (st ft³/sec)
$Q_g := Q_c$	volumetric flow rate of gas in system (st ft³/sec)

Basic standard and elevation atmospheric conditions

$p_{st} := 14.695$	API standard atmospheric pressure (psia)
$t_{st} := 60$	API standard atmospheric temperature (°F)
$alt := 4000.0$	surface location elevation altitude (ft)
$p_{at} := 12.685$	actual surface atmospheric pressure (psia)
$t_{at} := 44.74$	actual surface atmospheric temperature (°F)
$P_{st} := p_{st} \cdot 144$	convert actual atmospheric pressure to lb/ft²
$P_{st} = 2116.080$	(lb/ft² abs)
$T_{st} := t_{st} + 459.67$	convert actual atmospheric temperature to °R
$T_{st} = 519.670$	(°R)
$P_{at} := p_{at} \cdot 144$	convert actual atmospheric pressure to lb/ft² abs

$P_{at} = 1826.640$ (lb/ft^2 abs)

$T_{at} := t_{at} + 459.67$ convert actual atmospheric temperature to ^oR

$T_{at} = 504.410$ (^oR)

$P_{bp} := (p_{bp} + p_{at}) \cdot 144$ convert back presssure to lb/ft^2 abs

$P_{bp} = 13346.640$ (lb/ft^2 abs)

Initial foam quality at the blooey line

$Q_{bp} := \dfrac{P_{st} \cdot T_{at} \cdot Q_c}{P_{bp} \cdot T_{st}}$ volumetric flow rate of gas at back pressure valve (ft^3/sec)

$Q_{bp} = 5.438$ volumetric flow rate (ft^3/sec)

$\Gamma_{bp} := \dfrac{Q_{bp}}{Q_f + Q_{bp}}$ foam quality at the back pressure valve

$\Gamma_{bp} = 0.980$ foam quality at top of annulus

Constants

$SG_a := 1.0$ specific gravity of air

$SG_r := 2.7$ rock specific gravity

$k := 1.4$ air specific heat constant

$R := 53.36$ specific gas constant (ft-lb/lb-^oR)

$\gamma_w := 8.33$ specific weight of fresh water (lb/gal)

$\gamma_w := 62.4$ specific weight of fresh water (lb/ft^3)

$e_{cs} := 0.0005$ roughness of commercial steel (ft)

$g := 32.2$ gravitational constant (ft/sec^2)

$\beta := 0.01$ geothermal gradient (^oR/ft)

Borehole geometry

$D_c := 7000$ total depth of cased hole (ft)

$D_{oh} := 3000$ total depth of open hole (ft)

$H := D_c + D_{oh}$ total depth of well (ft)

$H = 10000.000$ (ft)

$ROP := 60$ estimated rate of penetration (ft/hr)

Preliminary calculations

$\gamma_{st} := \frac{P_{st} \cdot SG_a}{R \cdot T_{st}}$ specific weight of air at API conditions (lb/ft^3)

$\gamma_{st} = 0.076$ (lb/ft^3)

$\gamma_{at} := \frac{P_{at} \cdot SG_a}{R \cdot T_{at}}$ specific weight of air at surface atmos (lbs/ft^3)

$\gamma_{at} = 0.0679$ (lbs/ft^3)

$P_{bp} = 13346.640$ (lb/ft^2 abs)

$\gamma_{bp} := \frac{P_{bp} \cdot SG_a}{R \cdot T_{at}}$ specific weight of air at back pressure (lbs/ft^3)

$\gamma_{bp} = 0.4959$ (lbs/ft^3)

$rop := \frac{ROP}{60 \cdot 60}$ convert ROP to ft/sec

$rop = 0.0167$ (ft/sec)

$d_{bh} := 7.875$ diameter of borehole (inches)

$D_{bh} := \frac{d_{bh}}{12}$ convert diameter to ft

$D_{bh} = 0.656$ (ft)

$w_w := \gamma_w \cdot Q_f$ weight rate of flow of water (lb/sec)

$w_w = 6.951$ (lb/sec)

$w_g := \gamma_{st} \cdot Q_c$ weight rate of flow of gas at blooey line exit (lb/sec)

$w_g = 2.696$ (lb/sec)

$w_r := \left[\left(\frac{\pi}{4} \right) \cdot D_{bh}^{2} \cdot 62.4 \cdot SG_r \cdot rop \right]$ weight rate of solid flow out of blooey line (lb/sec)

$w_r = 0.950$ (lb/sec)

$w_t := w_g + w_w + w_r$ total weight rate of flow (gas+fluid+solids) (lb/sec)

$w_t = 10.598$ (lb/sec)

$t_{bh} := t_{at} + \beta \cdot H$ bottom hole temperature (°F)

$t_{bh} = 144.740$ (°F)

$t_{ave} := \frac{(t_{at} + t_{bh})}{2}$ average borehole temperature (°F)

$t_{ave} = 94.740$ (°F)

$T_{ave} := 459.67 + t_{ave}$ average borehole termperature (°R)

$T_{ave} = 554.410$ (°R)

$P_g := P_{at}$ define the gas pressure (lb/ft² abs)

$T_g := T_{at}$ define the gas temperature (°R)

Non-friction solution by trial and error to determine p_{bh}

$p_{bh} := 373.65$ bottom hole pressure (psig)

$P_{bh} := (p_{bh} + p_{at}) \cdot 144$ convert to lb/ft² abs

$P_{bh} = 55632.240$ (lb/ft^2 abs)

$$LS := Q_g P_g \cdot \left(\frac{T_{ave}}{T_g}\right) \cdot \ln\left(\frac{P_{bh}}{P_{bp}}\right) + Q_f \cdot P_{bh} - Q_f \cdot P_{bp}$$ left side of equation

$LS = 105975.904$

$RS := \left(w_g + w_w + w_r\right) \cdot H$ right side of equation

$RS = 105975.057$

$p_{bha} := p_{bh} + p_{at}$

$p_{bha} = 386.335$ bottom hole pressure (psia)

Illustrative Example 10.2 (USCS version) Friction Solution - Drilling Fluid is Stable Foam.
This is a trial and error solution. The upper limit of the first integral across each geometric cross section solution is the trial and error term that must be sought to allow the value left side of the right side intergral to be equalized.

Laboratory screening test data

$h_n := 500$ nominal beaker height (ml)

$\tau_n := 200$ nominal half-life time (sec)

$h_t := 610$ test height (ml)

$\tau_t := 280$ test half-life (sec)

$\Lambda := \left(\frac{h_t}{h_n}\right) \cdot \left(\frac{\tau_t}{\tau_n}\right)$ foam viscosity adjustment coefficient

$\Lambda = 1.7080$

Given input volumetric flow rates

$q_f := 50$ mud flow rate (gpm)

$q_c := 2120.0$ flow rate of a single compressor (scfm)

$p_{bp} := 80.0$ back pressure at top of annulus (psig)

$Q_f := \frac{q_f \cdot 231}{12^3 \cdot 60}$ convert water gpm to ft^3/sec

$Q_f = 0.1114$ volumetric flow rate of water (ft^3/sec)

$Q_c := \frac{q_c}{60}$ convert scfm to ft^3/sec

$Q_c = 35.3333$ air volumetric flow rate from compressor (s ft^3/sec)

$Q_g := Q_c$ volumetric flow rate of gas in system (s ft^3/sec)

Basic standard and elevation atmospheric conditions

$p_{st} := 14.695$ API standard atmospheric pressure (psia)

$t_{st} := 60$ API standard atmospheric temperature (°F)

$alt := 4000.0$ surface location elevation altitude (ft)

$p_{at} := 12.685$ pressure from table for surface elev. (psia)

$t_{at} := 44.74$ temperature from table for surface elevation (°F)

$P_{st} := p_{st} \cdot 144$ convert API standard pressure to lb/ft^2

$P_{st} = 2116.0800$ (lb/ft^2)

$T_{st} := t_{st} + 459.67$ convert API standard temp to °R

$T_{st} = 519.6700$ (°R)

$P_{at} := p_{at} \cdot 144$ convert actual atmospheric pressure to lb/ft^2

$P_{at} = 1826.6400$ (lb/ft^2)

$T_{at} := t_{at} + 459.67$ convert actual atmospheric temperature to °R

$T_{at} = 504.4100$ (°R)

$P_{bp} := (p_{bp} + p_{at}) \cdot 144$ convert back presssure to lb/ft^2

$P_{bp} = 13346.6400$ (lb/ft^2)

Initial foam quality at the blooey line

$$Q_{bp} := \frac{P_{st} \cdot T_{at} \cdot Q_c}{P_{bp} \cdot T_{st}}$$ volumetric flow rate of gas at back pressure valve (ft^3/sec)

$Q_{bp} = 5.4375$ volumetric flow rate (ft^3/sec)

$$\Gamma_{bp} := \frac{Q_{bp}}{Q_f + Q_{bp}}$$ foam quality at the back pressure valve

$\Gamma_{bp} = 0.980$ foam quality at top of annulus

Viscosities of gas and fluid

$\mu_{air} := 0.012$ viscosity of air (centipoises)

$\mu_g := \mu_{air} \cdot 2.089 \cdot 10^{-5}$ convert viscosity to lb-sec/ft^2

$\mu_g = 2.507 \times 10^{-7}$ (lb-sec/ft^2)

$\mu_w := 1.0$	viscosity of water (centipoises)
$\mu_f := \mu_w \cdot 2.089 \cdot 10^{-5}$	convert viscosity to lb-sec/ft^2
$\mu_f = 2.089 \times 10^{-5}$	(lb-sec/ft^2)

Constants

$SG_a := 1.0$	specific gravity of air
$k := 1.4$	air specific heat constant
$R := 53.36$	specific gas constant (ft-lb/lb-°R)
$\gamma_{fw} := 8.33$	specific weight of fresh water (lb/gal)
$\gamma_f := 62.4$	specific weight of fresh water (lb/ft^3)
$e_{cs} := 0.0005$	roughness of commercial steel (ft)
$g := 32.2$	gravitational constant (ft/sec^2)
$\beta := 0.01$	geothermal gradient (°R/ft)
$SG_r := 2.7$	specific gravity of sedimentary rock
$e_r := 0.01$	roughness of rock Table 8-1
$C := 0.81$	aerated fluid flow loss coefficient through nozzles

Borehole geometry

$H_c := 7000$	total depth of cased hole (ft)
$H_{oh} := 3000$	total depth of open hole (ft)
$ROP := 60$	estimated rate of penetration (ft/hr)
$d_{bh} := 7.875$	borehole (drill bit) diameter (in)
$od_{dc} := 6.25$	outer diameter of drill collar (in)
$id_{dc} := 2.8125$	inner diameter of drill collar (in)
$od_{dp} := 4.5$	outer diameter of drill pipe (in)

$id_{dp} := 3.826$	inner diameter of drill pipe (in)
$od_{tj} := 6.25$	outer diameter of drill pipe tool joint (in)
$id_{tj} := 2.75$	inner diameter of drill pipe tool joint (in)
$od_c := 8.625$	outer diameter of casing (in)
$id_c := 7.921$	inner diameter of casing to (in)
$L_r := 100$	length of return line (ft)
$id_r := 5.625$	inner diameter of return line (in)
$L_{dc} := 500$	total length of drill collars (ft)
$L_{tj} := 1.5$	average length of one tool joint (ft)
$L_{dp} := 30$	average length of one drill pipe section (ft)
$d_n := \frac{12}{32}$	nozzle diameter (in)
$N_n := 3$	number of nozzles in the bit
$D_{bh} := \frac{d_{bh}}{12}$	convert borehole diameter to ft
$D_{bh} = 0.6563$	(ft)
$ID_r := \frac{id_r}{12}$	convert return line diameter to ft
$ID_r = 0.4688$	(ft)
$OD_{dp} := \frac{od_{dp}}{12}$	convert drill pipe outer diameter to ft
$OD_{dp} = 0.3750$	(ft)
$ID_{dp} := \frac{id_{dp}}{12}$	convert drill pipe inner diameter to ft
$ID_{dp} = 0.3188$	(ft)
$ID_c := \frac{id_c}{12}$	convert casing inner diameter to ft

$ID_c = 0.6601$ (ft)

$OD_{dc} := \frac{od_{dc}}{12}$ convert drill collar outer diameter to ft

$OD_{dc} = 0.5208$ (ft)

$ID_{dc} := \frac{id_{dc}}{12}$ convert drill collar inner diameter to ft

$ID_{dc} = 0.2344$ (ft)

$OD_{tj} := \frac{od_{tj}}{12}$ convert tool joint outer diameter to ft

$OD_{tj} = 0.5208$ (ft)

$ID_{tj} := \frac{id_{tj}}{12}$ convert tool joint inner diameter to ft

$ID_{tj} = 0.2292$ (ft)

$d_n = 0.3750$ nozzle diameter (in)

$D_n := \frac{d_n}{12}$ nozzle diameter converted to ft

$D_n = 0.0313$ (ft)

Preliminary calculations

$\gamma_g := \frac{P_{st} \cdot SG_a}{R \cdot T_{st}}$ specific weight of air at surface (lb/ft^3)

$\gamma_g = 0.0763$ (lb/ft^3)

$\gamma_{bp} := \frac{P_{bp} \cdot SG_a}{R \cdot T_{at}}$ specific weight of air at back pressure (lb/ft^3)

$\gamma_{bp} = 0.4959$ (lb/ft^3)

$rop := \frac{ROP}{60 \cdot 60}$ convert ROP to ft/sec

$rop = 0.0167$ (ft/sec)

$w_f := \gamma_f \cdot Q_f$ weight rate of flow of water (lb/sec)

$w_f = 6.9514$ (lb/sec)

$w_g := \gamma_g \cdot Q_c$ weight rate of flow of gas blooey line exit (lb/sec)

$w_g = 2.6963$ (lb/sec)

$w_g := \gamma_{bp} \cdot Q_{bp}$ weight rate of flow of gas back pressure (lb/sec)

$w_g = 2.6963$ (lb/sec)

$w_s := \left[\left(\frac{\pi}{4}\right) \cdot D_{bh}^{\ 2} \cdot 62.4 \cdot SG_r \cdot rop\right]$ weight rate of solid flow out (lb/sec)

$w_s = 0.950$ (lb/sec)

$w_t := w_g + w_f + w_s$ total weight rate of flow (lb/sec)

$w_t = 10.5975$ (lb/sec)

$\mu_{usf} := \frac{\mu_f \cdot w_f + \mu_g \cdot w_g}{w_f + w_g}$ average absolute viscosity at BP valve (lb-sec/ft^2)

$\mu_{usf} = 1.512 \times 10^{-5}$ (lb-sec/ft^2)

$\rho_g := \frac{\gamma_g}{g}$ gas (air) density (lb-sec^2/ft^4)

$\rho_g = 2.370 \times 10^{-3}$ (lb-sec^2/ft^4)

$\nu_g := \frac{\mu_g}{\rho_g}$ gas (air) kinematic viscosity (ft^2/sec)

$\nu_g = 1.058 \times 10^{-4}$ (ft^2/sec)

$\rho_f := \frac{\gamma_f}{g}$ mud density (lb-sec^2/ft^4)

$\rho_f = 1.938 \times 10^{0}$ (lb-sec^2/ft^4)

$\nu_f := \frac{\mu_f}{\rho_f}$ liquid kinematic viscosity (ft^2/sec)

$\mu_f = 2.089 \times 10^{-5}$ (lb-sec/ft^2)

$\nu_f = 1.078 \times 10^{-5}$ (ft^2/sec)

Section height calculations

$H_1 := H_c - L_{tj} \cdot \left(\frac{H_c}{L_{dp}}\right)$ calculate length of drill pipe in cased section

$H_1 = 6650.0000$ (ft)

$H_2 := L_{tj} \cdot \left(\frac{H_c}{L_{dp}}\right)$ calculate length of tool joint in cased section

$H_2 = 350.0000$ (ft)

$H_3 := H_{oh} - L_{dc} - L_{tj} \cdot \left(\frac{H_{oh} - L_{dc}}{L_{dp}}\right)$ calculate length of drill pipe in open hole section

$H_3 = 2375.0000$ (ft)

$H_4 := L_{tj} \cdot \left(\frac{H_{oh} - L_{dc}}{L_{dp}}\right)$ calculate length of tool joints in open hole section

$H_4 = 125.0000$ (ft)

$H_5 := L_{dc}$ length of drill collar

$H_5 = 500.0000$ (ft)

Return line and Tee at top of annulus

$ID := ID_r$ set variable ID equal to return line's ID

$ID = 0.469$ (ft)

$A := \frac{\pi}{4} \cdot ID^2$ cross sectional area (ft^2)

$A = 0.173$ (ft^2)

$\gamma_{bp} := \frac{P_{bp} \cdot SG_a}{R \cdot T_{at}}$ specific weight of air upstream of back pressure valve (lb/ft^3)

$\gamma_{bp} = 0.4959$ (lb/ft^3)

$w_g := \gamma_{bp} \cdot Q_{bp}$ weight rate of flow of gas (lb/sec)

$w_g = 2.6963$ (lb/sec)

$Q_g := \frac{w_g}{\gamma_g}$ volumetric flow rate of gas (ft^3/sec)

$Q_g = 35.3333$ (ft^3/sec)

$q_g := Q_g \cdot 60$ convert flow to ft^3/min

$q_g = 2120.0000$ (ft^3/min)

$w_f := \gamma_f \cdot Q_f$ weight rate of flow of fluid (lb/sec)

$w_f = 6.9514$ (lb/sec)

$\gamma_{bp} := \frac{w_t}{\left(\frac{P_{at}}{P_{bp}}\right) \cdot \left(\frac{T_{at}}{T_{at}}\right) \cdot Q_g + Q_f}$ specific weight of foam just upstream of back pressure valve

$\gamma_{bp} = 2.1421$ (lb/ft^3)

$\rho_{bp} := \frac{\gamma_{bp}}{g}$ density of stable foam

$\rho_{bp} = 0.0665$ (lb-sec^2/ft^4)

$$n_{bp} := 0.095932 + 2.3654 \cdot \Gamma_{bp} - 10.46 \cdot \Gamma_{bp}^{2} + 12.955 \cdot \Gamma_{bp}^{3} + 14.467 \cdot \Gamma_{bp}^{4} - 39.673 \cdot \Gamma_{bp}^{5} + 20.625 \cdot \Gamma_{bp}^{6}$$

$n_{bp} = 0.3143$

$$K_{bp} := -0.15626 + 56.14 \cdot \Gamma_{bp} - 312.77 \cdot \Gamma_{bp}^{2} + 576.65 \cdot \Gamma_{bp}^{3} + 63.960 \cdot \Gamma_{bp}^{4} - 960.46 \cdot \Gamma_{bp}^{5} - 154.68 \cdot \Gamma_{bp}^{6} + 1$$

$K_{bp} = 3.0435$

$V_{bp} := \dfrac{Q_{bp} + Q_f}{\left(\frac{\pi}{4}\right) \cdot ID^2}$ average velocity of the foam just upstream of the back pressure valve

$V_{bp} = 32.1541$ (ft/sec)

$$\mu_e := \frac{\Lambda \cdot K_{bp}}{g} \cdot \left(\frac{3 \cdot n_{bp} + 1}{4 \cdot n_{bp}}\right)^{n_{bp}} \cdot \left(\frac{8 \cdot V_{bp}}{ID}\right)^{n_{bp}-1}$$

$\mu_e = 2.450 \times 10^{-3}$ (lb-sec/ft²)

$\mu_{bp}(\Gamma_{bp}) := \mu_e$ stable foam (lb-sec/ft²)

$\mu_{bp}(\Gamma_{bp}) = 2.450 \times 10^{-3}$ (lb-sec/ft²)

$\nu_{bp} := \dfrac{\mu_{bp}(\Gamma_{bp})}{\rho_{bp}}$ kinematic viscosity of the foam just upstream of the back pressure valve

$\nu_{bp} = 0.036825$ (ft²/sec)

$N_R := \dfrac{V_{bp} \cdot ID}{\nu_{bp}}$ Reynolds number calculation

$N_R = 409$ Reynolds number

$f_l := \dfrac{64}{N_R}$ friction factor from the Laminar equation

$f_l = 0.156366$ Laminar friction factor

$$f_h := \left[\frac{1}{-1.8\cdot\log\left[\left(\frac{\frac{e_{cs}}{ID}}{3.7}\right)^{1.11} + \left(\frac{6.9}{N_R}\right)\right]}\right]^2$$ friction factor from the Haaland correlation

$f_h = 0.0985$ Haaland friction factor

$f := f_l$ select friction factor equal to applicable factor

$$Q(P) := \frac{P_{at}}{P}\cdot\frac{T_{at}}{T_{at}}\cdot Q_g$$ flow rate as a function of pressure

$$\gamma_{mix}(P) := \frac{w_t}{Q(P) + Q_f}$$ specific weight as a function of pressure

$$V(P) := \frac{Q(P) + Q_f}{A}$$ velocity as a function of pressure

$p_r := 98.82$ trial pressure at entrance to return line (psia)

$P_r := p_r\cdot 144$ convert pressure to lb/ft^2 abs

$P_r = 14230.0800$ (lb/ft^2)

$$\int_{P_{bp}}^{P_r} \frac{1}{\gamma_{mix}(P)\cdot\left(V(P)^2\cdot\frac{f}{2\cdot g\cdot ID}\right)}\,dP = 100.0137$$ left hand side

$$\int_0^{L_r} 1\,dh = 100.0000$$ right hand side

$K_t := 25$ Tee constant from Figure 8-5 and 8-6

$K_v := 0.2$ loss factor for gate valves in blooey line

$$\gamma_{mix} := \frac{w_t}{\left(\frac{P_{at}}{P_r}\right)\cdot\left(\frac{T_{at}}{T_{at}}\right)\cdot Q_g + Q_f}$$ specific weight (lb/ft^3)

$\gamma_{mix} = 2.2805$ (lb/ft^3)

$$V := \left[\frac{\left(\frac{P_{at}}{P_r}\right)\cdot\left(\frac{T_{at}}{T_{at}}\right)\cdot Q_g + Q_f}{A}\right]$$ velocity (ft/sec)

$V = 26.9275$ (ft/sec)

$$\Delta P_T := \gamma_{mix}\cdot(2\cdot K_v + K_t)\cdot\frac{V^2}{2\cdot g}$$ pressure loss due to Tee and valves

$\Delta P_T = 652.1912$ (lb/ft^2)

$P_e := P_r + \Delta P_T$ sum pressure losses - entrance to return line

$P_e = 14882.2712$ (lb/ft^2)

$$p_e := \frac{P_e}{144}$$ convert from lb/ft^2 to psia

$p_e = 103.3491$ (psia)

$$V := \left[\frac{\left(\frac{P_{at}}{P_e}\right)\cdot\left(\frac{T_{at}}{T_{at}}\right)\cdot Q_g + Q_f}{A}\right]$$ velocity (ft/sec)

$V = 25.7757$ (ft/sec)

Annulus cased hole drill pipe body section

$ID := ID_c$ set inner diameter to casing ID

$OD := OD_{dp}$ set outer diameter to drill pipe OD

$P := P_e$ set pressure equal to previous section

$$A := \frac{\pi}{4}\cdot\left(ID^2 - OD^2\right)$$ cross sectional area (ft^2)

$T := T_{at} + \beta \cdot H_1$ calculate temperature at bottom of section

$T = 570.9100$ (°R)

$T_{ave1} := \frac{T_{at} + T}{2}$ calculate average temperature of section

$T_{ave1} = 537.6600$ (°R)

$T_{ave} := T_{ave1}$ set average temp to section average temp

$\gamma_g := \frac{P \cdot SG_a}{R \cdot T_{ave}}$ specific weight of the gas (lb/ft^3)

$\gamma_g = 0.5187$ (lb/ft^3)

$Q_{ga} := \frac{w_g}{\gamma_g}$ flow rate of gas (ft^3/sec)

$Q_{ga} = 5.1979$ (ft^3/sec)

$\Gamma := \frac{Q_{ga}}{Q_{ga} + Q_f}$ check foam quality of this section

$\Gamma = 0.9790$ foam quality

$$n := 0.095932 + 2.3654 \cdot \Gamma - 10.46 \cdot \Gamma^2 + 12.955 \cdot \Gamma^3 + 14.467 \cdot \Gamma^4 - 39.673 \cdot \Gamma^5 + 20.625 \cdot \Gamma^6$$

$n = 0.3121$ power law index

$$K := -0.15626 + 56.14 \cdot \Gamma - 312.77 \cdot \Gamma^2 + 576.65 \cdot \Gamma^3 + 63.960 \cdot \Gamma^4 - 960.46 \cdot \Gamma^5 - 154.68 \cdot \Gamma^6 + 1670.2 \cdot \Gamma^7 - 9$$

$K = 3.1150$ consistency index

$V := \frac{Q_{ga} + Q_f}{A}$ average velocity of the foam just upstream back pressure valve

$V = 22.9087$ (ft/sec)

$V_{a1} := V$ define the annulus velocity in Section 1

$$\mu_e := \frac{\Lambda \cdot K}{g} \cdot \left(\frac{2 \cdot n + 1}{3 \cdot n}\right)^n \cdot \left(\frac{12 \cdot V}{ID - OD}\right)^{n-1}$$ foam effective viscosity

$$\mu_e = 1.737 \times 10^{-3}$$ (lb-sec/ft^2)

$$\mu_{sf}(\Gamma) := \mu_e$$ stable foam viscosity (lb-sec/ft^2)

$$\mu_{sf}(\Gamma) = 1.737 \times 10^{-3}$$ (lb-sec/ft^2)

$$\mu_{sf1} := \mu_{sf}(\Gamma)$$

$$\gamma_{sf} := \frac{w_g + w_f}{\left(\frac{P_{at}}{P}\right) \cdot \left(\frac{T_{ave}}{T_{at}}\right) \cdot Q_g + Q_f}$$ specific weight of foam

$$\gamma_{sf} = 2.0379$$ (lb/ft^3)

$$\gamma_{a1} := \gamma_{sf}$$ define foam specific weight in Section 1

$$\rho_{sf} := \frac{\gamma_{sf}}{g}$$ density of stable foam

$$\rho_{sf} = 0.0633$$ (lb-sec^2/ft^4)

$$\nu_{sf}(\Gamma) := \frac{\mu_{sf}(\Gamma)}{\rho_{sf}}$$ kinematic viscosity of the foam

$$\nu_{sf}(\Gamma) = 0.027443$$ (ft^2/sec)

$$\nu_{sf1} := \nu_{sf}(\Gamma)$$

$$N_R := \frac{V \cdot (ID - OD)}{\nu_{sf}(\Gamma)}$$ Reynolds number calculation

$$N_R = 238$$ Reynolds number

$$f_l := \frac{64}{N_R}$$ friction factor from the Laminar equation

$$f_l = 0.268927$$ Laminar friction factor

$$f_h := \left[\frac{1}{-1.8\cdot\log\left[\left(\frac{\frac{e_{cs}}{ID-OD}}{3.7}\right)^{1.11} + \left(\frac{6.9}{N_R}\right)\right]}\right]^2$$ friction factor from the Haaland correlation

$f_h = 0.1311$

$f := f_l$ select friction factor equal to highest

$Q(P) := \frac{P_{at}}{P}\cdot\frac{T_{ave}}{T_{at}}\cdot Q_g$ flow rate as a function of pressure

$\gamma_{mix}(P) := \frac{w_t}{Q(P) + Q_f}$ specific weight as a function of pressure

$V(P) := \frac{Q(P) + Q_f}{A}$ velocity as a function of pressure

$p_1 := 683.37$ trial pressure at the bottom of the section (psia)

$P_1 := p_1\cdot 144$ convert pressure to lb/ft^2

$P_1 = 9.841 \times 10^4$ (lb/ft^2 abs)

$$\int_{P_e}^{P_1} \frac{1}{\gamma_{mix}(P)\cdot\left[1 + V(P)^2\cdot\frac{f}{2\cdot g\cdot(ID - OD)}\right]}\,dP = 6650.054$$ left hand side

$$\int_0^{H_1} 1\,dh = 6650.0000$$ right hand side

Annulus cased hole tool joint section

$ID := ID_c$ set inner diameter to casing ID

$OD := OD_{tj}$ set outer diameter to tool joint OD

$P := P_1$ set pressure equal to previous section

$A := \frac{\pi}{4} \cdot \left(ID^2 - OD^2\right)$ cross sectional area (ft^2)

$T_{temp} := T$ keep previous section temperature

$T := T_{temp} + \beta \cdot H_2$ calculate temperature at bottom of section

$T = 574.4100$ (°R)

$T_{ave2} := \frac{T_{temp} + T}{2}$ calculate average temperature of section

$T_{ave2} = 572.6600$ (°R)

$T_{ave} := T_{ave2}$ set average temp to section average temp

$\gamma_g := \frac{P \cdot SG_a}{R \cdot T_{ave}}$ specific weight of the gas (lb/ft^3)

$\gamma_g = 3.2204$ (lb/ft^3)

$Q_{ga} := \frac{w_g}{\gamma_g}$ flow rate of gas (ft^3/sec)

$Q_{ga} = 0.8373$ (ft^3/sec)

$\Gamma := \frac{Q_{ga}}{Q_{ga} + Q_f}$ check foam quality of this section

$\Gamma = 0.8826$ foam quality

$$n := 0.095932 + 2.3654 \cdot \Gamma - 10.46 \cdot \Gamma^2 + 12.955 \cdot \Gamma^3 + 14.467 \cdot \Gamma^4 - 39.673 \cdot \Gamma^5 + 20.625 \cdot \Gamma^6$$

$n = 0.2228$ power law index

$$K := -0.15626 + 56.14 \cdot \Gamma - 312.77 \cdot \Gamma^2 + 576.65 \cdot \Gamma^3 + 63.960 \cdot \Gamma^4 - 960.46 \cdot \Gamma^5 - 154.68 \cdot \Gamma^6 + 1670.2 \cdot \Gamma^7 -$$

$K = 4.9749$ consistency index

$$V := \frac{Q_{ga} + Q_f}{A}$$ average velocity of the foam just upstream of the back pressure valve

$V = 7.3454$ (ft/sec)

$V_{a2} := V$ define velocity in annulus Section 2

$$\mu_e := \frac{A \cdot K}{g} \cdot \left(\frac{2 \cdot n + 1}{3 \cdot n}\right)^n \cdot \left(\frac{12 \cdot V}{ID - OD}\right)^{n-1}$$

$\mu_e = 2.083 \times 10^{-3}$ (lb-sec/ft²)

$\mu_{sf}(\Gamma) := \mu_e$ stable foam viscosity (lb-sec/ft²)

$\mu_{sf}(\Gamma) = 2.083 \times 10^{-3}$ (lb-sec/ft²)

$\mu_{sf2} := \mu_{sf}(\Gamma)$

$$\gamma_{sf} := \frac{w_g + w_f}{\left(\frac{P_{at}}{P}\right) \cdot \left(\frac{T_{ave}}{T_{at}}\right) \cdot Q_g + Q_f}$$ specific weight of foam

$\gamma_{sf} = 11.2705$ (lb/ft³)

$\gamma_{a2} := \gamma_{sf}$ define foam specific weight in Section 2

$$\rho_{sf} := \frac{\gamma_{sf}}{g}$$ density of stable foam

$\rho_{sf} = 0.3500$ (lb-sec²/ft⁴)

$$\nu_{sf} := \frac{\mu_{sf}(\Gamma)}{\rho_{sf}}$$ kinematic viscosity of the foam

$\nu_{sf} = 0.005952$ (ft²/sec)

$\nu_{sf2} := \nu_{sf}$

$$N_R := \frac{V \cdot (ID - OD)}{\nu_{sf}}$$ Reynolds number calculation

$N_R = 172$ Reynolds number

$f_l := \frac{64}{N_R}$ friction factor from the Laminar equation

$f_l = 0.372401$ Laminar friction factor

$$f_h := \left[\frac{1}{-1.8\cdot\log\left[\left(\frac{\frac{e_{cs}}{ID-OD}}{3.7}\right)^{1.11} + \left(\frac{6.9}{N_R}\right)\right]}\right]^2$$

friction factor from the Haaland correlation

$f_h = 0.1594$

$f := f_l$ select friction factor equal to higher of the two

$Q(P) := \frac{P_{at}}{P}\cdot\frac{T_{ave}}{T_{at}}\cdot Q_g$ flow rate as a function of pressure

$\gamma_{mix}(P) := \frac{w_t}{Q(P) + Q_f}$ specific weight as a function of pressure

$V(P) := \frac{Q(P) + Q_f}{A}$ velocity as a function of pressure

$p_2 := 767.24$ trial pressure at the bottom of the section (psia)

$P_2 := p_2\cdot 144$ convert pressure to lb/ft^2

$P_2 = 1.105 \times 10^5$ (lb/ft^2)

$$\int_{P_1}^{P_2} \frac{1}{\gamma_{mix}(P)\cdot\left[1 + V(P)^2\cdot\frac{f}{2\cdot g\cdot(ID - OD)}\right]}\,dP = 350.053$$

left hand side

$$\int_{H_1}^{H_1+H_2} 1\,dh = 350.0000$$

right hand side

Annulus open hole drill pipe body section

$ID := D_{bh}$	set inner diameter to casing ID
$OD := OD_{dp}$	set outer diameter to drill pipe OD
$P := P_2$	pressure equal to previous section
$A := \frac{\pi}{4} \cdot \left(ID^2 - OD^2\right)$	cross sectional area (ft^2)
$T_{temp} := T$	keep previous section temperature
$T := T_{temp} + \beta \cdot H_3$	calculate temperature at bottom of section
$T = 598.1600$	(°R)
$T_{ave3} := \frac{T_{temp} + T}{2}$	calculate average temperature of section
$T_{ave3} = 586.2850$	(°R)
$T_{ave} := T_{ave3}$	set average temp to section average temp
$\gamma_g := \frac{P \cdot SG_a}{R \cdot T_{ave}}$	specific weight of the gas (lb/ft^3)
$\gamma_g = 3.5316$	(lb/ft^3)
$Q_{ga} := \frac{w_g}{\gamma_g}$	flow rate of gas (ft^3/sec)
$Q_{ga} = 0.7635$	(ft^3/sec)
$\Gamma := \frac{Q_{ga}}{Q_{ga} + Q_f}$	check foam quality of this section
$\Gamma = 0.8727$	foam quality

$$n := 0.095932 + 2.3654 \cdot \Gamma - 10.46 \cdot \Gamma^2 + 12.955 \cdot \Gamma^3 + 14.467 \cdot \Gamma^4 - 39.673 \cdot \Gamma^5 + 20.625 \cdot \Gamma^6$$

$n = 0.2247$	power law index

$K := -0.15626 + 56.14\cdot\Gamma - 312.77\cdot\Gamma^2 + 576.65\cdot\Gamma^3 + 63.960\cdot\Gamma^4 - 960.46\cdot\Gamma^5 - 154.68\cdot\Gamma^6 + 1670.2\cdot\Gamma^7 - 9$

$K = 4.8392$ consistency index

$V := \frac{Q_{ga} + Q_f}{A}$ average velocity of the foam just upstream of the back pressure valve

$V = 3.8407$ (ft/sec)

$V_{a3} := V$

$\mu_e := \frac{\Lambda\cdot K}{g}\cdot\left(\frac{2\cdot n + 1}{3\cdot n}\right)^n\cdot\left(\frac{12\cdot V}{ID - OD}\right)^{n-1}$ stable foam effective viscosity

$\mu_e = 5.850 \times 10^{-3}$ (lb-sec/ft²)

$\mu_{sf}(\Gamma) := \mu_e$ stable foam viscosity (lb-sec/ft²)

$\mu_{sf}(\Gamma) = 5.850 \times 10^{-3}$ (lb-sec/ft²)

$\mu_{sf3} := \mu_{sf}(\Gamma)$

$\gamma_{sf} := \frac{w_g + w_f}{\left(\frac{P_{at}}{P}\right)\cdot\left(\frac{T_{ave}}{T_{at}}\right)\cdot Q_g + Q_f}$ specific weight of foam

$\gamma_{sf} = 12.2061$ (lb/ft²)

$\gamma_{a3} := \gamma_{sf}$ define foam specific weight in Section 3

$\rho_{sf} := \frac{\gamma_{sf}}{g}$ density of stable foam

$\rho_{sf} = 0.3791$ (lb-sec²/ft⁴)

$\nu_{sf} := \frac{\mu_{sf}(\Gamma)}{\rho_{sf}}$ kinematic viscosity of the foam

$\nu_{sf} = 0.015431$ (ft²/sec)

$\nu_{sf3} := \nu_{sf}$

$N_R := \dfrac{V \cdot (ID - OD)}{\nu_{sf}}$ — Reynolds number calculation

$N_R = 70$ — Reynolds number

$e_{ave} := \dfrac{e_r \cdot ID^2 + e_{cs} \cdot OD^2}{ID^2 + OD^2}$ — calculate weighted average of roughness

$e_{ave} = 7.662 \times 10^{-3}$ — weighted average of roughness (ft)

$f_l := \dfrac{64}{N_R}$ — friction factor from the Laminar equation

$f_l = 0.914280$ — Laminar friction factor

$$f_h := \left[\frac{1}{-1.8 \cdot \log\left[\left(\dfrac{\frac{e_{ave}}{ID-OD}}{3.7}\right)^{1.11} + \left(\dfrac{6.9}{N_R}\right)\right]}\right]^2$$

friction factor from the Haaland correlation

$f_h = 0.3163$

$f := f_l$ — select friction factor equal to higher of the two

$Q(P) := \dfrac{P_{at}}{P} \cdot \dfrac{T_{ave}}{T_{at}} \cdot Q_g$ — flow rate as a function of pressure

$\gamma_{mix}(P) := \dfrac{w_t}{Q(P) + Q_f}$ — specific weight as a function of pressure

$V(P) := \dfrac{Q(P) + Q_f}{A}$ — velocity as a function of pressure

$p_3 := 1145.38$ — trial pressure at the bottom of the section (psia)

$P_3 := p_3 \cdot 144$ — convert pressure to lb/ft^2

$P_3 = 1.649 \times 10^5$ (lb/ft^2)

$$\int_{P_2}^{P_3} \frac{1}{\gamma_{mix}(P)\cdot\left[1 + V(P)^2\cdot\frac{f}{2\cdot g\cdot(ID - OD)}\right]}\,dP = 2375.016$$ left hand side

$$\int_{H_1+H_2}^{H_1+H_2+H_3} 1\,dh = 2375.0000$$ right hand side

Annulus open hole tool joint section

$ID := D_{bh}$ set inner diameter to casing ID

$OD := OD_{tj}$ set outer diameter to tool joint OD

$P := P_3$ set pressure equal to previous section

$A := \frac{\pi}{4}\cdot\left(ID^2 - OD^2\right)$ cross sectional area (ft^2)

$T_{temp} := T$ keep previous section temperature

$T := T_{temp} + \beta\cdot H_4$ calculate temperature at bottom of section

$T = 599.4100$ (°R)

$T_{ave4} := \frac{T_{temp} + T}{2}$ calculate average temperature of section

$T_{ave4} = 598.7850$ (°R)

$T_{ave} := T_{ave4}$ set average temp to section average temp

$\gamma_g := \frac{P\cdot SG_a}{R\cdot T_{ave}}$ specific weight of the gas (lb/ft^3)

$\gamma_g = 5.1621$ (lb/ft^3)

$Q_{ga} := \frac{w_g}{\gamma_g}$ flow rate of gas (ft^3/sec)

$Q_{ga} = 0.5223$ (ft³/sec)

$\Gamma := \frac{Q_{ga}}{Q_{ga} + Q_f}$ check foam quality of this section

$\Gamma = 0.8242$ foam quality

$n := 0.095932 + 2.3654 \cdot \Gamma - 10.46 \cdot \Gamma^2 + 12.955 \cdot \Gamma^3 + 14.467 \cdot \Gamma^4 - 39.673 \cdot \Gamma^5 + 20.625 \cdot \Gamma^6$

$n = 0.2455$ power law index

$K := -0.15626 + 56.14 \cdot \Gamma - 312.77 \cdot \Gamma^2 + 576.65 \cdot \Gamma^3 + 63.960 \cdot \Gamma^4 - 960.46 \cdot \Gamma^5 - 154.68 \cdot \Gamma^6 + 1670.2 \cdot \Gamma^7 - 9$

$K = 4.0405$ consistency index

$V := \frac{Q_{ga} + Q_f}{A}$ average velocity of the foam upstream of the back pressure valve

$V = 5.0622$ (ft/sec)

$V_{a4} := V$

$\mu_e := \frac{\Lambda \cdot K}{g} \cdot \left(\frac{2 \cdot n + 1}{3 \cdot n}\right)^n \cdot \left(\frac{12 \cdot V}{ID - OD}\right)^{n-1}$ effective viscosity

$\mu_e = 2.544 \times 10^{-3}$ (lb-sec/ft²)

$\mu_{sf}(\Gamma) := \mu_e$ stable foam viscosity (lb-sec/ft²)

$\mu_{sf}(\Gamma) = 2.544 \times 10^{-3}$ (lb-sec/ft²)

$\mu_{sf4} := \mu_{sf}(\Gamma)$

$\gamma_{sf} := \frac{w_g + w_f}{\left(\frac{P_{at}}{P}\right) \cdot \left(\frac{T_{ave}}{T_{at}}\right) \cdot Q_g + Q_f}$ specific weight of foam

$\gamma_{sf} = 16.7516$ (lb/ft²)

$\gamma_{a4} := \gamma_{sf}$ — define foam specific weight in Section 4

$\rho_{sf} := \frac{\gamma_{sf}}{g}$ — density of stable foam

$\rho_{sf} = 0.5202$ — ($lb\text{-}sec^2/ft^4$)

$\nu_{sf} := \frac{\mu_{sf}(\Gamma)}{\rho_{sf}}$ — kinematic viscosity of the foam

$\nu_{sf} = 0.004890$ — (ft^2/sec)

$\nu_{sf4} := \nu_{sf}$

$N_R := \frac{V \cdot (ID - OD)}{\nu_{sf}}$ — Reynolds number calculation

$N_R = 140$ — Reynolds number

$e_{ave} := \frac{e_r \cdot ID^2 + e_{cs} \cdot OD^2}{ID^2 + OD^2}$ — calculate weighted average of roughness between the steel and the rock

$e_{ave} = 6.329 \times 10^{-3}$ — weighted average of roughness

$f_l := \frac{64}{N_R}$ — friction factor from the Laminar equation

$f_l = 0.4565$ — Laminar friction factor

$$f_h := \left[\frac{1}{-1.8 \cdot \log\left[\left(\frac{\frac{e_{ave}}{ID-OD}}{3.7}\right)^{1.11} + \left(\frac{6.9}{N_R}\right)\right]}\right]^2$$

friction factor from the Haaland correlation

$f_h = 0.1995$

$f := f_l$ — select friction factor

$Q(P) := \frac{P_{at}}{P} \cdot \frac{T_{ave}}{T_{at}} \cdot Q_g$ — flow rate as a function of pressure

$\gamma_{mix}(P) := \dfrac{w_t}{Q(P) + Q_f}$	specific weight as a function of pressure
$V(P) := \dfrac{Q(P) + Q_f}{A}$	velocity as a function of pressure
$p_4 := 1179.04$	trial pressure at the bottom of the section (psia)
$P_4 := p_4 \cdot 144$	convert pressure to lb/ft^2
$P_4 = 1.698 \times 10^5$	(lb/ft^2 abs)

$$\int_{P_3}^{P_4} \frac{1}{\gamma_{mix}(P)\cdot\left[1 + V(P)^2 \cdot \dfrac{f}{2 \cdot g \cdot (ID - OD)}\right]} \, dP = 125.026$$

left hand side

$$\int_{H_1+H_2+H_3}^{H_1+H_2+H_3+H_4} 1 \, dh = 125.0000$$

right hand side

Annulus open hole drill collar section

$ID := D_{bh}$	set inner diameter to casing ID
$OD := OD_{dc}$	set outer diameter to drill collar OD
$P := P_4$	set pressure equal to previous section pressure
$A := \dfrac{\pi}{4} \cdot \left(ID^2 - OD^2\right)$	cross sectional area (ft^2)
$T_{temp} := T$	keep previous section temperature
$T := T_{temp} + \beta \cdot H_5$	calculate temperature at bottom of section
$T = 604.4100$	(°R)
$T_{ave5} := \dfrac{T_{temp} + T}{2}$	calculate average temperature of section

$T_{ave5} = 601.9100$ (°R)

$T_{ave} := T_{ave5}$ set average temp to section average temp

$\gamma_g := \dfrac{P \cdot SG_a}{R \cdot T_{ave}}$ specific weight of the gas (lb/ft³)

$\gamma_g = 5.2862$ (lb/ft³)

$Q_{ga} := \dfrac{w_g}{\gamma_g}$ flow rate of gas (ft³/sec)

$Q_{ga} = 0.5101$ (ft³/sec)

$\Gamma := \dfrac{Q_{ga}}{Q_{ga} + Q_f}$ check foam quality of this section

$\Gamma = 0.8207$ foam quality

$$n := 0.095932 + 2.3654 \cdot \Gamma - 10.46 \cdot \Gamma^2 + 12.955 \cdot \Gamma^3 + 14.467 \cdot \Gamma^4 - 39.673 \cdot \Gamma^5 + 20.625 \cdot \Gamma^6$$

$n = 0.2474$ power law index

$$K := -0.15626 + 56.14 \cdot \Gamma - 312.77 \cdot \Gamma^2 + 576.65 \cdot \Gamma^3 + 63.960 \cdot \Gamma^4 - 960.46 \cdot \Gamma^5 - 154.68 \cdot \Gamma^6 + 1670.2 \cdot \Gamma^7 -$$

$K = 3.9842$ consistency index

$V := \dfrac{Q_{ga} + Q_f}{A}$ average velocity of the foam upstream

$V = 4.9642$ (ft/sec)

$V_{a5} := V$

$$\mu_e := \frac{\Lambda \cdot K}{g} \cdot \left(\frac{2 \cdot n + 1}{3 \cdot n}\right)^n \cdot \left(\frac{12 \cdot V}{ID - OD}\right)^{n-1}$$

$\mu_e = 2.576 \times 10^{-3}$ (lb-sec/ft²)

$\mu_{sf}(\Gamma) := \mu_e$ stable foam viscosity (lb-sec/ft²)

$\mu_{sf}(\Gamma) = 2.576 \times 10^{-3}$ (lb-sec/ft²)

$\mu_{sf5} := \mu_{sf}(\Gamma)$

$\gamma_{sf} := \dfrac{w_g + w_f}{\left(\dfrac{P_{at}}{P}\right)\cdot\left(\dfrac{T_{ave}}{T_{at}}\right)\cdot Q_g + Q_f}$ specific weight of foam

$\gamma_{sf} = 17.0749$ (lb/ft³)

$\gamma_{a5} := \gamma_{sf}$ define foam specific weight in Section 5

$\rho_{sf} := \dfrac{\gamma_{sf}}{g}$ density of stable foam

$\rho_{sf} = 0.5303$ (lb-sec²/ft⁴)

$\nu_{sf} := \dfrac{\mu_{sf}(\Gamma)}{\rho_{sf}}$ kinematic viscosity of the foam

$\nu_{sf} = 0.004857$ (ft²/sec)

$\nu_{sf5} := \nu_{sf}$

$N_R := \dfrac{V\cdot(ID - OD)}{\nu_{sf}}$ Reynolds number calculation

$N_R = 138$ Reynolds number

$e_{ave} := \dfrac{e_r\cdot ID^2 + e_{cs}\cdot OD^2}{ID^2 + OD^2}$ calculate weighted average of roughness between the steel and the rock

$e_{ave} = 6.329 \times 10^{-3}$ weighted average of roughness

$f_l := \dfrac{64}{N_R}$ friction factor from the Laminar equation

$f_l = 0.4624$ Laminar friction factor

$$f_h := \left[\frac{1}{-1.8\cdot\log\left[\left(\frac{\frac{e_{ave}}{ID-OD}}{3.7}\right)^{1.11} + \left(\frac{6.9}{N_R}\right)\right]}\right]^2$$ friction factor from the Haaland correlation

$f_h = 0.2010$

$f := f_l$ select friction factor

$Q(P) := \frac{P_{at}}{P}\cdot\frac{T_{ave}}{T_{at}}\cdot Q_g$ flow rate as a function of pressure

$\gamma_{mix}(P) := \frac{w_t}{Q(P) + Q_f}$ specific weight as a function of pressure

$V(P) := \frac{Q(P) + Q_f}{A}$ velocity as a function of pressure

$p_5 := 1314.46$ trail pressure at the bottom of the section (psia)

$P_5 := p_5\cdot 144$ convert pressure to lb/ft²

$P_5 = 1.893 \times 10^5$ (lb/ft² abs)

$$\int_{P_4}^{P_5} \frac{1}{\gamma_{mix}(P)\cdot\left[1 + V(P)^2\cdot\frac{f}{2\cdot g\cdot(ID - OD)}\right]}\,dP = 500.024$$ left hand side

$$\int_{H_1+H_2+H_3+H_4}^{H_1+H_2+H_3+H_4+H_5} 1\,dh = 500.0000$$ right hand side

$\gamma_{bh} := \frac{P_5\cdot SG_a}{R\cdot T}$ specific weight of the gas (lb/ft³)

$\gamma_{bh} = 5.8690$ (lb/ft³)

$Q_{bh} := \frac{w_g}{\gamma_{bh}}$ flow rate of gas (ft³/sec)

$Q_{bh} = 0.4594$ (ft³/sec)

$$\Gamma_{bh} := \frac{Q_{bh}}{Q_{bh} + Q_f}$$ check foam quality of this section

$\Gamma_{bh} = 0.8048$ foam quality

Bottom hole through the bit nozzles

$$D_e := \left(N_n \cdot D_n^{\,2}\right)^{0.5}$$ equivalent single diameter (ft)

$D_e = 0.05413$ (ft)

$P := P_5$

1st Iteration to determine pressure inside drill string above nozzles

$$\gamma_{mix} := \frac{w_g + w_f}{Q_f + Q_g \cdot \left(\frac{P_{at}}{P}\right) \cdot \frac{T}{T_{at}}}$$ specific weight of mixture (lb/ft³)

$\gamma_{mix} = 18.5541$ (lb/ft³)

$$\Delta P_b := \frac{\left(w_g + w_f\right)^2}{2 \cdot C^2 \cdot g \cdot \gamma_{mix} \cdot \left(\frac{\pi}{4}\right)^2 \cdot D_e^{\,4}}$$ approximate the pressure change of aerated fluid through the drill bit

$\Delta P_b = 22424.9932$ (lb/ft²)

$$\Delta p_b := \frac{\Delta P_b}{144}$$ convert pressure to lb/in²

$\Delta p_b = 155.7291$ (lb/in²)

$P := P_5 + \Delta P_b$ sum pressures to get pressure above bit

$P = 211707.2332$ (lb/ft² abs)

$p := \frac{P}{144}$ convert pressure to psia

$p = 1470.1891$ (psia)

2nd Iteration to determine pressure inside drill string above nozzles

$\gamma_{mix} := \frac{w_g + w_f}{Q_f + Q_g \cdot \left(\frac{P_{at}}{P}\right) \cdot \frac{T}{T_{at}}}$ specific weight of mixture (lb/ft^3)

$\gamma_{mix} = 20.2385$ (lb/ft^3)

$\Delta P_b := \frac{(w_g + w_f)^2}{2 \cdot C^2 \cdot g \cdot \gamma_{mix} \cdot \left(\frac{\pi}{4}\right)^2 \cdot D_e^4}$ approximate the pressure change of aerated fluid through the drill bit

$\Delta P_b = 20558.5331$ (lb/ft^2)

$\Delta p_b := \frac{\Delta P_b}{144}$ convert pressure to lb/in^2

$\Delta p_b = 142.7676$ (lb/in^2)

$P := P_5 + \Delta P_b$ sum pressures to get pressure above bit

$P = 209840.7731$ lb/ft^2 abs

$p := \frac{P}{144}$ convert pressure to psia

$p = 1457.2276$ (psia)

3rd Iteration to determine pressure inside drill string above nozzles

$\gamma_{mix} := \frac{w_g + w_f}{Q_f + Q_g \cdot \left(\frac{P_{at}}{P}\right) \cdot \frac{T}{T_{at}}}$ specific weight of mixture (lb/ft^3)

$\gamma_{mix} = 20.1015$ (lb/ft^3)

$$\Delta P_b := \frac{(w_g + w_f)^2}{2 \cdot C^2 \cdot g \cdot \gamma_{mix} \cdot \left(\frac{\pi}{4}\right)^2 \cdot D_e^4}$$ approximate the pressure change of aerated fluid through the drill bit

$\Delta P_b = 20698.6611$ (lb/ft^2)

$$\Delta p_b := \frac{\Delta P_b}{144}$$ convert pressure to lb/in^2

$\Delta p_b = 143.7407$ (lb/in^2)

$P := P_5 + \Delta P_b$ sum pressures to get pressure above bit

$P = 209980.9011$ (lb/ft^2 abs)

$$p := \frac{P}{144}$$ convert pressure to psia

$p = 1458.2007$ (psia)

4th Iteration to determine pressure inside drill string above nozzles

$$\gamma_{mix} := \frac{w_g + w_f}{Q_f + Q_g \cdot \left(\frac{P_{at}}{P}\right) \cdot \frac{T}{T_{at}}}$$ specific weight of mixture (lb/ft^3)

$\gamma_{mix} = 20.1118$ (lb/ft^3)

$$\Delta P_b := \frac{(w_g + w_f)^2}{2 \cdot C^2 \cdot g \cdot \gamma_{mix} \cdot \left(\frac{\pi}{4}\right)^2 \cdot D_e^4}$$ approximate the pressure change of aerated fluid through the drill bit

$\Delta P_b = 20688.0543$ (lb/ft^2)

$$\Delta p_b := \frac{\Delta P_b}{144}$$ convert pressure to lb/in^2

$\Delta p_b = 143.6670$ (lb/in^2)

$P := P_5 + \Delta P_b$ sum pressures to get pressure above bit

$P = 209970.2943$ (lb/ft^2 abs)

$p := \frac{P}{144}$ convert pressure to psia

$p = 1458.1270$ (psia)

Convergence of pressure above the drill bit inside drill pipe

Open hole inside drill string drill collar section

$ID := ID_{dc}$ set inner diameter to casing ID

$A := \frac{\pi}{4} \cdot (ID^2)$ cross sectional area (ft^2)

$T_{ave} := T_{ave5}$ set ave temp to section ave temp

$\gamma_g := \frac{P \cdot SG_a}{R \cdot T_{ave}}$ specific weight of the gas (lb/ft^3)

$\gamma_g = 6.5375$ (lb/ft^3)

$\rho_g := \frac{\gamma_g}{g}$ gas (air) density ($lb\text{-}sec^2/ft^4$)

$\rho_g = 0.2030$ ($lb\text{-}sec^2/ft^4$)

$\nu_g := \frac{\mu_g}{\rho_g}$ gas (air) kinematic viscosity (ft^2/sec)

$\nu_g = 1.235 \times 10^{-6}$ (ft^2/sec)

$\nu_{ave} := \frac{w_g \cdot \nu_g + w_f \cdot \nu_f}{w_g + w_f}$ average kinematic viscosity of gas & mud

$\nu_{ave} = 8.112 \times 10^{-6}$ (ft^2/sec)

$$V := \left[\frac{\left(\frac{P_{at}}{P}\right)\cdot\left(\frac{T_{ave}}{T_{at}}\right)\cdot Q_g + Q_f}{A}\right]$$ velocity of flow in annulus (ft/sec)

$V = 11.0840$ (ft/sec)

$$N_R := \frac{V \cdot ID}{\nu_{ave}}$$ Reynolds number calculations

$N_R = 3.202 \times 10^5$ Reynolds number

$$f_h := \left[\frac{1}{-1.8\cdot\log\left[\left(\frac{\frac{e_{cs}}{ID}}{3.7}\right)^{1.11} + \left(\frac{6.9}{N_R}\right)\right]}\right]^2$$ friction factor from the Haaland correlation

$f_h = 0.0244$

$f := f_h$

$$Q(P) := \frac{P_{at}}{P}\cdot\frac{T_{ave}}{T_{at}}\cdot Q_g$$ flow rate as a function of pressure

$$\gamma_{mix}(P) := \frac{w_g + w_f}{Q(P) + Q_f}$$ specific weight as a function of pressure

$$V(P) := \frac{Q(P) + Q_f}{A}$$ velocity as a function of pressure

$p_4 := 1403.19$ trial pressure at the top of the section (psia)

$P_4 := p_4 \cdot 144$ convert pressure to lb/ft^2 abs

$P_4 = 2.021 \times 10^5$ (lb/ft^2 abs)

$$\int_{P_4}^{P} \frac{1}{\gamma_{mix}(P)\cdot\left[1 - V(P)^2\cdot\frac{f}{2\cdot g\cdot(ID)}\right]}\,dP = 500.029$$ left hand side

$$\int_{H_1+H_2+H_3+H_4}^{H_1+H_2+H_3+H_4+H_5} 1\,dh = 500.0000$$ right hand side

Open hole inside drill string tool joint section

$ID := ID_{tj}$ set inner diameter to casing ID

$A := \frac{\pi}{4} \cdot (ID^2)$ cross sectional area (ft^2)

$T_{ave} := T_{ave4}$ set average temp to section ave temp

$P := P_4$ set pressure to previous section pressure

$\gamma_g := \frac{P \cdot SG_a}{R \cdot T_{ave}}$ specific weight of the gas (lb/ft^3)

$\gamma_g = 6.3240$ (lb/ft^3)

$\rho_g := \frac{\gamma_g}{g}$ gas (air) density ($lb\text{-}sec^2/ft^4$)

$\rho_g = 0.1964$ ($lb\text{-}sec^2/ft^4$)

$\nu_g := \frac{\mu_g}{\rho_g}$ gas (air) kinematic viscosity (ft^2/sec)

$\nu_g = 1.276 \times 10^{-6}$ (ft^2/sec)

$\nu_{ave} := \frac{w_g \cdot \nu_g + w_f \cdot \nu_f}{w_g + w_f}$ average kinematic viscosity of gas & mud

$\nu_{ave} = 8.124 \times 10^{-6}$ (ft^2/sec)

$$V := \left[\frac{\left(\frac{P_{at}}{P}\right) \cdot \left(\frac{T_{ave}}{T_{at}}\right) \cdot Q_g + Q_f}{A} \right]$$ velocity of flow in annulus (ft/sec)

$V = 11.8937$ (ft/sec)

$N_R := \frac{V \cdot ID}{\nu_{ave}}$ Reynolds number calculations

$N_R = 3.355 \times 10^5$ Reynolds number

$f_h := \left[\frac{1}{-1.8 \cdot \log\left[\left(\frac{\frac{e_{cs}}{ID}}{3.7} \right)^{1.11} + \left(\frac{6.9}{N_R} \right) \right]} \right]^2$ friction factor from the Haaland correlation

$f_h = 0.0245$

$f := f_h$

$Q(P) := \frac{P_{at}}{P} \cdot \frac{T_{ave}}{T_{at}} \cdot Q_g$ flow rate as a function of pressure

$\gamma_{mix}(P) := \frac{w_g + w_f}{Q(P) + Q_f}$ specific weight as a function of pressure

$V(P) := \frac{Q(P) + Q_f}{A}$ velocity as a function of pressure

$p_3 := 1390.19$ trial pressure at the top of the section (psia)

$P_3 := p_3 \cdot 144$ convert pressure to lb/ft² abs

$P_3 = 2.002 \times 10^5$ (lb/ft² abs)

$$\int_{P_3}^{P_4} \frac{1}{\gamma_{mix}(P) \cdot \left[1 - V(P)^2 \cdot \frac{f}{2 \cdot g \cdot (ID)} \right]} \, dP = 125.081$$ left hand side

$$\int_{H_1+H_2+H_3}^{H_1+H_2+H_3+H_4} 1 \, dh = 125.0000$$ right hand side

Open hole inside drill string drill pipe body section

$ID := ID_{dp}$	set inner diameter to casing ID
$A := \frac{\pi}{4} \cdot \left(ID^2\right)$	cross sectional area (ft^2)
$T_{ave} := T_{ave3}$	set ave temp to section ave temp
$P := P_3$	set pressure to previous section pressure
$\gamma_g := \frac{P \cdot SG_a}{R \cdot T_{ave}}$	specific weight of the gas (lb/ft^3)
$\gamma_g = 6.3990$	(lb/ft^3)
$\rho_g := \frac{\gamma_g}{g}$	gas (air) density (lb-sec^2/ft^4)
$\rho_g = 0.1987$	(lb-sec^2/ft^4)
$\nu_g := \frac{\mu_g}{\rho_g}$	gas (air) kinematic viscosity (ft^2/sec)
$\nu_g = 1.261 \times 10^{-6}$	(ft^2/sec)
$\nu_{ave} := \frac{w_g \cdot \nu_g + w_f \cdot \nu_f}{w_g + w_f}$	average kinematic viscosity of gas & mud
$\nu_{ave} = 8.120 \times 10^{-6}$	(ft^2/sec)
$V := \left[\frac{\left(\frac{P_{at}}{P}\right) \cdot \left(\frac{T_{ave}}{T_{at}}\right) \cdot Q_g + Q_f}{A}\right]$	velocity of flow in annulus (ft/sec)
$V = 6.0889$	(ft/sec)
$N_R := \frac{V \cdot ID}{\nu_{ave}}$	Reynolds number calculations

$N_R = 2.391 \times 10^5$ Reynolds number

$$f_h := \left[\frac{1}{-1.8 \cdot \log\left[\left(\frac{\frac{e_{cs}}{ID}}{3.7} \right)^{1.11} + \left(\frac{6.9}{N_R} \right) \right]} \right]^2$$ friction factor from the Haaland correlation

$f_h = 0.0228$

$f := f_h$

$$Q(P) := \frac{P_{at}}{P} \cdot \frac{T_{ave}}{T_{at}} \cdot Q_g$$ flow rate as a function of pressure

$$\gamma_{mix}(P) := \frac{w_g + w_f}{Q(P) + Q_f}$$ specific weight as a function of pressure

$$V(P) := \frac{Q(P) + Q_f}{A}$$ velocity as a function of pressure

$p_2 := 1105.23$ trial pressure at the top of the section (psia)

$P_2 := p_2 \cdot 144$ convert pressure to lb/ft^2 abs

$P_2 = 1.592 \times 10^5$ (lb/ft^2 abs)

$$\int_{P_2}^{P_3} \frac{1}{\gamma_{mix}(P) \cdot \left[1 - V(P)^2 \cdot \frac{f}{2 \cdot g \cdot (ID)} \right]} \, dP = 2375.026$$ left hand side

$$\int_{H_1+H_2}^{H_1+H_2+H_3} 1 \, dh = 2375.0000$$ right hand side

Cased hole inside drill string tool joint section

$ID := ID_{tj}$ — set inner diameter to casing ID

$A := \frac{\pi}{4}\cdot\left(ID^2\right)$ — cross sectional area (ft^2)

$T_{ave} := T_{ave2}$ — set ave temp to section ave temp

$P := P_2$ — set pressure to previous section pressure

$\gamma_g := \frac{P\cdot SG_a}{R\cdot T_{ave}}$ — specific weight of the gas (lb/ft^3)

$\gamma_g = 5.2084$ — (lb/ft^3)

$\rho_g := \frac{\gamma_g}{g}$ — gas (air) density ($lb\text{-}sec^2/ft^4$)

$\rho_g = 0.1618$ — ($lb\text{-}sec^2/ft^4$)

$\nu_g := \frac{\mu_g}{\rho_g}$ — gas (air) kinematic viscosity (ft^2/sec)

$\nu_g = 1.550 \times 10^{-6}$ — (ft^2/sec)

$\nu_{ave} := \frac{w_g\cdot\nu_g + w_f\cdot\nu_f}{w_g + w_f}$ — average kinematic viscosity of gas & mud

$\nu_{ave} = 8.200 \times 10^{-6}$ — (ft^2/sec)

$V := \left[\frac{\left(\frac{P_{at}}{P}\right)\cdot\left(\frac{T_{ave}}{T_{at}}\right)\cdot Q_g + Q_f}{A}\right]$ — velocity of flow in annulus (ft/sec)

$V = 13.8628$ — (ft/sec)

$N_R := \frac{V\cdot ID}{\nu_{ave}}$ — Reynolds number calculations

$N_R = 3.874 \times 10^5$ Reynolds number

$$f_h := \left[\frac{1}{-1.8 \cdot \log\left[\left(\frac{\frac{e_{cs}}{ID}}{3.7}\right)^{1.11} + \left(\frac{6.9}{N_R}\right)\right]}\right]^2$$

friction factor from the Haaland correlation

$f_h = 0.0244$

$f := f_h$

$Q(P) := \frac{P_{at}}{P} \cdot \frac{T_{ave}}{T_{at}} \cdot Q_g$ flow rate as a function of pressure

$\gamma_{mix}(P) := \frac{w_g + w_f}{Q(P) + Q_f}$ specific weight as a function of pressure

$V(P) := \frac{Q(P) + Q_f}{A}$ velocity as a function of pressure

$p_1 := 1077.8$ trial pressure at the top of the section (psia)

$P_1 := p_1 \cdot 144$ convert pressure to lb/ft² abs

$P_1 = 1.552 \times 10^5$ (lb/ft²)

$$\int_{P_1}^{P_2} \frac{1}{\gamma_{mix}(P) \cdot \left[1 - V(P)^2 \cdot \frac{f}{2 \cdot g \cdot (ID)}\right]} \, dP = 350.034$$

left hand side

$$\int_{H_1}^{H_1+H_2} 1 \, dh = 350.0000$$

right hand side

Cased hole inside drill string drill pipe body section

$ID := ID_{dp}$ set inner diameter to casing ID

$A := \frac{\pi}{4} \cdot \left(ID^2\right)$ cross sectional area (ft^2)

$T_{ave} := T_{ave1}$ set ave temp to section ave temp

$P := P_1$ set pressure to previous section pressure

$\gamma_g := \frac{P \cdot SG_a}{R \cdot T_{ave}}$ specific weight of the gas (lb/ft^3)

$\gamma_g = 5.4097$ (lb/ft^3)

$\rho_g := \frac{\gamma_g}{g}$ gas (air) density (lb-sec^2/ft^4)

$\rho_g = 0.1680$ (lb-sec^2/ft^4)

$\nu_g := \frac{\mu_g}{\rho_g}$ gas (air) kinematic viscosity (ft^2/sec)

$\nu_g = 1.492 \times 10^{-6}$ (ft^2/sec)

$\nu_{ave} := \frac{w_g \cdot \nu_g + w_f \cdot \nu_f}{w_g + w_f}$ average kinematic viscosity of gas & mud

$\nu_{ave} = 8.184 \times 10^{-6}$ (ft^2/sec)

$V := \left[\frac{\left(\frac{P_{at}}{P}\right) \cdot \left(\frac{T_{ave}}{T_{at}}\right) \cdot Q_g + Q_f}{A}\right]$ velocity of flow in annulus (ft/sec)

$V = 6.9472$ (ft/sec)

$N_R := \frac{V \cdot ID}{\nu_{ave}}$ Reynolds number calculations

$N_R = 2.706 \times 10^5$ Reynolds number

$$f_h := \left[\frac{1}{-1.8\cdot\log\left[\left(\frac{\frac{e_{cs}}{ID}}{3.7}\right)^{1.11} + \left(\frac{6.9}{N_R}\right)\right]}\right]^2$$ friction factor from the Haaland correlation

$f_h = 0.0227$

$f := f_h$

$$Q(P) := \frac{P_{at}}{P}\cdot\frac{T_{ave}}{T_{at}}\cdot Q_g$$ flow rate as a function of pressure

$$\gamma_{mix}(P) := \frac{w_g + w_f}{Q(P) + Q_f}$$ specific weight as a function of pressure

$$V(P) := \frac{Q(P) + Q_f}{A}$$ velocity as a function of pressure

$p_{in} := 530.34$ trial pressure at the top of the section (psia)

$P_{in} := p_{in}\cdot 144$ convert pressure to lb/ft² abs

$P_{in} = 7.637 \times 10^4$ (lb/ft² abs)

$$\int_{P_{in}}^{P_1} \frac{1}{\gamma_{mix}(P)\cdot\left[1 - V(P)^2\cdot\frac{f}{2\cdot g\cdot(ID)}\right]}\,dP = 6650.051$$ left hand side

$$\int_0^{H_1} 1\,dh = 6650.0000$$ right hand side

Hole cleaning in the annulus

Cuttings details

$d_{cut} := 0.50$ cuttings approximate maximum size (inches)

$D_{cut} := \frac{d_{cut}}{12}$ convert to feet

$D_{cut} = 0.0417$ (ft)

$\gamma_{cut} := SG_r \cdot \gamma_f$ specific weight of cuttings (lb/ft^3)

$\gamma_{cut} = 168.4800$ (lb/ft^3)

Largest annulus cross-sectional area (cased hole drill pipe body section)

- Laminar terminal velocity

$$V_{tlama} := 0.0333 \cdot D_{cut}^2 \cdot \left[\frac{(\gamma_{cut} - \gamma_{a1})}{\mu_{sf1}}\right]$$

$V_{tlama} = 5.5402$ (ft/sec)

$$N_{cut} := \frac{(D_{cut} \cdot V_{tlama})}{\nu_{sf1}}$$

$N_{cut} = 8.4117$ No

- Transition terminal velocity

$$V_{ttrana} := 0.492 \cdot D_{cut} \cdot \left[\frac{(\gamma_{cut} - \gamma_{a1})^{\frac{2}{3}}}{(\gamma_{a1} \cdot \mu_{sf1})^{\frac{1}{3}}}\right]$$

$V_{ttrana} = 4.0702$ (ft/sec)

$$N_{cut} := \frac{(D_{cut} \cdot V_{ttrana})}{\nu_{sf1}}$$

$N_{cut} = 6.1798$ Yes

- *Turbulent terminal velocity*

$$V_{tturba} := 5.35\cdot\left[D_{cut}\cdot\left[\frac{(\gamma_{cut}-\gamma_{a1})}{\gamma_{a1}}\right]\right]^{\frac{1}{2}}$$

$V_{tturba} = 9.8693$ (ft/sec)

$$N_{cut} := \frac{(D_{cut}\cdot V_{tturba})}{\nu_{sf1}}$$

$N_{cut} = 14.9846$ No

- *Concentration factor*

$$C_{con} := \frac{rop}{(V_{a1} - V_{ttrana})}$$ Concentration factor

$C_{con} = 0.0009$ Concentration factor less than 0.04, so OK

$V_{cut1} := V_{ttrana}$

Cased hole drill pipe tool joint section

- *Laminar terminal velocity*

$$V_{tlama} := 0.0333\cdot D_{cut}^{2}\cdot\left[\frac{(\gamma_{cut}-\gamma_{a2})}{\mu_{sf2}}\right]$$

$V_{tlama} = 4.3629$ (ft/sec)

$$N_{cut} := \frac{(D_{cut}\cdot V_{tlama})}{\nu_{sf2}}$$

$N_{cut} = 30.5439$ No

- *Transition terminal velocity*

$$V_{ttrana} := 0.492 \cdot D_{cut} \cdot \left[\frac{(\gamma_{cut} - \gamma_{a2})^{\frac{2}{3}}}{(\gamma_{a2} \cdot \mu_{sf2})^{\frac{1}{3}}} \right]$$

$V_{ttrana} = 2.0854$ (ft/sec)

$$N_{cut} := \frac{(D_{cut} \cdot V_{ttrana})}{\nu_{sf2}}$$

$N_{cut} = 14.5994$ Yes

- *Turbulent terminal velocity*

$$V_{tturba} := 5.35 \cdot \left[D_{cut} \cdot \left[\frac{(\gamma_{cut} - \gamma_{a2})}{\gamma_{a2}} \right] \right]^{\frac{1}{2}}$$

$V_{tturba} = 4.0786$ (ft/sec)

$$N_{cut} := \frac{(D_{cut} \cdot V_{tturba})}{\nu_{sf2}}$$

$N_{cut} = 28.5539$ No

- *Concentration factor*

$$C_{con} := \frac{rop}{(V_{a2} - V_{ttrana})}$$ Concentration factor

$C_{con} = 0.0032$ Concentration factor less than 0.04, so OK

$V_{cut2} := V_{ttrana}$

Open hole drill pipe body section

- Laminar terminal velocity

$$V_{tlama} := 0.0333 \cdot D_{cut}^{2} \cdot \left[\frac{(\gamma_{cut} - \gamma_{a3})}{\mu_{sf3}} \right]$$

$V_{tlama} = 1.5445$ (ft/sec)

$$N_{cut} := \frac{(D_{cut} \cdot V_{tlama})}{\nu_{sf3}}$$

$N_{cut} = 4.1704$ No

- Transition terminal velocity

$$V_{ttrana} := 0.492 \cdot D_{cut} \cdot \left[\frac{(\gamma_{cut} - \gamma_{a3})^{\frac{2}{3}}}{(\gamma_{a3} \cdot \mu_{sf3})^{\frac{1}{3}}} \right]$$

$V_{ttrana} = 1.4337$ (ft/sec)

$$N_{cut} := \frac{(D_{cut} \cdot V_{ttrana})}{\nu_{sf3}}$$

$N_{cut} = 3.8711$ Yes

- Turbulent terminal velocity

$$V_{tturba} := 5.35 \cdot \left[D_{cut} \cdot \left[\frac{(\gamma_{cut} - \gamma_{a3})}{\gamma_{a3}} \right] \right]^{\frac{1}{2}}$$

$V_{tturba} = 3.9075$ (ft/sec)

$$N_{cut} := \frac{(D_{cut} \cdot V_{tturba})}{\nu_{sf3}}$$

$N_{cut} = 10.5509$ No

- *Concentration factor*

$$C_{con} := \frac{rop}{\left(V_{a3} - V_{ttrana}\right)}$$ Concentration factor

$C_{con} = 0.0069$ Concentration factor less than 0.04, so OK

$V_{cut3} := V_{ttrana}$

Open hole drill pipe tool joint section

- *Laminar terminal velocity*

$$V_{tlama} := 0.0333 \cdot D_{cut}^{2} \cdot \left[\frac{\left(\gamma_{cut} - \gamma_{a4}\right)}{\mu_{sf4}}\right]$$

$V_{tlama} = 3.4481$ (ft/sec)

$$N_{cut} := \frac{\left(D_{cut} \cdot V_{tlama}\right)}{\nu_{sf4}}$$

$N_{cut} = 29.3798$ No

- *Transition terminal velocity*

$$V_{ttrana} := 0.492 \cdot D_{cut} \cdot \left[\frac{\left(\gamma_{cut} - \gamma_{a4}\right)^{\frac{2}{3}}}{\left(\gamma_{a4} \cdot \mu_{sf4}\right)^{\frac{1}{3}}}\right]$$

$V_{ttrana} = 1.6696$ (ft/sec)

$$N_{cut} := \frac{\left(D_{cut} \cdot V_{ttrana}\right)}{\nu_{sf4}}$$

$N_{cut} = 14.2260$ Yes

- *Turbulent terminal velocity*

$$V_{tturba} := 5.35 \cdot \left[D_{cut} \cdot \left[\frac{(\gamma_{cut} - \gamma_{a4})}{\gamma_{a4}} \right] \right]^{\frac{1}{2}}$$

$V_{tturba} = 3.2867$ (ft/sec)

$$N_{cut} := \frac{(D_{cut} \cdot V_{tturba})}{\nu_{sf4}}$$

$N_{cut} = 28.0045$ No

- *Concentration factor*

$$C_{con} := \frac{rop}{(V_{a4} - V_{ttrana})}$$ Concentration factor

$C_{con} = 0.0049$ Concentration factor less than 0.04, so OK

$V_{cut4} := V_{ttrana}$

Open hole drill collar section

- *Laminar terminal velocity*

$$V_{tlama} := 0.0333 \cdot D_{cut}^{2} \cdot \left[\frac{(\gamma_{cut} - \gamma_{a5})}{\mu_{sf5}} \right]$$

$V_{tlama} = 3.3983$ (ft/sec)

$$N_{cut} := \frac{(D_{cut} \cdot V_{tlama})}{\nu_{sf5}}$$

$N_{cut} = 29.1508$ No

- Transition terminal velocity

$$V_{ttrana} := 0.492 \cdot D_{cut} \cdot \left[\frac{(\gamma_{cut} - \gamma_{a5})^{\frac{2}{3}}}{(\gamma_{a5} \cdot \mu_{sf5})^{\frac{1}{3}}} \right]$$

$V_{ttrana} = 1.6498$ (ft/sec)

$$N_{cut} := \frac{(D_{cut} \cdot V_{ttrana})}{\nu_{sf5}}$$

$N_{cut} = 14.1520$ Yes

- Turbulent terminal velocity

$$V_{tturba} := 5.35 \cdot \left[D_{cut} \cdot \left[\frac{(\gamma_{cut} - \gamma_{a5})}{\gamma_{a5}} \right] \right]^{\frac{1}{2}}$$

$V_{tturba} = 3.2519$ (ft/sec)

$$N_{cut} := \frac{(D_{cut} \cdot V_{tturba})}{\nu_{sf5}}$$

$N_{cut} = 27.8952$ No

- Concentration factor

$$C_{con} := \frac{rop}{(V_{a5} - V_{ttrana})}$$ Concentration factor

$C_{con} = 0.0050$ Concentration factor less than 0.04, so OK

$V_{cut5} := V_{ttrana}$

Time for cutting to reach surface

$$\tau_{a1} := \frac{H_1}{(V_{a1} - V_{cut1})}$$ time (sec)

$\tau_{a1} = 353.0004$ (sec)

$$\tau_{a2} := \frac{H_2}{(V_{a2} - V_{cut2})}$$ time (sec)

$\tau_{a2} = 66.5402$ (sec)

$$\tau_{a3} := \frac{H_3}{(V_{a3} - V_{cut3})}$$ time (sec)

$\tau_{a3} = 986.6959$ (sec)

$$\tau_{a4} := \frac{H_4}{(V_{a4} - V_{cut4})}$$ time (sec)

$\tau_{a4} = 36.8449$ (sec)

$$\tau_{a5} := \frac{H_5}{(V_{a5} - V_{cut5})}$$ time (sec)

$\tau_{a5} = 150.8552$ (sec)

- *Total time to surface*

$$\tau_{tot} := \frac{(\tau_{a1} + \tau_{a2} + \tau_{a3} + \tau_{a4} + \tau_{a5})}{60}$$ total time to surface (min)

$\tau_{tot} = 26.5656$ (min)

Illustrative Example 10.1: SI version, Non-Friction Solution - Drilling Fluid is Stable Foam. This is a closed form solution. However, it is solved by simple trial and error to match the left and right sides of the equation.

Given input conditions

$q_f := 189.25$	mud flow rate (liters/min)
$q_c := 1000.428$	flow rate single compressors (liters/sec)
$p_{bp} := 55.176$	back pressure at top of annulus (N/cm^2 gauge)
$Q_f := \frac{q_f}{1000 \cdot 60}$	convert water gpm to m^3/sec
$Q_f = 0.00315$	volumetric flow rate of water (m^3/sec)
$Q_c := \frac{q_c}{1000}$	convert scfm to m^3/sec
$Q_c = 1.00043$	air volumetric flow rate from compressor (st m^3/sec)
$Q_g := Q_c$	volumetric flow rate of gas in system (st ft^3/sec)

Basic standard and elevation atmospheric conditions

$p_{st} := 10.136$	API standard atmospheric pressure (N/cm^2 abs)
$t_{st} := 15.6$	API standard atmospheric temperature (°C)
$alt := 1219.0$	surface location elevation altitude (m)
$p_{at} := 8.749$	actual surface atmospheric pressure (N/cm^2 abs)
$t_{at} := 7.078$	actual surface atmospheric temperature (°C)
$P_{st} := p_{st} \cdot 10000.0$	convert actual atmospheric pressure to N/m^2
$P_{st} = 101360.000$	(N/m^2 abs)
$T_{st} := t_{st} + 273.15$	convert actual atmospheric temperature to °C
$T_{st} = 288.750$	(°C)
$P_{at} := p_{at} \cdot 10000.0$	convert actual atmospheric pressure to N/m^2 abs

$P_{at} = 87490.000$	(N/m² abs)
$T_{at} := t_{at} + 273.15$	convert actual atmospheric temperature to °C
$T_{at} = 280.228$	(°C)
$P_{bp} := (p_{bp} + p_{at}) \cdot 10000.0$	convert back presssure to N/m² abs
$P_{bp} = 639250.000$	(N/m² abs)

Initial foam quality at the blooey line

$Q_{bp} := \frac{P_{st} \cdot T_{at} \cdot Q_c}{P_{bp} \cdot T_{st}}$	volumetric flow rate of gas at back pressure valve (m³/sec)
$Q_{bp} = 0.154$	volumetric flow rate (m³/sec)
$\Gamma_{bp} := \frac{Q_{bp}}{Q_f + Q_{bp}}$	foam quality at the back pressure valve
$\Gamma_{bp} = 0.980$	foam quality at top of annulus

Constants

$SG_a := 1.0$	specific gravity of air
$SG_r := 2.7$	rock specific gravity
$k := 1.4$	air specific heat constant
$R := 29.31$	specific gas constant (N-m/N-°C)
$\rho_w := 1000.0$	density of fresh water (kg/m³)
$g := 9.81$	gravitational constant (m/sec²)
$\gamma_w := \rho_w \cdot g$	specific weight of fresh water (N/m³)
$\gamma_w = 9810.000$	(N/m³)
$e_{cs} := 0.000152939$	roughness of commercial steel (m)
$\beta := 0.01823$	geothermal gradient (°C/m)

Borehole geometry

$D_c := 2133.496$ total depth of cased hole (m)

$D_{oh} := 914.355$ total depth of open hole (m)

$H := D_c + D_{oh}$ total depth of well (m)

$H = 3047.851$ (m)

$ROP := 18.29$ estimated rate of penetration (m/hr)

Preliminary calculations

$\gamma_{st} := \frac{P_{st} \cdot SG_a}{R \cdot T_{st}}$ specific weight of air at API conditions (N/m^3)

$\gamma_{st} = 11.976$ (N/m^3)

$\gamma_{at} := \frac{P_{at} \cdot SG_a}{R \cdot T_{at}}$ specific weight of air at surface atmosphere (N/m^3)

$\gamma_{at} = 10.6520$ (N/m^3)

$P_{bp} = 639250.000$ (N/m^2 abs)

$\gamma_{bp} := \frac{P_{bp} \cdot SG_a}{R \cdot T_{at}}$ specific weight of air at back pressure (N/m^3)

$\gamma_{bp} = 77.8293$ (N/m^3)

$rop := \frac{ROP}{60 \cdot 60}$ convet ROP to m/sec

$rop = 0.0051$ (m/sec)

$d_{bh} := 200.03$ diameter of borehole (mm)

$D_{bh} := \frac{d_{bh}}{1000}$ convert diameter to m

$D_{bh} = 0.200$ (m)

$w_w := \gamma_w \cdot Q_f$ weight rate of flow of water (N/sec)

$w_w = 30.942$ (N/sec)

$w_g := \gamma_{st} \cdot Q_c$ weight rate of flow of gas at blooey line exit (N/sec)

$w_g = 11.982$ (N/sec)

$w_r := \left[\left[\left(\frac{\pi}{4}\right) \cdot D_{bh}^{\ 2} \cdot \gamma_w \cdot SG_r \cdot rop\right]\right]$ weight rate of solid flow out of blooey line (N/sec)

$w_r = 4.229$ (N/sec)

$w_t := w_g + w_w + w_r$ total weight rate of flow (gas+fluid+solids) (N/sec)

$w_t = 47.153$ (N/sec)

$t_{bh} := t_{at} + \beta \cdot H$ bottom hole temperature (°C)

$t_{bh} = 62.640$ (°C)

$t_{ave} := \frac{(t_{at} + t_{bh})}{2}$ average borehole temperature (°C)

$t_{ave} = 34.859$ (°C)

$T_{ave} := 273.15 + t_{ave}$ average borehole termperature (°C)

$T_{ave} = 308.009$ (°C)

$P_g := P_{at}$ define the gas pressure (N/m^2 abs)

$T_g := T_{at}$ define the gas temperature (°C)

Non-friction solution by trial and error to determine p_{bh}

$p_{bh} := 257.71$ bottom hole pressure (N/cm^2 guage)

$P_{bh} := (p_{bh} + p_{at}) \cdot 10000.0$ convert to N/m^2 abs

$P_{bh} = 2664590.000$ (N/m^2 abs)

$$LS := Q_g P_g \cdot \left(\frac{T_{ave}}{T_g}\right) \cdot \ln\left(\frac{P_{bh}}{P_{bp}}\right) + Q_f \cdot P_{bh} - Q_f \cdot P_{bp}$$ left side of equation

$LS = 143721.441$

$RS := (w_g + w_w + w_r) \cdot H$ right side of equation

$RS = 143714.824$

$p_{bha} := p_{bh} + p_{at}$

$p_{bha} = 266.459$ bottom hole pressure (N/cm^2 abs)

Illustrative Example 10.2; SI version, Friction Solution - Drilling Fluid is Stable Foam.
This is a trial and error solution. The upper limit of the first integral across each geometric cross section solution is the trial and error term that must be sought to allow the value left side of the right side intergral to be equalized.

Laboratory screening test data

$h_n := 500$ nominal beaker height (ml)

$\tau_n := 200$ nominal half-life time (sec)

$h_t := 610$ test height (ml)

$\tau_t := 280$ test half-life (sec)

$\Lambda := \left(\frac{h_t}{h_n}\right)\cdot\left(\frac{\tau_t}{\tau_n}\right)$ foam viscosity adjustment coefficient

$\Lambda = 1.7080$

Given input volumetric flow rates

$q_f := 189.25$ mud flow rate (lpm)

$q_c := 999.56$ flow rate of a single compressor (standard liter/sec)

$p_{bp} := 55.176$ back pressure at top of annulus (N/cm^2 gauge)

$Q_f := \frac{q_f}{1000\cdot 60}$ convert water lpm to m^3/sec

$Q_f = 0.0032$ volumetric flow rate of water (m^3/sec)

$Q_c := \frac{q_c}{1000}$ convert standard liter/sec to m^3/sec

$Q_c = 0.9996$ air volumetric flow rate from compressor (stand m^3/s

$Q_g := Q_c$ volumetric flow rate of gas in system (stand m^3/sec)

Basic standard and elevation atmospheric conditions

$p_{st} := 10.136$ API standard atmospheric pressure (N/cm^2 abs)

$t_{st} := 15.6$ API standard atmospheric temperature (^{o}C)

$alt := 1219$	surface location elevation altitude (m)
$p_{at} := 8.749$	pressure from table for surface elevation (N/cm^2 abs)
$t_{at} := 7.078$	temperature from table for surface elevation (°C)
$P_{st} := p_{st} \cdot 10000.0$	convert API standard pressure to N/cm^2
$P_{st} = 101360.0000$	(N/m^2 abs)
$T_{st} := t_{st} + 273.15$	convert API standard temperature to K
$T_{st} = 288.7500$	(K)
$P_{at} := p_{at} \cdot 10000.0$	convert actual atmospheric pressure to N/m^2
$P_{at} = 87490.0000$	(N/m^2 abs)
$T_{at} := t_{at} + 273.15$	convert actual atmospheric temperature to K
$T_{at} = 280.2280$	(K)
$P_{bp} := \left(p_{bp} + p_{at}\right) \cdot 10000.0$	convert back presssure to N/m^2
$P_{bp} = 639250.0000$	(N/m^2 abs)

Initial foam quality at the blooey line

$Q_{bp} := \dfrac{P_{st} \cdot T_{at} \cdot Q_c}{P_{bp} \cdot T_{st}}$	volumetric flow rate of gas at back pressure valve (m^3/sec)
$Q_{bp} = 0.1538$	volumetric flow rate (m^3/sec)
$\Gamma_{bp} := \dfrac{Q_{bp}}{Q_f + Q_{bp}}$	foam quality at the back pressure valve
$\Gamma_{bp} = 0.980$	foam quality at top of annulus

Viscosities of gas and fluid

$\mu_{air} := 0.012$	viscosity of air (centipoises)
$\mu_g := \mu_{air} \cdot 10^{-3}$	convert viscosity to $N\text{-}sec/m^2$
$\mu_g = 1.200 \times 10^{-5}$	($N\text{-}sec/m^2$)

$\mu_w := 1.0$ viscosity of water (centipoises)

$\mu_f := \mu_w \cdot 10^{-3}$ convert viscosity to N-sec/m^2

$\mu_f = 1.000 \times 10^{-3}$ (lb-sec/ft^2)

Constants

$SG_a := 1.0$ specific gravity of air

$k := 1.4$ air specific heat constant

$R := 29.31$ engineering gas constant (N-m/N-K)

$\rho_w := 1000.0$ density of fresh water (kg/m^3)

$g := 9.81$ gravitational constant (m/sec^2)

$\gamma_w := \rho_w \cdot g$ specific weight of water (N/m^3)

$\gamma_w = 9810.0$ (N/m^3)

$\gamma_{wg} := 9.81$ specific weight of fresh water (N/liter)

$e_{cs} := 0.000152939$ roughness of commercial steel (m)

$\beta := 0.01823$ geothermal gradient (°C/m)

$SG_r := 2.7$ specific gravity of sedimentary rock

$e_r := 0.003048$ roughness of rock from table (m)

$C := 0.81$ aerated fluid flow loss coefficient through nozzles

Borehole geometry

$H_c := 2133.5$ total depth of cased hole (m)

$H_{oh} := 914.36$ total depth of open hole (m)

$ROP := 18.29$ estimated rate of penetration (m/hr)

$d_{bh} := 200.03$ borehole (drill bit) diameter (mm)

$od_{dc} := 158.75$ outer diameter of drill collar (mm)

$id_{dc} := 71.438$ inner diameter of drill collar (mm)

$od_{dp} := 114.30$ outer diameter of drill pipe (mm)

$id_{dp} := 97.18$ inner diameter of drill pipe (mm)

$od_{tj} := 158.75$ outer diameter of drill pipe tool joint (mm)

$id_{tj} := 69.85$ inner diameter of drill pipe tool joint (mm)

$od_c := 219.08$ outer diameter of casing (mm)

$id_c := 201.19$ inner diameter of casing (mm)

$L_r := 30.479$ length of return line (m)

$id_r := 142.9$ inner diameter of return line (mm)

$L_{dc} := 152.4$ total length of drill collars (m)

$L_{tj} := 0.457$ average length of one tool joint (m)

$L_{dp} := 9.144$ average length of one drill pipe section (m)

$d_n := 9.525$ nozzle diameter (mm)

$N_n := 3$ number of nozzles in the bit

$D_{bh} := \frac{d_{bh}}{1000.0}$ convert borehole diameter to m

$D_{bh} = 0.2000$ (m)

$ID_r := \frac{id_r}{1000.0}$ convert return line diameter to m

$ID_r = 0.1429$ (m)

$OD_{dp} := \frac{od_{dp}}{1000.0}$ convert drill pipe outer diameter to m

$OD_{dp} = 0.1143$ (m)

$ID_{dp} := \frac{id_{dp}}{1000.0}$ convert drill pipe inner diameter to m

$ID_{dp} = 0.0972$ (m)

$ID_c := \frac{id_c}{1000.0}$ convert casing inner diameter to m

$ID_c = 0.2012$ (m)

$OD_{dc} := \frac{od_{dc}}{1000.0}$ convert drill collar outer diameter to m

$OD_{dc} = 0.1587$ (m)

$ID_{dc} := \frac{id_{dc}}{1000.0}$ convert drill collar inner diameter to m

$ID_{dc} = 0.0714$ (m)

$OD_{tj} := \frac{od_{tj}}{1000.0}$ convert tool joint outer diameter to m

$OD_{tj} = 0.1587$ (m)

$ID_{tj} := \frac{id_{tj}}{1000.0}$ convert tool joint inner diameter to m

$ID_{tj} = 0.0698$ (m)

$d_n = 9.5250$ nozzle diameter (mm)

$D_n := \frac{d_n}{1000.0}$ nozzle diameter converted to m

$D_n = 0.0095$ (m)

Preliminary calculations

$\gamma_g := \frac{P_{st} \cdot SG_a}{R \cdot T_{st}}$ specific weight of air at surface (N/m^3)

$\gamma_g = 11.9765$ (N/m^3)

$\gamma_{bp} := \frac{P_{bp} \cdot SG_a}{R \cdot T_{at}}$ specific weight of air at back pressure (N/m^3)

$\gamma_{bp} = 77.8293$ (N/m^3)

$rop := \frac{ROP}{60 \cdot 60}$	convert ROP to m/sec
$rop = 0.0051$	(m/sec)
$w_f := \gamma_w \cdot Q_f$	weight rate of flow of water (N/sec)
$w_f = 30.9424$	(N/sec)
$w_g := \gamma_g \cdot Q_c$	weight rate of flow of gas blooey line exit (N/sec)
$w_g = 11.9712$	(N/sec)
$w_g := \gamma_{bp} \cdot Q_{bp}$	weight rate of flow of gas back pressure (N/sec)
$w_g = 11.9712$	(N/sec)
$w_s := \left[\left(\frac{\pi}{4} \right) \cdot D_{bh}^{\ 2} \cdot \gamma_w \cdot SG_r \cdot rop \right]$	weight rate of solid flow out (N/sec)
$w_s = 4.229$	(N/sec)
$w_t := w_g + w_f + w_s$	total weight rate of flow (N/sec)
$w_t = 47.1424$	(N/sec)
$\mu_{usf} := \frac{\mu_f \cdot w_f + \mu_g \cdot w_g}{w_f + w_g}$	average absolute viscosity at BP valve (N-sec/m^2)
$\mu_{usf} = 7.244 \times 10^{-4}$	(N-sec/m^2)
$\rho_g := \frac{\gamma_g}{g}$	gas (air) density (N-sec^2/m^4)
$\rho_g = 1.221 \times 10^{0}$	(N-sec^2/m^4)
$\nu_g := \frac{\mu_g}{\rho_g}$	gas (air) kinematic viscosity (m^2/sec)

$\nu_g = 9.829 \times 10^{-6}$ (m²/sec)

$\rho_f := \frac{\gamma_w}{g}$ mud density ($N\text{-}sec^2/m^4$)

$\rho_f = 1.000 \times 10^{3}$ ($N\text{-}sec^2/m^4$)

$\nu_f := \frac{\mu_f}{\rho_f}$ liquid kinematic viscosity (m^2/sec)

$\mu_f = 1.000 \times 10^{-3}$ ($N\text{-}sec/m^2$)

$\nu_f = 1.000 \times 10^{-6}$ (m^2/sec)

Section height calculations

$H_1 := H_c - L_{tj} \cdot \left(\frac{H_c}{L_{dp}}\right)$ calculate length of drill pipe in cased section

$H_1 = 2026.8717$ (m)

$H_2 := L_{tj} \cdot \left(\frac{H_c}{L_{dp}}\right)$ calculate length of tool joint in cased section

$H_2 = 106.6283$ (m)

$H_3 := H_{oh} - L_{dc} - L_{tj} \cdot \left(\frac{H_{oh} - L_{dc}}{L_{dp}}\right)$ calculate length of drill pipe in open hole section

$H_3 = 723.8787$ (m)

$H_4 := L_{tj} \cdot \left(\frac{H_{oh} - L_{dc}}{L_{dp}}\right)$ calculate length of tool joints in open hole section

$H_4 = 38.0813$ (m)

$H_5 := L_{dc}$ length of drill collar

$H_5 = 152.4000$ (m)

Return line and Tee at top of annulus

$ID := ID_r$ set variable ID equal to return line's ID

$ID = 0.143$	(m)
$A := \frac{\pi}{4} \cdot ID^2$	cross sectional area (m^2)
$A = 0.016$	(m^2)
$\gamma_{bp} := \frac{P_{bp} \cdot SG_a}{R \cdot T_{at}}$	specific weight of air upstream of back pressure valve (N/m^3)
$\gamma_{bp} = 77.8293$	(N/m^3)
$w_g := \gamma_{bp} \cdot Q_{bp}$	weight rate of flow of gas (N/sec)
$w_g = 11.9712$	(N/sec)
$Q_g := \frac{w_g}{\gamma_g}$	volumetric flow rate of gas (m^3/sec)
$Q_g = 0.9996$	(m^3/sec)
$q_g := Q_g \cdot 60$	convert flow to m^3/min
$q_g = 59.9736$	(m^3/min)
$w_f := \gamma_w \cdot Q_f$	weight rate of flow of fluid (N/sec)
$w_f = 30.9424$	(N/sec)
$\gamma_{bp} := \frac{w_t}{\left(\frac{P_{at}}{P_{bp}}\right) \cdot \left(\frac{T_{at}}{T_{at}}\right) \cdot Q_g + Q_f}$	specific weight of foam just upstream of back pressure valve
$\gamma_{bp} = 336.8341$	(N/m^3)
$\rho_{bp} := \frac{\gamma_{bp}}{g}$	density of stable foam
$\rho_{bp} = 34.3358$	(N-sec^2/m^4)

$$n_{bp} := 0.095932 + 2.3654\cdot\Gamma_{bp} - 10.46\cdot\Gamma_{bp}^{2} + 12.955\cdot\Gamma_{bp}^{3} + 14.467\cdot\Gamma_{bp}^{4} - 39.673\cdot\Gamma_{bp}^{5} + 20.625\cdot\Gamma_{bp}^{6}$$

$n_{bp} = 0.3143$

$$K_{bp} := -0.15626 + 56.14\cdot\Gamma_{bp} - 312.77\cdot\Gamma_{bp}^{2} + 576.65\cdot\Gamma_{bp}^{3} + 63.960\cdot\Gamma_{bp}^{4} - 960.46\cdot\Gamma_{bp}^{5} - 154.68\cdot\Gamma_{bp}^{6} + 1$$

$K_{bp} = 3.0450$

$$V_{bp} := \frac{Q_{bp} + Q_f}{\left(\frac{\pi}{4}\right)\cdot ID^2}$$ average velocity of the foam just upstream of the back pressure valve

$V_{bp} = 9.7871$ (m/sec)

$$\mu_e := \frac{14.5727(\Lambda\cdot K_{bp})}{g}\cdot\left(\frac{3\cdot n_{bp} + 1}{4\cdot n_{bp}}\right)^{n_{bp}}\cdot\left(\frac{8\cdot V_{bp}}{ID}\right)^{n_{bp}-1}$$

$\mu_e = 1.173 \times 10^{-1}$ (N-sec/m²)

$\mu_{bp}(\Gamma_{bp}) := \mu_e$ stable foam (N-sec/m²)

$\mu_{bp}(\Gamma_{bp}) = 1.173 \times 10^{-1}$ (N-sec/m²)

$$\nu_{bp} := \frac{\mu_{bp}(\Gamma_{bp})}{\rho_{bp}}$$ kinematic viscosity of the foam just upstream of the back pressure valve

$\nu_{bp} = 0.003417$ (m²/sec)

$$N_R := \frac{V_{bp}\cdot ID}{\nu_{bp}}$$ Reynolds number calculation

$N_R = 409$ Reynolds number

$$f_l := \frac{64}{N_R}$$ friction factor from the Laminar equation

$f_l = 0.156367$ Laminar friction factor

$$f_h := \left[\frac{1}{-1.8\cdot\log\left[\left(\frac{\frac{e_{cs}}{ID}}{3.7}\right)^{1.11} + \left(\frac{6.9}{N_R}\right)\right]}\right]^2$$ friction factor from the Haaland correlation

$$f_h = 0.0985$$ Haaland friction factor

$$f := f_l$$ select friction factor equal to applicable factor

$$Q(P) := \frac{P_{at}}{P}\cdot\frac{T_{at}}{T_{at}}\cdot Q_g$$ flow rate as a function of pressure

$$\gamma_{mix}(P) := \frac{w_t}{Q(P) + Q_f}$$ specific weight as a function of pressure

$$V(P) := \frac{Q(P) + Q_f}{A}$$ velocity as a function of pressure

$$p_r := 68.15$$ trial pressure at entrance to return line (N/cm^2)

$$P_r := p_r\cdot 10000.0$$ convert pressure to N/m^2 abs

$$P_r = 681500.0000$$ (N/m^2)

$$\int_{P_{bp}}^{P_r} \frac{1}{\gamma_{mix}(P)\cdot\left(V(P)^2\cdot\frac{f}{2\cdot g\cdot ID}\right)}\, dP = 30.4864$$ left hand side

$$\int_0^{L_r} 1\, dh = 30.4790$$ right hand side

$$K_t := 25$$ Tee constant from Figure 8-5 and 8-6

$$K_v := 0.2$$ loss factor for gate valves in blooey line

$$\gamma_{mix} := \frac{w_t}{\left(\frac{P_{at}}{P_r}\right)\cdot\left(\frac{T_{at}}{T_{at}}\right)\cdot Q_g + Q_f}$$ specific weight (N/m³)

$\gamma_{mix} = 358.5624$ (N/m³)

$$V := \left[\frac{\left(\frac{P_{at}}{P_r}\right)\cdot\left(\frac{T_{at}}{T_{at}}\right)\cdot Q_g + Q_f}{A}\right]$$ velocity (m/sec)

$V = 8.1977$ (m/sec)

$$\Delta P_T := \gamma_{mix}\cdot(2\cdot K_v + K_t)\cdot\frac{V^2}{2\cdot g}$$ pressure loss due to Tee and valves

$\Delta P_T = 31195.0326$ (N/m²)

$P_e := P_r + \Delta P_T$ sum pressure losses - entrance to return line

$P_e = 712695.0326$ (N/m²)

$$p_e := \frac{P_e}{10000.0}$$ convert from N/m² to N/cm² abs

$p_e = 71.2695$ (N/cm² abs)

$$V := \left[\frac{\left(\frac{P_{at}}{P_e}\right)\cdot\left(\frac{T_{at}}{T_{at}}\right)\cdot Q_g + Q_f}{A}\right]$$ velocity (m/sec)

$V = 7.8475$ (m/sec)

Annulus cased hole drill pipe body section

$ID := ID_c$ set inner diameter to casing ID

$OD := OD_{dp}$ set outer diameter to drill pipe OD

$P := P_e$ set pressure equal to previous section

$A := \frac{\pi}{4} \cdot \left(ID^2 - OD^2\right)$ cross sectional area (m^2)

$T := T_{at} + \beta \cdot H_1$ calculate temperature at bottom of section

$T = 317.1779$ (K)

$T_{ave1} := \frac{T_{at} + T}{2}$ calculate average temperature of section

$T_{ave1} = 298.7029$ (K)

$T_{ave} := T_{ave1}$ set average temp to section average temp

$\gamma_g := \frac{P \cdot SG_a}{R \cdot T_{ave}}$ specific weight of the gas (N/m^3)

$\gamma_g = 81.4045$ (N/m^3)

$Q_{ga} := \frac{w_g}{\gamma_g}$ flow rate of gas (m^3/sec)

$Q_{ga} = 0.1471$ (m^3/sec)

$\Gamma := \frac{Q_{ga}}{Q_{ga} + Q_f}$ check foam quality of this section

$\Gamma = 0.9790$ foam quality

$$n := 0.095932 + 2.3654 \cdot \Gamma - 10.46 \cdot \Gamma^2 + 12.955 \cdot \Gamma^3 + 14.467 \cdot \Gamma^4 - 39.673 \cdot \Gamma^5 + 20.625 \cdot \Gamma^6$$

$n = 0.3120$ power law index

$$K := -0.15626 + 56.14 \cdot \Gamma - 312.77 \cdot \Gamma^2 + 576.65 \cdot \Gamma^3 + 63.960 \cdot \Gamma^4 - 960.46 \cdot \Gamma^5 - 154.68 \cdot \Gamma^6 + 1670.2 \cdot \Gamma^7 - 9$$

$K = 3.1162$ consistency index

$V := \frac{Q_{ga} + Q_f}{A}$ average velocity of the foam just upstream back pressure valve

$V = 6.9769$ (m/sec)

$V_{a1} := V$ define the annulus velocity in Section 1

$$\mu_e := \frac{14.5727\Lambda \cdot K}{g} \cdot \left(\frac{2 \cdot n + 1}{3 \cdot n}\right)^n \cdot \left(\frac{12 \cdot V}{ID - OD}\right)^{n-1}$$ foam effective viscosity

$\mu_e = 8.313 \times 10^{-2}$ (N-sec/m^2)

$\mu_{sf}(\Gamma) := \mu_e$ stable foam viscosity (N-sec/m^2)

$\mu_{sf}(\Gamma) = 8.313 \times 10^{-2}$ (N-sec/m^2)

$\mu_{sf1} := \mu_{sf}(\Gamma)$

$$\gamma_{sf} := \frac{w_g + w_f}{\left(\frac{P_{at}}{P}\right) \cdot \left(\frac{T_{ave}}{T_{at}}\right) \cdot Q_g + Q_f}$$ specific weight of foam

$\gamma_{sf} = 320.3718$ (N/m^3)

$\gamma_{a1} := \gamma_{sf}$ define foam specific weight in Section 1

$$\rho_{sf} := \frac{\gamma_{sf}}{g}$$ density of stable foam

$\rho_{sf} = 32.6577$ (N-sec^2/m^4)

$$\nu_{sf}(\Gamma) := \frac{\mu_{sf}(\Gamma)}{\rho_{sf}}$$ kinematic viscosity of the foam

$\nu_{sf}(\Gamma) = 0.002546$ (m^2/sec)

$\nu_{sf1} := \nu_{sf}(\Gamma)$

$$N_R := \frac{V \cdot (ID - OD)}{\nu_{sf}(\Gamma)}$$ Reynolds number calculation

$N_R = 238$ Reynolds number

$f_l := \frac{64}{N_R}$ friction factor from the Laminar equation

$f_l = 0.268745$ Laminar friction factor

$$f_h := \left[\frac{1}{-1.8 \cdot \log\left[\left(\frac{\frac{e_{cs}}{ID-OD}}{3.7} \right)^{1.11} + \left(\frac{6.9}{N_R} \right) \right]} \right]^2$$

friction factor from the Haaland correlation

$f_h = 0.1310$

$f := f_l$ select friction factor equal to highest

$Q(P) := \frac{P_{at}}{P} \cdot \frac{T_{ave}}{T_{at}} \cdot Q_g$ flow rate as a function of pressure

$\gamma_{mix}(P) := \frac{w_t}{Q(P) + Q_f}$ specific weight as a function of pressure

$V(P) := \frac{Q(P) + Q_f}{A}$ velocity as a function of pressure

$p_1 := 471.23$ trial pressure bottom of the section (N/cm^2 abs)

$P_1 := p_1 \cdot 10000.0$ convert pressure to N/m^2

$P_1 = 4.712 \times 10^6$ (N/m^2 abs)

$$\int_{P_e}^{P_1} \frac{1}{\gamma_{mix}(P) \cdot \left[1 + V(P)^2 \cdot \frac{f}{2 \cdot g \cdot (ID - OD)} \right]} \, dP = 2026.870$$

left hand side

$$\int_0^{H_1} 1 \, dh = 2026.8717$$

right hand side

Annulus cased hole tool joint section

$ID := ID_c$ set inner diameter to casing ID

$OD := OD_{tj}$ set outer diameter to tool joint OD

$P := P_1$ set pressure equal to previous section

$A := \frac{\pi}{4} \cdot \left(ID^2 - OD^2\right)$ cross sectional area (m²)

$T_{temp} := T$ keep previous section temperature

$T := T_{temp} + \beta \cdot H_2$ calculate temperature at bottom of section

$T = 319.1217$ (K)

$T_{ave2} := \frac{T_{temp} + T}{2}$ calculate average temperature of section

$T_{ave2} = 318.1498$ (K)

$T_{ave} := T_{ave2}$ set average temp to section average temp

$\gamma_g := \frac{P \cdot SG_a}{R \cdot T_{ave}}$ specific weight of the gas (N/m³)

$\gamma_g = 505.3421$ (N/m³)

$Q_{ga} := \frac{w_g}{\gamma_g}$ flow rate of gas (N³/sec)

$Q_{ga} = 0.0237$ (m³/sec)

$\Gamma := \frac{Q_{ga}}{Q_{ga} + Q_f}$ check foam quality of this section

$\Gamma = 0.8825$ foam quality

$$n := 0.095932 + 2.3654 \cdot \Gamma - 10.46 \cdot \Gamma^2 + 12.955 \cdot \Gamma^3 + 14.467 \cdot \Gamma^4 - 39.673 \cdot \Gamma^5 + 20.625 \cdot \Gamma^6$$

$n = 0.2228$ power law index

$$K := -0.15626 + 56.14 \cdot \Gamma - 312.77 \cdot \Gamma^2 + 576.65 \cdot \Gamma^3 + 63.960 \cdot \Gamma^4 - 960.46 \cdot \Gamma^5 - 154.68 \cdot \Gamma^6 + 1670.2 \cdot \Gamma^7 -$$

$K = 4.9740$ consistency index

$$V := \frac{Q_{ga} + Q_f}{A}$$

average velocity of the foam just upstream of the back pressure valve

$V = 2.2374$ (m/sec)

$V_{a2} := V$ define velocity in annulus Section 2

$$\mu_e := \frac{14.5729\Lambda \cdot K}{g} \cdot \left(\frac{2 \cdot n + 1}{3 \cdot n}\right)^n \cdot \left(\frac{12 \cdot V}{ID - OD}\right)^{n-1}$$

$\mu_e = 99.678 \times 10^{-3}$ (N-sec/m²)

$\mu_{sf}(\Gamma) := \mu_e$ stable foam viscosity (N-sec/m²)

$\mu_{sf}(\Gamma) = 9.968 \times 10^{-2}$ (N-sec/m²)

$\mu_{sf2} := \mu_{sf}(\Gamma)$

$$\gamma_{sf} := \frac{w_g + w_f}{\left(\frac{P_{at}}{P}\right) \cdot \left(\frac{T_{ave}}{T_{at}}\right) \cdot Q_g + Q_f}$$

specific weight of foam

$\gamma_{sf} = 1771.5548$ (N/m³)

$\gamma_{a2} := \gamma_{sf}$ define foam specific weight in Section 2

$$\rho_{sf} := \frac{\gamma_{sf}}{g}$$

density of stable foam

$\rho_{sf} = 180.5866$ (N-sec²/m⁴)

$$\nu_{sf} := \frac{\mu_{sf}(\Gamma)}{\rho_{sf}}$$

kinematic viscosity of the foam

$\nu_{sf} = 0.000552$ (m²/sec)

$\nu_{sf2} := \nu_{sf}$

$N_R := \frac{V\cdot(ID - OD)}{\nu_{sf}}$ Reynolds number calculation

$N_R = 172$ Reynolds number

$f_l := \frac{64}{N_R}$ friction factor from the Laminar equation

$f_l = 0.372027$ Laminar friction factor

$f_h := \left[\frac{1}{-1.8\cdot\log\left[\left(\frac{\frac{e_{cs}}{ID-OD}}{3.7}\right)^{1.11} + \left(\frac{6.9}{N_R}\right)\right]}\right]^2$ friction factor from the Haaland correlation

$f_h = 0.1593$

$f := f_l$ select friction factor equal to higher of the two

$Q(P) := \frac{P_{at}}{P}\cdot\frac{T_{ave}}{T_{at}}\cdot Q_g$ flow rate as a function of pressure

$\gamma_{mix}(P) := \frac{w_t}{Q(P) + Q_f}$ specific weight as a function of pressure

$V(P) := \frac{Q(P) + Q_f}{A}$ velocity as a function of pressure

$p_2 := 529.0$ trial pressure at the bottom of the section (psia)

$P_2 := p_2\cdot 10000.0$ convert pressure to N/m^2

$P_2 = 5.290 \times 10^6$ (N/m^2)

$$\int_{P_1}^{P_2} \frac{1}{\gamma_{mix}(P)\cdot\left[1 + V(P)^2\cdot\frac{f}{2\cdot g\cdot(ID - OD)}\right]}\,dP = 106.621$$ left hand side

$$\int_{H_1}^{H_1+H_2} 1\,dh = 106.6283$$ right hand side

Annulus open hole drill pipe body section

$ID := D_{bh}$ set inner diameter to casing ID

$OD := OD_{dp}$ set outer diameter to drill pipe OD

$P := P_2$ pressure equal to previous section

$A := \frac{\pi}{4}\cdot\left(ID^2 - OD^2\right)$ cross sectional area (m^2)

$T_{temp} := T$ keep previous section temperature

$T := T_{temp} + \beta\cdot H_3$ calculate temperature at bottom of section

$T = 332.3180$ (K)

$T_{ave3} := \frac{T_{temp} + T}{2}$ calculate average temperature of section

$T_{ave3} = 325.7199$ (K)

$T_{ave} := T_{ave3}$ set average temp to section average temp

$\gamma_g := \frac{P\cdot SG_a}{R\cdot T_{ave}}$ specific weight of the gas (N/m^3)

$\gamma_g = 554.1095$ (N/m^3)

$Q_{ga} := \frac{w_g}{\gamma_g}$ flow rate of gas (m^3/sec)

$Q_{ga} = 0.0216$ (m^3/sec)

$\Gamma := \frac{Q_{ga}}{Q_{ga} + Q_f}$ check foam quality of this section

$\Gamma = 0.8726$ foam quality

$$n := 0.095932 + 2.3654\cdot\Gamma - 10.46\cdot\Gamma^2 + 12.955\cdot\Gamma^3 + 14.467\cdot\Gamma^4 - 39.673\cdot\Gamma^5 + 20.625\cdot\Gamma^6$$

$n = 0.2247$ power law index

$$K := -0.15626 + 56.14\cdot\Gamma - 312.77\cdot\Gamma^2 + 576.65\cdot\Gamma^3 + 63.960\cdot\Gamma^4 - 960.46\cdot\Gamma^5 - 154.68\cdot\Gamma^6 + 1670.2\cdot\Gamma^7 - 9$$

$K = 4.8382$ consistency index

$V := \dfrac{Q_{ga} + Q_f}{A}$ average velocity of the foam just upstream of the back pressure valve

$V = 1.1698$ (m/sec)

$V_{a3} := V$

$\mu_e := \dfrac{14.5727\Lambda\cdot K}{g}\cdot\left(\dfrac{2\cdot n + 1}{3\cdot n}\right)^n\cdot\left(\dfrac{12\cdot V}{ID - OD}\right)^{n-1}$ stable foam effective viscosity

$\mu_e = 279.934 \times 10^{-3}$ (N-sec/m^2)

$\mu_{sf}(\Gamma) := \mu_e$ stable foam viscosity (N-sec/m^2)

$\mu_{sf}(\Gamma) = 2.799 \times 10^{-1}$ (N-sec/m^2)

$\mu_{sf3} := \mu_{sf}(\Gamma)$

$\gamma_{sf} := \dfrac{w_g + w_f}{\left(\dfrac{P_{at}}{P}\right)\cdot\left(\dfrac{T_{ave}}{T_{at}}\right)\cdot Q_g + Q_f}$ specific weight of foam

$\gamma_{sf} = 1918.4103$ (N/m^2)

$\gamma_{a3} := \gamma_{sf}$ define foam specific weight in Section 3

$\rho_{sf} := \dfrac{\gamma_{sf}}{g}$ density of stable foam

$\rho_{sf} = 195.5566$ (N-sec^2/m^4)

$\nu_{sf} := \dfrac{\mu_{sf}(\Gamma)}{\rho_{sf}}$ kinematic viscosity of the foam

$\nu_{sf} = 0.001431$ (m^2/sec)

$\nu_{sf3} := \nu_{sf}$

$N_R := \dfrac{V \cdot (ID - OD)}{\nu_{sf}}$ Reynolds number calculation

$N_R = 70$ Reynolds number

$e_{ave} := \dfrac{e_r \cdot ID^2 + e_{cs} \cdot OD^2}{ID^2 + OD^2}$ calculate weighted average of roughness

$e_{ave} = 2.335 \times 10^{-3}$ weighted average of roughness (m)

$f_l := \dfrac{64}{N_R}$ friction factor from the Laminar equation

$f_l = 0.913509$ Laminar friction factor

$$f_h := \left[\frac{1}{-1.8 \cdot \log\left[\left(\dfrac{\dfrac{e_{ave}}{ID-OD}}{3.7}\right)^{1.11} + \left(\dfrac{6.9}{N_R}\right)\right]}\right]^2$$ friction factor from the Haaland correlation

$f_h = 0.3161$

$f := f_l$ select friction factor equal to higher of the two

$Q(P) := \dfrac{P_{at}}{P} \cdot \dfrac{T_{ave}}{T_{at}} \cdot Q_g$ flow rate as a function of pressure

$\gamma_{mix}(P) := \dfrac{w_t}{Q(P) + Q_f}$ specific weight as a function of pressure

$V(P) := \dfrac{Q(P) + Q_f}{A}$ velocity as a function of pressure

$p_3 := 789.73$ trial pressure at the bottom of the section (psia)

$P_3 := p_3 \cdot 10000.0$ convert pressure to N/m²

$P_3 = 7.897 \times 10^6$ (N/m²)

$$\int_{P_2}^{P_3} \frac{1}{\gamma_{mix}(P) \cdot \left[1 + V(P)^2 \cdot \frac{f}{2 \cdot g \cdot (ID - OD)}\right]} dP = 723.873$$

left hand side

$$\int_{H_1+H_2}^{H_1+H_2+H_3} 1 \, dh = 723.8787$$

right hand side

Annulus open hole tool joint section

$ID := D_{bh}$ set inner diameter to casing ID

$OD := OD_{tj}$ set outer diameter to tool joint OD

$P := P_3$ set pressure equal to previous section

$A := \frac{\pi}{4} \cdot \left(ID^2 - OD^2\right)$ cross sectional area (m²)

$T_{temp} := T$ keep previous section temperature

$T := T_{temp} + \beta \cdot H_4$ calculate temperature at bottom of section

$T = 333.0122$ (K)

$T_{ave4} := \frac{T_{temp} + T}{2}$ calculate average temperature of section

$T_{ave4} = 332.6651$ (K)

$T_{ave} := T_{ave4}$ set average temp to section average temp

$\gamma_g := \frac{P \cdot SG_a}{R \cdot T_{ave}}$ specific weight of the gas (N/m³)

$\gamma_g = 809.9450$ (N/m³)

$Q_{ga} := \frac{w_g}{\gamma_g}$ flow rate of gas (m³/sec)

$Q_{ga} = 0.0148$ (m³/sec)

$\Gamma := \frac{Q_{ga}}{Q_{ga} + Q_f}$ check foam quality of this section

$\Gamma = 0.8241$ foam quality

$$n := 0.095932 + 2.3654 \cdot \Gamma - 10.46 \cdot \Gamma^2 + 12.955 \cdot \Gamma^3 + 14.467 \cdot \Gamma^4 - 39.673 \cdot \Gamma^5 + 20.625 \cdot \Gamma^6$$

$n = 0.2456$ power law index

$$K := -0.15626 + 56.14 \cdot \Gamma - 312.77 \cdot \Gamma^2 + 576.65 \cdot \Gamma^3 + 63.960 \cdot \Gamma^4 - 960.46 \cdot \Gamma^5 - 154.68 \cdot \Gamma^6 + 1670.2 \cdot \Gamma^7 - 9$$

$K = 4.0391$ consistency index

$V := \frac{Q_{ga} + Q_f}{A}$ average velocity of the foam upstream of the back pressure valve

$V = 1.5418$ (m/sec)

$V_{a4} := V$

$\mu_e := \frac{14.5727 \Lambda \cdot K}{g} \cdot \left(\frac{2 \cdot n + 1}{3 \cdot n}\right)^n \cdot \left(\frac{12 \cdot V}{ID - OD}\right)^{n-1}$ effective viscosity

$\mu_e = 121.758 \times 10^{-3}$ (N-sec/m²)

$\mu_{sf}(\Gamma) := \mu_e$ stable foam viscosity (N-sec/m²)

$\mu_{sf}(\Gamma) = 1.218 \times 10^{-1}$ (N-sec/m²)

$\mu_{sf4} := \mu_{sf}(\Gamma)$

$$\gamma_{sf} := \frac{w_g + w_f}{\left(\frac{P_{at}}{P}\right)\cdot\left(\frac{T_{ave}}{T_{at}}\right)\cdot Q_g + Q_f}$$ specific weight of foam

$\gamma_{sf} = 2632.7528$ (N/m^3)

$\gamma_{a4} := \gamma_{sf}$ define foam specific weight in Section 4

$$\rho_{sf} := \frac{\gamma_{sf}}{g}$$ density of stable foam

$\rho_{sf} = 268.3744$ (N-sec^2/m^4)

$$\nu_{sf} := \frac{\mu_{sf}(\Gamma)}{\rho_{sf}}$$ kinematic viscosity of the foam

$\nu_{sf} = 0.000454$ (m^2/sec)

$\nu_{sf4} := \nu_{sf}$

$$N_R := \frac{V\cdot(ID - OD)}{\nu_{sf}}$$ Reynolds number calculation

$N_R = 140$ Reynolds number

$$e_{ave} := \frac{e_r\cdot ID^2 + e_{cs}\cdot OD^2}{ID^2 + OD^2}$$ calculate weighted average of roughness between the steel and the rock

$e_{ave} = 1.929 \times 10^{-3}$ weighted average of roughness

$$f_l := \frac{64}{N_R}$$ friction factor from the Laminar equation

$f_l = 0.4562$ Laminar friction factor

$$f_h := \left[\frac{1}{-1.8\cdot\log\left[\left(\frac{\frac{e_{ave}}{ID-OD}}{3.7}\right)^{1.11} + \left(\frac{6.9}{N_R}\right)\right]}\right]^2$$ friction factor from the Haaland correlation

$f_h = 0.1994$

$f := f_l$ select friction factor

$Q(P) := \frac{P_{at}}{P} \cdot \frac{T_{ave}}{T_{at}} \cdot Q_g$ flow rate as a function of pressure

$\gamma_{mix}(P) := \frac{w_t}{Q(P) + Q_f}$ specific weight as a function of pressure

$V(P) := \frac{Q(P) + Q_f}{A}$ velocity as a function of pressure

$p_4 := 812.9$ trial pressure bottom of the section (N/cm^2 abs)

$P_4 := p_4 \cdot 10000.0$ convert pressure to N/m^2

$P_4 = 8.129 \times 10^6$ (N/m^2 abs)

$$\int_{P_3}^{P_4} \frac{1}{\gamma_{mix}(P) \cdot \left[1 + V(P)^2 \cdot \frac{f}{2 \cdot g \cdot (ID - OD)}\right]} dP = 38.058$$ left hand side

$$\int_{H_1+H_2+H_3}^{H_1+H_2+H_3+H_4} 1\, dh = 38.0813$$ right hand side

Annulus open hole drill collar section

$ID := D_{bh}$ set inner diameter to casing ID

$OD := OD_{dc}$ set outer diameter to drill collar OD

$P := P_4$ set pressure equal to previous section pressure

$A := \frac{\pi}{4}\cdot\left(ID^2 - OD^2\right)$ cross sectional area (m^2)

$T_{temp} := T$ keep previous section temperature

$T := T_{temp} + \beta\cdot H_5$ calculate temperature at bottom of section

$T = 335.7905$ (K)

$T_{ave5} := \frac{T_{temp} + T}{2}$ calculate average temperature of section

$T_{ave5} = 334.4014$ (K)

$T_{ave} := T_{ave5}$ set average temp to section average temp

$\gamma_g := \frac{P\cdot SG_a}{R\cdot T_{ave}}$ specific weight of the gas (N/m^3)

$\gamma_g = 829.3794$ (N/m^3)

$Q_{ga} := \frac{w_g}{\gamma_g}$ flow rate of gas (m^3/sec)

$Q_{ga} = 0.0144$ (m^3/sec)

$\Gamma := \frac{Q_{ga}}{Q_{ga} + Q_f}$ check foam quality of this section

$\Gamma = 0.8207$ foam quality

$$n := 0.095932 + 2.3654\cdot\Gamma - 10.46\cdot\Gamma^2 + 12.955\cdot\Gamma^3 + 14.467\cdot\Gamma^4 - 39.673\cdot\Gamma^5 + 20.625\cdot\Gamma^6$$

$n = 0.2475$ power law index

$$K := -0.15626 + 56.14\cdot\Gamma - 312.77\cdot\Gamma^2 + 576.65\cdot\Gamma^3 + 63.960\cdot\Gamma^4 - 960.46\cdot\Gamma^5 - 154.68\cdot\Gamma^6 + 1670.2\cdot\Gamma^7 -$$

$K = 3.9829$ consistency index

$V := \frac{Q_{ga} + Q_f}{A}$ average velocity of the foam upstream

$V = 1.5120$ (m/sec)

$V_{a5} := V$

$$\mu_e := \frac{14.5727\Lambda \cdot K}{g} \cdot \left(\frac{2 \cdot n + 1}{3 \cdot n}\right)^n \cdot \left(\frac{12 \cdot V}{ID - OD}\right)^{n-1}$$

$\mu_e = 123.275 \times 10^{-3}$ (N-sec/m²)

$\mu_{sf}(\Gamma) := \mu_e$ stable foam viscosity (N-sec/m²)

$\mu_{sf}(\Gamma) = 1.233 \times 10^{-1}$ (N-sec/m²)

$\mu_{sf5} := \mu_{sf}(\Gamma)$

$$\gamma_{sf} := \frac{w_g + w_f}{\left(\frac{P_{at}}{P}\right) \cdot \left(\frac{T_{ave}}{T_{at}}\right) \cdot Q_g + Q_f}$$ specific weight of foam

$\gamma_{sf} = 2683.4652$ (N/m³)

$\gamma_{a5} := \gamma_{sf}$ define foam specific weight in Section 5

$$\rho_{sf} := \frac{\gamma_{sf}}{g}$$ density of stable foam

$\rho_{sf} = 273.5439$ (N-sec²/m⁴)

$$\nu_{sf} := \frac{\mu_{sf}(\Gamma)}{\rho_{sf}}$$ kinematic viscosity of the foam

$\nu_{sf} = 0.000451$ (m²/sec)

$\nu_{sf5} := \nu_{sf}$

$$N_R := \frac{V \cdot (ID - OD)}{\nu_{sf}}$$ Reynolds number calculation

$N_R = 139$ Reynolds number

$$e_{ave} := \frac{e_r \cdot ID^2 + e_{cs} \cdot OD^2}{ID^2 + OD^2}$$ calculate weighted average of roughness between the steel and the rock

$$e_{ave} = 1.929 \times 10^{-3}$$ weighted average of roughness (m)

$$f_l := \frac{64}{N_R}$$ friction factor from the Laminar equation

$$f_l = 0.4621$$ Laminar friction factor

$$f_h := \left[\frac{1}{-1.8 \cdot \log\left[\left(\frac{\frac{e_{ave}}{ID - OD}}{3.7}\right)^{1.11} + \left(\frac{6.9}{N_R}\right)\right]}\right]^2$$ friction factor from the Haaland correlation

$$f_h = 0.2009$$

$$f := f_l$$ select friction factor

$$Q(P) := \frac{P_{at}}{P} \cdot \frac{T_{ave}}{T_{at}} \cdot Q_g$$ flow rate as a function of pressure

$$\gamma_{mix}(P) := \frac{w_t}{Q(P) + Q_f}$$ specific weight as a function of pressure

$$V(P) := \frac{Q(P) + Q_f}{A}$$ velocity as a function of pressure

trail pressure bottom of the section (N/cm^2 abs)

$$p_5 := 906.24$$

convert pressure to N/m^2

$$P_5 := p_5 \cdot 10000.0$$

$$P_5 = 9.062 \times 10^6$$ (N/m^2 abs)

$$\int_{P_4}^{P_5} \frac{1}{\gamma_{mix}(P) \cdot \left[1 + V(P)^2 \cdot \frac{f}{2 \cdot g \cdot (ID - OD)}\right]} dP = 152.404$$ left hand side

$$\int_{H_1+H_2+H_3+H_4}^{H_1+H_2+H_3+H_4+H_5} 1\,dh = 152.4000$$ right hand side

$$\gamma_{bh} := \frac{P_5 \cdot SG_a}{R \cdot T}$$ specific weight of the gas (N/m³)

$$\gamma_{bh} = 920.7867$$ (N/m³)

$$Q_{bh} := \frac{w_g}{\gamma_{bh}}$$ flow rate of gas (m³/sec)

$$Q_{bh} = 0.0130$$ (m³/sec)

$$\Gamma_{bh} := \frac{Q_{bh}}{Q_{bh} + Q_f}$$ check foam quality of this section

$$\Gamma_{bh} = 0.8048$$ foam quality

Bottom hole through the bit nozzles

$$D_e := \left(N_n \cdot D_n^{\,2}\right)^{0.5}$$ equivalent single diameter (m)

$$D_e = 0.01650$$ (m)

$$P := P_5$$

1st Iteration to determine pressure inside drill string above nozzles

$$\gamma_{mix} := \frac{w_g + w_f}{Q_f + Q_g \cdot \left(\frac{P_{at}}{P}\right) \cdot \frac{T}{T_{at}}}$$ specific weight of mixture (N/m³)

$$\gamma_{mix} = 2915.8308$$ (N/m³)

$$\Delta P_b := \frac{\left(w_g + w_f\right)^2}{2 \cdot C^2 \cdot g \cdot \gamma_{mix} \cdot \left(\frac{\pi}{4}\right)^2 \cdot D_e^{\,4}}$$ approximate the pressure change of aerated fluid through the drill bit

$\Delta P_b = 1073682.3461$ (N/m^2)

$$\Delta p_b := \frac{\Delta P_b}{10000.0}$$ convert pressure to N/cm^2

$\Delta p_b = 107.3682$ (N/cm^2)

$P := P_5 + \Delta P_b$ sum pressures to get pressure above bit

$P = 10136082.3461$ (N/m^2 abs)

$$p := \frac{P}{10000.0}$$ convert prssure to N/cm^2 abs

$p = 1013.6082$ (N/cm^2 abs)

2nd Iteration to determine pressure inside drill string above nozzles

$$\gamma_{mix} := \frac{w_g + w_f}{Q_f + Q_g \cdot \left(\frac{P_{at}}{P}\right) \cdot \frac{T}{T_{at}}}$$ specific weight of mixture (N/m^3)

$\gamma_{mix} = 3180.5307$ (N/m^3)

$$\Delta P_b := \frac{\left(w_g + w_f\right)^2}{2 \cdot C^2 \cdot g \cdot \gamma_{mix} \cdot \left(\frac{\pi}{4}\right)^2 \cdot D_e^4}$$ approximate the pressure change of aerated fluid through the drill bit

$\Delta P_b = 984325.0457$ (N/m^2)

$$\Delta p_b := \frac{\Delta P_b}{10000.0}$$ convert pressure to N/cm^2

$\Delta p_b = 98.4325$ (N/cm^2)

$P := P_5 + \Delta P_b$ sum pressures to get pressure above bit

$P = 10046725.0457$ N/m^2 abs

$$p := \frac{P}{10000.0}$$ convert prssure to N/cm^2 abs

$p = 1004.6725$ (N/cm^2 abs)

3rd Iteration to determine pressure inside drill string above nozzles

$$\gamma_{mix} := \frac{w_g + w_f}{Q_f + Q_g \cdot \left(\frac{P_{at}}{P}\right) \cdot \frac{T}{T_{at}}}$$ specific weight of mixture (N/m^3)

$\gamma_{mix} = 3159.0022$ (N/m^3)

$$\Delta P_b := \frac{(w_g + w_f)^2}{2 \cdot C^2 \cdot g \cdot \gamma_{mix} \cdot \left(\frac{\pi}{4}\right)^2 \cdot D_e^4}$$ approximate the pressure change of aerated fluid through the drill bit

$\Delta P_b = 991033.1990$ (N/m^2)

$$\Delta p_b := \frac{\Delta P_b}{144}$$ convert pressure to N/cm^2

$\Delta p_b = 6882.1750$ (N/cm^2)

$P := P_5 + \Delta P_b$ sum pressures to get pressure above bit

$P = 10053433.1990$ (N/m^2 abs)

$$p := \frac{P}{10000.0}$$ convert pressure to N/cm^2 abs

$p = 1005.3433$ (N/cm^2 abs)

4th Iteration to determine pressure inside drill string above nozzles

$$\gamma_{mix} := \frac{w_g + w_f}{Q_f + Q_g \cdot \left(\frac{P_{at}}{P}\right) \cdot \frac{T}{T_{at}}}$$ specific weight of mixture (N/m^3)

$\gamma_{mix} = 3160.6214$ (N/m³)

$\Delta P_b := \frac{(w_g + w_f)^2}{2 \cdot C^2 \cdot g \cdot \gamma_{mix} \cdot \left(\frac{\pi}{4}\right)^2 \cdot D_e^4}$ approximate the pressure change of aerated fluid through the drill bit

$\Delta P_b = 990525.4703$ (N/m²)

$\Delta p_b := \frac{\Delta P_b}{144}$ convert pressure to N/cm²

$\Delta p_b = 6878.6491$ (N/cm²)

$P := P_5 + \Delta P_b$ sum pressures to get pressure above bit

$P = 10052925.4703$ (N/m² abs)

$p := \frac{P}{10000.0}$ convert pressure to N/cm² abs

$p = 1005.2925$ (N/cm² abs)

Convergence of pressure above the drill bit inside drill pipe

Open hole inside drill string drill collar section

$ID := ID_{dc}$ set inner diameter to casing ID

$A := \frac{\pi}{4} \cdot \left(ID^2\right)$ cross sectional area (m²)

$T_{ave} := T_{ave5}$ set average temp to section average temp

$\gamma_g := \frac{P \cdot SG_a}{R \cdot T_{ave}}$ specific weight of the gas (N/m³)

$\gamma_g = 1025.6723$ (N/m³)

$\rho_g := \frac{\gamma_g}{g}$ gas (air) density (N-sec^2/m^4)

$\rho_g = 104.5537$ (N-sec^2/m^4)

$\nu_g := \frac{\mu_g}{\rho_g}$ gas (air) kinematic viscosity (m^2/sec)

$\nu_g = 1.148 \times 10^{-7}$ (m^2/sec)

$\nu_{ave} := \frac{w_g \cdot \nu_g + w_f \cdot \nu_f}{w_g + w_f}$ average kinematic viscosity of gas & mud

$\nu_{ave} = 7.531 \times 10^{-7}$ (m^2/sec)

$V := \left[\frac{\left(\frac{P_{at}}{P}\right) \cdot \left(\frac{T_{ave}}{T_{at}}\right) \cdot Q_g + Q_f}{A}\right]$ velocity of flow in annulus (m/sec)

$V = 3.3768$ (m/sec)

$N_R := \frac{V \cdot ID}{\nu_{ave}}$ Reynolds number calculations

$N_R = 3.203 \times 10^5$ Reynolds number

$f_h := \left[\frac{1}{-1.8 \cdot \log\left[\left(\frac{\frac{e_{cs}}{ID}}{3.7}\right)^{1.11} + \left(\frac{6.9}{N_R}\right)\right]}\right]^2$ friction factor from the Haaland correlation

$f_h = 0.0244$

$f := f_h$

$Q(P) := \frac{P_{at}}{P} \cdot \frac{T_{ave}}{T_{at}} \cdot Q_g$ flow rate as a function of pressure

$$\gamma_{mix}(P) := \frac{w_g + w_f}{Q(P) + Q_f}$$ specific weight as a function of pressure

$$V(P) := \frac{Q(P) + Q_f}{A}$$ velocity as a function of pressure

$p_4 := 967.40$ trial pressure top of the section (N/cm² abs)

$P_4 := p_4 \cdot 10000.0$ convert pressure to N/m² abs

$P_4 = 9.674 \times 10^6$ (N/m² abs)

$$\int_{P_4}^{P} \frac{1}{\gamma_{mix}(P) \cdot \left[1 - V(P)^2 \cdot \frac{f}{2 \cdot g \cdot (ID)}\right]} \, dP = 152.422$$ left hand side

$$\int_{H_1+H_2+H_3+H_4}^{H_1+H_2+H_3+H_4+H_5} 1 \, dh = 152.4000$$ right hand side

Open hole inside drill string tool joint section

$ID := ID_{tj}$ set inner diameter to casing ID

$A := \frac{\pi}{4} \cdot \left(ID^2\right)$ cross sectional area (m²)

$T_{ave} := T_{ave4}$ set average temp to section average temp

$P := P_4$ set pressure to previous section pressure

$$\gamma_g := \frac{P \cdot SG_a}{R \cdot T_{ave}}$$ specific weight of the gas (N/m³)

$\gamma_g = 992.1629$ (N/m³)

$$\rho_g := \frac{\gamma_g}{g}$$ gas (air) density (N-sec²/m⁴)

$\rho_g = 101.1379$ (N-sec²/m⁴)

$\nu_g := \dfrac{\mu_g}{\rho_g}$ gas (air) kinematic viscosity (m²/sec)

$\nu_g = 1.186 \times 10^{-7}$ (m²/sec)

$\nu_{ave} := \dfrac{w_g \cdot \nu_g + w_f \cdot \nu_f}{w_g + w_f}$ average kinematic viscosity of gas & mud

$\nu_{ave} = 7.541 \times 10^{-7}$ (m²/sec)

$V := \left[\dfrac{\left(\dfrac{P_{at}}{P}\right)\cdot\left(\dfrac{T_{ave}}{T_{at}}\right)\cdot Q_g + Q_f}{A}\right]$ velocity of flow in annulus (m/sec)

$V = 3.6236$ (m/sec)

$N_R := \dfrac{V \cdot ID}{\nu_{ave}}$ Reynolds number calculations

$N_R = 3.356 \times 10^5$ Reynolds number

$f_h := \left[\dfrac{1}{-1.8 \cdot \log\left[\left(\dfrac{\frac{e_{cs}}{ID}}{3.7}\right)^{1.11} + \left(\dfrac{6.9}{N_R}\right)\right]}\right]^2$ friction factor from the Haaland correlation

$f_h = 0.0245$

$f := f_h$

$Q(P) := \dfrac{P_{at}}{P}\cdot\dfrac{T_{ave}}{T_{at}}\cdot Q_g$ flow rate as a function of pressure

$\gamma_{mix}(P) := \dfrac{w_g + w_f}{Q(P) + Q_f}$ specific weight as a function of pressure

$V(P) := \frac{Q(P) + Q_f}{A}$ velocity as a function of pressure

$p_3 := 958.45$ trial pressure top of the section (N/cm^2 abs)

$P_3 := p_3 \cdot 10000.0$ convert pressure to N/m^2 abs

$P_3 = 9.585 \times 10^6$ (N/m^2 abs)

$$\int_{P_3}^{P_4} \frac{1}{\gamma_{mix}(P) \cdot \left[1 - V(P)^2 \cdot \frac{f}{2 \cdot g \cdot (ID)}\right]} dP = 38.059$$ left hand side

$$\int_{H_1+H_2+H_3}^{H_1+H_2+H_3+H_4} 1 \, dh = 38.0813$$ right hand side

Open hole inside drill string drill pipe body section

$ID := ID_{dp}$ set inner diameter to casing ID

$A := \frac{\pi}{4} \cdot \left(ID^2\right)$ cross sectional area (m^2)

$T_{ave} := T_{ave3}$ set average temp to section average temp

$P := P_3$ set pressure to previous section pressure

$\gamma_g := \frac{P \cdot SG_a}{R \cdot T_{ave}}$ specific weight of the gas (N/m^3)

$\gamma_g = 1003.9438$ (N/m^3)

$\rho_g := \frac{\gamma_g}{g}$ gas (air) density ($N\text{-}sec^2/m^4$)

$\rho_g = 102.3388$ ($N\text{-}sec^2/m^4$)

$\nu_g := \dfrac{\mu_g}{\rho_g}$ gas (air) kinematic viscosity (m^2/sec)

$\nu_g = 1.173 \times 10^{-7}$ (m^2/sec)

$\nu_{ave} := \dfrac{w_g \cdot \nu_g + w_f \cdot \nu_f}{w_g + w_f}$ average kinematic viscosity of gas & mud

$\nu_{ave} = 7.537 \times 10^{-7}$ (m^2/sec)

$V := \left[\dfrac{\left(\dfrac{P_{at}}{P}\right) \cdot \left(\dfrac{T_{ave}}{T_{at}}\right) \cdot Q_g + Q_f}{A} \right]$ velocity of flow in annulus (m/sec)

$V = 1.8551$ (m/sec)

$N_R := \dfrac{V \cdot ID}{\nu_{ave}}$ Reynolds number calculations

$N_R = 2.392 \times 10^5$ Reynolds number

$f_h := \left[\dfrac{1}{-1.8 \cdot \log\left[\left(\dfrac{\frac{e_{cs}}{ID}}{3.7} \right)^{1.11} + \left(\dfrac{6.9}{N_R} \right) \right]} \right]^2$ friction factor form the Haaland correlation

$f_h = 0.0228$

$f := f_h$

$Q(P) := \dfrac{P_{at}}{P} \cdot \dfrac{T_{ave}}{T_{at}} \cdot Q_g$ flow rate as a function of pressure

$\gamma_{mix}(P) := \dfrac{w_g + w_f}{Q(P) + Q_f}$ specific weight as a function of pressure

$V(P) := \dfrac{Q(P) + Q_f}{A}$ velocity as a function of pressure

$p_2 := 761.93$ trial pressure top of the section (N/cm^2 abs)

$P_2 := p_2 \cdot 10000.0$ convert pressure to N/m^2 abs

$P_2 = 7.619 \times 10^6$ (N/m^2 abs)

$$\int_{P_2}^{P_3} \frac{1}{\gamma_{mix}(P)\cdot\left[1 - V(P)^2 \cdot \frac{f}{2\cdot g\cdot (ID)}\right]} dP = 723.812$$ left hand side

$$\int_{H_1+H_2}^{H_1+H_2+H_3} 1\, dh = 723.8787$$ right hand side

Cased hole inside drill string tool joint section

$ID := ID_{tj}$ set inner diameter to casing ID

$A := \frac{\pi}{4}\cdot\left(ID^2\right)$ cross sectional area (m^2)

$T_{ave} := T_{ave2}$ set average temp to section average temp

$P := P_2$ set pressure to previous section pressure

$\gamma_g := \frac{P\cdot SG_a}{R\cdot T_{ave}}$ specific weight of the gas (N/m^3)

$\gamma_g = 817.0857$ (N/m^3)

$\rho_g := \frac{\gamma_g}{g}$ gas (air) density (N-sec^2/m^4)

$\rho_g = 83.2911$ (N-sec^2/m^4)

$\nu_g := \frac{\mu_g}{\rho_g}$ gas (air) kinematic viscosity (m²/sec)

$\nu_g = 1.441 \times 10^{-7}$ (m²/sec)

$\nu_{ave} := \frac{w_g \cdot \nu_g + w_f \cdot \nu_f}{w_g + w_f}$ average kinematic viscosity of gas & mud

$\nu_{ave} = 7.612 \times 10^{-7}$ (m²/sec)

$V := \left[\frac{\left(\frac{P_{at}}{P}\right) \cdot \left(\frac{T_{ave}}{T_{at}}\right) \cdot Q_g + Q_f}{A}\right]$ velocity of flow in annulus (m/sec)

$V = 4.2237$ (m/sec)

$N_R := \frac{V \cdot ID}{\nu_{ave}}$ Reynolds number calculations

$N_R = 3.876 \times 10^5$ Reynolds number

$f_h := \left[\frac{1}{-1.8 \cdot \log\left[\left(\frac{\frac{e_{cs}}{ID}}{3.7}\right)^{1.11} + \left(\frac{6.9}{N_R}\right)\right]}\right]^2$ friction factor from the Haaland correlation

$f_h = 0.0244$

$f := f_h$

$Q(P) := \frac{P_{at}}{P} \cdot \frac{T_{ave}}{T_{at}} \cdot Q_g$ flow rate as a function of pressure

$\gamma_{mix}(P) := \frac{w_g + w_f}{Q(P) + Q_f}$ specific weight as a function of pressure

$$V(P) := \frac{Q(P) + Q_f}{A}$$ velocity as a function of pressure

$$p_1 := 743.03$$ trial pressure top of the section (N/cm^2 abs)

$$P_1 := p_1 \cdot 10000.0$$ convert pressure to N/m^2 abs

$$P_1 = 7.430 \times 10^6$$ (N/m^2)

$$\int_{P_1}^{P_2} \frac{1}{\gamma_{mix}(P) \cdot \left[1 - V(P)^2 \cdot \frac{f}{2 \cdot g \cdot (ID)}\right]} dP = 106.608$$ left hand side

$$\int_{H_1}^{H_1+H_2} 1 \, dh = 106.6283$$ right hand side

Cased hole inside drill string drill pipe body section

$$ID := ID_{dp}$$ set inner diameter to casing ID

$$A := \frac{\pi}{4} \cdot \left(ID^2\right)$$ cross sectional area (m^2)

$$T_{ave} := T_{ave1}$$ set average temperature to section average temperature

$$P := P_1$$ set pressure to previous section pressure

$$\gamma_g := \frac{P \cdot SG_a}{R \cdot T_{ave}}$$ specific weight of the gas (N/m^3)

$$\gamma_g = 848.6938$$ (N/m^3)

$$\rho_g := \frac{\gamma_g}{g}$$ gas (air) density (N-sec^2/m^4)

$$\rho_g = 86.5131$$ (N-sec^2/m^4)

$$\nu_g := \frac{\mu_g}{\rho_g}$$ gas (air) kinematic viscosity (m^2/sec)

$\nu_g = 1.387 \times 10^{-7}$ (m²/sec)

$$\nu_{ave} := \frac{w_g \cdot \nu_g + w_f \cdot \nu_f}{w_g + w_f}$$ average kinematic viscosity of gas & mud

$\nu_{ave} = 7.597 \times 10^{-7}$ (m²/sec)

$$V := \left[\frac{\left(\frac{P_{at}}{P}\right) \cdot \left(\frac{T_{ave}}{T_{at}}\right) \cdot Q_g + Q_f}{A}\right]$$ velocity of flow in annulus (m/sec)

$V = 2.1166$ (m/sec)

$$N_R := \frac{V \cdot ID}{\nu_{ave}}$$ Reynolds number calculations

$N_R = 2.707 \times 10^5$ Reynolds number

$$f_h := \left[\frac{1}{-1.8 \cdot \log\left[\left(\frac{\frac{e_{cs}}{ID}}{3.7}\right)^{1.11} + \left(\frac{6.9}{N_R}\right)\right]}\right]^2$$ friction factor from the Haaland correlation

$f_h = 0.0227$

$f := f_h$

$$Q(P) := \frac{P_{at}}{P} \cdot \frac{T_{ave}}{T_{at}} \cdot Q_g$$ flow rate as a function of pressure

$$\gamma_{mix}(P) := \frac{w_g + w_f}{Q(P) + Q_f}$$ specific weight as a function of pressure

$$V(P) := \frac{Q(P) + Q_f}{A}$$ velocity as a function of pressure

$p_{in} := 365.51$ trial pressure top of the section (N/cm² abs)

$P_{in} := p_{in} \cdot 10000.0$ convert pressure to N/m2 abs

$P_{in} = 3.655 \times 10^{6}$ (N/m2 abs)

$$\int_{P_{in}}^{P_1} \frac{1}{\gamma_{mix}(P) \cdot \left[1 - V(P)^2 \cdot \frac{f}{2 \cdot g \cdot (ID)}\right]} \, dP = 2026.819$$ left hand side

$$\int_{0}^{H_1} 1 \, dh = 2026.8717$$ right hand side

Hole cleaning in the annulus

Cuttings details

$d_{cut} := 12.7$ cuttings approximate maximum size (mm)

$D_{cut} := \frac{d_{cut}}{1000}$ convert to meter

$D_{cut} = 0.0127$ (m)

$\gamma_{cut} := SG_r \cdot \gamma_w$ specific weight of cuttings (N/m3)

$\gamma_{cut} = 26487.0000$ (N/m3)

Largest annulus cross-sectional area (cased hole drill pipe body section)

- Laminar terminal velocity

$$V_{tlama} := 0.0333 \cdot D_{cut}^{2} \cdot \left[\frac{(\gamma_{cut} - \gamma_{a1})}{\mu_{sf1}}\right]$$

$V_{tlama} = 1.6905$ (m/sec)

$$N_{cut} := \frac{(D_{cut} \cdot V_{tlama})}{\nu_{sfl}}$$

$N_{cut} = 8.4341$ No

- *Transition terminal velocity*

$$V_{ttrana} := 0.331 \cdot D_{cut} \cdot \left[\frac{(\gamma_{cut} - \gamma_{al})^{\frac{2}{3}}}{(\gamma_{al} \cdot \mu_{sfl})^{\frac{1}{3}}} \right]$$

$V_{ttrana} = 1.2407$ (m/sec)

$$N_{cut} := \frac{(D_{cut} \cdot V_{ttrana})}{\nu_{sfl}}$$

$N_{cut} = 6.1897$ Yes

- *Turbulent terminal velocity*

$$V_{tturba} := 2.95 \cdot \left[D_{cut} \cdot \left[\frac{(\gamma_{cut} - \gamma_{al})}{\gamma_{al}} \right] \right]^{\frac{1}{2}}$$

$V_{tturba} = 3.0045$ (m/sec)

$$N_{cut} := \frac{(D_{cut} \cdot V_{tturba})}{\nu_{sfl}}$$

$N_{cut} = 14.9894$ No

- *Concentration factor*

$$C_{con} := \frac{rop}{(V_{al} - V_{ttrana})}$$ Concentration factor

$C_{con} = 0.0009$ Concentration factor less than 0.04, so OK

$V_{cut1} := V_{ttrana}$

Cased hole drill pipe tool joint section

- *Laminar terminal velocity*

$$V_{tlama} := 0.0333 \cdot D_{cut}^{2} \cdot \left[\frac{(\gamma_{cut} - \gamma_{a2})}{\mu_{sf2}} \right]$$

$V_{tlama} = 1.3317$ (m/sec)

$$N_{cut} := \frac{(D_{cut} \cdot V_{tlama})}{\nu_{sf2}}$$

$N_{cut} = 30.6417$ No

- *Transition terminal velocity*

$$V_{ttrana} := 0.331 \cdot D_{cut} \cdot \left[\frac{(\gamma_{cut} - \gamma_{a2})^{\frac{2}{3}}}{(\gamma_{a2} \cdot \mu_{sf2})^{\frac{1}{3}}} \right]$$

$V_{ttrana} = 0.6358$ (m/sec)

$$N_{cut} := \frac{(D_{cut} \cdot V_{ttrana})}{\nu_{sf2}}$$

$N_{cut} = 14.6280$ Yes

- *Turbulent terminal velocity*

$$V_{tturba} := 2.95 \cdot \left[D_{cut} \cdot \left[\frac{(\gamma_{cut} - \gamma_{a2})}{\gamma_{a2}} \right] \right]^{\frac{1}{2}}$$

$V_{tturba} = 1.2417$ (m/sec)

$$N_{cut} := \frac{(D_{cut} \cdot V_{tturba})}{\nu_{sf2}}$$

$N_{cut} = 28.5708$ No

- Concentration factor

$$C_{con} := \frac{rop}{(V_{a2} - V_{ttrana})}$$ Concentration factor

$C_{con} = 0.0032$ Concentration factor less than 0.04, so OK

$$V_{cut2} := V_{ttrana}$$

Open hole drill pipe body section

- Laminar terminal velocity

$$V_{tlama} := 0.0333 \cdot D_{cut}^{2} \cdot \left[\frac{(\gamma_{cut} - \gamma_{a3})}{\mu_{sf3}}\right]$$

$V_{tlama} = 0.4714$ (m/sec)

$$N_{cut} := \frac{(D_{cut} \cdot V_{tlama})}{\nu_{sf3}}$$

$N_{cut} = 4.1821$ No

- Transition terminal velocity

$$V_{ttrana} := 0.331 \cdot D_{cut} \cdot \left[\frac{(\gamma_{cut} - \gamma_{a3})^{\frac{2}{3}}}{(\gamma_{a3} \cdot \mu_{sf3})^{\frac{1}{3}}}\right]$$

$V_{ttrana} = 0.4371$ (m/sec)

$$N_{cut} := \frac{(D_{cut} \cdot V_{ttrana})}{\nu_{sf3}}$$

$N_{cut} = 3.8777$ Yes

- *Turbulent terminal velocity*

$$V_{tturba} := 2.95 \cdot \left[D_{cut} \cdot \left[\frac{(\gamma_{cut} - \gamma_{a3})}{\gamma_{a3}} \right] \right]^{\frac{1}{2}}$$

$V_{tturba} = 1.1897$ (m/sec)

$$N_{cut} := \frac{(D_{cut} \cdot V_{tturba})}{\nu_{sf3}}$$

$N_{cut} = 10.5552$ No

- *Concentration factor*

$$C_{con} := \frac{rop}{(V_{a3} - V_{ttrana})}$$ Concentration factor

$C_{con} = 0.0069$ Concentration factor less than 0.04, so OK

$V_{cut3} := V_{ttrana}$

Open hole drill pipe tool joint section

- *Laminar terminal velocity*

$$V_{tlama} := 0.0333 \cdot D_{cut}^{2} \cdot \left[\frac{(\gamma_{cut} - \gamma_{a4})}{\mu_{sf4}} \right]$$

$V_{tlama} = 1.0523$ (m/sec)

$$N_{cut} := \frac{(D_{cut} \cdot V_{tlama})}{\nu_{sf4}}$$

$N_{cut} = 29.4554$ No

- *Transition terminal velocity*

$$V_{ttrana} := 0.331 \cdot D_{cut} \cdot \left[\frac{(\gamma_{cut} - \gamma_{a4})^{\frac{2}{3}}}{(\gamma_{a4} \cdot \mu_{sf4})^{\frac{1}{3}}} \right]$$

$V_{ttrana} = 0.5090$ (m/sec)

$$N_{cut} := \frac{(D_{cut} \cdot V_{ttrana})}{\nu_{sf4}}$$

$N_{cut} = 14.2480$ Yes

- *Turbulent terminal velocity*

$$V_{tturba} := 2.95 \cdot \left[D_{cut} \cdot \left[\frac{(\gamma_{cut} - \gamma_{a4})}{\gamma_{a4}} \right] \right]^{\frac{1}{2}}$$

$V_{tturba} = 1.0007$ (m/sec)

$$N_{cut} := \frac{(D_{cut} \cdot V_{tturba})}{\nu_{sf4}}$$

$N_{cut} = 28.0123$ No

- *Concentration factor*

$$C_{con} := \frac{rop}{(V_{a4} - V_{ttrana})}$$ Concentration factor

$C_{con} = 0.0049$ Concentration factor less than 0.04, so OK

$V_{cut4} := V_{ttrana}$

Open hole drill collar section

- Laminar terminal velocity

$$V_{tlama} := 0.0333 \cdot D_{cut}^{2} \cdot \left[\frac{(\gamma_{cut} - \gamma_{a5})}{\mu_{sf5}} \right]$$

$V_{tlama} = 1.0371$ (m/sec)

$$N_{cut} := \frac{(D_{cut} \cdot V_{tlama})}{\nu_{sf5}}$$

$N_{cut} = 29.2263$ No

- Transition terminal velocity

$$V_{ttrana} := 0.331 \cdot D_{cut} \cdot \left[\frac{(\gamma_{cut} - \gamma_{a5})^{\frac{2}{3}}}{(\gamma_{a5} \cdot \mu_{sf5})^{\frac{1}{3}}} \right]$$

$V_{ttrana} = 0.5030$ (m/sec)

$$N_{cut} := \frac{(D_{cut} \cdot V_{ttrana})}{\nu_{sf5}}$$

$N_{cut} = 14.1740$ Yes

- Turbulent terminal velocity

$$V_{tturbe} := 2.95 \cdot \left[D_{cut} \cdot \left[\frac{(\gamma_{cut} - \gamma_{a5})}{\gamma_{a5}} \right] \right]^{\frac{1}{2}}$$

$V_{tturba} = 0.9901$ (m/sec)

$N_{cut} := \frac{(D_{cut} \cdot V_{tturba})}{\nu_{sf5}}$

$N_{cut} = 27.9031$ No

- *Concentration factor*

$C_{con} := \frac{rop}{(V_{a5} - V_{ttrana})}$ Concentration factor

$C_{con} = 0.0050$ Concentration factor less than 0.04, so OK

$V_{cut5} := V_{ttrana}$

Time for cutting to reach surface

$\tau_{a1} := \frac{H_1}{(V_{a1} - V_{cut1})}$ time (sec)

$\tau_{a1} = 353.3472$ (sec)

$\tau_{a2} := \frac{H_2}{(V_{a2} - V_{cut2})}$ time (sec)

$\tau_{a2} = 66.5746$ (sec)

$\tau_{a3} := \frac{H_3}{(V_{a3} - V_{cut3})}$ time (sec)

$\tau_{a3} = 987.9039$ (sec)

$\tau_{a4} := \frac{H_4}{(V_{a4} - V_{cut4})}$ time (sec)

$\tau_{a4} = 36.8712$ (sec)

$$\tau_{a5} := \frac{H_5}{(V_{a5} - V_{cut5})}$$ time (sec)

$\tau_{a5} = 151.0303$ (sec)

- *Total time to surface*

$$\tau_{tot} := \frac{(\tau_{a1} + \tau_{a2} + \tau_{a3} + \tau_{a4} + \tau_{a5})}{60}$$ total time to surface (min)

$\tau_{tot} = 26.5955$ (min)

Chapter 11 Illustrative Example MathCad™ Solutions

F

This appendix presents the detailed MathCad solutions for the illustrative examples in Chapter 11.

Illustrative Example 11.1: Hammer Performance (USCS)

Hammer No. 1

1) Basic input data

$l_s := 7.1$	length of piston chamber (in)
$e_s := 0.8$	stroke approximate ratio of length
$L_s := \frac{l_s}{12}$	convert chamber length to ft
$L_s = 0.592$	(ft)
$d_p := 5.9$	diameter of piston (in)
$d_b := 7.875$	diameter of drill bit (in)
$A_p := \left(\frac{\pi}{4}\right)\cdot\left(\frac{d_p}{12}\right)^2$	cross-sectional area of piston (ft^2)
$A_p = 0.19$	(ft^2)
$A_b := \left(\frac{\pi}{4}\right)\cdot\left(\frac{d_b}{12}\right)^2$	cross-sectional area of drill bit (ft^2)
$A_b = 0.338$	(ft^2)
$W_p := 97.0$	weight of piston (lb)
$g := 32.2$	acceleration of gravity (ft/sec^2)
$\gamma_{st} := 0.0763$	specific weight of standard air (lb/ft^3)
$\gamma_{at} := 0.0679$	specific weight of 4000 ft atmos air (lb/ft^3)
$q_{cat} := 2400$	compressor 4000 ft atmos air (acfm)
$q_{cst} := q_{cat}\cdot\left(\frac{\gamma_{at}}{\gamma_{st}}\right)$	equivalent standard air (scfm)
$q_{cst} = 2.136 \times 10^3$	(scfm)

$Q_{cst} := \frac{q_{cst}}{60}$	convert to ft^3/sec
$Q_{cst} = 35.596$	(ft^3/sec)
$H := 10000$	depth of drilling (ft)
$p_{at} := 12.685$	atmos pressure at 4000 ft (psia)
$P_{at} := p_{at} \cdot 144$	convert to lb/ft^2 abs
$P_{at} = 1.827 \times 10^3$	lb/ft^2 abs
$t_{at} := 44.74$	atmos temperature at 4000 ft (°F)
$T_{at} := t_{at} + 459.67$	convert to absolute temperature (°R)
$T_{at} = 504.41$	(°R)
$\beta := 0.01$	geothermal gradient (°F/ft)
$T_h := T_{at} + \beta \cdot H$	absolute temperature at depth (°R)
$T_h = 604.41$	(°R)
$p_{anbh} := 161.7$	bottom hole annulus pressure (psia)
$\Delta p_{DTH} := 340.0$	pressure drop through DTH (psi)
$p_{inaDTH} := \Delta p_{DTH} + p_{anbh}$	pressure above DTH (psia)
$p_{inaDTH} = 501.7$	(psia)
$P_{inaDTH} := p_{inaDTH} \cdot 144$	convert to lb/ft^2 abs
$P_{inaDTH} = 7.224 \times 10^4$	(lb/ft^2 abs)

$R := 53.36$ engineering gas constant (lb-ft/lb-°R)

$S_a := 1.0$ specific gravity of air

2) Pressure and weight rate of flow

$w_a := \gamma_{st} \cdot Q_{cst}$ weight rate of flow of air (lb/sec)

$w_a = 2.716$ (lb/sec)

$\gamma_{inaDTH} := \frac{P_{inaDTH} \cdot S_a}{R \cdot T_h}$ specific weight of air entering (lb/ft^3)

$\gamma_{inaDTH} = 2.24$ (lb/ft^3)

$Q_{inaDTH} := \frac{w_a}{\gamma_{inaDTH}}$ volumetric flow through DTH (ft^3/sec)

$Q_{inaDTH} = 1.212$ (ft^3/sec)

3) Piston motion and impact velocity

$Vol_{DTH} := e_s \cdot L_s \cdot A_p$ volume of DTH per stroke (ft^3)

$Vol_{DTH} = 0.09$ (ft^3)

$N_{ps} := \frac{Q_{inaDTH} \cdot 60}{2Vol_{DTH}}$ number of strokes per minute (spm)

$N_{ps} = 404.756$ (spm)

$\Delta\tau_{ts} := \frac{1}{\left(\frac{N_{ps}}{60}\right)}$ period of stroke (sec)

$\Delta\tau_{ts} = 0.148$ (sec)

$$V_p := \left[2\cdot\left[\frac{P_{inaDTH}\cdot A_p + W_p}{\left(\frac{W_p}{g}\right)}\right]\cdot(e_s\cdot L_s)\right]^{0.5}$$ velocity of piston at shank (ft/sec)

$V_p = 65.885$ (ft/sec)

$$\Delta\tau_{dn} := \frac{2\cdot e_s\cdot L_s}{V_p}$$ time of down stroke (sec)

$\Delta\tau_{dn} = 0.014$ (sec)

4) Kinetic energy terms

$$KE_{ps} := \left(\frac{1}{2}\right)\cdot\left(\frac{W_p}{g}\right)\cdot V_p^{\,2}$$ piston KE at shank (lb-ft)

$KE_{ps} = 6.538 \times 10^3$ (lb-ft)

$$ke_b := \frac{KE_{ps}}{A_b}$$ KE per drill bit area (lb-ft/ft^2)

$ke_b = 1.933 \times 10^4$ (lb-ft/ft^2)

$KE_{total} := KE_{ps}\cdot N_{ps}$ total KE to rock per minute (lb-ft/min)

$KE_{total} = 2.646 \times 10^6$ (lb-ft/min)

Hammer No. 2

1) Basic input data

$l_s := 4.6876$ length of piston chamber (in)

$e_s := 0.80$ stroke approximate ratio of length

$$L_s := \frac{l_s}{12}$$ convert chamber length to ft

$L_s = 0.391$ (ft)

$d_p := 6.00$ diameter of piston (in)

$d_b := 7.875$ diameter of drill bit (in)

$A_p := \left(\frac{\pi}{4}\right)\cdot\left(\frac{d_p}{12}\right)^2$ cross-sectional area of piston (ft^2)

$A_p = 0.196$ (ft^2)

$A_b := \left(\frac{\pi}{4}\right)\cdot\left(\frac{d_b}{12}\right)^2$ cross-sectional area of drill bit (ft^2)

$A_b = 0.338$ (ft^2)

$W_p := 78.0$ weight of piston (lb)

$g := 32.2$ acceleration of gravity (ft/sec^2)

$\gamma_{st} := 0.0763$ specific weight of standard air (lb/ft^3)

$\gamma_{at} := 0.0679$ specific weight of 4000 ft atmos air (lb/ft^3)

$q_{cat} := 2400.0$ compressor 4000 ft atmos air (acfm)

$q_{cst} := q_{cat}\cdot\left(\frac{\gamma_{at}}{\gamma_{st}}\right)$ equivalent standard air (scfm)

$q_{cst} = 2.136 \times 10^3$ (scfm)

$Q_{cst} := \frac{q_{cst}}{60}$ convert to ft^3/sec

$Q_{cst} = 35.596$ (ft^3/sec)

$H := 10000$ depth of drilling (ft)

$\beta := 0.01$ geothermal gradient (°F/ft)

$T_b := T_{at} + \beta\cdot H$ absolute temperature at depth (°R)

$T_h = 604.41$ (°R)

$p_{anbh} := 161.7$ bottom hole annulus pressure (psia)

$\Delta p_{DTH} := 550.0$ pressure drop through DTH (psi)

$p_{inaDTH} := \Delta p_{DTH} + p_{anbh}$ pressure above DTH (psia)

$p_{inaDTH} = 711.7$ (psia)

$P_{inaDTH} := p_{inaDTH} \cdot 144$ convert to lb/ft² abs

$P_{inaDTH} = 1.025 \times 10^5$ (lb/ft² abs)

$R := 53.36$ engineering gas constant (lb-ft/lb-°R)

$S_a := 1.0$ specific gravity of air

2) Pressure and weight rate of flow

$w_a := \gamma_{st} \cdot Q_{cst}$ weight rate of flow of air (lb/sec)

$w_a = 2.716$ (lb/sec)

$\gamma_{inaDTH} := \frac{P_{inaDTH} \cdot S_a}{R \cdot T_h}$ specific weight of air entering (lb/ft³)

$\gamma_{inaDTH} = 3.178$ (lb/ft³)

$Q_{inaDTH} := \frac{w_a}{\gamma_{inaDTH}}$ volumetric flow through DTH (ft³/sec)

$Q_{inaDTH} = 0.855$ (ft³/sec)

3) Piston motion and impact velocity

$$Vol_{DTH} := e_s \cdot L_s \cdot A_p$$ volume of DTH per stroke (ft^3)

$$Vol_{DTH} = 0.061$$ (ft^3)

$$N_{ps} := \frac{Q_{inaDTH} \cdot 60}{2 Vol_{DTH}}$$ number of strokes per minute (spm)

(spm)

$$N_{ps} = 417.878$$

period of stroke (sec)

$$\Delta\tau_{ts} := \frac{1}{\left(\frac{N_{ps}}{60}\right)}$$

(sec)

$$\Delta\tau_{ts} = 0.144$$

$$V_p := \left[2 \cdot \left[\frac{P_{inaDTH} \cdot A_p + W_p}{\left(\frac{W_p}{g}\right)}\right] \cdot \left(e_s \cdot L_s\right)\right]^{0.5}$$ velocity of piston at shank (ft/sec)

$$V_p = 72.195$$ (ft/sec)

$$\Delta\tau_{dn} := \frac{2 \cdot e_s \cdot L_s}{V_p}$$ time of down stroke (sec)

$$\Delta\tau_{dn} = 8.657 \times 10^{-3}$$ (sec)

4) Kinetic energy terms

$$KE_{ps} := \left(\frac{1}{2}\right) \cdot \left(\frac{W_p}{g}\right) \cdot V_p^{\,2}$$ piston KE at shank (lb-ft)

$$KE_{ps} = 6.313 \times 10^{3}$$ (lb-ft)

$ke_b := \frac{KE_{ps}}{A_b}$ KE per drill bit area (lb-ft/ft^2)

$ke_b = 1.866 \times 10^4$ (lb-ft/ft^2)

$KE_{total} := KE_{ps} \cdot N_{ps}$ total KE to rock per minute (lb-ft/min)

$KE_{total} = 2.638 \times 10^6$ (lb-ft/min)

Illustrative Example 11.2 PDM (USCS version) - Three Dresser Clark CFB-4 Compressors - Drilling Fluid is Air. Determining the bottom hole and injection pressures while drilling at 10,000 ft of depth for the drilling operations described in Illustrative Examples 8.1, 8.2, and 8.3.

Compressor data - Dresser Clark Model CFB-4 with Caterpillar Model D398 prime mover

$q_c := 1200$ flow rate produced from one compressor (acfm)

$num_c := 2$ number of compressors (required for min KE)

$\varepsilon_m := 0.8$ mechanical efficiency

$n_s := 4$ number of stages

$HP_{max} := 760$ max sea level Hp rating of prime mover (HP)

$c := 0.02$ clearance volume ratio for compressor

Basic standard and elevation atomospheric conditions

$p_{st} := 14.696$ API standard atmospheric pressure (psia)

$t_{st} := 60.0$ API standard atmospheric temperature (°F)

$alt := 4000$ surface elevation of drilling site (ft)

$p_{at} := 12.685$ pressure from table for surface elevation (psia)

$t_{at} := 44.74$ temperature from table for surface elevation (°F)

$P_{st} := p_{st} \cdot 144$ convert API standard pressure to lb/ft^2

$P_{st} = 2116.224$ (lb/ft^2 abs)

$T_{st} := t_{st} + 459.67$ convert API standard temp to °R

$T_{st} = 519.67$ (°R)

$P_{at} := p_{at} \cdot 144$ convert actual atm pressure to lb/ft^2

$P_{at} = 1826.64$ (lb/ft^2 abs)

$T_{at} := t_{at} + 459.67$ convert actual atm temp to ^oR

Borehole geometry

$H_c := 7000.0$ total depth of cased hole (ft)

$H_{oh} := 3000.0$ total depth of open hole (ft)

$ROP := 60$ estimated rate of penetration (ft/hr)

$d_{bh} := 7.875$ borehole (drill bit) diameter (in)

$od_{dc} := 6.25$ outer diameter in drill collar (in)

$id_{dc} := 2.8125$ inner diameter in drill collar (in)

$od_{dp} := 4.5$ outer diameter in drill pipe (in)

$id_{dp} := 3.826$ inner diameter in drill pipe (in)

$od_{tj} := 6.25$ outer diameter in drill pipe tool joint (in)

$id_{tj} := 2.75$ inside diameter in drill pipe tool joint (in)

$od_c := 8.625$ outer diameter in drill casing (in)

$id_c := 7.921$ inside diameter in drill casing (in)

$od_{bl} := 8.625$ outer diameter in blooey line (in)

$id_{bl} := 8.097$ inside diameter in blooey line (in)

$L_{bl} := 200$ length of blooey line (ft)

$e_{cs} := 0.0005$ roughness of commercial steel (ft)

$e_r := 0.01$	roughness of rock table (ft)
$L_{dc} := 500$	total length of drill collars (ft)
$L_{tj} := 1.5$	average length of one tool joint (ft)
$L_{dp} := 30$	average length of one drill pipe section (ft)
$d_n := \frac{10}{32}$	this diameter is selected to obtain 200 rpm in the PDM (in)
$N_n := 3$	number of nozzles in the bit
$\beta := 0.01$	temperature gradient (°F/ft)

Section height calculations

$H_1 := H_c - L_{tj} \cdot \left(\frac{H_c}{L_{dp}}\right)$	calculate length of drill pipe in cased section
$H_1 = 6650$	(ft)
$H_2 := L_{tj} \cdot \left(\frac{H_c}{L_{dp}}\right)$	calculate length of tool joint in cased section
$H_2 = 350$	(ft)
$H_3 := H_{oh} - L_{dc} - L_{tj} \cdot \left(\frac{H_{oh} - L_{dc}}{L_{dp}}\right)$	calculate length of drill pipe in open hole section
$H_3 = 2375$	(ft)
$H_4 := L_{tj} \cdot \left(\frac{H_{oh} - L_{dc}}{L_{dp}}\right)$	calculate length of tool joints in open hole section
$H_4 = 125$	(ft)
$H_5 := L_{dc}$	length of drill collar
$H_5 = 500$	(ft)

Formation natural gas and water influx

$q_w := 0.5$	injected "mist" water (bbl/hr)
$q_{ng} := 0.0$	natural gas influx (scfm)
$q_{fw} := 0.00$	formation water influx (bbl/hr)

Constants

$SG_a := 1.0$	specific gravity of air
$SG_w := 1.0$	specific gravity of fresh water
$SG_{ng} := 0.70$	specific gravity of natural gas
$SG_{fw} := 1.07$	specific gravity of formation water
$SG_r := 2.7$	specific gravity of rock
$SG_{df} := 0.8156$	specific gravity of diesel fuel
$k := 1.4$	gas ratio of specific heats
$R := 53.36$	engineering gas constant (ft-lb/lb-°R)
$\gamma_w := 62.4$	specific weight of fresh water (lb/ft³)
$\gamma_{wg} := 8.33$	specific weight of water (lb/gal)
$g := 32.2$	gravitational constant [ft/sec^2]

Preliminary calculations

$D_{bh} := \frac{d_{bh}}{12}$	convert borehole diameter to ft

$D_{bh} = 0.6563$ (ft)

$rop := \frac{ROP}{60 \cdot 60}$ convert ROP to ft/sec

$rop = 0.0167$ (ft/sec)

$w_s := \left[\left(\frac{\pi}{4}\right) \cdot D_{bh}^{\ 2} \cdot 62.4 \cdot SG_r \cdot rop\right]$ weight rate of solid flow out of blooey line (lb/sec)

$w_s = 0.9498$ (lb/sec)

$q_a := q_c \cdot num_c$ air flow rate from all compressors (acfm)

$Q_a := \frac{q_a}{60}$ convert air flow rate to ft^3/ sec

$Q_a = 40$ (ft^3/sec)

$Q_w := \frac{(q_w \cdot 42 \cdot 231)}{60 \cdot 60 \cdot 12^3}$ convert "mist" water flow to ft^3/sec

$Q_w = 0.0008$ (ft^3/sec)

$Q_{ng} := \frac{q_{ng}}{60}$ convert natural gas flow rate to ft^3/sec

$Q_{ng} = 0$ (ft^3/sec)

$Q_{gt} := Q_a + Q_{ng}$ total flow of the gas (ft^3/sec)

$Q_{gt} = 40$ (ft^3/sec)

$Q_{fw} := \frac{(q_{fw} \cdot 42 \cdot 231)}{60 \cdot 60 \cdot 12^3}$ convert formation water flow to ft^3/sec

$Q_{fw} = 0$ (ft^3/sec)

$\gamma_a := \frac{(P_{at} \cdot SG_a)}{R \cdot T_{at}}$ specific weight of air (lb/ft^3)

$\gamma_a = 0.0679$ (lb/ft^3)

$\gamma_{ng} := \frac{(P_{at} \cdot SG_{ng})}{R \cdot T_{at}}$ specific weight of natural gas (lb/ft^3)

$\gamma_{ng} = 0.0475$ (lb/ft^3)

$w_a := \gamma_a \cdot Q_a$ weight rate of flow of air (lb/sec)

$w_a = 2.7146$ (lb/sec)

$w_w := SG_w \cdot \gamma_w \cdot Q_w$ weight rate of flow of "mist" water (lb/sec)

$w_w = 0.0487$ (lb/sec)

$w_{ng} := \gamma_{ng} \cdot Q_{ng}$ weight rate of flow of natural gas (lb/sec)

$w_{ng} = 0$ (lb/sec)

$w_{fw} := SG_{fw} \cdot \gamma_w \cdot Q_{fw}$ weight rate of flow of formation water (lb/sec)

$w_{fw} = 0$ (lb/sec)

$w_{gt} := w_a + w_{ng}$ total weight rate of flow of gas (lb/sec)

$w_{gt} = 2.7146$ (lb/sec)

Blooey line

$ID := \frac{id_{bl}}{12}$ convert blooey line inner diameter to ft

$ID = 0.6747$ (ft)

$A := \frac{\pi}{4} \cdot ID^2$ cross sectional area (ft²)

$A = 0.3576$ (ft²)

$f := \left(\frac{1}{2 \cdot \log\left(\frac{ID}{e_{cs}}\right) + 1.14} \right)^2$ friction factor using von Karman correlation

$f = 0.0183$

$\frac{id_c}{od_{dp}} = 1.7602$ compute D1/D3 ratio for Fig. 8-5

$\frac{id_c}{id_{bl}} = 0.9783$ compute D1/D2 ratio for Fig. 8-5

Figure 8.5 gives the following flow resistance coefficient for the blind Tee

$K_t := 25$ Tee resistance coefficient

$K_v := 0.2$ gate valve resistance coefficient

$P_b := \left[\left[\left[\left(f \cdot \frac{L_{bl}}{ID} \right) + K_t + 2 \cdot K_v \right] \cdot \left[\frac{\left(w_{gt}^{\,2} \cdot R \cdot T_{at} \right)}{g \cdot A^2} \right] + P_{at}^{\,2} \right]^{0.5} \right]$ calculate pressure of the gas at the entrance to the blooey line

$P_b = 2195.6749$ (lb/ft²)

$p_b := \frac{P_b}{144}$ convert to lb/in²

$p_b = 15.2477$ (psia)

Annulus cased hole drill pipe body section

$H := H_1$ (ft)

$ID := \frac{id_c}{12}$ convert inner diameter of casing to ft

$ID = 0.6601$ (ft)

$OD := \frac{od_{dp}}{12}$ convert outer diameter of drill pipe to ft

$OD = 0.375$ (ft)

$A := \frac{\pi}{4}\cdot\left(ID^2 - OD^2\right)$ annulus area between casing and drill pipe (ft^2)

$A = 0.2318$ (ft^2)

$a_a := \frac{SG_a}{R}\cdot\left[1 + \frac{\left(w_s + w_w + w_{fw}\right)}{w_{gt}}\right]$ calculate constant

$a_a = 0.0256$

$f := \left[\frac{1}{2\cdot\log\left[\frac{(ID - OD)}{e_{cs}}\right] + 1.14}\right]^2$ friction factor using von Karman correlation

$f = 0.0226$

$b_a := \left[\left[\frac{f}{2\cdot g\cdot(ID - OD)}\right]\cdot\left(\frac{R}{SG_a}\right)^2\cdot\frac{w_{gt}^{\ 2}}{\left(\frac{\pi}{4}\right)^2\cdot\left(ID^2 - OD^2\right)^2}\right]$ calculate constant

$b_a = 480.8635$

$T := T_{at} + \beta\cdot H_1$ temperature at bottom of section (°R)

$T = 570.91$ (°R)

$T_{ave1} := \frac{(T_{at} + T)}{2}$ average temperature of section (°R)

$T_{ave1} = 537.66$ (°R)

$T_{ave} := T_{ave1}$

$$P_1 := \left[\left[\left(P_b^{\ 2} + b_a \cdot T_{ave}^{\ 2}\right) \cdot e^{\left[\frac{(2 \cdot a_a \cdot H)}{T_{ave}}\right]}\right] - b_a \cdot T_{ave}^{\ 2}\right]^{0.5}$$ calculate pressure

$P_1 = 11495.7701$ (lb/ft²)

$p_1 := \frac{P_1}{144}$ convert to psia

$p_1 = 79.8317$ (psia)

$\gamma := \frac{P_1 \cdot SG_a}{R \cdot T}$ specific weight of gas (lb/ft³)

$\gamma = 0.3774$ (lb/ft³)

$V := \frac{w_{gt}}{\gamma \cdot A}$ velocity (ft/sec)

$V = 31.04$ (ft/sec)

$KE_1 := \frac{1}{2} \cdot \frac{\gamma}{g} \cdot V^2$ kinetic energy of gas flow (lb-ft/ft³)

$KE_1 = 5.6456$ (lb-ft/ft³)

Annulus cased hole tool joint section

$H := H_2$

$ID := \frac{id_c}{12}$ convert inner diameter of casing to ft

$ID = 0.6601$ (ft)

$OD := \frac{od_{tj}}{12}$ convert outer diameter of tool joint to ft

$OD = 0.5208$ (ft)

$OD_t := \frac{od_{dp}}{12}$ use drill pipe OD for kinetic energy to ft

$OD_t = 0.375$ (ft)

$A := \frac{\pi}{4}\left(ID^2 - OD_t^{\,2}\right)$ annulus area for tool joint KE (ft^2)

$A = 0.2318$ (ft^2)

$a_a := \frac{SG_a}{R}\cdot\left[1 + \frac{\left(w_s + w_w + w_{fw}\right)}{w_{gt}}\right]$ calculate constant

$a_a = 0.0256$

$f := \left[\frac{1}{2\cdot\log\left[\frac{(ID - OD)}{e_{cs}}\right] + 1.14}\right]^2$ friction factor using von Karman correlation

$f = 0.0275$

$b_a := \left[\left[\frac{f}{2\cdot g\cdot(ID - OD)}\right]\cdot\left(\frac{R}{SG_a}\right)^2\cdot\frac{w_{gt}^{\,2}}{\left(\frac{\pi}{4}\right)^2\cdot\left(ID^2 - OD^2\right)^2}\right]$ calculate constant

$b_a = 3858.1981$

$T_{temp} := T$

$T := T_{temp} + \beta \cdot H_2$ temperature at bottom of section (°R)

$T = 574.41$ (°R)

$T_{ave2} := \frac{(T_{temp} + T)}{2}$ average temperature of section (°R)

$T_{ave2} = 572.66$ (°R)

$T_{ave} := T_{ave2}$

$$P_2 := \left[\left[\left(P_1^{\ 2} + b_a \cdot T_{ave}^{\ 2}\right) \cdot e^{\left[\frac{(2 \cdot a_a \cdot H)}{T_{ave}}\right]}\right] - b_a \cdot T_{ave}^{\ 2}\right]^{0.5}$$

calculate pressure

$P_2 = 13290.2784$ (lb/ft²)

$p_2 := \frac{P_2}{144}$ convert to psia

$p_2 = 92.2936$ (psia)

$\gamma := \frac{P_2 \cdot SG_a}{R \cdot T}$ specific weight of gas (lb/ft³)

$\gamma = 0.4336$ (lb/ft³)

$V := \frac{w_{gt}}{\gamma \cdot A}$ velocity of gas (ft/sec)

$V = 27.0135$ (ft/sec)

$KE_2 := \frac{1}{2} \cdot \frac{\gamma}{g} \cdot V^2$ kinetic energy of gas flow (lb-ft/ft³)

$KE_2 = 4.9133$ (lb-ft/ft^3)

Annulus of open hole drill pipe section

$H := H_3$

$ID := D_{bh}$ open hole diameter of borehole (ft)

$ID = 0.6563$ (ft)

$OD := \frac{od_{dp}}{12}$ convert diameter of drill pipe to ft

$OD = 0.375$ (ft)

$A := \frac{\pi}{4}\left(ID^2 - OD^2\right)$ area between borehole and drill pipe (ft^2)

$A = 0.2278$ (ft^2)

$a_a := \frac{SG_a}{R}\cdot\left[1 + \frac{\left(w_s + w_w + w_{fw}\right)}{w_{gt}}\right]$ calculate constant

$a_a = 0.0256$

$e_{ave} := \frac{\left(e_r ID^2 + e_{cs} OD^2\right)}{ID^2 + OD^2}$ weighted average roughness (ft)

$e_{ave} = 0.0077$ (ft)

$f := \left[\frac{1}{2\cdot\log\left[\frac{(ID - OD)}{e_{ave}}\right] + 1.14}\right]^2$ friction factor using von Karman correlation

$f = 0.0549$

$$b_a := \left[\left[\frac{f}{2\cdot g\cdot(ID - OD)}\right]\cdot\left(\frac{R}{SG_a}\right)^2\cdot\frac{w_{gt}^{\;2}}{\left(\frac{\pi}{4}\right)^2\cdot\left(ID^2 - OD^2\right)^2}\right]$$ calculate constant

$b_a = 1224.6798$

$T_{temp} := T$

$T := T_{temp} + \beta\cdot H_3$ temperature at bottom of section (°R)

$T = 598.16$

$$T_{ave3} := \frac{\left(T_{temp} + T\right)}{2}$$ average temperature of section (°R)

$T_{ave3} = 586.285$ (°R)

$T_{ave} := T_{ave3}$

$$P_3 := \left[\left[\left(P_2^{\;2} + b_a\cdot T_{ave}^{\;2}\right)\cdot e^{\left[\frac{(2\cdot a_a\cdot H)}{T_{ave}}\right]}\right] - b_a\cdot T_{ave}^{\;2}\right]^{0.5}$$ calculate pressure

$P_3 = 17736.0029$ (lb/ft²)

$$p_3 := \frac{P_3}{144}$$ convert to psia

$p_3 = 123.1667$ (psia)

$\gamma := \frac{P_3 \cdot SG_a}{R \cdot T}$ specific weight of gas (lb/ft^3)

$\gamma = 0.5557$ (lb/ft^3)

$V := \frac{w_{gt}}{\gamma \cdot A}$ velocity of gas (ft/sec)

$V = 21.4459$ (ft/sec)

$KE_3 := \frac{1}{2} \cdot \frac{\gamma}{g} \cdot V^2$ kinetic energy of gas flow (lb-ft/ft^3)

$KE_3 = 3.9685$ (lb-ft/ft^3)

Annulus open hole tool joint section

$H := H_4$

$ID := D_{bh}$ open hole diameter of borehole (ft)

$ID = 0.6563$ (ft)

$OD := \frac{od_{tj}}{12}$ convert tool joint diameter to ft

$OD = 0.5208$ (ft)

$OD_t := \frac{od_{dp}}{12}$ drill pipe diameter (ft)

$OD_t = 0.375$ (ft)

$A := \frac{\pi}{4}\left(ID^2 - OD_t^2\right)$ annulus area for tool joint KE (ft^2)

$A = 0.2278$ (ft^2)

$$a_a := \frac{SG_a}{R} \cdot \left[1 + \frac{(w_s + w_w + w_{fw})}{w_{gt}}\right]$$ calculate constant

$a_a = 0.0256$

$$e_{ave} := \frac{\left(e_r ID^2 + e_{cs} OD^2\right)}{ID^2 + OD^2}$$ weighted average roughness (ft)

$e_{ave} = 0.0063$ (ft)

$$f := \left[\frac{1}{2 \cdot \log\left[\frac{(ID - OD)}{e_{ave}}\right] + 1.14}\right]^2$$ friction factor using von Karman correlation

$f = 0.0692$

$$b_a := \left[\left[\frac{f}{2 \cdot g \cdot (ID - OD)}\right] \cdot \left(\frac{R}{SG_a}\right)^2 \cdot \frac{w_{gt}^{\,2}}{\left(\frac{\pi}{4}\right)^2 \cdot \left(ID^2 - OD^2\right)^2}\right]$$ calculate constant

$b_a = 10627.4659$

$T_{temp} := T$

$T := T_{temp} + \beta \cdot H_4$ temperature at bottom of section (°R)

$T = 599.41$

$$T_{ave4} := \frac{(T_{temp} + T)}{2}$$ average temperature of section (°R)

$T_{ave4} = 598.785$ (°R)

$T_{ave} := T_{ave4}$

$$P_4 := \left[\left[\left(P_3^{\,2} + b_a \cdot T_{ave}^{\,2}\right) \cdot e^{\left[\frac{(2 \cdot a_a \cdot H)}{T_{ave}}\right]}\right] - b_a \cdot T_{ave}^{\,2}\right]^{0.5}$$ calculate pressure

$P_4 = 18945.9611$ (lb/ft²)

$$p_4 := \frac{P_4}{144}$$ convert to psia

$p_4 = 131.5692$ (psia)

$$\gamma := \frac{P_4 \cdot SG_a}{R \cdot T}$$ specific weight of gas (lb/ft³)

$\gamma = 0.5923$ (lb/ft³)

$$V := \frac{w_{gt}}{\gamma \cdot A}$$ velocity of gas (ft/sec)

$V = 20.1183$ (ft/sec)

$$KE_4 := \frac{1}{2} \cdot \frac{\gamma}{g} \cdot V^2$$ kinetic energy of gas flow (lb-ft/ft³)

$KE_4 = 3.7228$ (ft-lb/ft³)

Annulus open hole drill collar section

$H := H_5$

$ID := D_{bh}$ open hole diameter of borehole (ft)

$ID = 0.6563$ (ft)

$$OD := \frac{od_{dc}}{12}$$ convert drill collar diameter to ft

$OD = 0.5208$ (ft)

$$A := \frac{\pi}{4}\left(ID^2 - OD^2\right)$$ drill pipe diameter (ft)

$A = 0.1252$ (ft)

$$a_a := \frac{SG_a}{R}\cdot\left[1 + \frac{\left(w_s + w_w + w_{fw}\right)}{w_{gt}}\right]$$ calculate constant

$a_a = 0.0256$

$$e_{ave} := \frac{\left(e_r ID^2 + e_{cs} OD^2\right)}{ID^2 + OD^2}$$ weighted average roughness (ft)

$e_{ave} = 0.0063$ (ft)

$$f := \left[\frac{1}{2\cdot\log\left[\frac{(ID - OD)}{e_{ave}}\right] + 1.14}\right]^2$$ friction factor using von Karman correlation

$f = 0.0692$

$$b_a := \left[\left[\frac{f}{2\cdot g\cdot(ID - OD)}\right]\cdot\left(\frac{R}{SG_a}\right)^2\cdot\frac{w_{gt}^2}{\left(\frac{\pi}{4}\right)^2\cdot\left(ID^2 - OD^2\right)^2}\right]$$ calculate constant

$b_a = 10627.4659$

$T_{temp} := T$

$T := T_{temp} + \beta\cdot H_5$ temperature at bottom of section (°R)

$T = 604.41$

$T_{ave5} := \frac{(T_{temp} + T)}{2}$ average temperature of section (°R)

$T_{ave5} = 601.91$ (°R)

$T_{ave} := T_{ave5}$

$$P_5 := \left[\left[\left(P_4{}^2 + b_a \cdot T_{ave}{}^2\right) \cdot e^{\left[\frac{(2 \cdot a_a \cdot H)}{T_{ave}}\right]} \right] - b_a \cdot T_{ave}{}^2 \right]^{0.5}$$

$P_5 = 23282.5904$ (lb/ft²)

$P_{bh} := P_5$ annulus bottom hole pressure (lb/ft²)

$p_5 := \frac{P_5}{144}$ convert to psia

$p_5 = 161.6847$ (psia)

$p_{bh} := p_5$ annulus bottom hole pressure (psia)

$\gamma := \frac{P_5 \cdot SG_a}{R \cdot T}$ specific weight of gas (lb/ft³)

$\gamma = 0.7219$ (lb/ft³)

$V := \frac{w_{gt}}{\gamma \cdot A}$ velocity of gas (ft/sec)

$V = 30.0373$ (ft/sec)

$KE_5 := \frac{1}{2} \cdot \frac{\gamma}{g} \cdot V^2$ kinetic energy of gas flow (lb-ft/ft³)

$KE_5 = 10.1139$ (ft-lb/ft³)

Drill bit nozzles

$d_n := \frac{10}{32}$ trial and error nozzle diameter (32nds of an inch)

$d_n = 0.3125$ nozzle diameter (in)

$D_n := \frac{d_n}{12}$ convert nozzle diameter to ft

$D_n = 0.026$ (ft)

$A_n := N_n \cdot \left(\frac{\pi}{4}\right) \cdot D_n^2$ summed area of all nozzles in bit (ft^2)

$A_n = 0.00160$ (ft^2)

$$P := \frac{w_{gt} \cdot T^{0.5}}{A_n \cdot \left[\left(\frac{g \cdot k \cdot SG_a}{R}\right) \cdot \left(\frac{2}{k+1}\right)^{\left(\frac{k+1}{k-1}\right)}\right]^{0.5}}$$ calculate pressure

$P = 78521.805$ (lb/ft^2)

$P_{sonic} := \frac{P_{bh}}{0.528}$ verify sonic flow

$P_{sonic} = 44095.8152$ (lb/ft^2) - flow is sonic since $P > P_{sonic}$

$p := \frac{P}{144}$ convert to lb/in^2

$p = 545.2903$ (psia)

Positive displacement motor performance

$q_{mm} := 400$ graph mud flow rate through PDM (gpm)

$N_{mm} := 200$ graph PDM (rpm)

$s := \frac{231.016 \cdot q_{mm}}{N_{mm}}$ calculate specific displacement of PDM

$s = 462.032$ (in^3)

$p_{bm} := p$ pressure at bottom of motor (psia)

$p_{bm} = 545.2903$ (psia)

From mud motor graph pressure drop across motor for 25 horsepower output is 220 psi.

$\Delta p_{mm} := 220$ pressure drop (psi)

$p_{am} := p_{bm} + \Delta p_{mm}$ pressure above the motor

$p_{am} = 765.2903$ (psia)

$P_{am} := p_{am} \cdot 144$ convert to consistent units (lb/ft^2)

$P_{am} = 110201.805$ (lb/ft^2)

$p_{mav} := \frac{p_{bm} + p_{am}}{2}$ average pressure in the motor (psia)

$p_{mav} = 655.2903$ (psia)

$P_{mav} := p_{mav} \cdot 144$ convert to consistent units

$P_{mav} = 94361.805$ (lb/ft^2)

$\gamma := \frac{P_{mav} \cdot SG_a}{R \cdot T}$ specific weight of air in motor cavity

$\gamma = 2.9258$ (lb/ft^3)

$Q_{mav} := \frac{w_{gt}}{\gamma}$ average volumetric flow rate of air through motor

$Q_{mav} = 0.9278$ (ft^3/sec)

$q_{mav} := \frac{Q_{mav} \cdot 12^3 \cdot 60}{231.016}$ convert to gallons per minute (gpm)

$q_{mav} = 416.4066$ (gpm)

$N_m := \frac{231.016 \cdot q_{mav}}{s}$ bit speed of motor when operating on air (rpm)

$N_m = 208.2$ verified that selected nozzle diameter is correct to control PDM bit speed at required ~200 rpm "close enough"

$P := P_{am}$ set pressure for inside drill string (lb/ft^2)

$p := \frac{P}{144}$ convert to lb/ft^2

$p = 765.2903$ (psia)

Open hole inside drill string drill collar section

$D := \frac{id_{dc}}{12}$ convert diameter to ft

$D = 0.2344$ (ft)

$a_i := \left(\frac{SG_a}{R}\right) \cdot \left(1 + \frac{w_w}{w_{gt}}\right)$ calculate constant

$a_i = 0.0191$

$f := \left(\frac{1}{2 \cdot \log\left(\frac{D}{e_{cs}}\right) + 1.14}\right)^2$ friction factor using von Karman correlation

$f = 0.0238$

$$b_i := \left(\frac{f}{2\cdot g\cdot D}\right)\cdot\left(\frac{R}{SG_a}\right)^2\cdot\frac{w_{gt}^{\,2}}{\left(\frac{\pi}{4}\right)^2\cdot D^4}$$ calculate constant

$b_i = 17776.0049$

$$P := \left[\frac{\left[P^2 + b_i\cdot T_{ave5}^{\,2}\cdot\left[e^{\frac{(2\cdot a_i\cdot H_5)}{T_{ave5}}} - 1\right]\right]}{e^{\frac{(2\cdot a_i\cdot H_5)}{T_{ave5}}}}\right]^{(0.5)}$$ calculate pressure

$P = 109391.4312$ (lb/ft^2)

$$p := \frac{P}{144}$$ convert to psia

$p = 759.6627$ (psia)

Open hole inside drill string tool joint section

$$D := \frac{id_{tj}}{12}$$ convert diameter to ft

$D = 0.2292$ (ft)

$$a_i := \left(\frac{SG_a}{R}\right)\cdot\left(1 + \frac{w_w}{w_{gt}}\right)$$ calculate constant

$a_i = 0.0191$

$$f := \left(\frac{1}{2\cdot\log\left(\frac{D}{e_{cs}}\right) + 1.14}\right)^2$$ friction factor using von Karman correlation

$f = 0.0239$

$$b_i := \left(\frac{f}{2\cdot g\cdot D}\right)\cdot\left(\frac{R}{SG_a}\right)^2\cdot\frac{w_{gt}^{\,2}}{\left(\frac{\pi}{4}\right)^2\cdot D^4}$$ calculate constant

$b_i = 20010.2715$

$$P := \left[\frac{\left[P^2 + b_i\cdot T_{ave4}^{\,2}\cdot\left[e^{\frac{(2\cdot a_i\cdot H_4)}{T_{ave4}}} - 1\right]\right]}{e^{\frac{(2\cdot a_i\cdot H_4)}{T_{ave4}}}}\right]^{(0.5)}$$ calculate pressure

$P = 109217.5377$ (lb/ft^2)

$$p := \frac{P}{144}$$ convert to psia

$p = 758.4551$ (psia)

Open hole inside drill string drill pipe body section

$$D := \frac{id_{dp}}{12}$$ convert diameter to ft

$D = 0.3188$ (ft)

$$a_i := \left(\frac{SG_a}{R}\right)\cdot\left(1 + \frac{w_w}{w_{gt}}\right)$$ calculate constant

$a_i = 0.0191$

$$f := \left(\frac{1}{2 \cdot \log\left(\frac{D}{e_{cs}}\right) + 1.14} \right)^2$$ friction factor using von Karman correlation

$f = 0.022$

$$b_i := \left(\frac{f}{2 \cdot g \cdot D}\right) \cdot \left(\frac{R}{SG_a}\right)^2 \cdot \frac{w_{gt}^{\,2}}{\left(\frac{\pi}{4}\right)^2 \cdot D^4}$$ calculate constant

$b_i = 3519.4219$

$$P := \left[\frac{\left[P^2 + b_i \cdot T_{ave3}^{\,2} \cdot \left[e^{\frac{(2 \cdot a_i \cdot H_3)}{T_{ave3}}} - 1 \right] \right]}{e^{\frac{(2 \cdot a_i \cdot H_3)}{T_{ave3}}}} \right]^{(0.5)}$$ calculate pressure

$P = 101948.5294$ (lb/ft2)

$$p := \frac{P}{144}$$ convert to psia

$p = 707.9759$ (psia)

Cased hole inside drill string tool joint section

$$D := \frac{id_{tj}}{12}$$ convert diameter to ft

$D = 0.2292$ (ft)

$$a_i := \left(\frac{SG_a}{R}\right) \cdot \left(1 + \frac{w_w}{w_{gt}}\right)$$ calculate constant

$a_i = 0.0191$

$$f := \left(\frac{1}{2\cdot\log\left(\frac{D}{e_{cs}}\right) + 1.14}\right)^2$$ friction factor using von Karman correlation

$f = 0.0239$

$$b_i := \left(\frac{f}{2\cdot g\cdot D}\right)\cdot\left(\frac{R}{SG_a}\right)^2\cdot\frac{w_{gt}^{\,2}}{\left(\frac{\pi}{4}\right)^2\cdot D^4}$$ calculate constant

$b_i = 20010.2715$

$$P := \left[\frac{\left[P^2 + b_i\cdot T_{ave2}^{\,2}\cdot\left[e^{\frac{(2\cdot a_i\cdot H_2)}{T_{ave2}}} - 1\right]\right]}{e^{\frac{(2\cdot a_i\cdot H_2)}{T_{ave2}}}}\right]^{(0.5)}$$ calculate pressure

$P = 101514.5061$ (lbs/ft^2)

$$p := \frac{P}{144}$$ convert to psia

$p = 704.9618$ (psia)

Cased hole inside drill string pipe body section

$$D := \frac{id_{dp}}{12}$$ convert diameter to ft

$D = 0.3188$ (ft)

$$a_i := \left(\frac{SG_a}{R}\right)\cdot\left(1 + \frac{w_w}{w_{gt}}\right)$$ calculate constant

$a_i = 0.0191$

$$f := \left(\frac{1}{2 \cdot \log\left(\frac{D}{e_{cs}} \right) + 1.14} \right)^2$$

friction factor using von Karman correlation

$f = 0.022$

$$b_i := \left(\frac{f}{2 \cdot g \cdot D} \right) \cdot \left(\frac{R}{SG_a} \right)^2 \cdot \frac{w_{gt}^2}{\left(\frac{\pi}{4} \right)^2 \cdot D^4}$$

calculate constant

$b_i = 3519.4219$

$$P := \left[\frac{\left[P^2 + b_i \cdot T_{ave1}^2 \cdot \left[e^{\frac{(2 \cdot a_i \cdot H_1)}{T_{ave1}}} - 1 \right] \right]}{e^{\frac{(2 \cdot a_i \cdot H_1)}{T_{ave1}}}} \right]^{0.5}$$

calculate pressure

$P = 82530.6518$ (lb/ft^2)

$$p := \frac{P}{144}$$

convert to psia

$p = 573.1295$ (psia)

$p_{in} := p$

$p_{in} = 573.1295$ drilling string injection pressure (psia)

Kinetic energy summary of each section (KE greater than 3.0 lb-ft/ft^3)

$KE_1 = 5.6456$ (lb-ft/ft^3)

$KE_2 = 4.9133$ (lb-ft/ft^3)

$KE_3 = 3.9685$ (lb-ft/ft^3)

$KE_4 = 3.7228$ (lb-ft/ft^3)

$KE_5 = 10.1139$ (lb-ft/ft^3)

Important design parameter summary

$q_a = 2400$ air compressor flow rate (acfm)

$p_{in} = 573.1295$ required surface air injection pressure (psia)

$p_{bh} = 161.6847$ bottom hole pressure (psia)

Illustrative Example 11.1: Hammer Performance (SI)

Hammer No. 1

1) Basic input data

$l_s := 180.34$	length of piston chamber (mm)
$e_s := 0.8$	stroke approximate ratio of length
$L_s := \frac{l_s}{1000}$	convert chamber length to m
$L_s = 0.18$	(m)
$d_p := 149.86$	diameter of piston (mm)
$d_b := 200.03$	diameter of drill bit (mm)
$A_p := \left(\frac{\pi}{4}\right)\cdot\left(\frac{d_p}{1000}\right)^2$	cross-sectional area of piston (m^2)
$A_p = 0.018$	(m^2)
$A_b := \left(\frac{\pi}{4}\right)\cdot\left(\frac{d_b}{1000}\right)^2$	cross-sectional area of drill bit (m^2)
$A_b = 0.031$	(m^2)
$W_p := 431.5$	weight of piston (N)
$g := 9.81$	acceleration of gravity (m/sec^2)
$\gamma_{st} := 11.976$	specific weight of standard air (N/m^3)
$\gamma_{at} := 10.652$	specific weight of 1219 m atmos air (N/m^3
$q_{cat} := 1131.6$	compressor 1219 m atmos air (actual liter/sec)
$q_{cst} := q_{cat}\cdot\left(\frac{\gamma_{at}}{\gamma_{st}}\right)$	equivalent standard air (stand liters/sec)

$q_{cst} = 1.006 \times 10^3$	(stand liters/sec)
$Q_{cst} := \frac{q_{cst}}{1000}$	convert to m^3/sec
$Q_{cst} = 1.006$	(stand m^3/sec)
$H := 3047.9$	depth of drilling (m)
$p_{at} := 8.749$	atmos pressure at 1219 m (N/cm^2 abs)
$P_{at} := p_{at} \cdot 10000.0$	convert to N/m^2 abs
$P_{at} = 8.749 \times 10^4$	N/m^2 abs
$t_{at} := 7.078$	atmos temperature at 1219 m (°C)
$T_{at} := t_{at} + 273.15$	convert to absolute temperature (K)
$T_{at} = 280.228$	(K)
$\beta := 0.01823$	geothermal gradient (°C/m)
$T_h := T_{at} + \beta \cdot H$	absolute temperature at depth (K)
$T_h = 335.791$	(K)
$p_{anbh} := 111.52$	bottom hole annulus pressure (N/cm^2 abs)
$\Delta p_{DTH} := 234.50$	pressure drop through DTH (N/cm^2)
$p_{inaDTH} := \Delta p_{DTH} + p_{anbh}$	pressure above DTH (N/cm^2 abs)
$p_{inaDTH} = 346.02$	(N/cm^2 abs)
$P_{inaDTH} := p_{inaDTH} \cdot 10000.0$	convert to N/m^2 abs

$P_{inaDTH} = 3.46 \times 10^{6}$ (N/m^2 abs)

$R := 29.31$ engineering gas constant (N-m/N-K)

$S_a := 1.0$ specific gravity of air

2) Pressure and weight rate of flow

$w_a := \gamma_{st} \cdot Q_{cst}$ weight rate of flow of air (N/sec)

$w_a = 12.054$ (N/sec)

$\gamma_{inaDTH} := \dfrac{P_{inaDTH} \cdot S_a}{R \cdot T_h}$ specific weight of air entering (N/m^3)

$\gamma_{inaDTH} = 351.573$ (N/m^3)

$Q_{inaDTH} := \dfrac{w_a}{\gamma_{inaDTH}}$ volumetric flow through DTH (m^3/sec)

$Q_{inaDTH} = 0.034$ (m^3/sec)

3) Piston motion and impact velocity

$Vol_{DTH} := e_s \cdot L_s \cdot A_p$ volume of DTH per stroke (m^3)

$Vol_{DTH} = 2.545 \times 10^{-3}$ (m^3)

$N_{ps} := \dfrac{Q_{inaDTH} \cdot 60}{2Vol_{DTH}}$ number of strokes per minute (spm)

$N_{ps} = 404.19$ (spm)

$\Delta\tau_{ts} := \dfrac{1}{\left(\dfrac{N_{ps}}{60}\right)}$ period of stroke (sec)

$\Delta\tau_{ts} = 0.148$ (sec)

$$V_p := \left[2\cdot\left[\frac{P_{inaDTH}\cdot A_p + W_p}{\left(\frac{W_p}{g}\right)}\right]\cdot(e_s\cdot L_s)\right]^{0.5}$$ velocity of piston at shank (m/sec)

$V_p = 20.08$ (m/sec)

$$\Delta\tau_{dn} := \frac{2\cdot e_s\cdot L_s}{V_p}$$ time of down stroke (sec)

$\Delta\tau_{dn} = 0.014$ (sec)

4) Kinetic energy terms

$$KE_{ps} := \left(\frac{1}{2}\right)\cdot\left(\frac{W_p}{g}\right)\cdot V_p^{\,2}$$ piston KE at shank (N-m)

$KE_{ps} = 8.868 \times 10^3$ (N-m)

$$ke_b := \frac{KE_{ps}}{A_b}$$ KE per drill bit area (N-m/m²)

$ke_b = 2.822 \times 10^5$ (N-m/m²)

$KE_{total} := KE_{ps}\cdot N_{ps}$ total KE to rock per minute (N-m/min)

$KE_{total} = 3.584 \times 10^6$ (N-m/min)

Hammer No. 2

1) Basic input data

$l_s := 119.065$ length of piston chamber (mm)

$e_s := 0.80$ stroke approximate ratio of length

$L_s := \frac{l_s}{1000}$	convert chamber length to m
$L_s = 0.119$	(m)
$d_p := 152.4$	diameter of piston (mm)
$d_b := 200.03$	diameter of drill bit (mm)
$A_p := \left(\frac{\pi}{4}\right)\cdot\left(\frac{d_p}{1000}\right)^2$	cross-sectional area of piston (m^2)
$A_p = 0.018$	(m^2)
$A_b := \left(\frac{\pi}{4}\right)\cdot\left(\frac{d_b}{1000}\right)^2$	cross-sectional area of drill bit (m^2)
$A_b = 0.031$	(m^2)
$W_p := 346.983$	weight of piston (N)
$g := 9.81$	acceleration of gravity (m/sec^2)
$\gamma_{st} := 11.976$	specific weight of standard air (N/m^3)
$\gamma_{at} := 10.652$	specific weight of 1219 m atmos air (N/m^3)
$q_{cat} := 1131.60$	compressor 1219 m atmos air (actual liters/sec)
$q_{cst} := q_{cat}\cdot\left(\frac{\gamma_{at}}{\gamma_{st}}\right)$	equivalent standard air (stand liters/sec)
$q_{cst} = 1.006 \times 10^3$	(stand liters/sec)
$Q_{cst} := \frac{q_{cst}}{1000}$	convert to m^3/sec
$Q_{cst} = 1.006$	(m^3/sec)

$H := 3047.9$	depth of drilling (m)
$\beta := 0.01823$	geothermal gradient (°C/m)
$T_h := T_{at} + \beta \cdot H$	absolute temperature at depth (°C)
$T_h = 335.791$	(°C)
$p_{anbh} := 111.52$	bottom hole annulus pressure (N/cm² abs)
$\Delta p_{DTH} := 379.34$	pressure drop through DTH (N/cm² abs)
$p_{inaDTH} := \Delta p_{DTH} + p_{anbh}$	pressure above DTH (N/cm² abs)
$p_{inaDTH} = 490.86$	(N/cm² abs)
$P_{inaDTH} := p_{inaDTH} \cdot 10000.0$	convert to N/m² abs
$P_{inaDTH} = 4.909 \times 10^{6}$	(N/m² abs)
$R := 29.31$	engineering gas constant (N-m/N-K)
$S_a := 1.0$	specific gravity of air

2) Pressure and weight rate of flow

$w_a := \gamma_{st} \cdot Q_{cst}$	weight rate of flow of air (N/sec)
$w_a = 12.054$	(N/sec)
$\gamma_{inaDTH} := \dfrac{P_{inaDTH} \cdot S_a}{R \cdot T_h}$	specific weight of air entering (N/m³)
$\gamma_{inaDTH} = 498.738$	(N/m³)
$Q_{inaDTH} := \dfrac{w_a}{\gamma_{inaDTH}}$	volumetric flow through DTH (m³/sec)

$Q_{inaDTH} = 0.02417$ (m³/sec)

3) Piston motion and impact velocity

$Vol_{DTH} := e_s \cdot L_s \cdot A_p$ volume of DTH per stroke (m^3)

$Vol_{DTH} = 1.738 \times 10^{-3}$ (m^3)

$$N_{ps} := \frac{Q_{inaDTH} \cdot 60}{2 Vol_{DTH}}$$ number of strokes per minute (spm)

$N_{ps} = 417.291$ (spm)

$$\Delta\tau_{ts} := \frac{1}{\left(\frac{N_{ps}}{60}\right)}$$ period of stroke (sec)

$\Delta\tau_{ts} = 0.144$ (sec)

$$V_p := \left[2 \cdot \left[\frac{P_{inaDTH} \cdot A_p + W_p}{\left(\frac{W_p}{g}\right)}\right] \cdot (e_s \cdot L_s)\right]^{0.5}$$ velocity of piston at shank (m/sec)

$V_p = 22.003$ (ft/sec)

$$\Delta\tau_{dn} := \frac{2 \cdot e_s \cdot L_s}{V_p}$$ time of down stroke (sec)

$\Delta\tau_{dn} = 8.658 \times 10^{-3}$ (sec)

4) Kinetic energy terms

$$KE_{ps} := \left(\frac{1}{2}\right) \cdot \left(\frac{W_p}{g}\right) \cdot V_p^{\,2}$$ piston KE at shank (N-m)

$KE_{ps} = 8.562 \times 10^3$ (N-m)

$ke_b := \frac{KE_{ps}}{A_b}$ KE per drill bit area (N-m/m^2)

$ke_b = 2.725 \times 10^5$ (N-m/m^2)

$KE_{total} := KE_{ps} \cdot N_{ps}$ total KE to rock per minute (N-m/min)

$KE_{total} = 3.573 \times 10^6$ (N-m/min)

Illustrative Example 11.2 PDM (SI version) - Three Dresser Clark CFB-4 Compressors - Drilling Fluid is Air. Determining the bottom hole and injection pressures while drilling at 3047.9 m of depth for the drilling operations described in Illustrative Examples 8.1, 8.2, and 8.3.

Compressor data - Dresser Clark Model CFB-4 with Caterpillar Model D398 prime mover

$q_c := 566.254$	flow rate produced from one compressor (actual liters/sec)
$num_c := 2$	number of compressors (required for SF above min KE)
$\varepsilon_m := 0.8$	mechanical efficiency
$n_s := 4$	number of stages
$HP_{max} := 566.7$	max sea level power rating of compressor prime mover (kW)
$c := 0.02$	clearance volume ratio for compressor

Basic standard and elevation atomospheric conditions

$p_{st} := 10.136$	API standard atmospheric pressure (N/cm^2 abs)
$t_{st} := 15.6$	API standard atmospheric temperature (°C)
$alt := 1219$	surface elevation of drilling site (m)
$p_{at} := 8.749$	pressure from table for surface elevation (N/cm^2 abs)
$t_{at} := 7.078$	temperature from table for surface elevation (°C)
$P_{st} := p_{st} \cdot 10000.0$	convert API standard pressure to N/m^2
$P_{st} = 101360$	(N/m^2 abs)
$T_{st} := t_{st} + 273.15$	convert API standard temp to K
$T_{st} = 288.75$	(K)

$P_{at} := p_{at} \cdot 10000.0$ convert actual atm pressure to N/m^2

$P_{at} = 87490$ (N/m^2 abs)

$T_{at} := t_{at} + 273.15$ convert actual atm temp to K

$T_{at} = 280.228$ (K)

Borehole geometry

$H_c := 2133.5$ total depth of cased hole (m)

$H_{oh} := 914.36$ total depth of open hole (m)

$ROP := 18.29$ estimated rate of penetration (m/hr)

$d_{bh} := 200.03$ borehole (drill bit) diameter (mm)

$od_{dc} := 158.75$ outer diameter in drill collar (mm)

$id_{dc} := 71.438$ inner diameter in drill collar (mm)

$od_{dp} := 114.30$ outer diameter in drill pipe (mm)

$id_{dp} := 97.18$ inner diameter in drill pipe (mm)

$od_{tj} := 158.75$ outer diameter in drill pipe tool joint (mm)

$id_{tj} := 69.85$ inside diameter in drill pipe tool joint (mm)

$od_c := 219.08$ outer diameter in drill casing (mm)

$id_c := 201.19$ inside diameter in drill casing (mm)

$od_{bl} := 219.08$ outer diameter in blooey line (mm)

$id_{bl} := 205.66$ inside diameter in blooey line (mm)

$L_{bl} := 60.957$ length of blooey line (m)

$e_{cs} := 0.000152939$ roughness of commercial steel (m)

$e_r := 0.003048$	roughness of rock table (m)
$L_{dc} := 152.39$	total length of drill collars (m)
$L_{tj} := 0.457$	average length of one tool joint (m)
$L_{dp} := 9.144$	average length of one drill pipe section (m)
$d_n := 7.9375$	nozzle diameter is selected to obtain 200 rpm in the PDM (mm)
$N_n := 3$	number of nozzles in the bit
$\beta := 0.01823$	temperature gradient (°C/m)

Section height calculations

$H_1 := H_c - L_{tj} \cdot \left(\frac{H_c}{L_{dp}} \right)$	calculate length of drill pipe in cased section
$H_1 = 2026.8717$	(m)
$H_2 := L_{tj} \cdot \left(\frac{H_c}{L_{dp}} \right)$	calculate length of tool joint in cased section
$H_2 = 106.6283$	(m)
$H_3 := H_{oh} - L_{dc} - L_{tj} \cdot \left(\frac{H_{oh} - L_{dc}}{L_{dp}} \right)$	calculate length of drill pipe in open hole section
$H_3 = 723.8882$	(m)
$H_4 := L_{tj} \cdot \left(\frac{H_{oh} - L_{dc}}{L_{dp}} \right)$	calculate length of tool joints in open hole section
$H_4 = 38.0818$	(m)
$H_5 := L_{dc}$	length of drill collar
$H_5 = 152.39$	(m)

Formation natural gas and water influx

$q_w := 0.5$	injected "mist" water (bbl/hr)
$q_{ng} := 0.0$	natural gas influx (liters/sec)
$q_{fw} := 0.00$	formation water influx (bbl/hr)

Constants

$SG_a := 1.0$	specific gravity of air
$SG_w := 1.0$	specific gravity of fresh water
$SG_{ng} := 0.70$	specific gravity of natural gas
$SG_{fw} := 1.07$	specific gravity of formation water
$SG_r := 2.7$	specific gravity of rock
$SG_{df} := 0.8156$	specific gravity of diesel fuel
$k := 1.4$	gas ratio of specific heats
$R := 29.31$	engineering gas constant (ft-lb/lb-°R)
$\rho_w := 1000.0$	density of fresh water (kg/m²)
$g := 9.81$	gravitational constant (m/sec²)
$\gamma_w := \rho_w \cdot g$	specific weight of fresh water (N/m³)
$\gamma_w = 9810$	(N/m³)
$\gamma_{wg} := 9.81$	specific weight of water (N/liter)

Preliminary calculations

$D_{bh} := \frac{d_{bh}}{1000}$ convert borehole diameter to m

$D_{bh} = 0.20003$ (m)

$rop := \frac{ROP}{60 \cdot 60}$ convert ROP to m/sec

$rop = 0.00508$ (m/sec)

$w_s := \left[\left(\frac{\pi}{4}\right) \cdot D_{bh}{}^2 \cdot \gamma_w \cdot SG_r \cdot rop\right]$ weight rate of solid flow out of blooey line (N/sec)

$w_s = 4.22887$ (N/sec)

$q_a := q_c \cdot num_c$ air flow rate from all compressors (actual liters/sec)

$Q_a := \frac{q_a}{1000}$ convert air flow rate to m^3/ sec

$Q_a = 1.13251$ (m^3/sec)

$Q_w := \frac{(q_w \cdot 42 \cdot 0.003785)}{60 \cdot 60}$ convert "mist" water flow to m^3/sec

$Q_w = 0.00002$ (m^3/sec)

$Q_{ng} := \frac{q_{ng}}{1000}$ convert natural gas flow rate to m^3/sec

$Q_{ng} = 0$ (m^3/sec)

$Q_{gt} := Q_a + Q_{ng}$ total flow of the gas (m^3/sec)

$Q_{gt} = 1.13251$ (m^3/sec)

$Q_{fw} := \frac{(q_{fw} \cdot 42 \cdot 0.003785)}{60 \cdot 60}$ convert formation water flow to m^3/sec

$Q_{fw} = 0$ (m^3/sec)

$\gamma_a := \frac{(P_{at} \cdot SG_a)}{R \cdot T_{at}}$ specific weight of air (N/m^3)

$\gamma_a = 10.652$ (N/m^3)

$\gamma_{ng} := \frac{(P_{at} \cdot SG_{ng})}{R \cdot T_{at}}$ specific weight of natural gas (N/m^3)

$\gamma_{ng} = 7.4564$ (N/m^3)

$w_a := \gamma_a \cdot Q_a$ weight rate of flow of air (N/sec)

$w_a = 12.06347$ (N/sec)

$w_w := SG_w \cdot \gamma_w \cdot Q_w$ weight rate of flow of "mist" water (N/sec)

$w_w = 0.2166$ (N/sec)

$w_{ng} := \gamma_{ng} \cdot Q_{ng}$ weight rate of flow of natural gas (N/sec)

$w_{ng} = 0$ (N/sec)

$w_{fw} := SG_{fw} \cdot \gamma_w \cdot Q_{fw}$ weight rate of flow of formation water (N/sec

$w_{fw} = 0$ (N/sec)

$w_{gt} := w_a + w_{ng}$ total weight rate of flow of gas (N/sec)

$w_{gt} = 12.06347$ (N/sec)

Blooey line

$ID := \frac{id_{bl}}{1000}$ convert blooey line inner diameter to m

$ID = 0.2057$ (m)

$A := \frac{\pi}{4} \cdot ID^2$ cross sectional area (m²)

$A = 0.0332$ (m²)

$f := \left(\frac{1}{2 \cdot \log\left(\frac{ID}{e_{cs}}\right) + 1.14} \right)^2$ friction factor using von Karman correlation

$f = 0.0183$

$\frac{id_c}{od_{dp}} = 1.7602$ compute D1/D3 ratio for Fig.8-5

$\frac{id_c}{id_{bl}} = 0.9783$ compute D1/D2 ratio for Fig. 8-5

Figure 8.5 gives the following flow resistance coefficient for the blind Tee

$K_t := 25$ Tee resistance coefficient

$K_v := 0.2$ gate valve resistance coefficient

$$P_b := \left[\left[\left[\left(f \cdot \frac{L_{bl}}{ID} \right) + K_t + 2 \cdot K_v \right] \cdot \left[\frac{\left(w_{gt}^2 \cdot R \cdot T_{at} \right)}{g \cdot A^2} \right] + P_{at}^2 \right]^{0.5} \right]$$

calculate pressure of the gas at the entrance to the blooey line

$P_b = 105152.6953$ (N/m²)

$p_b := \frac{P_b}{10000.0}$ convert to N/cm²

$p_b = 10.5153$ (N/cm² abs)

Annulus cased hole drill pipe body section

$H := H_1$ (m)

$ID := \frac{id_c}{1000}$ convert inner diameter of casing to m

$ID = 0.2012$ (m)

$OD := \frac{od_{dp}}{1000}$ convert outer diameter of drill pipe to ft

$OD = 0.1143$ (ft)

$A := \frac{\pi}{4}\cdot\left(ID^2 - OD^2\right)$ annulus area between casing and drill pipe (m^2)

$A = 0.0215$ (m^2)

$a_a := \frac{SG_a}{R}\cdot\left[1 + \frac{\left(w_s + w_w + w_{fw}\right)}{w_{gt}}\right]$ calculate constant

$a_a = 0.0467$

$f := \left[\frac{1}{2\cdot\log\left[\frac{(ID - OD)}{e_{cs}}\right] + 1.14}\right]^2$ friction factor using von Karman correlation

$f = 0.0226$

$b_a := \left[\left[\frac{f}{2\cdot g\cdot(ID - OD)}\right]\cdot\left(\frac{R}{SG_a}\right)^2\cdot\frac{w_{gt}^{\;2}}{\left(\frac{\pi}{4}\right)^2\cdot\left(ID^2 - OD^2\right)^2}\right]$ calculate constant

$b_a = 3578631.4513$

$T := T_{at} + \beta\cdot H_1$ temperature at bottom of section (K)

$T = 317.1779$ (K)

$T_{ave1} := \frac{(T_{at} + T)}{2}$ average temperature of section (K)

$T_{ave1} = 298.7029$ (K)

$T_{ave} := T_{ave1}$

$$P_1 := \left[\left[\left(P_b^{\,2} + b_a \cdot T_{ave}^{\,2}\right) \cdot e^{\left[\frac{(2 \cdot a_a \cdot H)}{T_{ave}}\right]}\right] - b_a \cdot T_{ave}^{\,2}\right]^{0.5}$$ calculate pressure

$P_1 = 550678.3399$ (N/m^2)

$p_1 := \frac{P_1}{10000.0}$ convert to N/cm^2 abs

$p_1 = 55.0678$ (N/cm^2 abs)

$\gamma := \frac{P_1 \cdot SG_a}{R \cdot T}$ specific weight of gas (N/m^3)

$\gamma = 59.2351$ (N/m^3)

$V := \frac{w_{gt}}{\gamma \cdot A}$ velocity (m/sec)

$V = 9.4591$ (m/sec)

$KE_1 := \frac{1}{2} \cdot \frac{\gamma}{g} \cdot V^2$ kinetic energy of gas flow ($N\text{-}m/m^3$)

$KE_1 = 270.1319$ ($N\text{-}m/m^3$)

Annulus cased hole tool joint section

$H := H_2$

$ID := \frac{id_c}{1000}$ convert inner diameter of casing to m

$ID = 0.2012$ (m)

$OD := \frac{od_{tj}}{1000}$ convert outer diameter of tool joint to m

$OD = 0.1587$ (m)

$OD_t := \frac{od_{dp}}{1000}$ use drill pipe OD for kinetic energy to m

$OD_t = 0.1143$ (m)

$A := \frac{\pi}{4}\left(ID^2 - OD_t^2\right)$ annulus area for tool joint KE (m^2)

$A = 0.0215$ (m^2)

$a_a := \frac{SG_a}{R}\cdot\left[1 + \frac{\left(w_s + w_w + w_{fw}\right)}{w_{gt}}\right]$ calculate constant

$a_a = 0.0467$

$f := \left[\frac{1}{2\cdot\log\left[\frac{(ID - OD)}{e_{cs}}\right] + 1.14}\right]^2$ friction factor using von Karman correlation

$f = 0.0275$

$b_a := \left[\left[\frac{f}{2\cdot g\cdot(ID - OD)}\right]\cdot\left(\frac{R}{SG_a}\right)^2\cdot\frac{w_{gt}^{\,2}}{\left(\frac{\pi}{4}\right)^2\cdot\left(ID^2 - OD^2\right)^2}\right]$ calculate constant

$b_a = 28719630.1504$

$T_{temp} := T$

$T := T_{temp} + \beta \cdot H_2$ temperature at bottom of section (K)

$T = 319.1217$ (K)

$$T_{ave2} := \frac{(T_{temp} + T)}{2}$$ average temperature of section (K)

$T_{ave2} = 318.1498$ (K)

$T_{ave} := T_{ave2}$

$$P_2 := \left[\left[\left(P_1^{\,2} + b_a \cdot T_{ave}^{\,2}\right) \cdot e^{\left[\frac{(2 \cdot a_a \cdot H)}{T_{ave}}\right]} \right] - b_a \cdot T_{ave}^{\,2} \right]^{0.5}$$ calculate pressure

$P_2 = 636636.2166$ (N/m^2)

$$p_2 := \frac{P_2}{10000.0}$$ convert to N/cm^2 abs

$p_2 = 63.6636$ N/cm^2 abs

$$\gamma := \frac{P_2 \cdot SG_a}{R \cdot T}$$ specific weight of gas (N/m^3)

$\gamma = 68.0643$ (N/m^3)

$$V := \frac{w_{gt}}{\gamma \cdot A}$$ velocity of gas (m/sec)

$V = 8.232$ (m/sec)

$$KE_2 := \frac{1}{2} \cdot \frac{\gamma}{g} \cdot V^2$$ kinetic energy of gas flow ($N\text{-}m/m^3$)

$KE_2 = 235.091$ ($N\text{-}m/m^3$)

Annulus of open hole drill pipe section

$H := H_3$

$ID := D_{bh}$ open hole diameter of borehole (m)

$ID = 0.20003$ (m)

$OD := \frac{od_{dp}}{1000}$ convert diameter of drill pipe to m

$OD = 0.1143$ (m)

$A := \frac{\pi}{4}\left(ID^2 - OD^2\right)$ area between borehole and drill pipe (m^2)

$A = 0.02116$ (m^2)

$a_a := \frac{SG_a}{R} \cdot \left[1 + \frac{\left(w_s + w_w + w_{fw}\right)}{w_{gt}}\right]$ calculate constant

$a_a = 0.04669$

$e_{ave} := \frac{\left(e_r ID^2 + e_{cs} OD^2\right)}{ID^2 + OD^2}$ weighted average roughness (m)

$e_{ave} = 0.00234$ (m)

$f := \left[\frac{1}{2 \cdot \log\left[\frac{(ID - OD)}{e_{ave}}\right] + 1.14}\right]^2$ friction factor using von Karman correlation

$f = 0.05486$

$$b_a := \left[\left[\frac{f}{2\cdot g\cdot(ID - OD)}\right]\cdot\left(\frac{R}{SG_a}\right)^2\cdot\frac{w_{gt}^{\ 2}}{\left(\frac{\pi}{4}\right)^2\cdot\left(ID^2 - OD^2\right)^2}\right]$$ calculate constant

$b_a = 9102577.0212$

$T_{temp} := T$

$T := T_{temp} + \beta\cdot H_3$ temperature at bottom of section (K)

$T = 332.3182$

$$T_{ave3} := \frac{\left(T_{temp} + T\right)}{2}$$ average temperature of section (K)

$T_{ave3} = 325.7199$ (K)

$T_{ave} := T_{ave3}$

$$P_3 := \left[\left[\left(P_2^{\ 2} + b_a\cdot T_{ave}^{\ 2}\right)\cdot e^{\left[\frac{\left(2\cdot a_a\cdot H\right)}{T_{ave}}\right]}\right] - b_a\cdot T_{ave}^{\ 2}\right]^{0.5}$$ calculate pressure

$P_3 = 849422.4024$ (N/m^2)

$$p_3 := \frac{P_3}{10000.0}$$ convert to N/cm^2 abs

$p_3 = 84.9422$ (N/cm^2 abs)

$$\gamma := \frac{P_3\cdot SG_a}{R\cdot T}$$ specific weight of gas (N/m^3)

$\gamma = 87.20749$ (N/m^3)

$V := \frac{w_{gt}}{\gamma \cdot A}$ velocity of gas (m/sec)

$V = 6.536$ (m/sec)

$KE_3 := \frac{1}{2} \cdot \frac{\gamma}{g} \cdot V^2$ kinetic energy of gas flow (lb-ft/ft^3)

$KE_3 = 189.87808$ (N-m/m^3)

Annulus open hole tool joint section

$H := H_4$

$ID := D_{bh}$ open hole diameter of borehole (m)

$ID = 0.2$ (m)

$OD := \frac{od_{tj}}{1000}$ convert tool joint diameter to m

$OD = 0.1587$ (m)

$OD_t := \frac{od_{dp}}{1000}$ drill pipe diameter (m)

$OD_t = 0.1143$ (m)

$A := \frac{\pi}{4}\left(ID^2 - OD_t^2\right)$ annulus area for tool joint KE (m^2)

$A = 0.02116$ (m^2)

$a_a := \frac{SG_a}{R} \cdot \left[1 + \frac{\left(w_s + w_w + w_{fw}\right)}{w_{gt}}\right]$ calculate constant

$a_a = 0.04669$

$e_{av} := \frac{\left(e_r ID^2 + e_{cs} OD^2\right)}{ID^2 + OD^2}$ weighted average roughness (m)

$e_{ave} = 0.00193$ (m)

$$f := \left[\frac{1}{2\cdot\log\left[\frac{(ID - OD)}{e_{ave}}\right] + 1.14}\right]^2$$ friction factor using von Karman correlation

$f = 0.0692$

$$b_a := \left[\left[\frac{f}{2\cdot g\cdot(ID - OD)}\right]\cdot\left(\frac{R}{SG_a}\right)^2\cdot\frac{w_{gt}^{\,2}}{\left(\frac{\pi}{4}\right)^2\cdot\left(ID^2 - OD^2\right)^2}\right]$$ calculate constant

$b_a = 78975169.4556$

$T_{temp} := T$

$T := T_{temp} + \beta\cdot H_4$ temperature at bottom of section (K)

$T = 333.0124$

$$T_{ave4} := \frac{\left(T_{temp} + T\right)}{2}$$ average temperature of section (K)

$T_{ave4} = 332.6653$ (K)

$T_{ave} := T_{ave4}$

$$P_4 := \left[\left[\left(P_3^{\,2} + b_a\cdot T_{ave}^{\,2}\right)\cdot e^{\left[\frac{\left(2\cdot a_a\cdot H\right)}{T_{ave}}\right]}\right] - b_a\cdot T_{ave}^{\,2}\right]^{0.5}$$ calculate pressure

$P_4 = 907304.7005$ (N/m^2)

$$p_4 := \frac{P_4}{10000.0}$$ convert to N/cm^2 abs

$p_4 = 90.7305$ (N/cm^2 abs)

$\gamma := \frac{P_4 \cdot SG_a}{R \cdot T}$ specific weight of gas (N/m³)

$\gamma = 92.9559$ (N/m³)

$V := \frac{w_{gt}}{\gamma \cdot A}$ velocity of gas (m/sec)

$V = 6.1318$ (m/sec)

$KE_4 := \frac{1}{2} \cdot \frac{\gamma}{g} \cdot V^2$ kinetic energy of gas flow (N-m/m³)

$KE_4 = 178.136$ (N-m/m³)

Annulus open hole drill collar section

$H := H_5$

$ID := D_{bh}$ open hole diameter of borehole (m)

$ID = 0.2$ (m)

$OD := \frac{od_{dc}}{1000}$ convert drill collar diameter to m

$OD = 0.1587$ (m)

$A := \frac{\pi}{4}\left(ID^2 - OD^2\right)$ drill pipe diameter (m)

$A = 0.0116$ (m)

$a_a := \frac{SG_a}{R} \cdot \left[1 + \frac{\left(w_s + w_w + w_{fw}\right)}{w_{gt}}\right]$ calculate constant

$a_a = 0.04669$

$e_{av} := \frac{\left(e_r ID^2 + e_{cs} OD^2\right)}{ID^2 + OD^2}$ weighted average roughness (m)

$e_{ave} = 0.00193$ (m)

$$f := \left[\frac{1}{2\cdot\log\left[\frac{(ID - OD)}{e_{ave}}\right] + 1.14}\right]^2$$ friction factor using von Karman correlation

$f = 0.06923$

$$b_a := \left[\left[\frac{f}{2\cdot g\cdot(ID - OD)}\right]\cdot\left(\frac{R}{SG_a}\right)^2\cdot\frac{w_{gt}^{\,2}}{\left(\frac{\pi}{4}\right)^2\cdot\left(ID^2 - OD^2\right)^2}\right]$$ calculate constant

$b_a = 78975169.4556$

$T_{temp} := T$

$T := T_{temp} + \beta\cdot H_5$ temperature at bottom of section (K)

$T = 335.7905$

$$T_{ave5} := \frac{\left(T_{temp} + T\right)}{2}$$ average temperature of section (K)

$T_{ave5} = 334.4015$ (K)

$T_{ave} := T_{ave5}$

$$P_5 := \left[\left[\left(P_4^{\,2} + b_a\cdot T_{ave}^{\,2}\right)\cdot e^{\left[\frac{(2\cdot a_a\cdot H)}{T_{ave}}\right]}\right] - b_a\cdot T_{ave}^{\,2}\right]^{0.5}$$

$P_5 = 1114861.8506$ (N/m^2)

$P_{bh} := P_5$ annulus bottom hole pressure (N/m^2)

$$p_5 := \frac{P_5}{10000.0}$$ convert to N/cm^2 abs

$p_5 = 111.48619$ (N/cm^2 abs)

$p_{bh} := p_5$ annulus bottom hole pressure (N/cm^2 abs)

$$\gamma := \frac{P_5 \cdot SG_a}{R \cdot T}$$ specific weight of gas (N/m^3)

$\gamma = 113.2757$ (N/m^3)

$$V := \frac{w_{gt}}{\gamma \cdot A}$$ velocity of gas (m/sec)

$V = 9.1554$ (m/sec)

$$KE_5 := \frac{1}{2} \cdot \frac{\gamma}{g} \cdot V^2$$ kinetic energy of gas flow ($lb\text{-}ft/ft^3$)

$KE_5 = 483.94194$ ($N\text{-}m/m^3$)

Drill bit nozzles

$d_n := 7.9375$ trial and error nozzle diameter (mm)

$d_n = 7.9375$ nozzle dia (mm)

$$D_n := \frac{d_n}{1000}$$ convert nozzle diameter to m

$D_n = 0.00794$ (m)

$$A_n := N_n \cdot \left(\frac{\pi}{4}\right) \cdot D_n^2$$ summed area of all nozzles in bit (m^2)

$A_n = 0.00015$ (m^2)

$$P := \frac{w_{gt} \cdot T^{0.5}}{A_n \cdot \left[\left(\frac{g \cdot k \cdot SG_a}{R}\right) \cdot \left(\frac{2}{k+1}\right)^{\left(\frac{k+1}{k-1}\right)}\right]^{0.5}}$$ calculate pressure

$P = 3759075.5757$ (N/m^2)

$P_{sonic} := \frac{P_{bh}}{0.528}$ verify sonic flow

$P_{sonic} = 2111480.7776$ (lb/ft^2) - flow is sonic since P > P_{sonic}

$p := \frac{P}{10000.0}$ convert to N/cm^2 abs

$p = 375.9076$ (N/cm^2 abs)

Positive displacement motor performance

$q_{mm} := 1514.0$ graph mud flow rate through PDM (liters/min)

$N_{mm} := 200$ graph PDM (rpm)

$s := \frac{q_{mm}}{N_{mm} \cdot 1000.0}$ calculate specific displacement of PDM

$s = 0.00757$ (m^3)

$p_{bm} := p$ pressure at bottom of motor (N/cm^2 abs)

$p_{bm} = 375.9076$ (N/cm^2 abs)

From mud motor graph pressure drop across motor for 25 horsepower output is 220 psi.

$\Delta p_{mm} := 151.734$ pressure drop (N/cm^2)

$p_{am} := p_{bm} + \Delta p_{mm}$ pressure above the motor

$p_{am} = 527.6416$ (N/cm^2 abs)

$P_{am} := p_{am} \cdot 10000.0$ convert to consistent units (N/m^2)

$P_{am} = 5276415.5757$ (N/m^2)

$p_{mav} := \frac{p_{bm} + p_{am}}{2}$ average pressure in the motor (N/cm^2 abs)

$p_{mav} = 451.7746$ (N/cm^2 abs)

$P_{mav} := p_{mav} \cdot 10000.0$ convert to consistent units

$P_{mav} = 4517745.5757$ (N/m^2)

$\gamma := \frac{P_{mav} \cdot SG_a}{R \cdot T}$ specific weight of air in motor cavity

$\gamma = 459.0263$ (N/m^3)

$Q_{mav} := \frac{w_{gt}}{\gamma}$ average volumetric flow rate of air through motor

$Q_{mav} = 0.0263$ (m^3/sec)

$q_{mav} := Q_{mav} \cdot 1000 \cdot 60$ convert to gallons per minute (liters/min)

$q_{mav} = 1576.8342$ (gpm)

$N_m := \frac{q_{mav}}{s \cdot 1000}$ bit speed of motor when operating on air (rpm)

$N_m = 208.3$ verified that selected nozzle diameter is correct to control PDM bit speed at required ~200 rpm "close enough"

$P := P_{am}$ set pressure for inside drill string (N/m^2)

$p := \frac{P}{10000.0}$ convert to N/m^2

$p = 527.6416$ (N/cm^2 abs)

Open hole inside drill string drill collar section

$$D := \frac{id_{dc}}{1000}$$ convert diameter to m

$$D = 0.0714$$ (m)

$$a_i := \left(\frac{SG_a}{R}\right)\cdot\left(1 + \frac{w_w}{w_{gt}}\right)$$ calculate constant

$$a_i = 0.0347$$

$$f := \left(\frac{1}{2\cdot\log\left(\frac{D}{e_{cs}}\right) + 1.14}\right)^2$$ friction factor using von Karman correlation

$$f = 0.0238$$

$$b_i := \left(\frac{f}{2\cdot g\cdot D}\right)\cdot\left(\frac{R}{SG_a}\right)^2\cdot\frac{w_{gt}^{\;2}}{\left(\frac{\pi}{4}\right)^2\cdot D^4}$$ calculate constant

$$b_i = 132269290.907$$

$$P := \left[\frac{\left[P^2 + b_i\cdot T_{ave5}^{\;2}\cdot\left[e^{\frac{(2\cdot a_i\cdot H_5)}{T_{ave5}}} - 1\right]\right]}{e^{\frac{(2\cdot a_i\cdot H_5)}{T_{ave5}}}}\right]^{(0.5)}$$ calculate pressure

$$P = 5237743.3339$$ (N/m^2)

$$p := \frac{P}{10000.0}$$ convert to N/cm^2 abs

$$p = 523.7743$$ (N/cm^2 abs)

Open hole inside drill string tool joint section

$$D := \frac{id_{tj}}{1000}$$ convert diameter to m

$D = 0.0698$ (m)

$$a_i := \left(\frac{SG_a}{R}\right)\cdot\left(1 + \frac{w_w}{w_{gt}}\right)$$ calculate constant

$a_i = 0.0347$

$$f := \left(\frac{1}{2\cdot\log\left(\frac{D}{e_{cs}}\right) + 1.14}\right)^2$$ friction factor using von Karman correlation

$f = 0.024$

$$b_i := \left(\frac{f}{2\cdot g\cdot D}\right)\cdot\left(\frac{R}{SG_a}\right)^2\cdot\frac{w_{gt}^2}{\left(\frac{\pi}{4}\right)^2\cdot D^4}$$ calculate constant

$b_i = 148900137.9185$

$$P := \left[\frac{\left[P^2 + b_i\cdot T_{ave4}^2\cdot\left[e^{\frac{(2\cdot a_i\cdot H_4)}{T_{ave4}}} - 1\right]\right]}{e^{\frac{(2\cdot a_i\cdot H_4)}{T_{ave4}}}}\right]^{(0.5)}$$ calculate pressure

$P = 5229453.6324$ (N/m^2)

$$p := \frac{P}{10000.0}$$ convert to N/cm^2 abs

$p = 522.9454$ (N/cm^2 abs)

Open hole inside drill string drill pipe body section

$$D := \frac{id_{dp}}{1000}$$ convert diameter to m

$D = 0.0972$ (m)

$$a_i := \left(\frac{SG_a}{R}\right)\cdot\left(1 + \frac{w_w}{w_{gt}}\right)$$ calculate constant

$a_i = 0.0347$

$$f := \left(\frac{1}{2\cdot\log\left(\frac{D}{e_{cs}}\right) + 1.14}\right)^2$$ friction factor using von Karman correlation

$f = 0.022$

$$b_i := \left(\frac{f}{2\cdot g\cdot D}\right)\cdot\left(\frac{R}{SG_a}\right)^2\cdot\frac{w_{gt}^{\,2}}{\left(\frac{\pi}{4}\right)^2\cdot D^4}$$ calculate constant

$b_i = 26188180.1607$

$$P := \left[\frac{\left[P^2 + b_i\cdot T_{ave3}^{\,2}\cdot\left[e^{\frac{(2\cdot a_i\cdot H_3)}{T_{ave3}}} - 1\right]\right]}{e^{\frac{(2\cdot a_i\cdot H_3)}{T_{ave3}}}}\right]^{(0.5)}$$ calculate pressure

$P = 4881873.4711$ (N/m^2)

$$p := \frac{P}{10000.0}$$ convert to N/cm^2 abs

$p = 488.1873$ (N/cm^2 abs)

Cased hole inside drill string tool joint section

$$D := \frac{id_{tj}}{1000}$$ convert diameter to m

$D = 0.0698$ (m)

$$a_i := \left(\frac{SG_a}{R}\right)\cdot\left(1 + \frac{w_w}{w_{gt}}\right)$$ calculate constant

$a_i = 0.0347$

$$f := \left(\frac{1}{2\cdot\log\left(\frac{D}{e_{cs}}\right) + 1.14}\right)^2$$ friction factor using von Karman correlation

$f = 0.024$

$$b_i := \left(\frac{f}{2\cdot g\cdot D}\right)\cdot\left(\frac{R}{SG_a}\right)^2\cdot\frac{w_{gt}^{\,2}}{\left(\frac{\pi}{4}\right)^2\cdot D^4}$$ calculate constant

$b_i = 148900137.9185$

$$P := \left[\frac{\left[P^2 + b_i\cdot T_{ave2}^{\,2}\cdot\left[e^{\frac{(2\cdot a_i\cdot H_2)}{T_{ave2}}} - 1\right]\right]}{e^{\frac{(2\cdot a_i\cdot H_2)}{T_{ave2}}}}\right]^{(0.5)}$$ calculate pressure

$P = 4861181.4522$ (N/m^2)

$$p := \frac{P}{10000.0}$$ convert to N/cm^2 abs

$p = 486.1181$ (N/cm^2 abs)

Cased hole inside drill string pipe body section

$D := \frac{id_{dp}}{1000.0}$ convert diameter to m

$D = 0.0972$ (m)

$a_i := \left(\frac{SG_a}{R}\right)\cdot\left(1 + \frac{w_w}{w_{gt}}\right)$ calculate constant

$a_i = 0.0347$

$f := \left(\frac{1}{2\cdot\log\left(\frac{D}{e_{cs}}\right) + 1.14}\right)^2$ friction factor using von Karman correlation

$f = 0.022$

$b_i := \left(\frac{f}{2\cdot g\cdot D}\right)\cdot\left(\frac{R}{SG_a}\right)^2\cdot\frac{w_{gt}^2}{\left(\frac{\pi}{4}\right)^2\cdot D^4}$ calculate constant

$b_i = 26188180.1607$

$P := \left[\frac{\left[P^2 + b_i\cdot T_{ave1}^2\cdot\left[e^{\frac{(2\cdot a_i\cdot H_1)}{T_{ave1}}} - 1\right]\right]}{e^{\frac{(2\cdot a_i\cdot H_1)}{T_{ave1}}}}\right]^{0.5}$ calculate pressure

$P = 3953220.1539$ (N/m2)

$p := \frac{P}{10000.0}$ convert to N/cm2 abs

$p = 395.322$ (N/cm2 abs)

$p_{in} := p$

$p_{in} = 395.322$ drill string injection pressure (N/cm^2 abs)

Kinetic energy summary of each section (KE greater than 143.5 $N\text{-}m/m^3$)

$KE_1 = 270.1319$ ($N\text{-}m/m^3$)

$KE_2 = 235.091$ ($N\text{-}m/m^3$)

$KE_3 = 189.8781$ ($N\text{-}m/m^3$)

$KE_4 = 178.136$ ($N\text{-}m/m^3$)

$KE_5 = 483.9419$ ($N\text{-}m/m^3$)

Important design parameter summary

$q_a = 1132.508$ air compressor flow rate (actual liters/sec)

$p_{in} = 395.322$ required surface air injection pressure (N/cm^2 abs)

$p_{bh} = 111.4862$ bottom hole pressure (N/cm^2 abs)

APPENDIX

Direct Circulation Minimum Volumetric Flow Rates

G

This appendix gives direct circulation minimum volumetric flow rates for deep air drilled boreholes. For API Mechanical Equipment standard conditions, the figures that follow give the approximate minimum volumetric flow rates as a function of drilling depth for the drilling rates of 0, 30, 60, 90, and 120 ft/hr. Figures are developed from the calculations presented in Chapter 8.

Figure G-1: Borehole $4^{1}/_{2}$ inches, drill pipe $2^{3}/_{8}$ inches
Figure G-2: Borehole $4^{3}/_{4}$ inches, drill pipe $2^{3}/_{8}$ inches
Figure G-3: Borehole $6^{1}/_{4}$ inches, drill pipe $2^{7}/_{8}$ inches
Figure G-4: Borehole $6^{3}/_{4}$ inches, drill pipe $2^{7}/_{8}$ inches
Figure G-5: Borehole $7^{3}/_{8}$ inches, drill pipe $2^{7}/_{8}$ inches
Figure G-6: Borehole $6^{1}/_{4}$ inches, drill pipe $3^{1}/_{2}$ inches
Figure G-7: Borehole $6^{3}/_{4}$ inches, drill pipe $3^{1}/_{2}$ inches
Figure G-8: Borehole $7^{3}/_{8}$ inches, drill pipe $3^{1}/_{2}$ inches
Figure G-9: Borehole $7^{7}/_{8}$ inches, drill pipe $3^{1}/_{2}$ inches
Figure G-10: Borehole $7^{3}/_{8}$ inches, drill pipe 4 inches
Figure G-11: Borehole $7^{7}/_{8}$ inches, drill pipe 4 inches
Figure G-12: Borehole $8^{3}/_{4}$ inches, drill pipe 4 inches
Figure G-13: Borehole $7^{7}/_{8}$ inches, drill pipe $4^{1}/_{2}$ inches
Figure G-14: Borehole $8^{3}/_{4}$ inches, drill pipe $4^{1}/_{2}$ inches
Figure G-15: Borehole 9 inches, drill pipe $4^{1}/_{2}$ inches
Figure G-16: Borehole $8^{3}/_{4}$ inches, drill pipe 5 inches
Figure G-17: Borehole 9 inches, drill pipe 5 inches
Figure G-18: Borehole $9^{7}/_{8}$ inches, drill pipe 5 inches
Figure G-19: Borehole 11 inches, drill pipe 5 inches
Figure G-20: Borehole $9^{7}/_{8}$ inches, drill pipe $5^{1}/_{2}$ inches
Figure G-21: Borehole 11 inches, drill pipe $5^{1}/_{2}$ inches
Figure G-22: Borehole $12^{1}/_{4}$ inches, drill pipe $5^{1}/_{2}$ inches
Figure G-23: Borehole 11 inches, drill pipe $6^{5}/_{8}$ inches
Figure G-24: Borehole $12^{1}/_{4}$ inches, drill pipe $6^{5}/_{8}$ inches
Figure G-25: Borehole 15 inches, drill pipe $6^{5}/_{8}$ inches
Figure G-26: Borehole $17^{1}/_{2}$ inches, drill pipe $6^{5}/_{8}$ inches

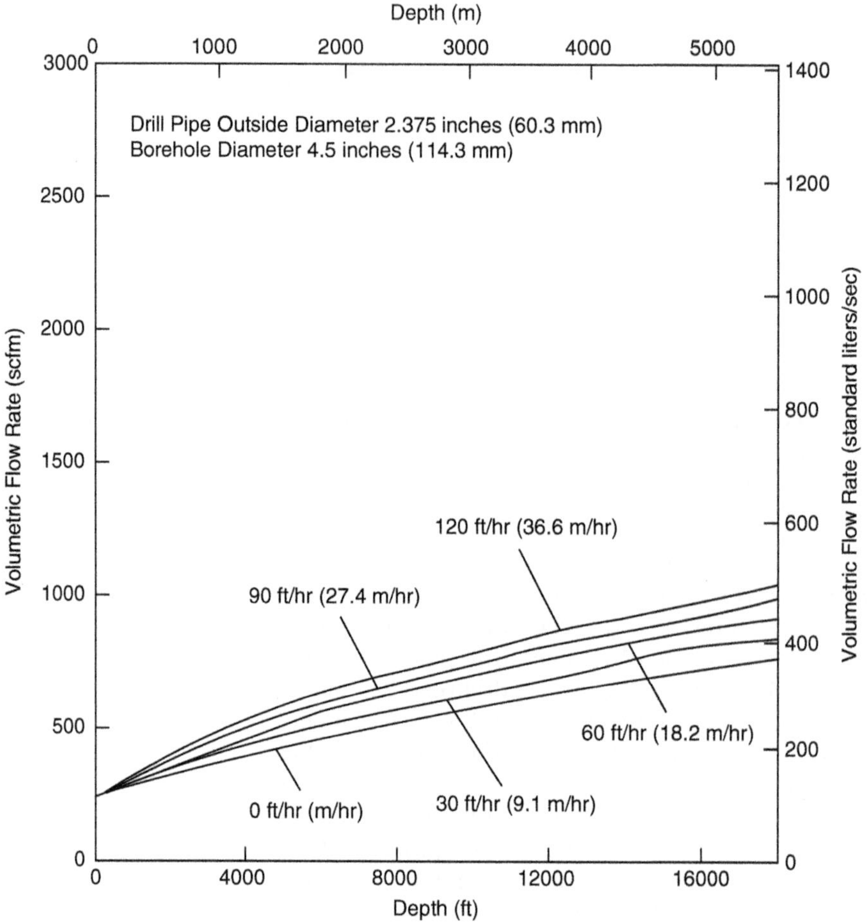

FIGURE G-1. Minimum volumetric flow rate of air at API standard conditions for $2^3/_8$-inch drill pipe and $4^1/_2$-inch borehole.

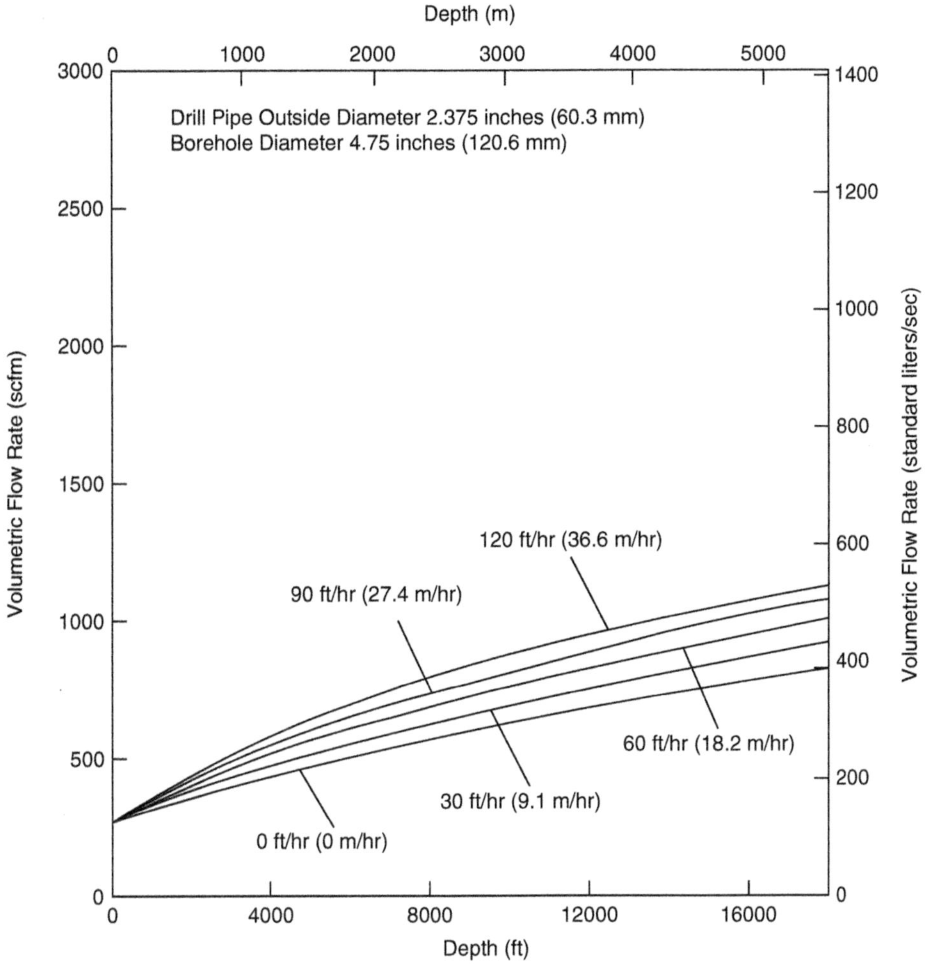

FIGURE G-2. Minimum volumetric flow rate of air at API standard conditions for $2^3/_8$-inch drill pipe and $4^3/_4$-inch borehole.

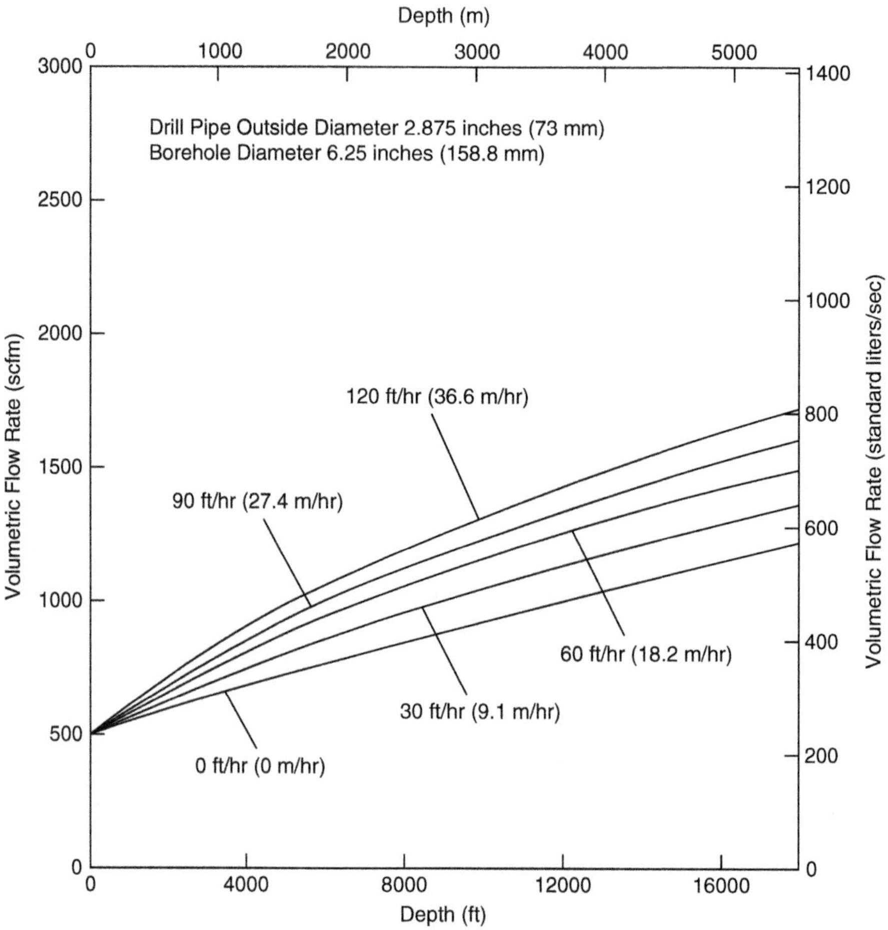

FIGURE G-3. Minimum volumetric flow rate of air at API standard conditions for $2^7/_8$-inch drill pipe and $6^1/_4$-inch borehole.

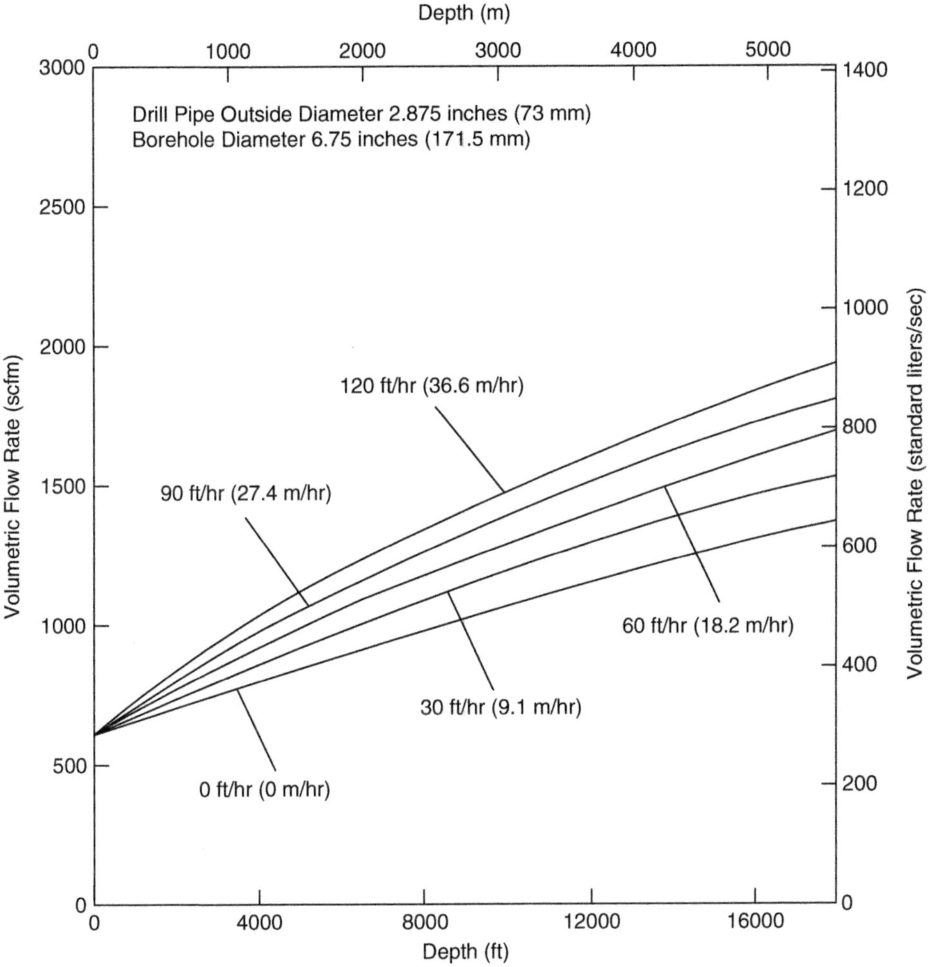

FIGURE G-4. Minimum volumetric flow rate of air at API standard conditions for $2^7/_8$-inch drill pipe and $6^3/_4$-inch borehole.

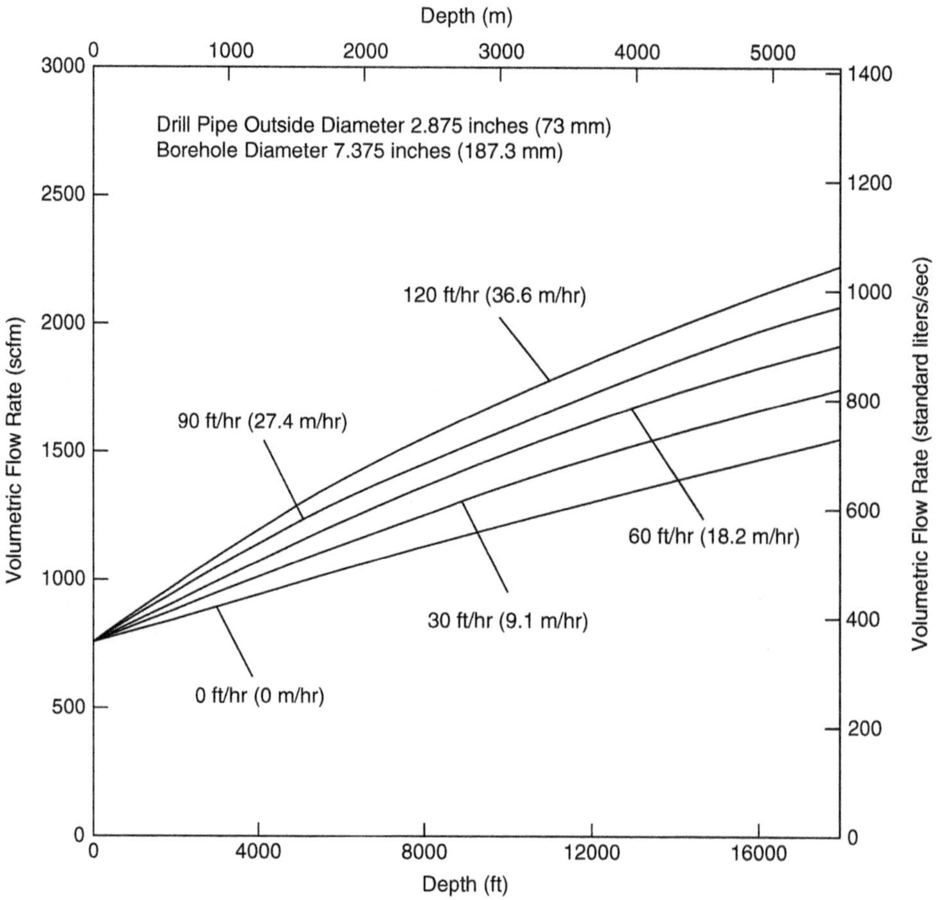

FIGURE G-5. Minimum volumetric flow rate of air at API standard conditions for $2^7/_8$-inch drill pipe and $7^3/_8$-inch borehole.

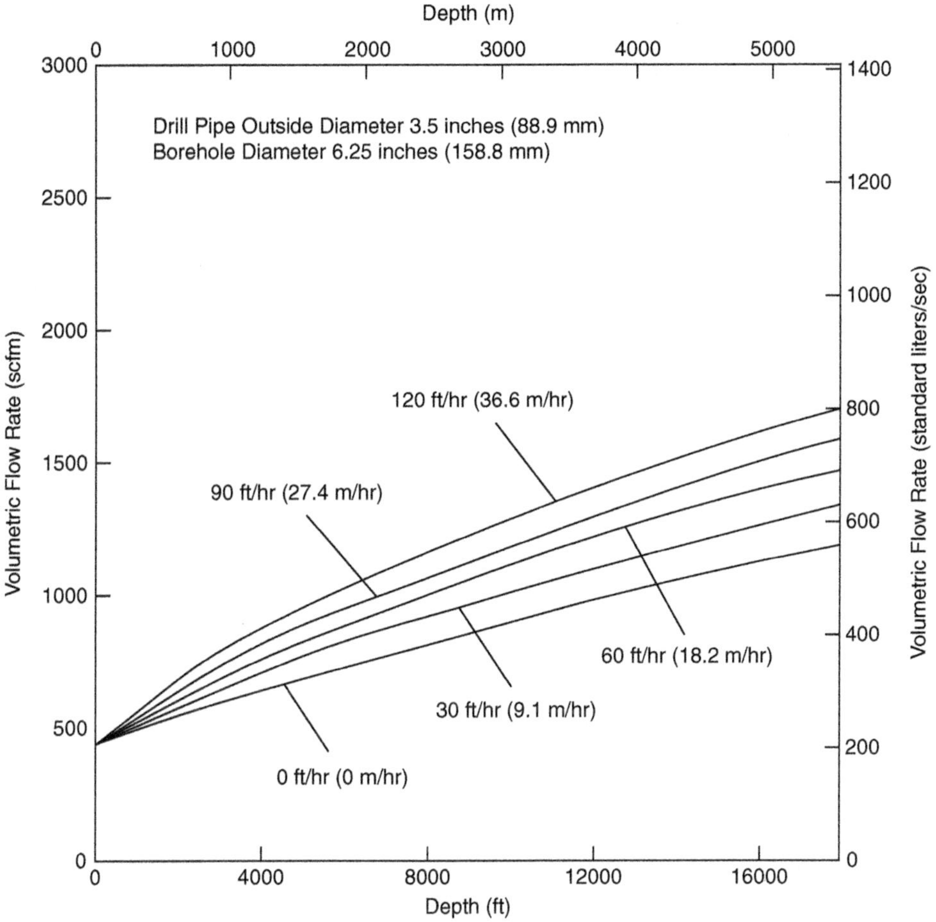

FIGURE G-6. Minimum volumetric flow rate of air at API standard conditions for $3^1/_2$-inch drill pipe and $6^1/_4$-inch borehole.

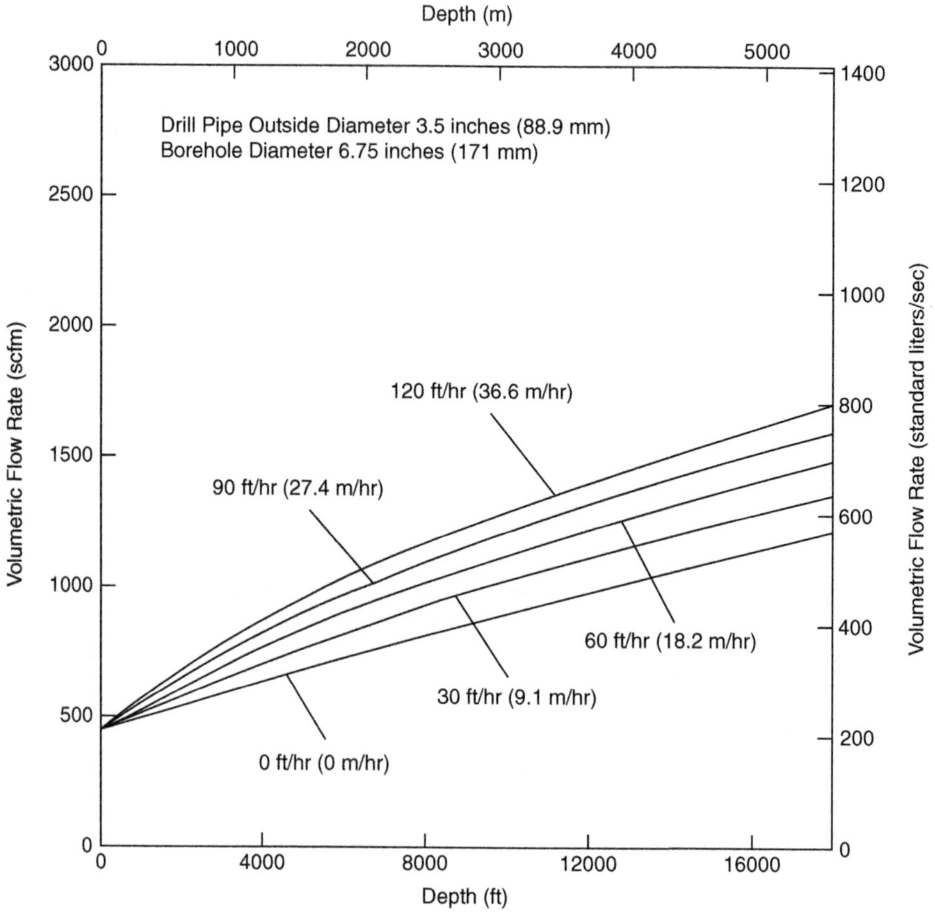

FIGURE G-7. Minimum volumetric flow rate of air at API standard conditions for $3^1/_2$-inch drill pipe and $6^3/_4$-inch borehole.

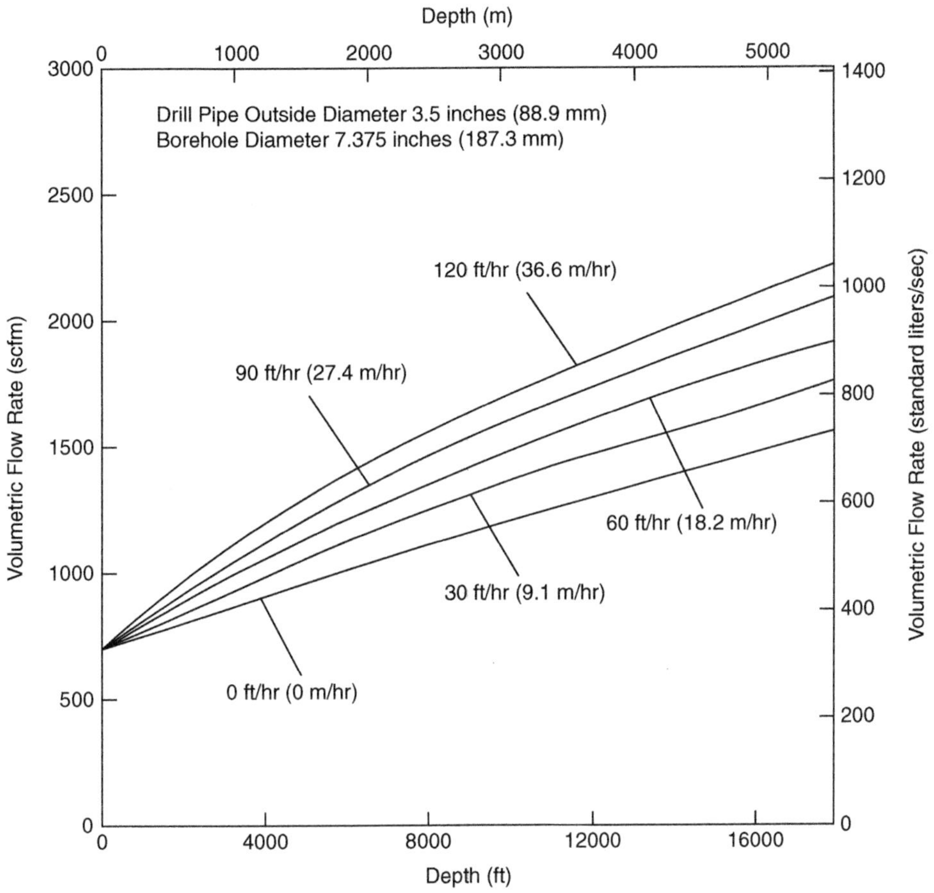

FIGURE G-8. Minimum volumetric flow rate of air at API standard conditions for $3^1/_2$-inch drill pipe and $7^3/_8$-inch borehole.

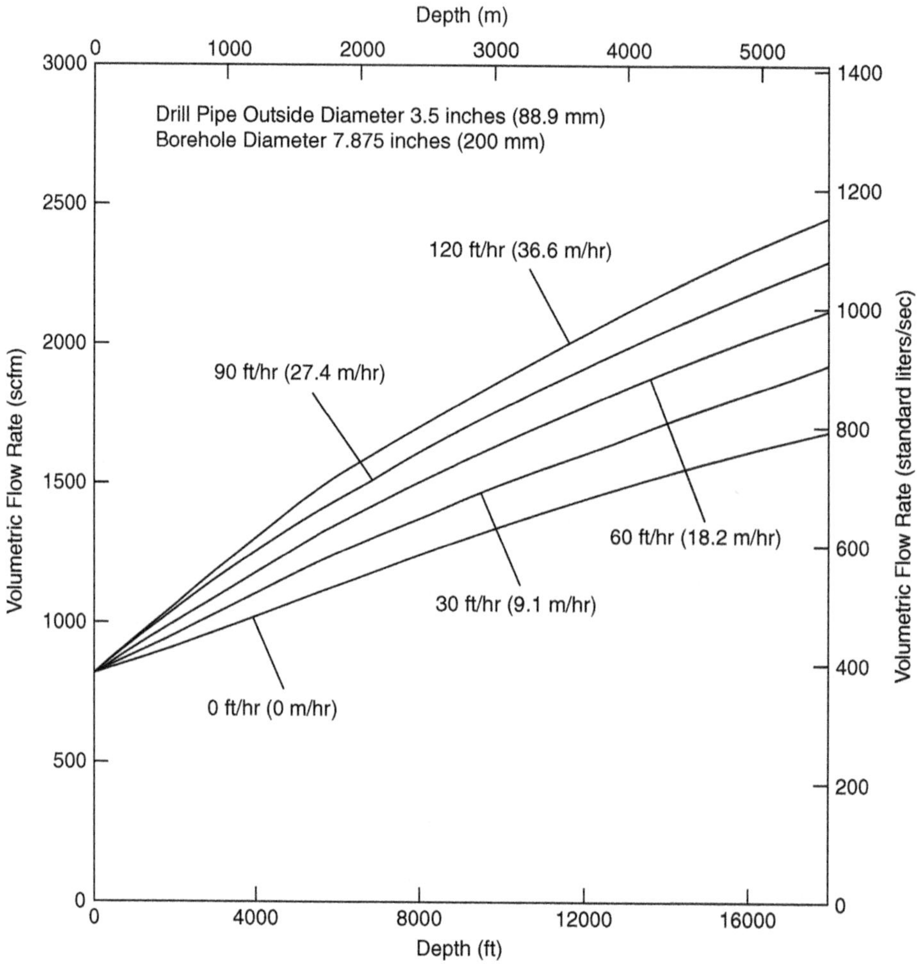

FIGURE G-9. Minimum volumetric flow rate of air at API standard conditions for $3^1/_2$-inch drill pipe and $7^7/_8$-inch borehole.

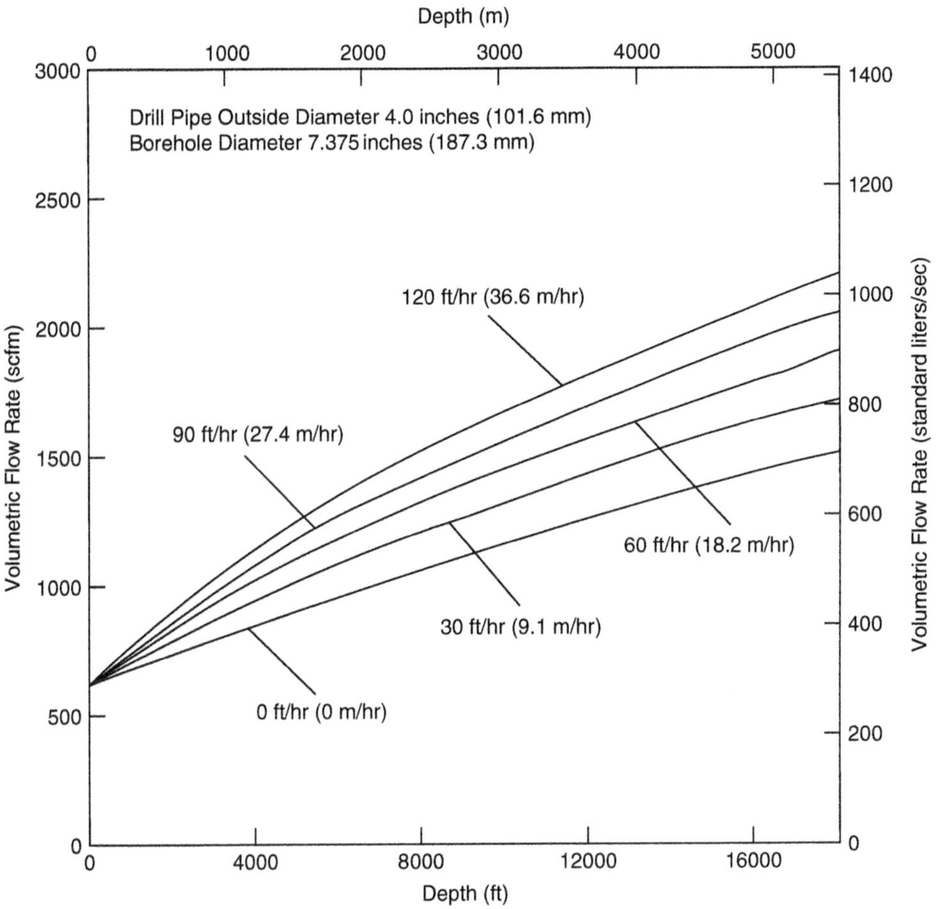

FIGURE G-10. Minimum volumetric flow rate of air at API standard conditions for 4-inch drill pipe and $7^3/_8$-inch borehole.

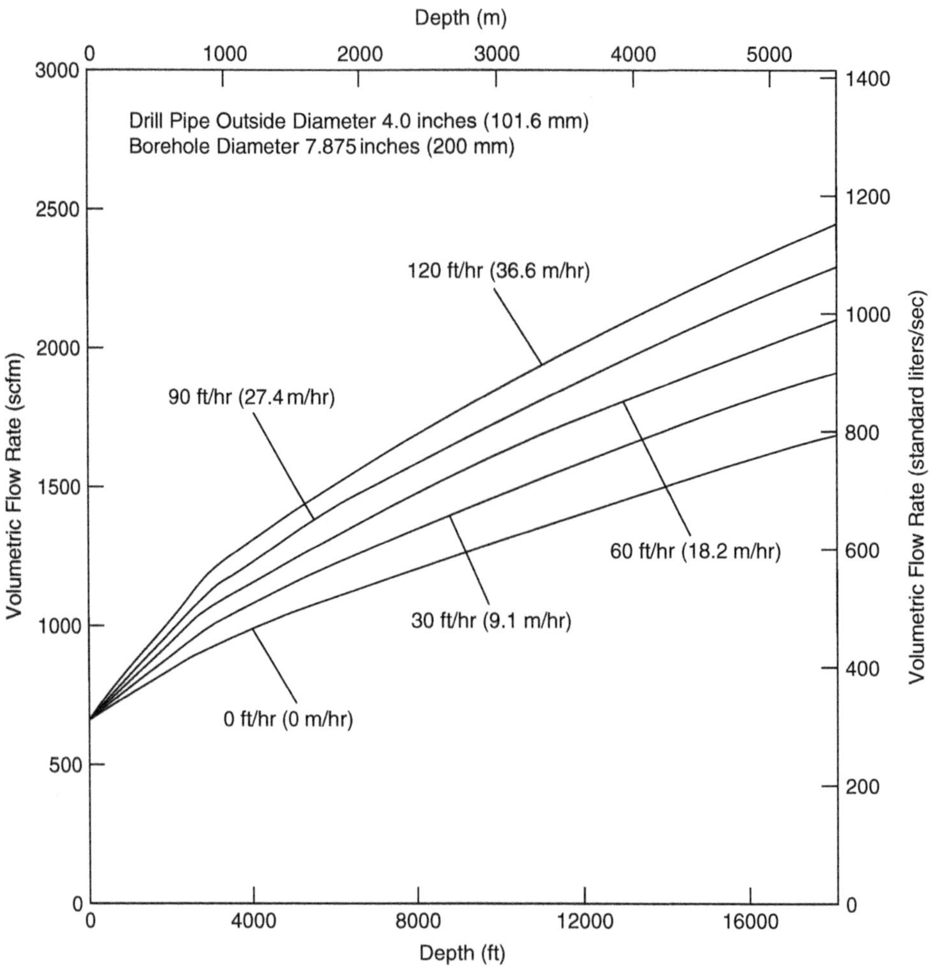

FIGURE G-11. Minimum volumetric flow rate of air at API standard conditions for 4-inch drill pipe and $7^{7}/_{8}$-inch borehole.

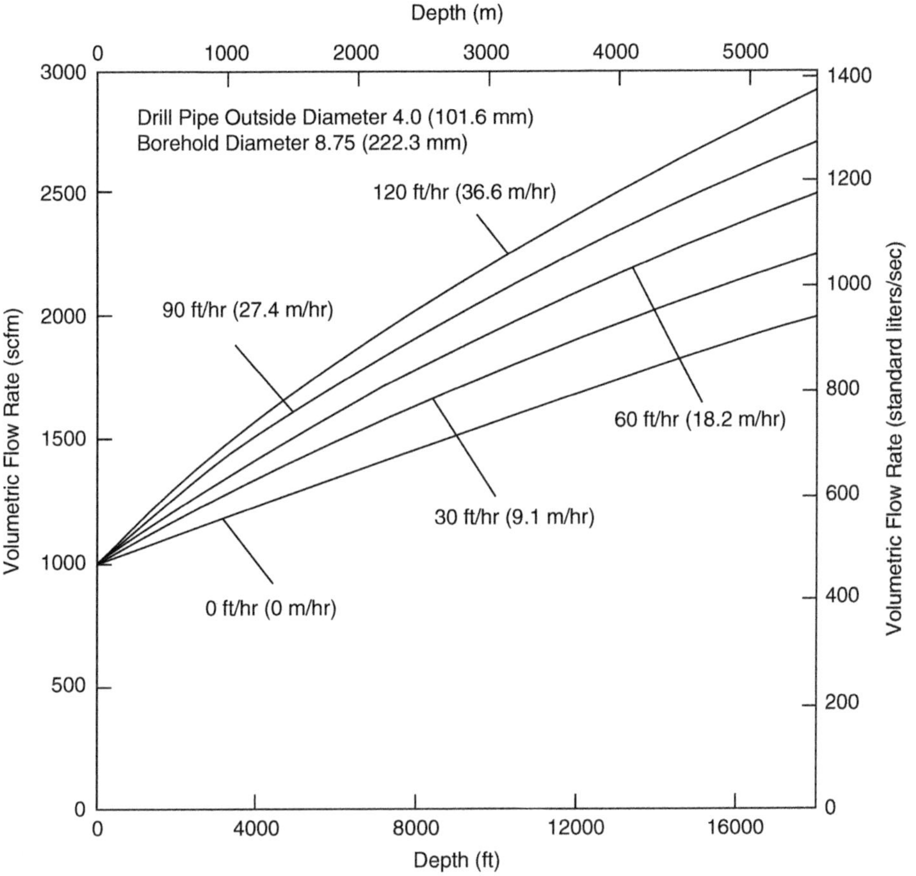

FIGURE G-12. Minimum volumetric flow rate of air at API standard conditions for 4-inch drill pipe and $8^3/_4$-inch borehole.

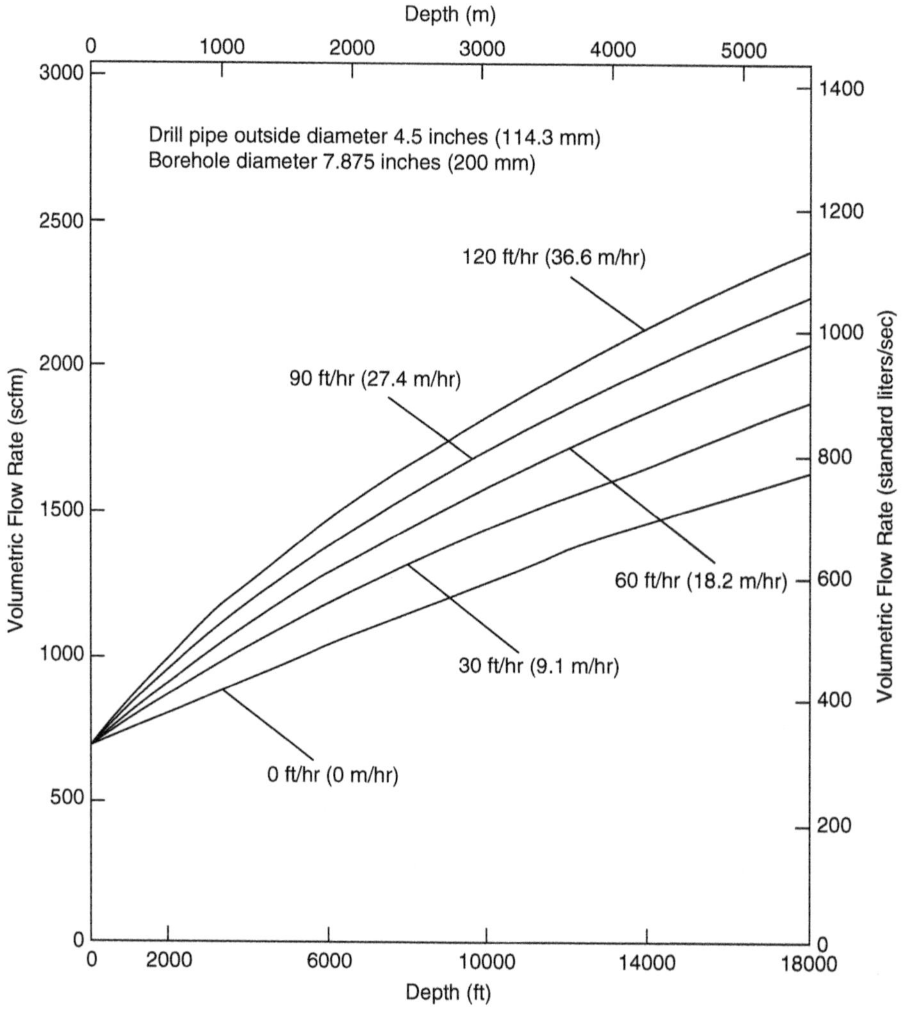

FIGURE G-13. Minimum volumetric flow rate of air at API standard conditions for $4^1/_2$-inch drill pipe and $7^7/_8$-inch borehole.

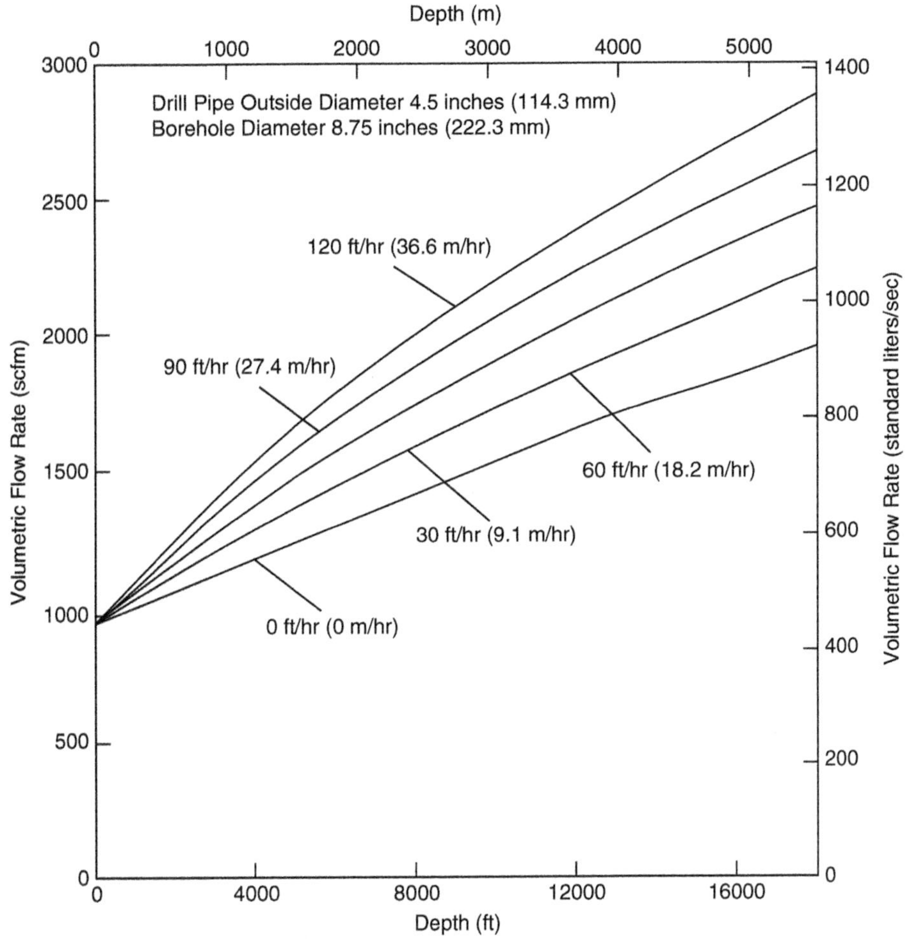

FIGURE G-14. Minimum volumetric flow rate of air at API standard conditions for $4^1/_2$-inch drill pipe and $8^3/_4$-inch borehole.

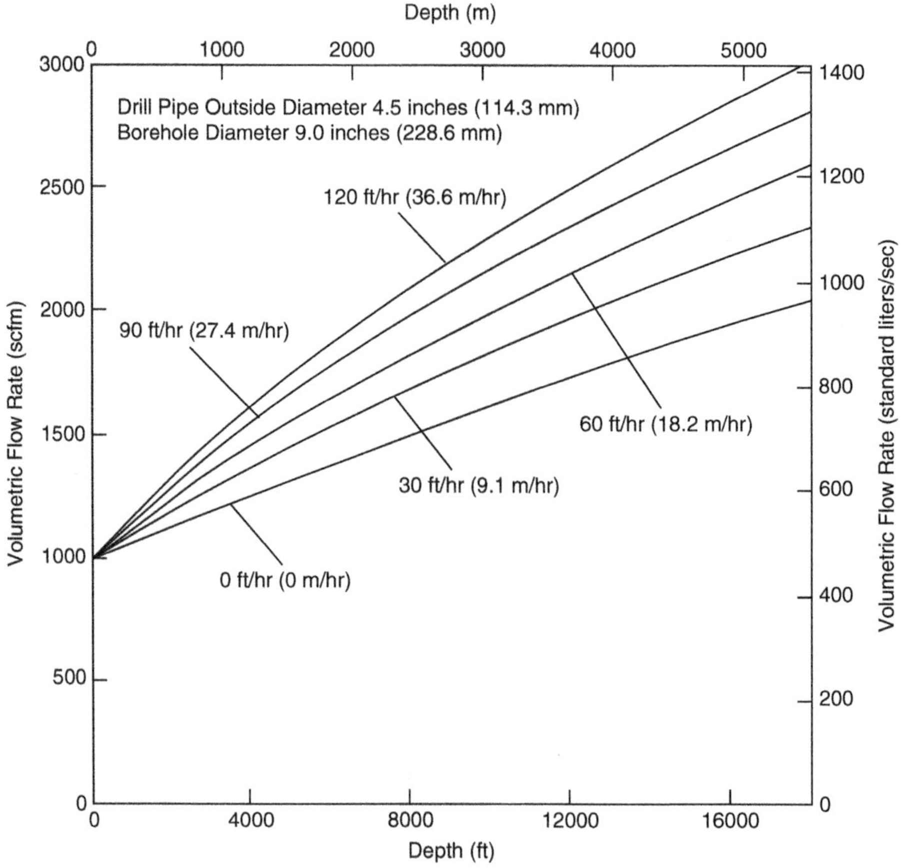

FIGURE G-15. Minimum volumetric flow rate of air at API standard conditions for $4^1/_2$-inch drill pipe and 9-inch borehole.

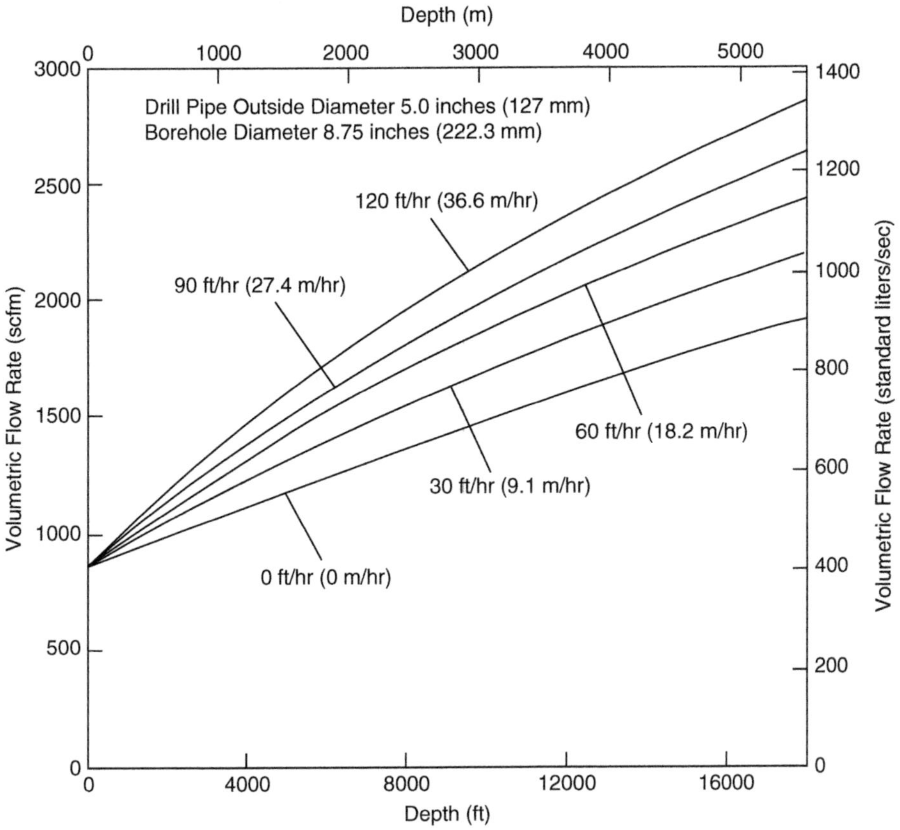

FIGURE G-16. Minimum volumetric flow rate of air at API standard conditions for 5-inch drill pipe and $8^3/_4$-inch borehole.

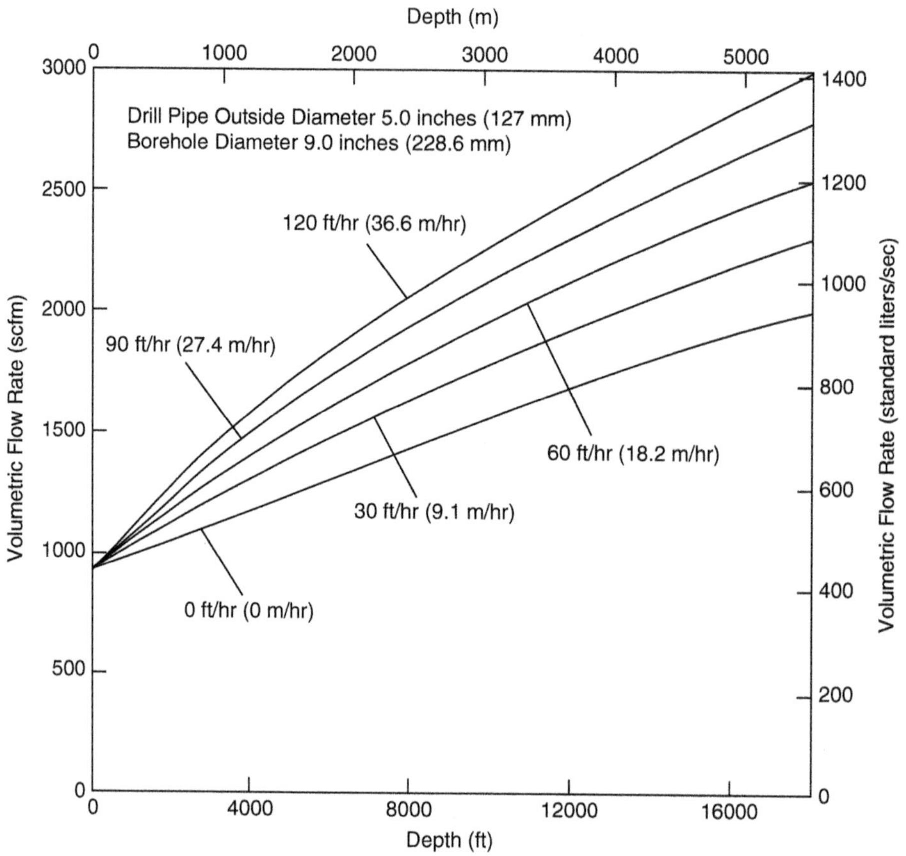

FIGURE G-17. Minimum volumetric flow rate of air at API standard conditions for 5-inch drill pipe and 9-inch borehole.

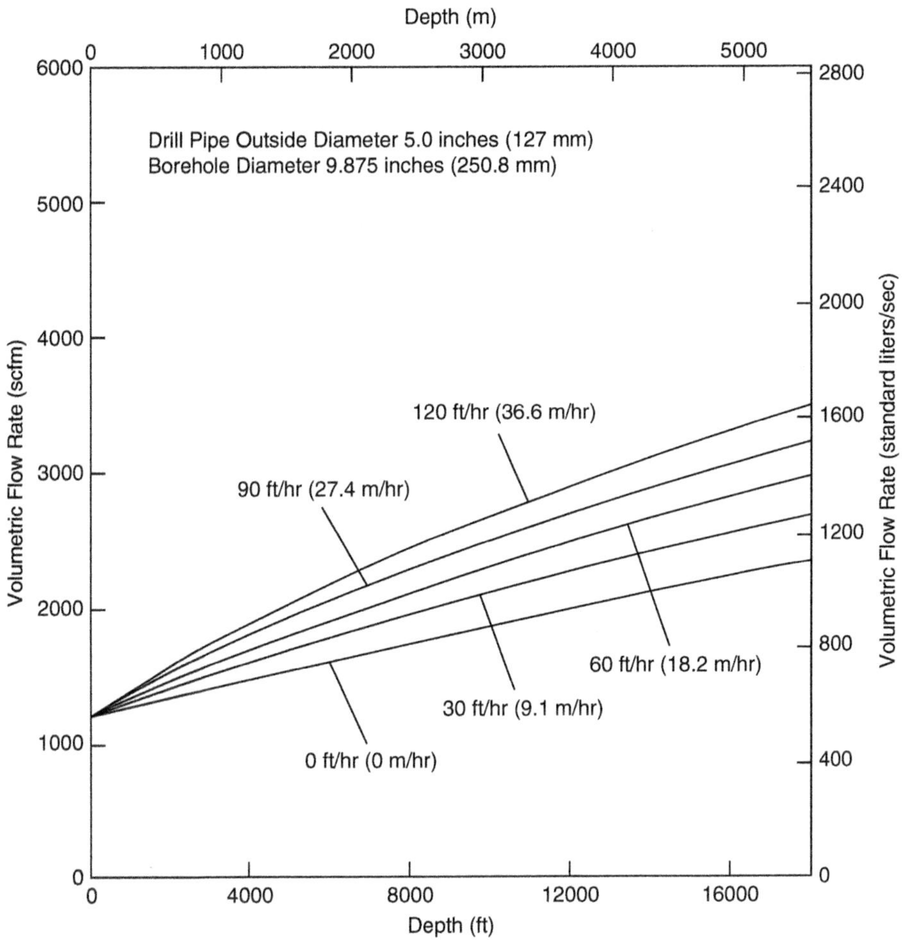

FIGURE G-18. Minimum volumetric flow rate of air at API standard conditions for 5-inch drill pipe and $9^7/_8$-inch borehole.

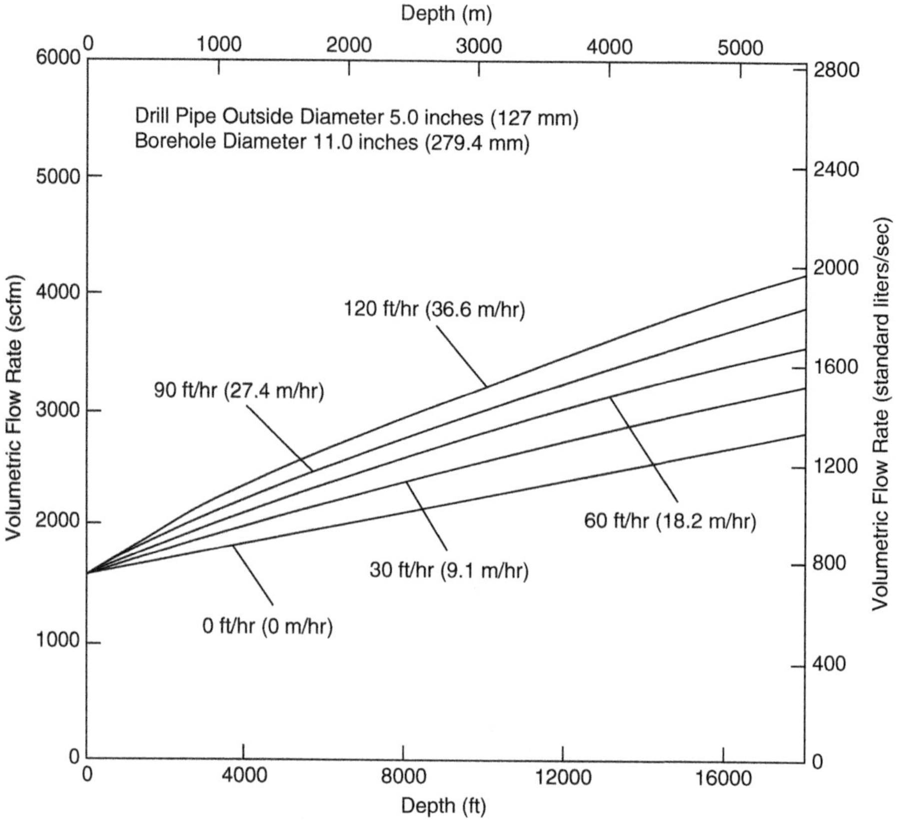

FIGURE G-19. Minimum volumetric flow rate of air at API standard conditions for 5-inch drill pipe and 11-inch borehole.

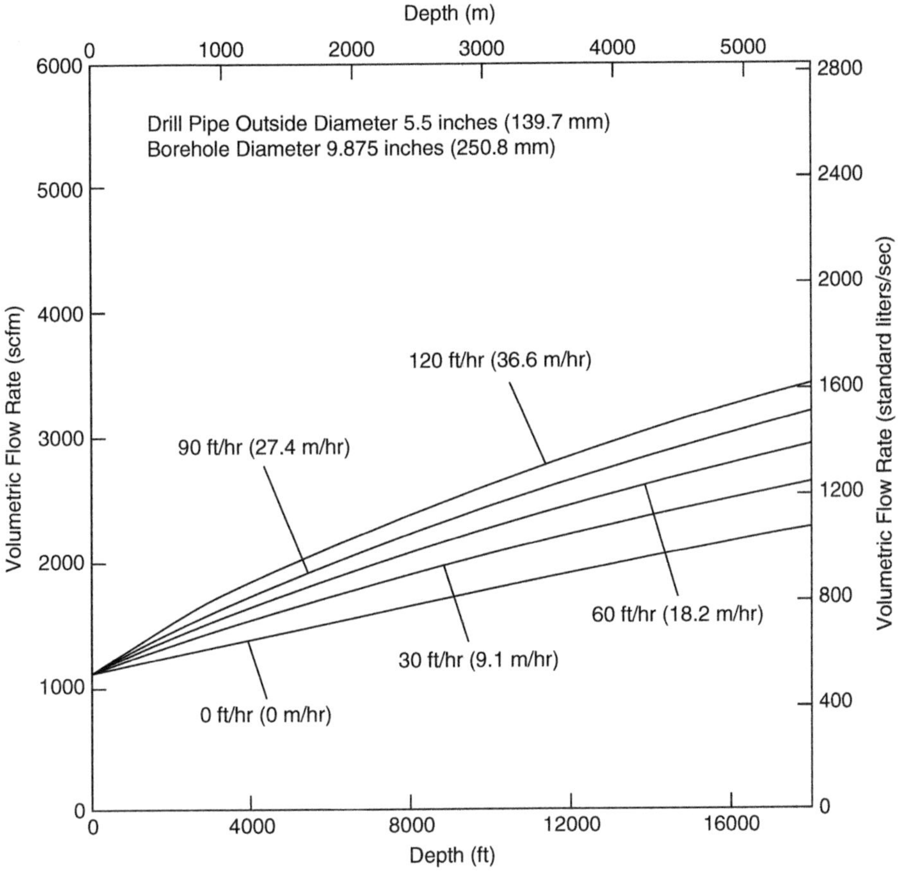

FIGURE G-20. Minimum volumetric flow rate of air at API standard conditions for $5^1/_2$-inch drill pipe and $9^7/_8$-inch borehole.

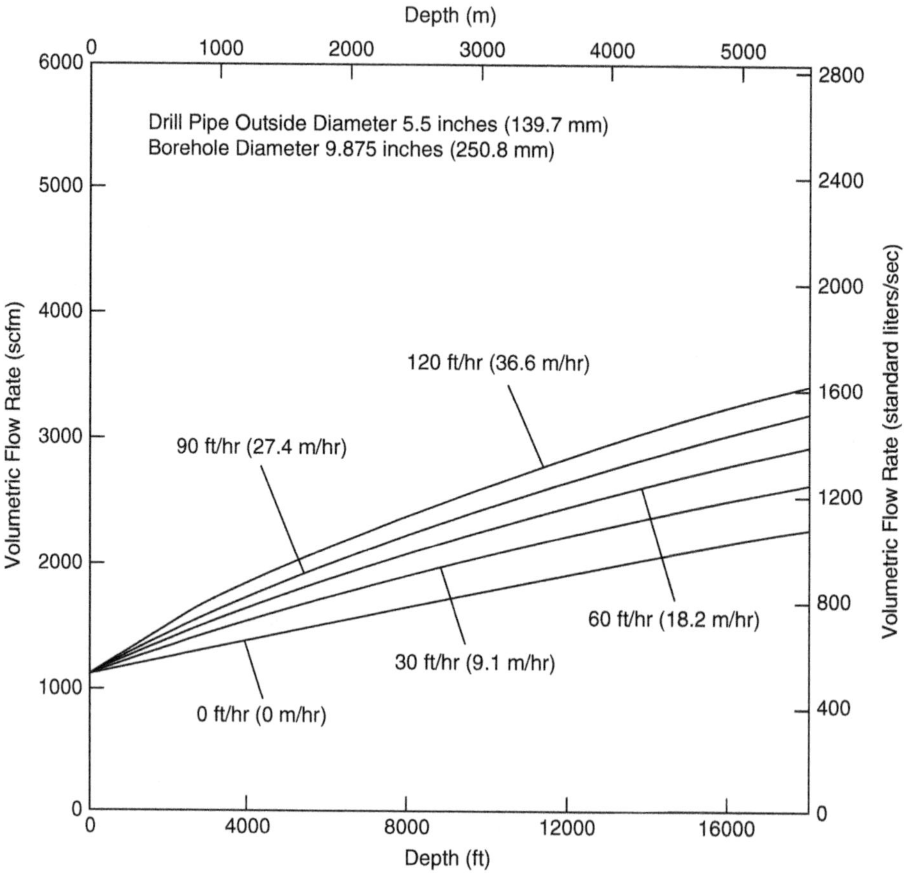

FIGURE G-21. Minimum volumetric flow rate of air at API standard conditions for $5^1/_2$-inch drill pipe and 11-inch borehole.

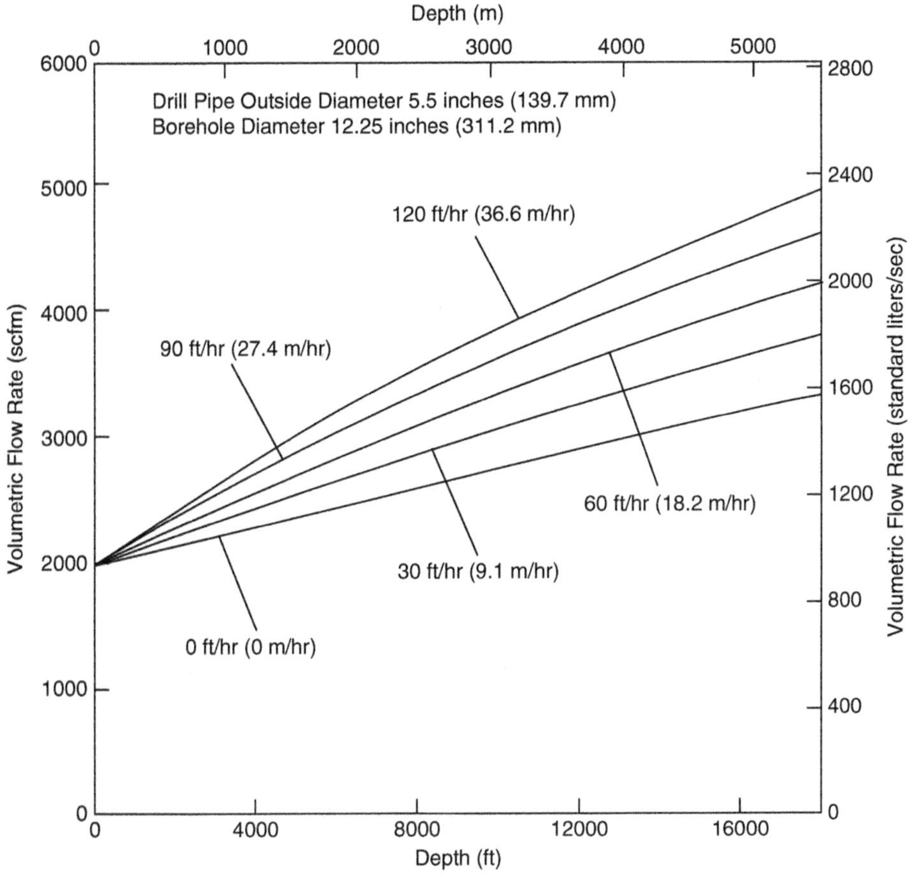

FIGURE G-22. Minimum volumetric flow rate of air at API standard conditions for $5^1/_2$-inch drill pipe and $12^1/_4$-inch borehole.

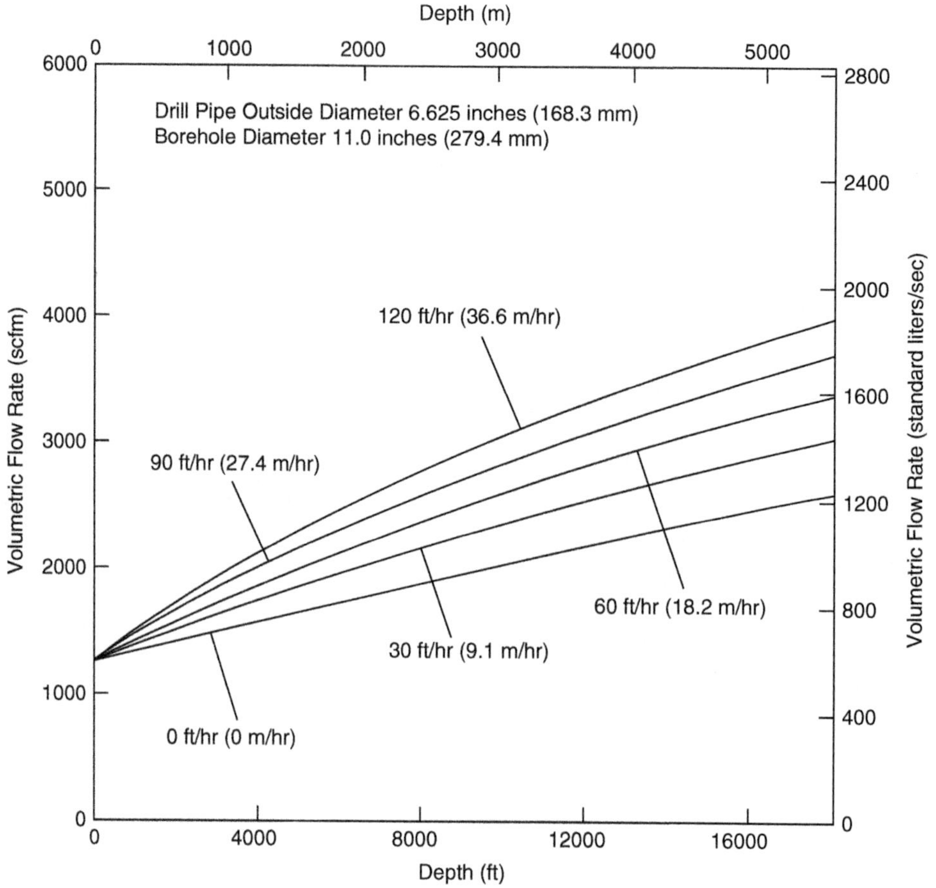

FIGURE G-23. Minimum volumetric flow rate of air at API standard conditions for $6^5/_8$-inch drill pipe and 11-inch borehole.

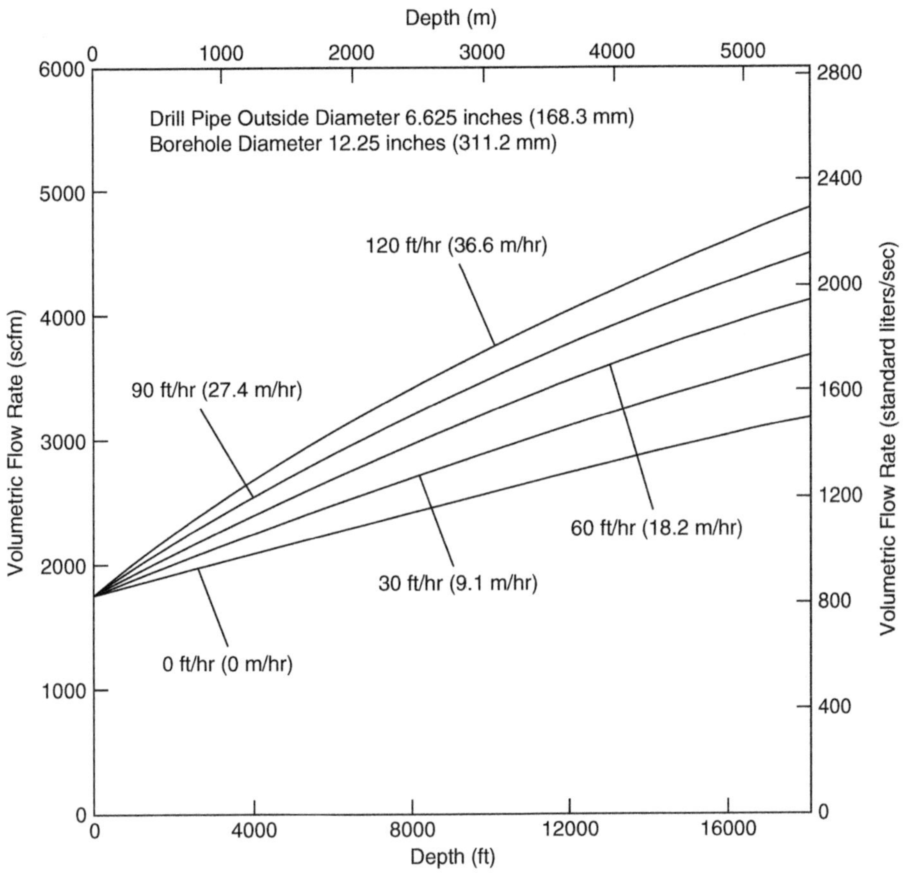

FIGURE G-24. Minimum volumetric flow rate of air at API standard conditions for $6^5/_8$-inch drill pipe and $12^1/_4$-inch borehole.

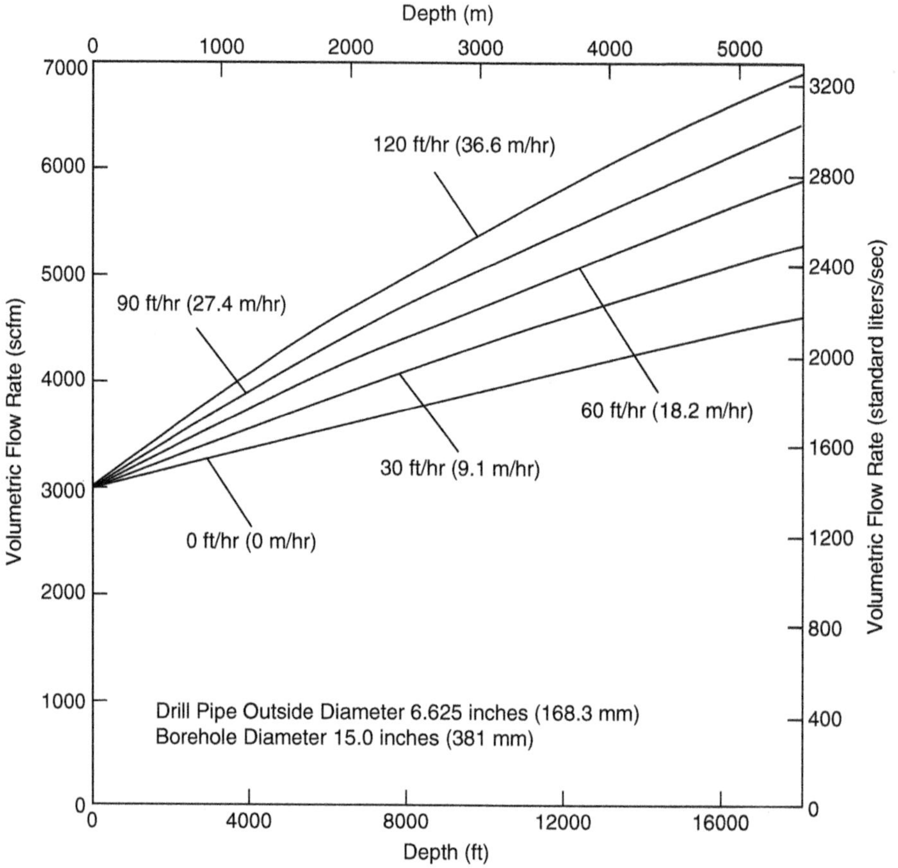

FIGURE G-25. Minimum volumetric flow rate of air at API standard conditions for $6^5/_8$-inch drill pipe and 15-inch borehole.

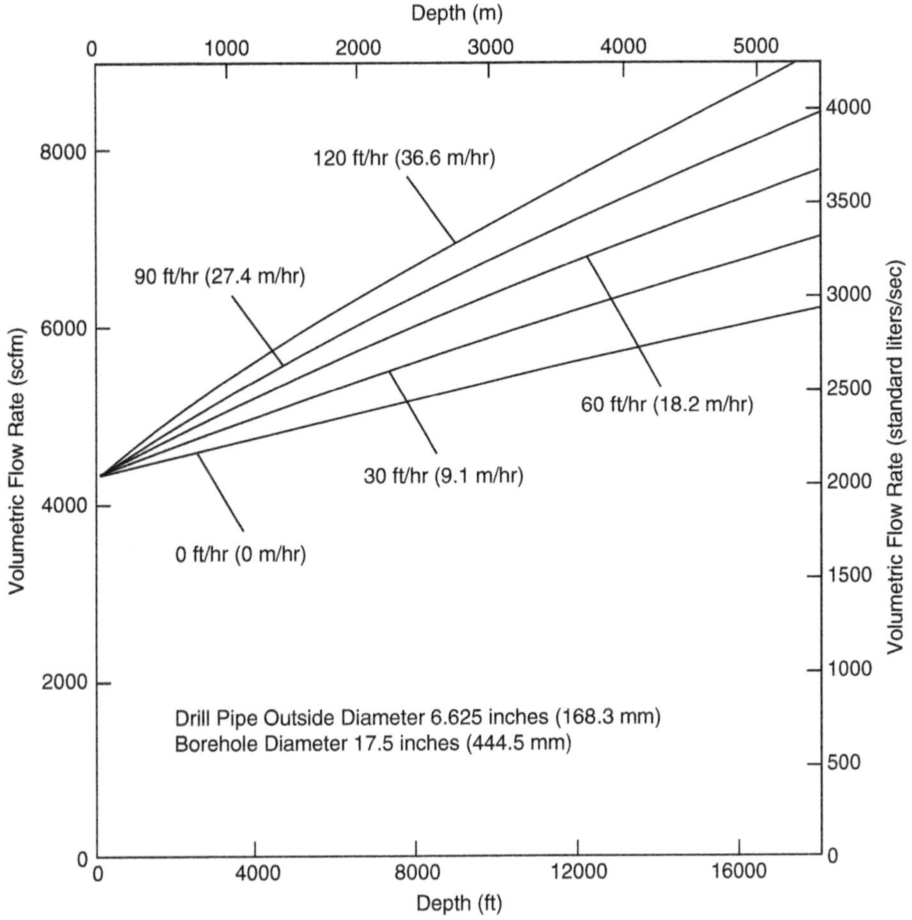

FIGURE G-26. Minimum volumetric flow rate of air at API standard conditions for $6^5/_8$-inch drill pipe and $17^1/_2$-inch borehole.

INDEX

Printed and bound by CPI Group (UK) Ltd, Croydon, CR0 4YY
03/10/2024
01040317-0015